advancing learning changing lives

D1393430

Edexcel IGCSE
Mathematics B

Student Book

D A Turner, I A Potts, W R J Waite, B V Hony

A PEARSON COMPANY

Published by Pearson Education Limited, a company incorporated in England and Wales, having its registered office at Edinburgh Gate, Harlow, Essex, CM20 2JE. Registered company number: 872828

www.pearsonschoolsandfecolleges.co.uk

Edexcel is a registered trade mark of Edexcel Limited

Text © Pearson Education Ltd 2010

First published 2010

14 13 12 11 10
10 9 8 7 6 5 4 3 2 1

ISBN 978 0 435044 10 7

Edited by Ros Davis and Deborah Dobson
Designed by Richard Ponsford and Creative Monkey
Typeset by Tech-Set Ltd, Gateshead
Original illustrations © Pearson Education Ltd 2010
Illustrated by Tech-Set Ltd, Gateshead
Cover design by Creative Monkey
Cover photo © Shutterstock.com/Ansem
Printed by Multivista Global Ltd

Acknowledgements
The author and publisher would like to thank the following individuals and organisations for permission to reproduce photographs:

(Key: b-bottom; c-centre; l-left; t-top)

Pearson Education Ltd: Brand X Pictures / Philip Coblentz 73, Brofsky Studio Inc 228, Comstock Images 339, Corbis 143, 184, Creatas 302, Digital Stock 345, Digital Vision 32, 37, 182, 352, Digital Vision / Robert Harding World Imagery / Jim Reed 47, Gareth Boden 208, Imagestate / John Foxx Collection 106, 109, 164, Jules Selmes 148, Marc Lansky / Illinois Bureau of Tourism 145, Mark Bassett 226, MindStudio 64t, Naki Kouyioumtzis 96, Photodisc 14t, 57, 85, 438, Photodisc / Alan D. Carey 170, 242, Photodisc / Getty Images 120, Photodisc / Karl Weatherly 338, Photodisc / Keith Brofsky 19b, Photodisc / Lawrence M. Sawyer 100t, Photodisc / Life File / Emma Lee 146, Photodisc / Life File / Jeremy Hoare 461, Photodisc / Neil Beer 135c, Photodisc / Photolink 3b, 14b, 52, 80l, 151, 176c, 176b, 235, 360, 456, Photodisc / Photolink / Rim Light 176t, Photodisc / Ryan McVay 80r, Photodisc / Sami Sarkis 178, Photodisc / Steve Cole 139, Photodisc / StockTrek 5, Rob Judges 107, Shenval / Alamy 197, Steve Shott 324, Stockbyte 62, 346, 406b, Tudor Photography 12, 306; **Shutterstock:** Alexander Sakhatovsky 142, Anyka 81b, James Steidl 58, Julie Phipps 81t, Monkey Business Images 144, Neale Cousland 3t, Sailorr 21, STILLFX 36

All other images © Pearson Education Ltd 2010.

The author and publisher would also like to thank the following for permission to reproduce copyright material:

Every effort has been made to contact copyright holders of material reproduced in this book. Any omissions will be rectified in subsequent printings if notice is given to the publishers.

Websites
The websites used in this book were correct and up to date at the time of publication. It is essential for tutors to preview each website before using it in class so as to ensure that the URL is still accurate, relevant and appropriate. We suggest that tutors bookmark useful websites and consider enabling students to access them through the school/college intranet.

Disclaimers
This material has been published on behalf of Edexcel and offers high-quality support for the delivery of Edexcel qualifications.

This does not mean that the material is essential to achieve any Edexcel qualification, nor does it mean that this is the only suitable material available to support any Edexcel qualification. Edexcel material will be used verbatim in setting any Edexcel examination or assessment. Any resource lists produced by Edexcel shall include this and other appropriate resources.

Copies of official specifications for all Edexcel qualifications may be found on the Edexcel website, www.edexcel.com.

Contents

Contents

iv

Course structure

This book is written for students following the IGCSE Specification B for the Edexcel examination board, and contains numerous exercises and examples as well as much consolidation material. The authors are very experienced teachers, and most of the material has been thoroughly tested in the classroom.

The book contains ten **units** of work, with each unit containing the following five **sections.**

Number
Algebra
Functions and graphs
Mensuration, geometry and trigonometry
Sets, statistics, probatility, vectors and matrices

In each section there are:

- **Concise explanations** and worked examples, with highlighted **Key Points.**

- **Activities** which lead pupils to discover mathematical principles for themselves.

- **Investigations** which prepare students for independent thought.

- **Parallel exercises** with the second exercise (starred) being more demanding than he first, allowing students to consolidate basic principles before attempting more difficult questions. Starred exercises are designed to challenge students working towards IGCSE grades A/A All exercises are carefully graded in difficulty to give students confidence. Real data is used wherever possible to make questions meaningful to students. At the end of some exercises there are more challenging questions that are marked in blue.

- Revision exercises at the end.

Each unit ends with a multiple choice paper and a self-assessment paper.
Unit 10 is a **consolidation unit**.

Simplifying fractions

> **Remember**
>
> A fraction has been simplified when the numerator (the top) and the denominator (the bottom) are expressed as whole numbers cancelled down as far as possible.
>
> $\frac{28}{42} = \frac{14}{21} = \frac{2}{3}$ $\frac{0.8}{1.6} = \frac{8}{16} = \frac{1}{2}$

Fractions are important when working with probabilities and retios. They are also used in many other calculations in everyday life.

> **Remember**
>
> To change a fraction or a decimal into a percentage, multiply it by 100.
> To change a percentage into a fraction, rewrite % as division by 100.
> To write a fraction as a decimal, divide the top number by the bottom number.

Exercise 1

Simplify these.

1 $\frac{8}{12}$ 2 $\frac{15}{45}$ 3 $\frac{0.6}{1.2}$

Copy and complete this table, giving the fractions in their lowest terms.

	Fraction	Decimal	Percentage
4		0.75	
5			25%
6	$\frac{3}{20}$		
7		0.35	

Change each of these to a mixed number.

8 $\frac{8}{3}$ 9 $\frac{17}{5}$

Change each of these to an improper fraction.

10 $2\frac{1}{3}$ 11 $1\frac{5}{6}$

Exercise 1*

Simplify and write each of these as a single fraction.

1 $\frac{6}{21}$ 2 $\frac{15}{90}$ 3 $\frac{0.7}{1.4}$ 4 $\frac{0.9}{1.2}$

5 $5 \times \frac{3}{18}$ 6 $4 \times \frac{7}{42}$ 7 $\frac{15}{27} \times 0.8$ 8 $0.3 \times \frac{7}{12}$

Simplify and write each of these as an ordinary number.

9 $68 \div 0.1$ 10 765×0.001 11 $\frac{7.8}{0.2}$

12 $\frac{36}{1.5}$ 13 $25 \times \frac{105}{100}$ 14 $46 \times \frac{91}{100}$

Directed numbers

This section will remind you of how to work with negative numbers, and why it is important to do calculations in the correct order.

> **Remember**
>
> - Directed numbers
>
> $$3 + (-4) = 3 - 4 = -1$$
> $$3 - (-\quad) = 3 + 4 = -7$$
> $$(-3) + (-4) = -3 - 4 = -7$$
> $$(-3) - (-4) = -3 + 4 = 1$$
> $$6 \times (-2) = -12$$
> $$6 \div (-2) = -3$$
> $$(-6) \div (-2) = 3$$
> $$(-6) \times (-2) = 12$$
>
> - Order of operations
>
>   The mnemonic BIDMAS may help you to remember the correct priority of operations when doing calculations working from left to right:
>   Brackets
>   Indices
>   Division and/or Multiplication
>   Addition and /or Subtraction

Exercise 2

Calculate these.

1 $(-5) + 10$ **2** $10 - (-3)$ **3** $17 + (-4)$

4 $(-7) - (-4)$ **5** $(-4) \times 3$ **6** $12 \div (-2)$

7 $\dfrac{(-16)}{8}$ $\dfrac{(-24)}{(-8)}$ **9** $4 \times 3 - 2$

Exercise 2*

Calculate these.

1 $4 - 2 \times 3$ **2** $20 \div 1 \quad 3$ **3** $16 \quad 4 \div 2$

4 $16 + \dfrac{1}{2}$ **5** $16 \div 2 + 4$ **6** $\dfrac{16}{2} + 4$

7 $(2 \times 3)^2$ **8** $(4 + 3)^2$ **9** $(2 + 2) \times 6^2$

Percentages

Percentages are used to compare quantities. The unit of comparison is 100, and this is why the term 'per cent' is used ('per' means divide, and 'cent' means 100).

Example 1

Pacific Airlines increased ticket prices by 8%. Calculate the new price of a $2450 ticket.

🌀 PACIFIC AIRLINES		
Boarding Closes 15 mins Before Boarding		Carrier PACIFIC AIRLINES
Carrier PACIFIC AIRLINES	Flight Date	Name MISS K WIGLEY
Name MISS K WIGLEY	PA 21 18 MAY	Flight PA 21 Date 18 MAY
From To	Seat 36A	Seat 36A
AROUND THE WORLD-5 STOPS		From To
Price	Boarding Time Gate	AROUND THE WORLD -5 STOPS
$2646,00	21:05 12	
PACIFIC-AIRLINES.com	SEQ NBR Class	SEQ NBR 40 Class LUXURY
	40 LUXURY FLYER	FLYER

$100\% + 8\% = 108\%.$

So multiplying factor $= \dfrac{108}{108} = 1.08$

New price
$= \$2450 \quad 1.08$
$= \$2646$

Example 2

In 1348–49, the population of England was 4 million.
The Black Death reduced the population by 37.5%. Find the new population.

$100\% - 37.5\% = 62.5\%$

So multiplying factor $= \dfrac{62.5}{100} = 0.625$

New population
$= 4 \text{ million} \times 0.625$
$= 2.5 \text{ million}$

Exercise 3

For Questions 1–3, find:

1 5% of 36   **2** 12% of 46 m   **3** 9% of 7.6

4 Increase $30 by 6%.   **5** Reduce $50 by 9%.

6 Arun buys a car for $15 000, and sells it for $12 750. What is his percentage loss?

7 A bicycle is bought for $250, and sold for $275. What is the percentage profit?

8 Gembira throws the javelin 34 m. Then she improves this by 1.7 m. What is her percentage improvement?

9 A rare stamp is bought for $7800 and increases in value by $468.
Show that the increase in value is 6%.

10 Using percentages, comment on these figures.

	1980	2000	2010
Life expectancy of men in the Caribbean	70 years	72 years	75 years

Exercise 3*

For Questions 1–3, find:

1 1.5% of 50   **2** 5.7% of $3000   **3** 7.5% of 700

4 Increase 67 km by 1.5%.   **5** Decrease $87 by 8%.

6 Marta buys a bicycle for $350, reduced from $402.50. What is her percentage saving?

7 A necklace is bought for $34 and sold for $38.25. Show that the percentage profit is 12.5%.

8 What is the percentage error if I use a value of $3\frac{1}{7}$ for π?

9 A transatlantic airline ticket costs $320 in the US, and $360 in Europe. As a percentage, how much cheaper is the ticket in the US? As a percentage, how much more expensive is the ticket in Europe?

10 Use the data below to calculate the percentage difference due to gender and comment.

Events (world records)	1975	1985	1995	2005
Men's 100 m	9.95 s	9.93 s	9.84 s	9.77 s
Women's 100 m	11.07 s	10.76 s	10.49 s	10.49 s
Men's 1500 m	3 min 32.2 s	3 min 29.7 s	3 min .82 s	3 min 26.00 s
Women's 1500 m	4 min 1.4 s	3 min 52.4 s	3 min 50.46 s	3 min 50.46 s

Standard form (positive indices)

You can write the very large number 100 000 000 more simply as 1×10^8 using **standard form**.
All numbers can be written in standard form, for example:

$$2904 = 2.904 \times 1000 = 2.904 \times 10^3$$

A standard form number can be converted back to an ordinary number:

$$5.6 \times 10^5 = 5.6 \times 100\,000 = 560\,000$$

Key Points
- In **standard form**, a million is written as 1×10^6.
- **Standard form** is always written as $a \times 10^b$, where a is between 1 and 10, but never equal to 10, and b is an integer (a whole number).

Example 3

Convert 549 into standard form.

$549 \rightarrow$ *divide by 100* $\rightarrow 5.49$
So *multiply by 100* to compensate.
$549 = 5.49 \times 100 = 5.49 \times 10^2$

Example 4

Convert 7 670 000 into standard form.

7 6 · 7 0 0 0 0.   *move the decimal point*
↑ ↑ ↑ ↑ ↑ ↑   *six places to make 7.67*
So, $7670000 = 7.67 \times 10^6$

N.B. 'Moving the decimal point one place to the left' divides the number by 10.

Activity 1

In the human brain, there are about 100 000 000 000 neurons, and over the human lifespan 1000 000 000 000 000 neural connections are made.

- Write these numbers in standard form.
- Calculate the approximate number of neural connections made per second in an average human lifespan of 75 years.

Exercise 4

Calculate these, and write each answer in standard form.

1 $10^2 \times 10^3$	**2** $10^7 \times 10^3$	**3** $10^1 \times 10^2$	**4** $10^5 \times 10^8$
5 $10^5 \div 10^3$	**6** $\dfrac{10^7}{10^4}$	**7** $10^6 \div 10^3$	**8** $10^{10} \div 10^9$

Write each of these in standard form.

9 456	**10** 123.45	**11** 568	**12** 706.05

Write each of these as an ordinary number.

13 4×10^3	**14** 4.09×10^6	**15** 5.6×10^2	**16** 7.97×10^6

17 The area of the surface of the largest known star is about 10^{15} square miles. The area of the surface of the Earth is about 10^{11} square miles. How many times greater is the star's area?

18 Calculate $(2 \times 10^4) \times (4.2 \times 10^5)$ and write the answer in standard form.

Exercise 4*

Write each of these in standard form.

1 45 089   **2** 29.83 million

Calculate these, and write each answer in standard form.

3 10×10^2	**4** $\dfrac{10^9}{10^4}$	**5** $10^{12} \times 10^9$	**6** $10^7 \div 10^7$

Calculate these, and write each answer in standard form.

7 $(5.6 \times 10^5) + (5.6 \times 10^6)$ **8** $(3.6 \times 10^4) \div (9 \times 10^2)$

Calculate these, and write each answer in standard form.

9 $(4.5 \times 10^5)^3$ **10** $10^{12} \div (4 \times 10^7)$ **11** $10^9 - (3.47 \times 10^7)$

You will need the information in this table to answer Questions 12 and 13.

Celestial body	Approximate distance from Earth (km)
Sun	1.5×10^8
Saturn	1.5×10^9
Andromeda Galaxy (nearest major galaxy)	1.5×10^{19}

Copy and complete these sentences.

12 The Andromeda Galaxy is ... times further away from the Earth than Saturn.

13 To make a scale model showing the distances of the four bodies from the Earth, a student marks the Sun 1 cm from the Earth.

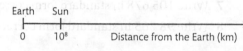

How far along the line should the other three celestial bodies be placed?

Significant figures and decimal places

It is often useful to simplify numbers by writing them either correct to so many **significant figures** (s.f.) or correct to so many **decimal places** (d.p.).

Example 5

Write 672 900 correct to 3 significant figures.
672 900 = 673 000 (to 3 s.f.)

4th s.f. = 9; 9 > 5. So 2 rounds up to 3.
(672 900 is closer in value to 673 000 than to 672 000.)

Example 6

Write 0.007645 correct to 2 significant figures.
0.007645 = 0.0076 (to 2 s.f.)

3rd s.f. = 4; 4 < 5. So 6 is not rounded up to 7.
(0.007 645 is closer in value to 0.0076 than to 0.0077.)

Example 7

Write 6.4873 correct to 2 decimal places.
6.4873 = 6.49 (to 2 d.p.)

3rd d.p. = 7; 7 > 5. So 8 rounds up to 9.
(6.4873 is closer in value to 6.49 than to 6.48.)

Example 8

Write 23.428 correct to 1 decimal place.
23.428 = 23.4 (to 1 d.p.)

2nd d.p. = 2; 2 < 5. So 4 is not rounded up to 5.
(23.428 is closer in value to 23.4 than to 23.5.)

Exercise 5

1 Write 783 correct to 1 significant figure.

2 Write 3783 correct to 3 significant figures.

3 Correct 0.439 to 2 significant figures.

4 Correct 0.5057 to 3 significant figures.

5 Write 34.777 to 2 decimal places.

6 Write 3.009 to 1 decimal place.

7 Write 105 678 in standard form correct to 1 s.f.

8 Write 98 765 in standard form correct to 1 s.f.

Exercise 5*

1 Write 10.49 correct to 1 significant figure.

2 Write 45.703 correct to 3 significant figures.

3 Correct 0.0688 to 2 significant figures.

4 Correct 0.049 549 to 3 significant figures.

5 Write 8.997 to 2 decimal places.

6 Write 6.96 to 1 decimal place.

7 Write 105 678 in standard form correct to 3 s.f.

8 Write 98 765 in standard form correct to 3 s.f.

Exercise 6 (Revision)

Write each of these as a fraction in its lowest terms.

1 $\frac{4}{12}$ 2 $\frac{4}{14}$ 3 $\frac{6}{30}$

4 $\frac{12}{96}$ 5 0.5   6 0.25

7 0.2   8 0.75   9 0.1

10 0.3   11 $2 \times \frac{3}{15}$ 12 $\frac{4}{32} \times 6$

Calculate these.

13 $5 \div 0.1$ 14 $12 \div 0.2$ 15 $3 \div 0.01$

16 $(-4) + 12$ 17 $(-4) - 12$ 18 $(-4) \times 12$

19 $(-4) \div 12$ 20 $(-4) \times (-12)$

Write the following percentages as fractions in their lowest terms.

21 25%   22 10%   23 75%

24 60%   25 35%

26 Find 10% of 1500 m.

27 Find 15% of $2400.

28 Increase 1500 m by 10%.

29 Decrease $2400 by 15%.

30 A mobile phone from Pineapple Net is advertised at a 25% reduction in the New Year sales. Its price before New Year is $120. What is the sale price?

31 Zack's pocket money is increased from €15 per week by 15%. How much does he receive per year after the increase?

32 Sami buys a computer game for $36 after the cost has been reduced by 36%. What was the original cost?

Calculate these, and write each answer in standard form.

33 $(3 \times 10^4) \times (2 \times 10^6)$ 34 $(8 \times 10^7) \div (2 \times 10^5)$ 35 $(7 \times 10^7) \times (8 \times 10^8)$

Write each of these correct to 3 significant figures.

36 1234   **37** 1235   **38** 1236

39 54321   **40** 54399

Write each of these correct to 3 decimal places.

41 1.2344   **42** 1.2345   **43** 1.2305

44 1.2035   **45** 1.2007

Exercise 6* (Revision)

Write each of these as a fraction in its lowest terms.

1 $\dfrac{0.4}{4.4}$ **2** $3.6 \times \dfrac{3}{72}$ **3** $\dfrac{3}{70} \times 21$ **4** $\dfrac{1.2}{3.2} \times \dfrac{2.4}{7.2}$

5 Find 10% of 5% of 8400 g.

6 Alec buys a model boat for $120, then sells it for $75. What is his percentage loss?

7 Lonice buys a painting for $1250, then sells it for $1400. What is her percentage profit?

8 Leandra's fastest time for the 400 m is 70 seconds. In his next race he improves by 10% and in the race after that he improves from this time by 10% again.
   a What is his new fastest time?
   b What is his overall percentage improvement?

9 Nina was 1.25 m tall. One year later she was 10% taller and in the next year her height increased a further 12% from her new height.
   a What is her height after both increases?
   b What is her overall percentage height increase over the two years?

10 Leo has a salary of €120 000. Calculate his new salary if it is
   a increased by 10% then decreased by 10%.
   b increased by x% then decreased by x%.

Calculate these.

11 $74.5 \div (0.1)^2$ **12** 74.5×0.001 **13** $74.5 \div 0.001$ **14** $74.5 \div (0.001)^2$

Calculate these and write each answer in standard form correct to 2 significant figures.

15 4321×1234 **16** $(3.5 \times 10^8) \times (2.5 \times 10^6)$

17 $(3.6 \times 10^8) \div (7.2 \times 10^6)$ **18** $(3.6 \times 10^8) + (7.2 \times 10^6)$

19 $(3.6 \times 10^8) - (7.2 \times 10^6)$ **20** $(2.5 \times 10^5)^2$

Write each of these correct to 3 significant figures.

21 0.2005   **22** 0.002005   **23** 3075.7   **24** 47555

Write each of these as a decimal correct to 3 decimal places.

25 0.0785   **26** 0.0715   **27** $\left(\dfrac{1}{4}\right)^2$ **28** $\left(\dfrac{1}{3}\right)^2$

Unit 1 : Algebra

Algebraic expressions contain letters which stand for numbers but can be treated in the same way as expressions containing numbers.

Activity 2

Think of a number. Add 7 and then double the answer. Subtract 10, halve the result, and then subtract the number you originally thought of. Algebra can show you why the answer is always 2.

Think of a number:	x
Add 7:	$x + 7$
Double the result:	$2x + 14$
Subtract 10:	$2x + 4$
Halve the result:	$x + 2$
Subtract the original number:	2

- Make up two magic number tricks of your own, one like the one above and another that is longer. Check that they work using algebra. Then test them on a friend.

- Think of a number. Double it, add 12, halve the result, and then subtract the original number.
  - Use algebra to find the answer.
  - If you add a number other than 12, the answer will change. Work out the connection between the number you add and the answer.

Simplifying algebraic expressions

You will often find it useful to simplify algebraic expressions before using them.

Exercise 7

Simplify these as much as possible.

1 $2a + 3a$

2 $6ab - 4ab$

3 $2a + 3b$

4 $3xy + 4yx$

5 $9ab - 5ab$

6 $5xy + 2yx$

7 $4pq - 7qp$

8 $2xy + y - 3xy$

9 $x - 3x + 2 - 4x$

10 $7cd - 8dc + 3cd$

11 $6xy - 12xy + 2xy$

12 $4ab + 10bc - 2ab - 5cb$

13 $3ba - ab + 3ab - 5ab$

14 $4gh - 5jk - 2gh + 7$

Exercise 7*

Simplify these as much as possible.

1 $2ab + 3ba$

2 $4xy - 2xz$

3 $4x + 3 - x$

4 $2y + 3z + y - z$

5 $7xy + 5xy - 13xy$

6 $7ab - b - 3ab$

7 $2ab - 3ba + 7ab$

8 $12ab - 6ba + ba - 7ab$

Key Points

You can only add or subtract **like** terms.

- $3ab + 2ab = 5ab$, but the terms in $3ab + b$ cannot be added together.

- $3a^2 + 2a^2 = 5a^2$, but the terms in $3a^2 + 2a$ cannot be added together.

You can check **your** simplifications by substituting numbers.

9 $4ab + 10bc - ba - 7cb$

10 $q^2 + q^3 + 2q^2 - q^3$

11 $x^2 - 5x + 4 - x^2 + 6x - 3$

12 $5a^2 + a^3 - 3a^2 + a$

13 $h^3 + 5h - 3 - 4h^2 - 2h + 7 + 5h^2$

14 $3a^2b - 2ab + 4ba^2 - ba$

Simplifying algebraic products

The multiplication sign is often left out.

Key Point

$3ab$ means $3 \times a \times b$.

Exercise 8

Simplify these.

1 $3 \times 2a$

2 $2x \times x$

3 $3x \times x^2$

4 $5a^3 \times 3a^2$

5 $2t \times 3s$

6 $4r \times s^2$

7 $2a^2 \times b^2$

8 $2y \times 2y \times y$

9 $2x^2 \times 3 \times 2x$

10 $(2a)^2 \times 5a$

Exercise 8*

Simplify these.

1 $2x \times 5y$

2 $3x^2 \times 4x^3$

3 $8a \times a^2$

4 $2x \times 4x \times 3x$

5 $5x^3 \times 3y^2 \times x$

6 $a^2 \times 2a^4 \times 3a$

7 $(3y)^2 \times 2y$

8 $6xy^2 \times 2x^3 \times 3xy$

9 $5abc \times 2ab^2c^3 \times 3ac$

10 $7x \times 2y^2 \times (2y)^2$

Simplifying algebraic expressions with brackets

To simplify an expression with brackets, multiply each term inside the bracket by the term outside the bracket.

Example 1

Simplify $2(3 + x)$.

$2(3 + x) = 2 \times 3 + 2 \times x = 6 + 2x$

The diagram helps show that $2(3 + x) = 6 + 2x$.

The area of the whole rectangle is $2(3 + x)$.

The area of rectangle A is 6.

The area of rectangle B is $2x$.

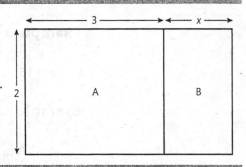

Key Points

    $3(x + y)$ means $3 \times (x + y) = 3 \times x + 3 \times y = 3x + 3y$

Be very careful with negative signs outside a bracket.

    $-2(a - 3)$ means $-2 \times (a - 3) = (-2) \times (a) + (-2) \times (-3) = -2a + 6$

When multiplying, the number 1 is usually left out.

    $-(2x + 3)$ means $-1 \times (2x + 3) = (-1) \times (2x) + (-1) \times (3) = -2x - 3$

Exercise 9

Remove the brackets and simplify these if possible.

1 $4(x + y)$ **2** $3(a - b)$ **3** $5(2 + 3a)$

4 $2(b - 4c)$ **5** $-3(2a + 8)$ **6** $-4(3 - x)$

7 $-(a - 2b)$ **8** $3a + 2(a + 2b)$ **9** $3(t - 4) - 6$

10 $7x - (x - y)$

Exercise 9*

Remove the brackets and simplify these if possible.

1 $5(x + 2y)$ **2** $3(2x - 5y)$ **3** $4(3m - 2)$

4 $2(x - y + z)$ **5** $5(3a + b - 4c)$ **6** $\frac{1}{2}(4x - 6y + 8)$

7 $5x - 3(2x - y)$ **8** $-4(x + y - z)$ **9** $0.4(2 - x) - (x + 3)$

10 $\frac{3}{4}(4x - 8y) - \frac{3}{5}(15x - 5y)$

Solving equations

If is often easier to solve mathematical problems using algebra. Let the unknown quantity be a letter, usually x, and then write down the facts in the form of an equation.

There are six basic types of equation:

$$x + 3 = 12 \qquad x - 3 = 12 \qquad 3 - x = 12$$

$$3x = 12 \qquad \frac{x}{3} = 12 \qquad \frac{3}{x} = 12$$

Solving an equation means getting x on its own, on one side of the equation.

R member

T solve equations, do the same thing to both sides.

Always check your answer.

Example 2

$x + 3 = 12$ (Subtract 3 from both sides)

   $x = 9$ (Check: $9 + 3 = 12$)

Example 3

$x - 3 = 12$ (Add 3 to both sides)

   $x = 15$ (Check: $15 - 3 = 12$)

Example 4

  $3 - x = 12$ (Add x to both sides)

     $3 = 12 + x$ (Subtract 12 from both sides)

$-12 + 3 = x$

     $x = -9$ (Check: $3 - (-9) = 12$)

Example 5

$3x = 12$ (Divide both sides by 3)

  $x = 4$ (Check: $3 \times 4 = 12$)

Example 6

$\frac{x}{3} = 12$ (Multiply both sides by 3)

$x = 36$ (Check: $36 \div 3 = 12$)

Example 7

$\frac{3}{x} = 12$ (Multiply both sides by x)

$3 = 12x$ (Divide both sides by 12)

$\frac{1}{4} = x$ (Check: $3 \div \frac{1}{4} = 12$)

Exercise 10

Solve these for x.

1 $5x = 20$

2 $x + 5 = 20$

3 $x - 5 = 20$

4 $\frac{x}{5} = 20$

5 $3 = \frac{36}{x}$

6 $20 - x = 5$

7 $5x = 12$

8 $x - 3.8 = 9.7$

9 $3.8 = \frac{x}{7}$

10 $x + 9.7 = 11.1$

11 $13.085 - x = 12.1$

12 $\frac{34}{x} = 5$

Exercise 10*

Solve these for x, giving each answer correct to 3 significant figures.

1 $23.5 + x = 123.4$

2 $7.6x = 39$

3 $39.6 = x - 1.064$

4 $45.7 = \frac{x}{12.7}$

5 $7.89 = \frac{67}{x}$

6 $40.9 - x = 2.06$

Example 8

$3x - 5 = 7$ (Add 5 to both sides)

$\quad 3x = 12$ (Divide both sides by 3)

$\quad\quad x = 4$ (Check: $3 \times 4 - 5 = 7$)

Example 9

$4(x + 3) = 20$ (Divide both sides by 4)

$\quad x + 3 = 5$ (Subtract 3 from both sides)

$\quad\quad x = 2$ (Check: $4(2 + 3) = 20$)

Example 10

$2(x + 3) = 9$ (Multiply out the bracket)

$2x + 6 = 9$ (Subtract 6 from both sides)

$\quad 2x = 3$ (Divide both sides by 2)

$\quad\quad x = \frac{3}{2}$ (Check: $2\left(\frac{3}{2} + 3\right) = 9$)

Exercise 11

Solve these for x.

1 $2x + 4 = 10$

2 $4x + 5 = 1$

3 $12x - 8 = -32$

4 $2(x + 3) = 10$

5 $5(x - 2) = 30$

6 $5 - x = 4$

7 $9 = 3 - x$

8 $2(6 - 3x) = 6$

9 $4(2 - x) = 16$

10 $3(x - 5) = -13$

11 $9(x + 4) = 41$

12 $5(10 - 3x) = 30$

13 $7(2 - 5x) = 49$

14 The sum of two consecutive numbers is 477.
What are the numbers? (Let the first number be x.)

15 Find x and the size of each angle in this triangle.

16 AB is a straight line.
Find x and the size of each angle.

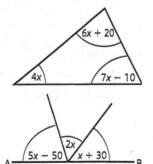

17 The formula for converting degrees Fahrenheit (F) to degrees Celsius (C) is $F = 32 + 1.8C$.
Find C when F is 5.

Exercise 11*

Solve these for x.

1 $5x - 3 = 17$ **2** $27 = 3(x - 2)$ **3** $7(x - 3) = -35$

4 $12(x + 5) = 0$ **5** $-7 = 9 + 4x$ **6** $5 - 4x = -15$

7 $34 = 17(2 - x)$

8 The sum of three consecutive even numbers is 222. Find the numbers.

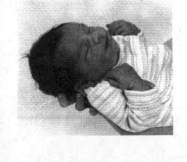

9 Zul and Amelia have a baby daughter, Shanise. Zul is 23 kg heavier than Amelia, who is four times heavier than Shanise. Their combined weight is 122 kg. How heavy is each person?

10 Emma buys some cans of cola at 28p each, and twice as many cans of orange at 22p each. She also buys ten fewer cans of lemonade than orange at 25p each. She spends £14.58. How many cans of cola did she buy?

11 Solve for x: $1.4(x - 3) + 0.2(2x - 1) = 0.1$.

12 Hama is training by running to a post across a field and then back. She runs the outward leg at 7 m/s and the return leg at 5 m/s. She takes 15.4 s. Find the distance to the post.

13 A piece of wire 30 cm long is cut into two pieces. One of these is bent into a circle, and the other is bent into a square enclosing the circle, as shown in the diagram. What is the diameter of the circle? (Remember that the circumference of a circle $= 2\pi r$.)

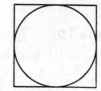

14 Maureen sets off on a walk at 6 km/h. Ten minutes later, her brother John sets off after her on his bicycle at 15 km/h. How far must John go to catch up with Maureen?

Equations with x on both sides

Sometimes x appears on both sides of an equation.

Example 11

Solve this for x.

$$7x - 3 = 3x + 5 \quad \text{(Subtract } 3x \text{ from both sides)}$$
$$7x - 3 - 3x = 5 \quad \text{(Add 3 to both sides)}$$
$$4x = 5 + 3$$
$$4x = 8 \quad \text{(Divide both sides by 4)}$$
$$x = 2 \quad \text{(Check: } 7 \times 2 - 3 = 3 \times 2 + 5)$$

Exercise 12

Solve these for x.

1 $8x - 3 = 4x + 1$ **2** $2x + 5 = 5x - 1$ **3** $7x - 5 = 9x - 13$ **4** $2x + 7 = 5x + 16$

5 $5x + 1 = 8 - 2x$ **6** $14 - 3x = 10 - 7x$ **7** $6 + 2x = 6 - 3x$ **8** $8x + 9 = 6x + 8$

9 Find the value of x and the perimeter of this rectangle.

4x − 2

4x + 5

x + 4

10 The result of adding 36 to a certain number is the same as multiplying that number by 5. What is the number?

Exercise 12*

Solve these for x.

1 $3x + 8 = 7x - 8$ **2** $7x + 5 = 5x + 1$ **3** $5x + 7 = 9x + 1$

4 $4x + 3 = 7 - x$ **5** $15x - 4 = 10 - 3x$ **6** $5(x + 1) = 4(x + 2)$

7 $8(x + 5) = 10(x + 3)$ **8** $3(x - 5) = 7(x + 4) - 7$ **9** $3.1(4.8x - 1) - 3.9 = x + 1$

10 If $\frac{x}{3} - 1$ is twice as large as $\frac{x}{4} - 3$, what is the value of x?

11 A father is three times as old as his son. In 14 years' time, he will be twice as old as his son. How old is the father now?

Negative signs outside brackets

> **Key Point**
>
> $-(2x - 5)$ means $-1 \times (2x - 5) = (-1) \times (2x) + (-1) \times (-5) = -2x + 5$

Example 12

Solve this for x.

$2(3x + 1) - (2x - 5) = 15$	(Remove brackets)
$6x + 2 - 2x + 5 = 15$	(Simplify)
$6x - 2x = 15 - 2 - 5$	(Subtract 2 and 5 from both sides)
$4x = 8$	(Divide both sides by 4)
$x = 2$	(Check: $2(3 \times 2 + 1) - (2 \times 2 - 5) = 15$)

Exercise 13

Solve these for x.

1 $3(x - 2) - 2(x + 1) = 5$ **2** $3(2x + 1) - 2(2x - 1) = 11$

3 $2(5x - 7) - 6(2x - 3) = 0$ **4** $4(3x - 1) - (x - 2) = 42$

5 $4(3 - 5x) - 7(5 - 4x) + 3 = 0$

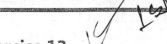

Exercise 13*

Solve these for x.

1 $5(x - 3) - 4(x + 1) = -11$ **2** $4(3x + 5) - 5(2x + 6) = 0$

3 $3(3x + 1) - 8(2x - 3) + 1 = 0$ **4** $-2(x + 3) - 6(2x - 4) + 108 = 0$

5 $7(5x - 3) - 10 = 2(3x - 5) - 3(5 - 7x)$

6 Anna is shooting at a target at a fair. If she hits the target she receives 50c, but if she misses she has to pay 20c for the shot. After 15 shots, Anna finds she has made a profit of $1.20. How many hits has she had?

7 Michael is doing a multiple-choice test with 20 questions. He scores 3 marks for a correct answer and loses 1 mark if the answer is incorrect. Michael answers all the questions and scores 40 marks. How many questions has he got right?

8 Freddie the frog is climbing up a well. Every day he climbs up 3 m but some nights he falls asleep and slips back 4 m. At the start of the sixteenth day, he has climbed a total of 29 m. On how many nights was he asleep?

Exercise 14 (Revision)

Simplify these as much as possible.

1 $x + 2x + 3 - 5$ 2 $3ba - ab + 3ab - 4ba$ 3 $2a \times 3$

4 $2a \times a$ 5 $a^2 \times a$ 6 $2a^2 \times a^2$

7 $2a \times 2a \times a^2$ 8 $7a - 4a(b + 3)$ 9 $4(x + y) - 3(x - y)$

Solve these equations.

10 $2(x - 1) = 12$ 11 $7x - 5 = 43 - 3x$ 12 $5 - (x + 1) = 3x - 4$

13 Find three consecutive numbers whose sum is 438.

14 The perimeter of a rectangle is 54 cm. One side is x cm long, and the other is 6 cm longer.
   a Form an equation involving x.
   b Solve the equation, and write down the length of each of the sides.

Exercise 14* (Revision)

Simplify these as much as possible.

1 $6xy^2 - 3x^2y - 2y^2x$ 2 $2xy^2 \times x^2y$

3 $p - (p - (p - (p - 1)))$ 4 $xy(x^2 + xy + y^2) - x^2(y^2 - xy - x^2)$

Solve these equations.

5 $4 = \frac{x}{5}$ 6 $4 = \frac{5}{x}$

7 $43 - 2x = 7 - 8x$ 8 $1.3 - 0.3x = 0.2x + 0.3$

9 $0.6(x + 1) + 0.2(6 - x) = x - 0.6$

10 The length of a conference room is one and a half times its width. A carpet that is twice as long as it is wide is placed in the centre of the room, leaving a 3 m wide border round the carpet.
Find the area of the carpet.

11 Two years ago, my age was four times that of my son. Eight years ago, my age was ten times that of my son. Find the age of my son now.

12 A river flows at 2 m/s. What is the speed through the water of a boat that can go twice as fast downstream as upstream?

13 Matt goes to buy a television. If he pays cash, he gets a discount of 7%. If he pays by instalments, he has to pay an extra 10% in interest. The difference between the two methods is $2176. Find the cost of the television.

Gradient of a straight line

The slope of a line is its **gradient**. This is usually represented by the letter m.

For a straight line,

$$m = \frac{\text{change in the } y\text{-coordinates}}{\text{change in the } x\text{-coordinates}} = \frac{\text{'rise'}}{\text{'run'}}$$

Example 1

Find the gradient of the straight line joining A(1, 2) to B(3, 6).

First draw a diagram.
Mark in the changes in coordinates.

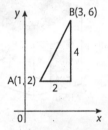

For AB, the change in the y-coordinate is $6 - 2 = 4$, and the change in the x-coordinate is $3 - 1 = 2$. The gradient is $\frac{4}{2} = 2$ (a positive gradient, running *uphill*).

Example 2

Find the gradient of the straight line joining C(6, 4) to D(12, 1).

First draw a diagram.
Mark in the changes in coordinates.

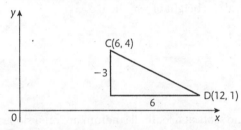

For CD, the change in the y-coordinate is $1 - 4 = -3$, and the change in the x-coordinate is $12 - 6 = 6$. The gradient is $\frac{6}{-3} = -\frac{1}{2}$ (a negative gradient, running *downhill*).

Key Points

- Gradient $= \dfrac{\text{'rise'}}{\text{'run'}}$
- Lines like this have a *positive* gradient.
- Lines like this have a *negative* gradient.
- Parallel lines have the same gradient.
- Always draw a diagram.

Investigate

- Find the gradient of the line AB.

- Investigate the gradient of AB as point B moves closer and closer to point C. Tabulate your results. What is the gradient of the horizontal line AC?

What is the gradient of any horizontal line?

- Investigate the gradient of AB as point A moves closer and closer to point C. Tabulate your results. What is the gradient of the vertical line BC?

What is the gradient of any vertical line?

Exercise 15

For Questions 1−5, find the gradient of the straight line joining A to B when

1
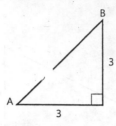

2
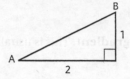

3 A is (1, 3) and B is (2, 6)

4 A is (−4, −1) and B is (4, 1)

5 A is (−2, 2) and B is (2, 1)

6 A hill has a gradient of 0.1. What is the value of *h*?

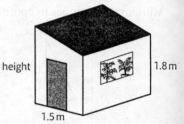

gradient ≐ 0.1

7 A ladder reaches 6 m up a vertical wall and has a gradient of 4.
How far is the foot of the ladder from the wall?

8 The roof of a lean-to garden shed has a gradient of 0.35.
Find the height of the shed.

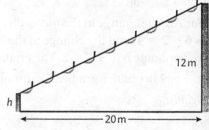

9 The seats at a football stadium are on a slope with
a gradient of $\frac{1}{2}$.
What is the height *h* of the bottom seats?

10 A road has a gradient of $\frac{1}{15}$ for 90 m. Then there is a horizontal section 130 m long.
The final section has a gradient of $\frac{1}{25}$ for 200 m.

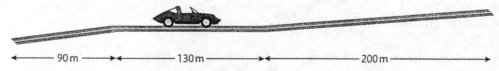

a Find the total height gained from start to finish.
b What is the average gradient from start to finish?

11 The masts for London's O$_2$ Arena were held up
during erection by wire ropes. The top of a mast, A,
is 106 m above the ground, and C is vertically below A.
The gradient of one wire rope, AB, is 1, and CD is 53 m.
a Find the gradient of AD.
b Find the length of BD.

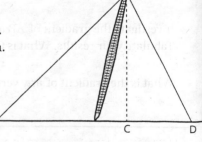

Exercise 15*

For Questions 1 and 2, find the gradients of the straight lines joining A to B when

1 A is $(-4, -1)$, B is $(4, 2)$ **2** A is $(-2, 4)$, B is $(2, 1)$

3 The line joining $A(1, 4)$ to $B(5, p)$ has a gradient of 12. Find the value of p.

For Questions 4 and 5, find, by calculating gradients, whether or not the opposite sides of quadrilateral ABCD are parallel.

4 A is $(2, 1)$, B is $(14, 9)$, C is $(24, 23)$, D is $(10, 13)$

5 A is $(2, 1)$, B is $(14, 7)$, C is $(20, 19)$, D is $(8, 13)$

6 Alexander enjoys mountain biking. He has found that the maximum gradient which he can cycle up is 0.3, and the maximum gradient that he can safely descend is 0.5. His map has a scale of 2 cm to 1 km, with contours every 25 m.
   a What is the minimum distance between the contours on his map that allows him to go uphill?
   b What is the minimum distance between the contours on his map that allows him to go downhill?

7 One of the world's tallest roller coasters is in Blackpool, England. It has a maximum height of 72 m, and gives white-knuckle rides at up to 140 km per hour. The maximum drop is 65 m over a horizontal distance of 65 m in two sections. The first section has a gradient of 3, and the second section has a gradient of $\frac{1}{2}$.
How high above the ground is point A?

8 The line joining $(3, p)$ to $(7, -4p)$ is parallel to the line joining $(-1, -3)$ to $(3, 7)$. Find p.

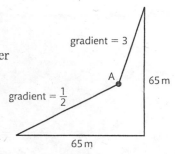

Straight-line graphs
Graphs of the form $y = mx + c$

Activity 3

- For each of these equations, copy and complete this table of values.

x	-2	0	2
y			

    $y = x + 1$ $y = -x + 1$ $y = 2x - 1$

    $y = -2x + 1$ $y = 3x - 1$ $y = \frac{1}{2}x + 2$

- Draw **one** set of axes, with the x-axis labelled from -2 to 2 and the y-axis from -7 to 5. Plot the graphs of all six equations on this set of axes.
- Copy and complete this table.

Equation	Gradient	y-intercept
$y = x + 1$		
$y = -x + 1$		
$y = 2x - 1$		
$y = -2x + 1$		
$y = 3x + 1$		
$y = \frac{1}{2}x + 2$		
$y = mx + c$		

Can you see a connection between the number in front of x and the gradient?
The **y-intercept** is the value of y where the line crosses the y-axis.

Can you see a connection between the number at the end of the equation and the y intercept?

Key Points
The graph of
$y = mx + c$
is a straight line with
gradient m and
y intercept c.

Sketching a straight line means showing the approximate position and slope of the line *without* plotting the line. If you know the gradient and intercept you can sketch the straight line easily.

Example 3

Sketch these two lines.

$y = 2x - 1 \qquad y = -\frac{1}{2}x + 3$

$y = 2x - 1$ is a straight line with gradient 2 and intercept -1.

$y = -\frac{1}{2}x + 3$ is a straight line with gradient $-\frac{1}{2}$ and intercept 3.

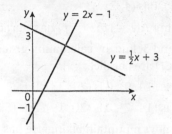

Exercise 16

For Questions 1–8, write down the gradient and y intercept and then sketch the graph of the equation.

1 $y = x + 1$ **2** $y = \frac{1}{2}x + 4$ **3** $y = 3x + 5$ **4** $y = x - 7$

5 $y = \frac{1}{3}x + 2$ **6** $y = -\frac{1}{2}x + 5$ **7** $y = -\frac{1}{3}x - 2$ **8** $y = 4 - 2x$

For Questions 9 and 10, write down the equations of the lines with gradient

9 with gradient 2, passing through (0, 1)

10 with gradient -1, passing through (0, 2)

For Questions 11–13, write down the equations of the lines that are parallel to

11 $y = 2x - 7$, passing through (0, 4)   **12** $y = 4 - 5x$, passing through (0, -1)

13 Write down possible equations for these sketch graphs.

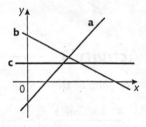

Exercise 16*

For Questions 1–8, write down the gradient and y intercept and then sketch the graph of each equation.

1 $y = 3x + 2$ **2** $y = -2x$ **3** $y = 5x + \frac{1}{2}$ **4** $y = -\frac{3}{4}$

5 $y = -3x + \frac{5}{2}$ **6** $y = 6x - \frac{3}{2}$ **7** $x = -1.5$ **8** $y = -\frac{2}{3}x - \frac{5}{3}$

For Questions 9 and 10, write down the equations of the lines with gradient

9 2.5, passing through (0, -2.3)   **10** $\frac{1}{4}$, passing through (4, 2)

For Questions 11–13, write down the equations of the lines that are parallel to

11 $2y = 5x + 7$, passing through (0, -3.5)   **12** $7x + 6y = 13$, passing through (6, 7)

13 Write down possible equations for these sketch graphs.

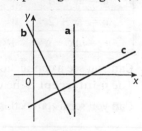

Activity 4

Equipment needed: a cylinder with a diameter of $5-10$ cm (a drinks can or cardboard tube is ideal), a length of string about 30 times as long as the diameter of the cylinder, a ruler and graph paper.

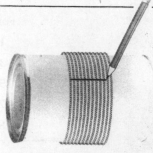

- Wrap the string tightly around the cylinder, keeping the turns close together. Ask a friend to draw a straight line across the string while you hold the ends of the string.
- Unwind the string. Measure the distance of each mark from the first mark. Enter your results in a table.
- Plot these points on a graph of D against M. (Plot D on the vertical axis and M on the horizontal axis.) Draw the best straight line through these points.
- Calculate the gradient of the line, and then write down the equation of the line.
- The gradient should equal πd, where d is the diameter of the cylinder. Use your gradient to work out an estimate for π.
- Repeat the activity with various cylinders, and obtain further estimates for π.

Mark M	Distance D (cm)
1	0
2	
3	
4	
5	
6	

Graphs of the form $ax + by = c$

The graph of $3x + 4y = 12$ is a straight line.

The equation can be rearranged as $y = -\frac{3}{4}x + 3$, showing that the graph is a straight line with gradient $-\frac{3}{4}$ and y intercept $(0, 3)$.

An easy way to draw or sketch this graph is to find where the graph crosses the axes.

Example 4

Sketch the graph of $x + 2y = 8$.

Substituting $y = 0$ gives $x = 8$, which shows that $(8, 0)$ lies on the line.

Substituting $x = 0$ gives $y = 4$, which shows that $(0, 4)$ lies on the line.

Exercise 17

For Questions $1-7$, find where the graph crosses the axes and sketch the graph.

1 $x + y = 5$
2 $3x + y = 6$
3 $2x + y = 6$
4 $3x + 2y = 12$
5 $4x + 5y = 20$
6 $x - 2y = 4$
7 $4y - 3x = 24$

8 A firm selling CDs finds that the number sold (N thousand) is related to the price (£P) by the formula $6P + N = 90$.
   a Draw the graph of N against P for $0 \le N \le 90$ (the vertical axis should be the P axis, and the horizontal axis should be the N axis).
   b Use your graph to find the price when 30 000 CDs are sold.
   c Use your graph to find the number sold if the price of a CD is set at £8.
   d Use your graph to find the price if 90 000 CDs are sold. Is this a sensible value?

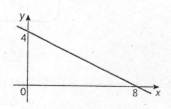

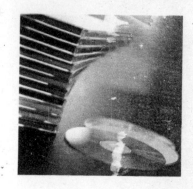

Exercise 17*

For Questions 1−7, find where the graph crosses the axes and sketch the graph.

1 $4x + y = 12$

2 $x - 5y = 10$

3 $6x + 3y = 36$

4 $6x + 4y = 21$

5 $4x - 5y = 30$

6 $7y - 2x = 21$

7 $6x - 7y = -21$

8 Courtney has started playing golf. To try to reduce her handicap she has lessons with a professional. She keeps a record of her progress.

Week (W)	5	10	20	30
Handicap (H)	22	21	20	19

a Plot these points on a graph of H against W. Draw in the best straight line.

b What was Courtney's handicap before she started having lessons?

c Find the gradient and intercept of the line.
Write down the equation of the line in the form $ax + by = c$.

d To have a trial for the youth team, Courtney needs to have a handicap of less than 12. Use your equation to find how many weeks it will take Courtney to reduce her handicap to 12. Do you think this is a reasonable time?

Activity 5

Your aim is to find the equation of the straight line joining two points.

- Plot the points A(1, 3) and B(5, 5) on a graph. Find the gradient of AB.

- Calculate where the straight line passing through AB will intercept the y-axis.

- Write down the equation of the straight line passing through A and B.

- Use this method to find the equation of the straight line joining these pairs of points:
  $(-2, 1)$ and $(-1, 4)$ $(-3, 4)$ and $(6, 1)$ $(-2, -1)$ and $(4, 3)$

Exercise 18 (Revision)

1 Find the gradient of the straight line joining A to B when
  a A is $(3, 4)$, B is $(5, 8)$ **b** A is $(-1, 2)$, B is $(1, 0)$

2 The foot of a ladder is 1.5 m from the base of a vertical wall. The gradient of the ladder is 3. How far does the ladder reach up the wall?

3 Write down the gradient and y intercept of the graph of
  a $y = 3x - 2$ **b** $y = -2x + 5$

4 Write down the equations of the lines with
  a gradient 2, passing through $(0, -1)$ **b** gradient -3, passing through $(0, 2)$

5 Sketch the following graphs.
  a $y = 2x - 3$ **b** $y = 4 - x$ **c** $2x + 5y = 10$

6 Which of these lines are parallel?
  $y = 2x + 4$ $x - 3y = 1$ $4x = 2y + 7$ $9y = 3x + 4$
  $4x - 3y = 12$ $3x - 4y = 12$ $3y = 4x - 1$ $4y = 3x + 7$

7 Find the gradients of the lines parallel to AB when
  a A is $(1, 2)$, B is $(4, 4)$ **b** A is $(-2, 1)$, B is $(2, -1)$

Exercise 18* (Revision)

1 Find the gradients of the lines parallel to AB when

   a A is $(-2, -1)$, B is $(4, 2)$ **b** A is $(-1, 4)$, B is $(1, -1)$

2 The Leaning Tower of Pisa is 55 m high, and the gradient of its lean is 11. By how much does the top overhang the bottom?

3 Sketch the following graphs

   a $y = 3x - 2$ **b** $y = 3 - 2x$ **c** $2y = 5 - x$ **d** $5x + 3y = 10$

4 Find b such that the line from the origin to $(3, 4b)$ is parallel to the line from the origin to $(b, 3)$.

5 Find the equation of the lines passing through $(6, 4)$ that are parallel and perpendicular to $3y = x + 21$.

6 A temperature F in degrees Fahrenheit is related to the temperature C in degrees Celsius by the formula $F = \frac{9}{5}C + 32$.

   a Draw a graph of F against C for $-50 \leqslant C \leqslant 40$.

   b Use your graph to estimate 80 °F and -22 °F in degrees Celsius and 25 °C in degrees Fahrenheit.

   c Use your graph to find which temperature has the same value in both degrees Fahrenheit and degrees Celsius.

Write down the equations that will produce these patterns.

7

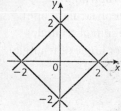

8

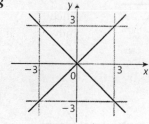

Basic principles

Triangles

(A dashed line indicates an axis of symmetry.)

Isosceles triangle

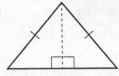

Acute, obtuse and right angles are possible.

Equilateral triangle

The rotational symmetry is of order 3.

Angle properties

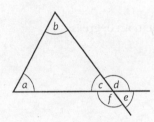

$$a + b + c = 180° \quad \text{(Angle sum of triangle)}$$
$$c + d = 180° \quad \text{(Angles on straight line are supplementary)}$$
$$c = e \quad \text{(Vertically opposite angles)}$$
$$c + d + e + f = 360° \quad \text{(Angles at a point)}$$
$$\text{Since} \quad d = 180° - c \quad \text{(Angles on straight line)}$$
$$\text{and} \quad a + b = 180° - c \quad \text{(Angle sum of triangle)}$$
$$d = a + b \quad \text{(Exterior angle of triangle)}$$

Parallel lines

Alternate angles are equal.

Co-interior angles $a + b = 180°$

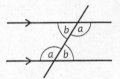

Corresponding angles are equal.

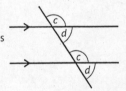

Quadrilaterals

Square

Rotational symmetry of order 4

Rhombus

Rotational symmetry of order 2

Rectangle

Rotational symmetry of order 2

Parallelogram

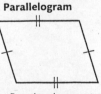

Rotational symmetry of order 2

Arrowhead

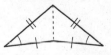

Kite

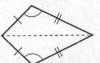

Acute, obtuse and right angles are possible.

Trapezium

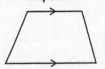

There is no symmetry. Right angles are possible.

Isosceles trapezium

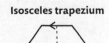

Activity 6

Copy and complete this table to show which properties are true for each type of quadrilateral.

Property	Square	Rectangle	Rhombus	Parallelogram	Arrowhead	Trapezium	Kite
The diagonals are equal in length.				No			
The diagonals bisect each other.				Yes			
The diagonals are perpendicular.				No			
The diagonals bisect the angles at the corners.				No			
Both pairs of opposite angles are equal.				Yes			

Angles of a regular polygon

Interior angles of an *n*-sided polygon

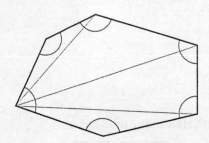

Exterior angles of an *n*-sided polygon

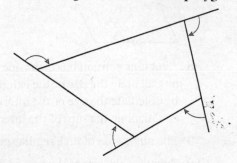

The polygon can be divided into $(n - 2)$ triangles. Therefore the angle sum = $(n - 2) \times 180°$.

The angles add up to one complete turn. Therefore they sum to 360°.

Interior and exterior angles add up together to $180n°$.

In a *regular* polygon, all the interior angles are equal and all the exterior angles are equal.

Each interior angle = $\dfrac{(n - 2) \times 180°}{n}$

Each exterior angle = $\dfrac{360°}{n}$

Example 1

Find the angle sum of a polygon with seven sides.

$n = 7$

Angle sum = $(7 - 2) \times 180°$

$= 5 \times 180° = 900°$

Example 2

A regular polygon has ten sides. Find the size of each interior and each exterior angle.

$n = 10$

Interior angle = $\dfrac{(10 - 2) \times 180°}{10}$

$= 8 \times 18° = 144°$

Exterior angle = $180° - 144° = 36°$

Or, find the exterior angle first.

Exterior angle = $\dfrac{360°}{10} = 36°$

Interior angle = $180° - 36° = 144°$

Exercise 19

Calculate the size of each lettered angle.

1
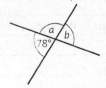

2
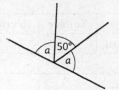

3
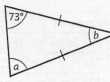

4
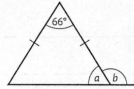

5
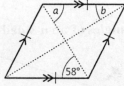

6
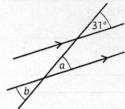

7
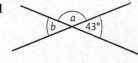

8 A regular polygon has eight sides.
   a Calculate the size of the exterior angles.
   b Calculate the size of the interior angles.
   c Calculate the sum of the interior angles.

9 The angle sum of an irregular polygon is 1260°. How many sides has it?

10 Calculate the size of the two unknown angles.

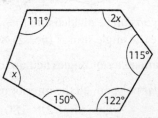

Exercise 19*

Calculate the size of each lettered angle.

1
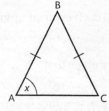

2

3
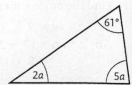

4 Express ∠ABC in terms of x.

5 Find the size of angle x.
   Write out a 'solution' giving a
   reason for each step.

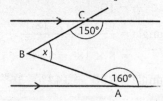

6 ABCD is a rectangle. Find angles *a* and *b*, giving reasons with each step.

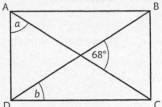

7 Find angles *a* and *b*, giving reasons with each step.

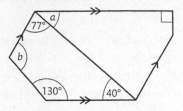

8 Find, giving reasons, the size of angle *a*.

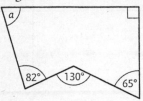

9 Find angles *a* and *b*, giving reasons with each step.

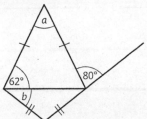

Constructions

The properties of triangles and quadrilaterals are used in the standard ruler and compass constructions.

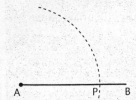 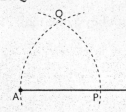 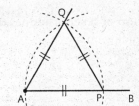
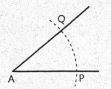 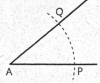 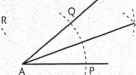 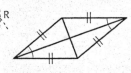

Remember

Constructing a perpendicular bisector of a line (diagonals of a rhombus)

- Draw arcs from A, with the same radius, above and below the line.
- With the same radius, draw arcs from B to intersect those from A, above and below the line. Label these two intersections P and Q.
- Draw the line PQ. **PQ is the perpendicular bisector of AB.**
  (Note: R is the mid-point of AB.)

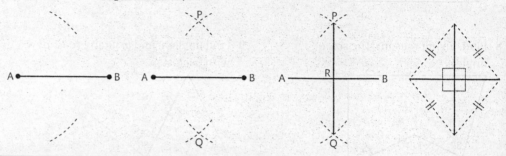

Remember

Constructing a perpendicular from a point X on the line

- With the same radius, draw arcs from the point X to cut the line at either side. Label these points A and B.
- The perpendicular bisector of this part AB of the line will pass through X.

A locus is the position of a set of points that obey a particular rule. It can be a line, curve or region, depending on the rule.

Remember

Common loci

Points on the **angle bisector** of ∠BAC are equidistant from the lines AB and AC.

Points on the **perpendicular bisector** of PQ are equidistant from P and Q.

Points on the **circle** with centre at X are equidistant from X.

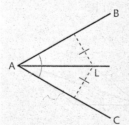

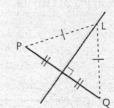

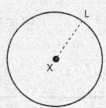

Example 3

A 10 km road race attracts so many runners that it is decided to split up the beginning of the race and have three different starts in a park.

The top diagram shows the three starting positions, A, B and C.
On an accurate scale drawing, show the point P where the three routes must converge so that they are all of the same distance.
Measure AP, and hence calculate the distance from each starting position to P.

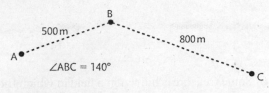

Choose a scale 1 cm : 50 m. As P is to be equidistant from A and B, it must be on the perpendicular bisector of AB. Draw this.

As P is also to be equidistant from B and C, it must be on the perpendicular bisector of BC. Draw this.

The point of intersection is equidistant from A and B and C.

AP = 19 cm. Therefore the distance from each starting position to P is 950 m.

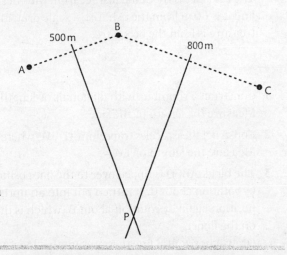

Exercise 20

Questions 1–6 should be done on plain paper. Protractors can be used, and all construction arcs are to be shown. Make a rough sketch of the figure before you begin a construction.

1 Construct triangle ABC, where AB = 8 cm, ∠A = 60°, and ∠B = 45°.
   Measure the length of AC.

2 A fierce dog is tethered by a rope 10 m long to a post 6 m from a straight path. If the path is 2 m wide, draw a scale diagram to illustrate the area of path along which a walker would be in danger.

3 P, Q and R represent the positions of three radio beacons. Signals from P have a range of 300 km, Q has a range of 350 km and R has a range of 200 km.

   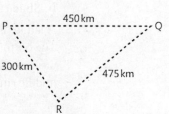

   a Reproduce the diagram and shade the region in which all three signals can be received.

   b Measure the shortest distance from Q to this region.

4 Gas rig Beta is 7 km from gas rig Gamma on a bearing of 210°. Bearings are measured from North in a clockwise direction. The region less than 4 km from gas rig Beta is an exclusion zone for ships.

   a Using a scale of 1 cm to 1 km, draw a scale diagram showing the positions of the gas rigs, and shade the region that represents the exclusion zone.

   b A boat sails so that it is always the same distance from Gamma and Beta.
      Draw the route taken by the boat.

   c For what distance is the boat within 4 km of oil rig Beta?

5 PQ is a breakwater, 750 m long, with a lighthouse at Q.
Using a scale drawing, find the closest distance to the breakwater from a
ship which is 280 m from the lighthouse at Q and 190 m from P.

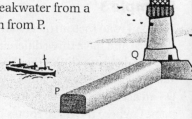

6 Some treasure is hidden in a field in which there are three trees: an ash A,
a beech B and a chestnut C. BC = 300 m, CA = 210 m and AB = 165 m.
The treasure is the same distnce from the chestnut as from the beech,
and it is 60 m from the ash. Use a scale drawing to find out how far the
treasure is from the beech tree.

Exercise 20*

1 Construct a rhombus with diagonals of length 9 cm and 6 cm.
Measure the length of the side.

2 Construct the isosceles trapezium TUVW, where TU = 8.5 cm, VW = 5 cm and ∠UTW = 75°.
Measure the length of TW.

3 The block ABCD is tipped over to the flat position
by rotation about C. It is then put into an upright
position again by rotation about B (which is then
on the floor).
Draw a horizontal line to represent the floor, and then
draw the locus of A during these two movements.

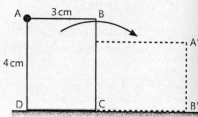

4 The diagram shows a sheep pen that is in the middle of a field. A sheepdog is tethered at
the corner C by a rope 6 m long.

a Draw a scale diagram of the pen, and shade the region that the dog can cover if
he is outside the pen.
b Shade the region that he can cover if he is inside the pen.

5 A ladder is 15 m long. It is resting almost vertically against a wall.
The bottom of the ladder is pulled out from the wall and allowed to slide into the horizontal
position. Draw x- and y-axes from 0 to 15, and make a scale drawing of the locus of the
middle rung of the ladder. (A 15 cm ruler may be useful.)

Exercise 21 (Revision)

Use a compass and ruler to draw the following. Remember to show all your construction arcs.

1 An equilateral triangle of sides 7 cm.

2 A triangle of sides 7 cm, 8 cm and 9 cm.

3 The perpendicular bisector of the line AB where AB = 8 cm.

4 Angles of 30°, 60° and 45°.

5 In a party game, a valuable prize is hidden within a triangle formed by an Oak tree (O), an Apple tree (A) and a Plum tree (P).

  a Given that OA = 16 m, AP = 18 m and OP = 20 m construct the triangle OAP using a scale of 1 cm = 2 m.

  b The prize is equidistant from the Apple tree and the Plum tree and 12 m from the Oak tree. By careful construction find the distance of the prize from the Plum tree.

Calculate the size of each lettered angle.

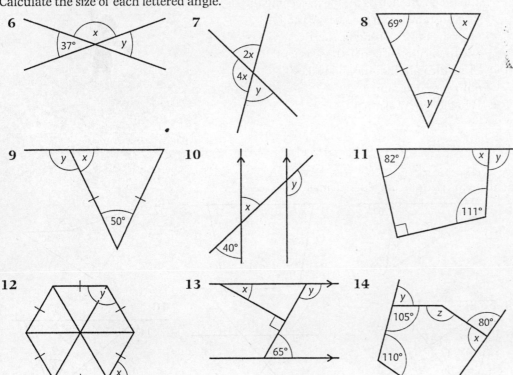

6
37° x y

7
2x 4x y

8
69° x y

9
y x 50°

10
y x 40°

11
82° x y 111°

12
y x

13
x y 65°

14
y 105° z 80° x 110°

Exercise 21* (Revision)

1 A regular decagon has ten sides. Calculate

  a the size of the exterior angles.

  b the size of the interior angles.

  c the sum of the interior angles.

2 The angle sum of a regular polygon is 3600°. Calculate

  a the number of sides of this polygon.

  b the exterior angle of this polygon.

3 The rectangle ABCD represents a map of an area 30 m × 60 m.
A mobile phone mast, M, is to be placed such that it is equidistant from A and B and 20 m from point E, such that BE : EC = 1 : 2.

  a Draw the map using a scale drawing of 1 cm = 5 m.

  b Showing your construction lines clearly, find the shortest distance of M from D.

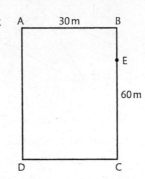

A — 30 m — B
E
60 m
D — C

4 The diagram represents a rectangular lawn. There is a water
sprinkler at the point E, halfway between C and D.
The sprinkler wets the area within 15 m from E.
a Using a scale of 1 cm to 5 m, draw a diagram of the garden,
and shade the area wetted by the sprinkler.
b A child is playing on the lawn. She starts at A,
and then runs across the lawn, keeping the
same distance from the sides AD and AB
until she is 10 m from the side DC.
She then runs straight to the corner B.
Draw the path that the child takes
onto your diagram.
c What length of her path is wet?

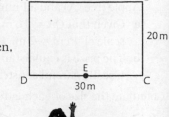

Calculate the size of each lettered angle.

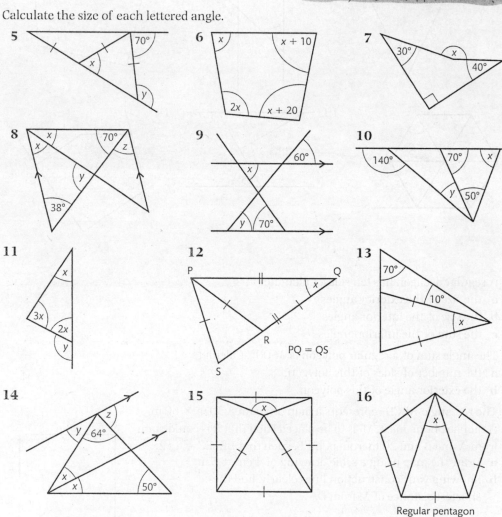

5 70°, x, y

6 x, x + 10, 2x, x + 20

7 30°, x, 40°

8 x, x, 70°, z, y, 38°

9 60°, x, y, 70°

10 140°, 70°, x, y, 50°

11 x, 3x, 2x, y

12 P, Q, x, R, S, PQ = QS

13 70°, 10°, x

14 z, y, 64°, x, x, 50°

15 x

16 x, Regular pentagon

Basic ideas of sets

The concept of a set is a simple but powerful idea. The theory of sets is mainly due to the work of the German mathematician, Cantor. It led to arguments and controversy, but by the 1920s his ideas were generally accepted and led to great advances in mathematics.

The objects can be numbers, animals, ideas, colours, in fact anything you can imagine. A set can be described by **listing** all the members of the set, or by giving a **rule** to describe the members. The list or rule is enclosed by **braces** { }.

Example 1

A set described by a list:

{Maureen, John, Louice} is the set consisting of the three people called Maureen, John and Louice.

Example 2

A set described by a rule:

{even numbers between 1 and 11} is the set consisting of the five numbers 2, 4, 6, 8, 10.

> **Key Point**
> A **set** is a collection of objects, which are called the elements or members of the set.

Sets are often labelled by a single capital letter. A = {odd numbers between 2 and 10} means A is the set consisting of the four numbers 3, 5, 7, 9.

Sets can be infinite in size, for example the set of prime numbers.

> **Key Point**
> The **number of elements** in the set A is written as $n(A)$.

Example 3

If E = {2, 8, 4, 6, 10} and F = {even numbers between 1 and 11}, then:

$n(E) = 5$, $n(F) = 5$; in other words both E and F have the same number of elements

$3 \notin$ E means 3 is not a member of the set E

$6 \in$ F means 6 is a member of the set F

E = F because both E and F have the same members. The order in which the members are listed does not matter.

> **Key Point**
> **Membership** of a set is indicated by the symbol $\in$ and non-membership by the symbol $\notin$.

The concept of the empty set might seem strange, but it is very useful.

> **Key Point**
> The **empty set** is the set with no members. It is denoted by the symbol $\varnothing$ or { }.

Example 4

Give two examples of the empty set.

a The set of people you know over 4 m tall.
b The set of odd numbers divisible by two.

Exercise 22

1 Write down two more members of each of these sets.

a {carrot, potato, pea, ...}
b {red, green, blue, ...}
c {a, b, c, d, ...}
d {1, 3, 5, 7, ...}

2 List these sets.

   a {days of the week}   **b** {square numbers less than 101}

   c {subjects you study at school}   **d** {prime numbers less than 22}

3 Describe these sets by a rule.

   a {a, b, c, d}   **b** {Tuesday, Thursday}

   c {1, 4, 9, 16}   **d** {2, 4, 6, 8, ...}

4 Which of these statements are true?

   a cat $\in$ {animals with two legs}   **b** Square $\notin$ {parallelograms}

   c $1 \in$ {prime numbers}   **d** $2 \notin$ {odd numbers}

5 . Which of these are examples of the empty set?

   a The set of men with no teeth

   b The set of months of the year with 32 days

   c The set of straight lines drawn on the surface of a sphere

   d The set of prime numbers between 35 and 43.

Exercise 22*

1 Write down two more members of each of these sets.

   a {Venus, Earth, Mars, ...}   **b** {triangle, square, hexagon, ...}

   c {hydrogen, iron, aluminium, ...}   **d** {1, 4, 9, 16, ...}

2 List these sets.

   a {all possible means of any two elements of 1, 3, 5}

   b {different digits of 11^4}

   c {all factors of 35}

   d {powers of 10 less than one million}

3 Describe these sets by a rule.

   a {spring, summer, autumn, winter}   **b** {circle, ellipse, parabola, hyperbola}

   c {1, 2, 4, 8, 16}   **d** {(3, 4, 5), (5, 12, 13), (7, 24, 25), ...}

4 Which of these statements are true?

   a Everest $\in$ {mountains over 2000 m high}

   b $2000 \notin$ {leap years}

   c $2x + 3y = 5 \in$ {straight-line graphs}

   d $-2 \in$ {solutions of $x^3 - 2x^2 = 0$}

5 Which of these are examples of the empty set?

   a The set of three-legged kangaroos

   b The set that has the numeral zero as its only member

   c The set of common factors of 11 and 13

   d The set of solutions of $x^2 = -1$.

Venn diagrams

Sets can be shown in a diagram called a **Venn diagram** after the English mathematician John Venn (1834–1923). The members of the set are shown within a closed curve.

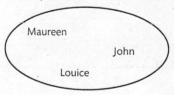

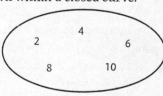

When the number of elements in a set is so large that they cannot all be shown, then a simple closed curve is drawn to indicate the set. If T = {all tabby cats} then this is shown on a Venn diagram as

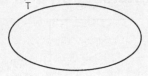

If C = {all cats in the world}, then T and C can be shown on a Venn diagram as

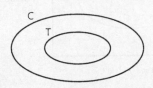

The set T is shown inside the set C because every member of T is also a member of C.

Key Point

If T is inside C, then T is called a **subset** of C. This is written as T ⊂ C.

Example 5

A = {1, 2, 3, 4, 5, 6, 7, 8, 9}
a List the subset O = {odd numbers}.
b List the subset P = {prime numbers}.
c Is Q = {8, 4, 6} a subset of A?
d Is R = {0, 1, 2, 3} a subset of A?

Answers
a O = {1, 3, 5, 7, 9}
b P = {2, 3, 5, 7}
c Q is a subset of A (Q ⊂ A) because every member of Q is also a member of A.
d R is not a subset of A (R ⊄ A) because the element 0 is a member of R but is not a member of A.

If the problem was only about cats in this world and wasn't concerned about cats outside this world, then it is more usual to call the set C the **universal set**, denoted by ℰ. The universal set contains all the elements being discussed in a particular problem, and is shown as a rectangle, where T = {all tabby cats}.

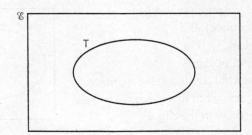

If the problem was only about cats in Rome then ℰ = {all cats in Rome}; the Venn diagram does not change. The cats outside T are all non-tabby cats. This set is denoted by T′ and is known as the **complement** of T.

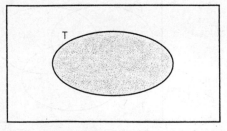

T shown shaded

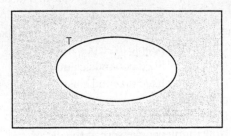

T′ shown shaded

Unit 1 : Sets

33

Intersection and union

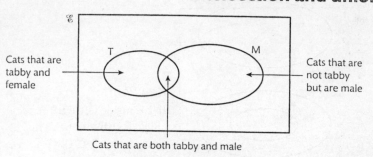

Cats that are tabby and female

Cats that are both tabby and male

Cats that are not tabby but are male

Sets can overlap. Let M = {all male cats}. T and M overlap because some cats are both tabby and male. T and M are shown on this Venn diagram:

The set of cats that are both tabby and male is where the sets T and M overlap.

Key Point
Where T and M overlap is called the **intersection** of the two sets T and M, and is written T ∩ M.

Example 6

$\mathscr{E}$ = {all positive integers less than 10}, P = {prime numbers less than 10} and O = {odd numbers less than 10}.

a Illustrate these sets on a Venn diagram.

b Find the set P ∩ O and n(P ∩ O).

c List P′.

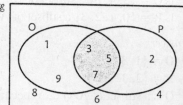

Answers

a The set P ∩ O is shown shaded on the Venn diagram.

b From the Venn diagram, P ∩ O = {3, 5, 7} and n(P ∩ O) = 3.

c P′ is every element not in P, so P′ = {1, 4, 6, 8, 9}.

Exercise 23

1 On the Venn diagram, $\mathscr{E}$ = {pupils in a class},
C = {pupils who like chocolate} and
T = {pupils who like toffee}.

a How many pupils like chocolate?

b Find n(T) and express what this means in words.

c Find n(C ∩ T) and express what this means in words.

d How many pupils are there in the class?

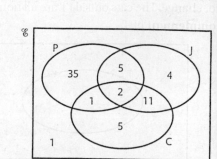

2 $\mathscr{E}$ = {all cars in the world}, P = {pink cars}, R = {Rolls-Royce cars}.

a Describe the set P ∩ R in words.

b If P ∩ R = ∅, describe what this means.

3 On the Venn diagram, $\mathscr{E}$ = {people at a disco},
P = {people who like pop music}, C = {people who like classical music} and J = {people who like jazz}.

a How many people liked pop music only?

b How many liked pop music and classical music?

c How many liked jazz and classical music, but not pop music?

d How many liked all three types of music?

e How many people were at the disco?

Unit 1 : Sets

4 On the Venn diagram, $\mathscr{E}$ = {ice-creams in a shop},
C = {ice-creams containing chocolate},
N = {ice-creams containing nuts} and
R = {ice-creams containing raisins}.

   a How many ice-creams contain both chocolate
     and nuts?

   b How many ice-creams contain all three ingredients?

   c How many ice-creams contain just raisins?

   d How many ice-creams contain chocolate and raisins but not nuts?

   e How many different types of ice-creams are there in the shop?

Exercise 23*

1 $\mathscr{E}$ = {all positive integers less than 12}, A = {2, 4, 6, 8, 10}, B = {4, 5, 6, 7, 8}.

   a Illustrate this information on a Venn diagram.

   b List A ∩ B and find n(A ∩ B).

   c Does A ∩ B = B ∩ A?

   d List (A ∩ B)′.

   e Is A ∩ B a subset of A?

2 $\mathscr{E}$ = {all positive integers less than 12}, E = {1, 2, 3, 4}, F = {5, 6, 7, 8}.

   a Illustrate this information on a Venn diagram.

   b List E ∩ F.

   c If E ∩ F = ∅, what does this imply about the sets E and F?

3 $\mathscr{E}$ = {letters of the alphabet}, V = {vowels}, A = {a, b, c, d, e}, B = {d, e, u}.

   a Illustrate this information on a Venn diagram.

   b List the sets V ∩ A, V ∩ B′, A′ ∩ B.

   c List the set V ∩ A ∩ B.

4 $\mathscr{E}$ = {all positive integers}, F = {4, 8, 12, 16, 20, 24}, S = {6, 12, 18, 24}.

   a Illustrate this information on a Venn diagram.

   b List F ∩ S.

   c What is the smallest member of F ∩ S?

   d F is the set of the multiples of 4, S is the set of the multiples of 6. What is the LCM of
     4 and 6? How is this related to the set F ∩ S?

   e Use this method to find the LCM of   **(i)** 6 and 8   **(ii)** 8 and 10.

5 Show that a set of three elements has eight subsets, including ∅. Find a rule giving the
   number of subsets (including ∅) for a set of n elements.

The union of two sets is the set of elements that belong to A or to B or to both A and B.

Example 7

$\mathscr{E}$ = {all positive integers less than 10},
P = {prime numbers less than 10} and
O = {odd numbers less than 10}.

   a Illustrate these sets on a Venn diagram.

   b Find the set P ∪ O and n(P ∪ O).

Answers

   a The set P ∪ O is shown shaded in the Venn diagram.

   b From the Venn diagram, P ∪ O = {1, 2, 3, 5, 7, 9} and n(P ∪ O) = 6.

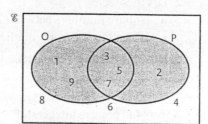

> **Key Point**
>
> The **union** of two sets
> A and B is the set of
> elements that belong to
> both sets, and is written
> A ∪ B.

Exercise 24

1 $\mathscr{E}$ = {all positive integers less than 10}, A = {1, 3, 5, 7, 9}, B = {3, 4, 5, 6}.
   a Illustrate this information on a Venn diagram.
   b List A $\cup$ B and find n(A $\cup$ B).
   c Does A $\cup$ B = B $\cup$ A?
   d List (A $\cup$ B)'.
   e Is A $\cup$ B a subset of A?

2 $\mathscr{E}$ = {pack of 52 playing cards}, B = {black cards}, C = {clubs}, K = {kings}.
   a Draw a Venn diagram to show the sets B, C and K.
   b Describe the set B $\cup$ K.
   c Describe the set B $\cup$ K $\cup$ C.
   d Describe the set B' $\cup$ K.

3 $\mathscr{E}$ = {all triangles}, E = {equilateral triangles}, I = {isosceles triangles} and
   R = {right-angled triangles}.
   a Draw a Venn diagram to show the sets E, I and R.
   b Sketch a member of I $\cap$ R.
   c Describe the sets I $\cup$ E and I $\cup$ R.
   d Describe the sets I $\cap$ E and E $\cap$ R.

Exercise 24*

1 $\mathscr{E}$ = {all positive integers less than 10}, E = {2, 4, 6, 8}, O = {1, 3, 5, 7, 9}.
   a Illustrate this information on a Venn diagram.
   b List E $\cup$ O.
   c If n(E) + n(O) = n(E $\cup$ O), what does this imply about the sets E and O?
   d If (E $\cup$ O)' = $\varnothing$, what does this tell you about E and O?

2 In Joe's Pizza Parlour, H is the set of pizzas containing ham and C is the set of pizzas
   containing cheese.
   a Describe the set H $\cup$ C in words.
   b Describe the set H $\cap$ C in words.
   c If (H $\cup$ C)' = $\varnothing$, what can you say?

3 If n(A) = n(A $\cup$ B), what can you say about the sets A and B?

Exercise 25 (Revision)

1 Write down two more members of these sets.
   a {salt, pepper, thyme, ...}   b {cat, dog, rabbit, ...}
   c {apple, banana, orange, ...}   d {red, black, blue, ...}

2 List these sets.
   a {square numbers between 2 and 30}   b {all factors of 24}
   c {vowels in the word 'mathematics'}   d {months of the year containing 30 days}

3 Describe these sets by a rule.
   a {2, 3, 5, 7}   b {32, 34, 36, 38}
   c {Saturday, Sunday}   d {a, e, i, o, u}

4 $\mathscr{E}$ = {all positive integers}, P = {prime numbers}, E = {even numbers}, O = {odd numbers}.
   Say which of these are true or false.
   a 51 $\in$ P   b P is a subset of O
   c E $\cap$ O = $\varnothing$ d E $\cup$ O = $\mathscr{E}$

5 ℰ = {positive integers less than 11}

A = {multiples of 2}   B = {multiples of 4}

  a Illustrate this information on a Venn diagram.

  b List the set A′ and describe it in words.

  c What is $n(B')$?

  d Is B ⊂ A? Explain your answer.

6 ℰ = {positive integers less than 15}, A = {5, 7, 11, 13}, B = {6, 7, 9}, C = {multiples of 3}.

  a List C.

  b Draw a Venn diagram to illustrate the sets ℰ, A, B and C.

  c List A ∪ B.

  d List B ∩ C.

  e What is A ∩ C?

7 Draw Venn diagrams to illustrate these statements.

  a A ∩ B = ∅   **b** A ∩ B ≠ ∅

  c A ∩ B = A   **d** A ∪ B = A

8 ℰ = {members of an expedition to the South Pole}, A = {people born in Africa},

F = {females}, C = {people born in China}.

  a Describe A ∩ F.

  b What is A ∩ C?

  c Amber ∈ A ∪ C. What can you say about Amber?

  d Illustrate the sets ℰ, A, F and C on a Venn diagram.

9 ℰ = {Suzy's clothes}, D = {Dresses}, R = {Red clothes} and G = {Green clothes}

  a D ∩ R = ∅. Describe what this means in words.

  b D ⊂ G. Describe what this means in words.

  c Illustrate all this information on a Venn diagram.

10 The following information was obtained about all the fast food restaurants in a town.
Six sold burgers and pizzas, four sold pizzas only, nine sold burgers, while two served
neither burgers nor pizzas.

  a Draw a Venn diagram to represent all of this information

  b How many fast food restaurants are there in the town?

Exercise 25* (Revision)

1 List these sets.

  a {multiples of 4 less than 20}

  b {colours of the rainbow}

  c {arrangements of the letters CAT}

  d {all pairs of products of 1, 2, 3}

2 Describe these sets by a rule.

  a {1, 2, 3, 4, 6, 12}

  b {1, 1, 2, 3, 5}

  c {hearts, clubs, diamonds, spades}

  d {tetrahedron, cube, octahedron, dodecahedron, icosahedron}

3 a A and B are two sets. A contains 12 members, B contains 17 members and A ∪ B
contains 26 members. How many members of A are not in A ∩ B?

  b Draw a Venn diagram with circles representing three sets A, B and C, such that these
are true:

A ∩ C = ∅, B ∩ C ≠ ∅, A ∩ B = ∅.

4 $\mathscr{E}$ = {pack of 52 playing cards}, A = {aces}, B = {black cards}, D = {diamonds}.
  a Describe A ∩ D.
  b Describe B ∩ D.
  c Describe A ∪ D.
  d Find n(A ∩ B).
  e Illustrate the sets $\mathscr{E}$, A, B and D on a Venn diagram.

5 $\mathscr{E}$ = {triangles}, R = {right-angled triangles}, I = {isosceles triangles},
  E = {equilateral triangles}.
  a Describe I ∩ R.
  b Describe I ∩ E.
  c Describe R ∩ E.
  d Draw a Venn diagram to illustrate the sets $\mathscr{E}$, R, I and E.

6 $\mathscr{E}$ = {positive integers less than 30}, P = {multiples of 4}, Q = {multiples of 5},
  R = {multiples of 6}.
  a List P ∩ Q.
  b $x \in$ P ∩ R. List the possible values of x.
  c Is it true that Q ∩ R = ∅? Explain your answer.

7 $\mathscr{E}$ = {even numbers less than 15}, A = {multiples of 4}
  B satisfies A ∩ B = ∅ and n(B) = 4.
  What is A ∪ B?

8 $n(\mathscr{E})$ = 17, n(B′) = 9 and n(A′ ∩ B) = 6
  a Find n(B)
  b Find n(A ∩ B)
  c Draw a Venn diagram to illustrate this information.

9 A class of 30 students was asked to choose **at least** one option subject from list A and list B
  Two students forgot to hand their forms in. Of the rest, twenty two chose list A and twenty
  five chose list B.
  a Draw a Venn diagram to illustrate this information.
  b How many students chose both options?

10 There are 30 Widgets, and every Widget is a Woodle.
  There are 20 Wopets, half of which are Woodles. No Wopet is a Widget.
  Half of all Woodles are Widgets.
  a Draw a Venn diagram to represent this information.
  b How many Woodles are neither Widgets nor Wopets?

1 The number 437 600 in standard form correct to 3 s.f. is
   A 4.375×10^5
   B 43.8×10^4
   C 4.38×10^5
   D 4.37×10^5

2 If $3x + 4 = 28 + x$, the value of x is
   A 8
   B 16
   C 12
   D 26

3 A rectangular field is $(2x + 1)$ m long and x m wide. If the perimeter is 14 m, the area is
   A $5\,m^2$
   B $10\,m^2$
   C $14\,m^2$
   D $15\,m^2$

4 A radio decreases in value by 55%. If it is now valued at $157.50, its value before the decrease was
   A $350
   B $1575
   C $3500
   D $286

5 The equation of the straight line with a gradient of $\frac{1}{3}$ passing through point (0, 2) is
   A $3y = x + 6$
   B $y = 2x + \frac{1}{3}$
   C $3y + x = 6$
   D $y = \frac{1}{3}x$

6 The equation of the straight line through points A(0, 10) and B(10, 2) is
   A $5y - 4x = 5$
   B $x + y = 10$
   C $y = 4x + 50$
   D $5y + 4x = 50$

7 The internal angle of a regular octagon is
   A $158°$
   B $140°$
   C $120°$
   D $135°$

8 The larger angle between the hands of a clock at 18:05 is nearest to
   A $210°$
   B $200°$
   C $180°$
   D $170°$

9 In a class of 20 pupils, 14 like reggae music, 3 like classical music and 4 do not like either. The number of pupils who like both reggae and classical is
   A 0
   B 1
   C 2
   D 3

10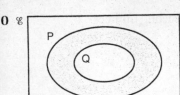

The shaded region is described by
   A Q'
   B $P \cup Q'$
   C $P \cap Q$
   D $P \cap Q'$

1 **a** Write the number 385 000 in standard form.
    b Write the number 3.25×10^3 as an ordinary number.

2 **a** Write the number 36.5782 correct to 2 decimal places.
    b Write the number 36.5782 correct to 2 significant figures.

3 **a** Calculate $(4.8 \times 10^5) \div (1.2 \times 10^3)$, giving your answer in standard form to 3 significant figures.
    b Calculate $(4.8 \times 10^5) + (1.2 \times 10^3)$, giving your answer in standard form to 3 significant figures.

4 **a** Calculate $(3 \times 10^3) \times (4 \times 10^2)$, giving your answer in standard form to 3 significant figures.
    b Calculate $(3 \times 10^3) - (4 \times 10^2)$, giving your answer in standard form to 3 significant figures.

5 **a** Find 30% of $15.40.
    b Decrease $48.60 by 12%.

6 **a** Simplify the expression $3ba + 4ab - ba + 2ab$.
    b Simplify the expression $4xy - x(y - 3)$.
    c Simplify the expression $3ab \times a^2$.

7 **a** Solve the equation $2(3x + 1) = 20$.
    b Solve the equation $2(4a - 3) - (2a + 5) = 10$.
    c Solve the equation $7 - 2x = 3x - 8$.

8 Three consecutive numbers sum to 525.
    a If the first of the consecutive numbers is x, what is the second number?
    b Write down an equation in x.
    c Solve your equation to find x.

9 Which two of the following lines are parallel?
    $x - 3y = 12$, $3y + x = 5$, $y = 3x - 2$, $6y - 2x = 7$

10 Find the gradient of the line through these pairs of points.
    a $P = (1, 5)$ and $Q = (3, 9)$ **b** $C = (4, 4)$ and $D = (1, 16)$

11 **a** A road has a gradient of $\frac{1}{20}$. What is the value of d?
    b The line joining $A = (2, 5)$ to $B = (6, q)$ has a gradient of $\frac{1}{2}$. Find the value of q.

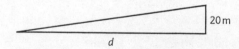

12 **a** Construct triangle ABC, with AB = 7 cm, angle BAC = 75° and angle ABC = 60°.
    b Measure AC.

13 **a** Construct triangle PQR such that PQ = 8 cm, $\angle$PQR = 50°, and $\angle$RPQ = 80°.
    b Construct the perpendicular from R to intersect PQ at S.
    c Measure RS, and hence calculate the area of $\triangle$PQR.

14 In the diagram, ABC and AED are straight lines.
BE and CD are parallel. Angle BAE = 32° and angle EDC = 68°.
Work out the value of p.

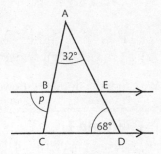

15 a Draw axes on graph paper with $-8 \leqslant x \leqslant 8$ and $-8 \leqslant y \leqslant 8$.
b Draw the graph of the line $y = 3x + 2$.
c Draw the graph of the line $x + 2y = 8$.
d Write down the coordinates of the point where the two lines intersect.

16 Write down the gradient, intercept and
equation of the lines shown on the diagram.

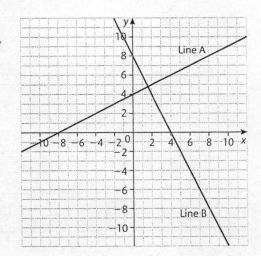

17 The Venn diagram shows four sets
A, B, C, and D.

$$A \subset B \qquad C \cup D = D$$
$$A \cap B \neq \varnothing \qquad A \cup C = \mathscr{E}$$
$$C \cup D = C \qquad A \cap C = \varnothing$$

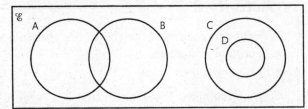

Choose a statement from the box that correctly describes the relationship between
a A and C **b** D and C **c** A and B

18 $n(\mathscr{E}) = 20$, $n(B') = 7$, $n(A' \cap B) = 6$
a Find $n(B)$. **b** Find $n(A \cap B)$.

19 $\mathscr{E}$ = {all triangles}, A = {isosceles triangles}, B = {right angled triangles}
a Draw a Venn diagram to illustrate the sets A and B.
b Calculate the three angles of a member of $A \cap B$
C = {equilateral triangles}
c Add set C to your Venn diagram.

20 $\mathscr{E}$ = {positive whole numbers less than 19}, A = {multiples of 2}, B = {multiples of 3}
a List the set $A \cap B$. **b** Describe the set $A \cap B$.
c Describe the set A'. **d** List the set $A' \cap B$.
e Describe the set $A' \cap B$.

Standard form (negative indices)

We can write numbers, however small, in **standard form**.

Key Point

$$10^{-n} = \frac{1}{10^n}$$

Example 1

$$10^{-2} = \frac{1}{10^2} = \frac{1}{100} = 0.01$$

$$10^{-6} = \frac{1}{10^6} = \frac{1}{1\,000\,000} = 0.000\,001$$

Activity 7

Copy and complete the table.

Decimal form	Fraction form or multiples of 10	Standard form
0.1	$\frac{1}{10} = \frac{1}{10^1}$	1×10^{-1}
	$\frac{1}{100} = \frac{1}{10^2}$	
0.001	=	
0.0001	=	
	=	1×10^{-5}

Activity 8

Write down the mass of each of the first three objects in grams
- in ordinary numbers
- in standard form.

Copy and complete these statements:
- A house mouse is ... times heavier than a pigmy shrew.
- A pigmy shrew is ... times heavier than a grain of sand.
- A grain of sand is 100 000 times lighter than a ....
- A pigmy shrew is 10 000 times heavier than a ....
- A ... is 100 million times heavier than a ....
- A house mouse is ... 10 000 billion times heavier than a ....

House mouse

10^{-2} kg

Grain of sand

10^{-7} kg

Pigmy shrew

10^{-3} kg

Staphylococcus bacterium

10^{-15} kg

Example 2

Write 0.987 in standard form.

$$0.987 = 9.87 \times \frac{1}{10} = 9.87 \times 10^{-1}$$

To display this on your calculator, press

Exercise 26

For Questions 1–4, write each number in standard form.

1 0.1   **2** 0.001   **3** $\frac{1}{1000}$ **4** 10

For Questions 5–8, write each one as an ordinary number.

5 10^{-3} **6** 1.2×10^{-3} **7** 10^{-6} **8** 4.67×10^{-2}

For Questions 9–12, write each number in standard form.

9 0.543   **10** 0.007   **11** 0.67   **12** 100

For Questions 13–17, write each one as an ordinary number.
Check your answers with a calculator.

13 $10^{-2} \times 10^4$ **14** $10^2 \div 10^{-2}$

15 $(3.2 \times 10^{-2}) \times (4 \times 10^3)$ **16** $(6 \times 10^{-1}) \div (2 \times 10^{-2})$

17 $(2 \times 10^{-2}) \times (9 \times 10^{-1}))$

Exercise 26*

For Questions 1–4, write each answer as an ordinary number.

1 $10^3 \times 10^{-2}$ **2** $10^{-2} + 10^{-3}$ **3** $10^{-4} \times 10^2$ **4** $10^{-3} + 10^{-4}$

For Questions 5–8, write each answer in standard form.

5 $10 \div 10^{-2}$ **6** $10^{-1} \div 10^{-2}$ **7** $10^3 \div 10^{-1}$ **8** $10^{-2} \div 10^{-4}$

You will need this information to answer Questions 9 and 10.

Cough virus   Human hair   Pin

9.144×10^{-6} mm diameter 5×10^{-2} mm diameter 6×10^{-1} mm diamete

9 How many viruses, to the nearest thousand, can be placed in a straight line across the width of a human hair?

10 How many viruses, to the nearest thousand, can be placed in a straight line across the width of a pin?

11 A molecule of water is a very small thing, so small that its volume is 10^{-27} m^3.

     a How many molecules are there in 1 m^3 of water?
        If you wrote your answer in full, how many zero digits would there be?

     b If you assume that a water molecule is in the form of a cube, show that its side length is 10^{-9} m.

     c If a number of water molecules were placed touching each other in a straight line, how many would there be in a line 1 cm long?

     d The volume of a cup of tea is 200 cm^3.
        How many molecules of water would the cup hold?
        If all these were placed end to end in a straight line, how long would the line be?
        Take the circumference of the Earth to be 40 000 km.
        How many times would the line of molecules go around the Earth?

Four rules of fractions

This section will give you practice in using fractions. The questions will help you to understand much of the algebra in this unit.

Addition and subtraction

A common denominator is required.

Example 3

Addition

$\frac{3}{4} + \frac{1}{6}$

$= \frac{9}{12} + \frac{2}{12} = \frac{9+2}{12} = \frac{11}{12}$

Example 4

Subtraction

$\frac{3}{4} - \frac{2}{5}$

$= \frac{15}{20} - \frac{8}{20} = \frac{15-8}{20} = \frac{7}{20}$

Example 5

With mixed fractions

$3\frac{1}{3} - 1\frac{3}{4}$

$= \frac{10}{3} - \frac{7}{4}$

$= \frac{40-21}{12} = \frac{19}{12} = 1\frac{7}{12}$

Multiplication and division

Convert mixed fractions into improper fractions.

Example 6

Multiplication

$1\frac{3}{4} \times \frac{3}{5}$

$= \frac{7}{4} \times \frac{3}{5} = \frac{7 \times 3}{4 \times 5} = \frac{21}{20} = 1\frac{1}{20}$

Example 7

Cancelling

$1\frac{5}{9} \times 2\frac{1}{7}$

$= \frac{14}{9} \times \frac{15}{7} = \frac{14 \times 15}{9 \times 7}$

Divide top and bottom by 7, and by 3.

$= \frac{2 \times 5}{3 \times 1} = \frac{10}{3} = 3\frac{1}{3}$

Example 8

Division

$\frac{3}{4} \div \frac{5}{11}$

Turn the divisor upside down and multiply.

$= \frac{3}{4} \times \frac{11}{5} = \frac{3 \times 11}{4 \times 5} = \frac{33}{20} = 1\frac{13}{20}$

Example 9

Multiplying with a whole number.

$\frac{3}{4} \times 7$

Change the whole number into a fraction.

$= \frac{3}{4} \times \frac{7}{1} = \frac{3 \times 7}{4 \times 1} = \frac{21}{4} = 5\frac{1}{4}$

Example 10

Dividing into a whole number.

$8 \div 1\frac{1}{2} = \frac{8}{1} \div \frac{3}{2} = \frac{8}{1} \times \frac{2}{3} = \frac{8 \times 2}{1 \times 3} = \frac{16}{3} = 5\frac{1}{3}$

Exercise 27

Work these out.

1. $\frac{2}{7} + \frac{4}{7}$
2. $\frac{3}{10} + \frac{1}{10}$
3. $\frac{7}{9} + \frac{4}{9}$
4. $\frac{5}{6} - \frac{1}{3}$
5. $\frac{3}{8} + \frac{7}{12}$
6. $3\frac{1}{4} + 1\frac{1}{6}$
7. $\frac{5}{6} \times \frac{1}{3}$
8. $\frac{3}{4} \div \frac{7}{8}$
9. $4 \times \frac{3}{20}$
10. $\frac{12}{25} \div 4$
11. $2\frac{1}{7} \times 1\frac{2}{5}$
12. $1\frac{1}{8} \div \frac{3}{4}$
13. $2\frac{5}{6} + 1\frac{3}{4}$
14. $5\frac{3}{10} - 2\frac{11}{20}$
15. $1\frac{3}{5} \times 3$
16. $1\frac{4}{5} \div 6$

Exercise 27*

Work these out.

1. $\frac{1}{3} + \frac{5}{12}$
2. $\frac{5}{6} - \frac{7}{30}$
3. $\frac{1}{5} + \frac{3}{10} + \frac{9}{20}$
4. $\frac{1}{8} + \frac{1}{8} + \frac{1}{8} + \frac{1}{8} + \frac{1}{8} + \frac{1}{8}$
5. $\frac{5}{8} \times \frac{8}{25}$
6. $\frac{2}{3} \div \frac{20}{21}$
7. $\frac{2}{3} \div 3$
8. $\frac{1}{2} \times \frac{15}{16} \times \frac{4}{5}$
9. $4\frac{1}{2} + 3\frac{1}{6}$
10. $7\frac{2}{3} - 1\frac{1}{6}$
11. $7\frac{2}{3} - \frac{8}{9}$
12. $3\frac{1}{7} \times \frac{7}{15}$
13. $4 \times \frac{2}{3}$
14. $2\frac{1}{3} \div 2\frac{4}{5}$
15. $14 \div 3\frac{1}{9}$
16. $\left(2\frac{1}{8} + 2\frac{1}{4}\right) \div 2\frac{1}{3}$

Ratio

Example 11

A marinade in a recipe contains rice wine and soy sauce in the ratio $2 : 3$.

How much of each ingredient is needed for 100 ml of the mixture?
(Add the ratios together: $2 + 3 = 5$.)

Then the parts are in the ratio $\frac{2}{5} : \frac{3}{5}$.

Amount of rice wine $= \frac{2}{5}$ of $100 = 40$ ml. Amount of soy sauce $= \frac{3}{5}$ of $100 = 60$ ml.

Exercise 28

1 Divide $392 in the ratio of $3 : 4$. 　　2 Divide 752 kg in the ratio of $1 : 7$.

3 Divide 984 in the ratio of $7 : 5$. 　　4 Divide 13.5 in the ratio of $3 : 2$.

Example 12

Divide £1170 in the ratio of $2 : 3 : 4$. 　　(Add the ratios together: $2 + 3 + 4 = 9$.)

Then the first part $= \frac{2}{9}$ of £1170 $=$ £260.

The second part $\quad = \frac{3}{9}$ of £1170 $=$ £390.

The third part $\quad = \frac{4}{9}$ of £1170 $=$ £520. 　　(Check: £260 + £390 + £520 = £1170.)

Exercise 28*

1 Divide $120 in the ratio $3 : 5$.

2 The fuel for a lawn mower is a mixture of 8 parts petrol to one part oil. How much oil is required to make 1 litre of fuel?

3 Mr Chan has three daughters, An, Lien and Tao, aged 7, 8 and 10 years respectively. He shares $100 between them in the ratio of their ages. How much does Lien receive?

4 A breakfast cereal contains the vitamins thiamin, riboflavin and niacin in the ratio $2 : 3 : 25$. A bowl of cereal contains 10 mg of these vitamins. Calculate the amount of riboflavin in a bowl of cereal.

Positive integer powers of numbers

Powers are used to write certain numbers in a convenient way. To help you understand how the rules of indices work, study the table carefully.

Operation	Example	Rule
Multiplying	$3^4 \times 3^2 = (3 \times 3 \times 3 \times 3) \times (3 \times 3)$ $= 3^6 = 3^{4+2} = 729$	**Add** the indices $(a^m \times a^n = a^{m+n})$
Dividing	$3^4 \div 3^2 = \dfrac{3 \times 3 \times 3 \times 3}{3 \times 3} = 3^2 = 3^{4-2} = 9$	**Subtract** the indices $(a^m \div a^n = a^{m-n})$
Raising to a power	$(3^4)^2 = (3 \times 3 \times 3 \times 3) \times (3 \times 3 \times 3 \times 3)$ $= 3^8 = 3^{4 \times 2} = 6561$	**Multiply** the indices $(a^m)^n = a^{mn}$

Example 13

Write 30^5 in standard form

$30^5 = (3 \times 10)^5 = 3^5 \times 10^5 = 243 \times 10^5 = 2.43 \times 10^7$

Write 4^5 as a power of 2

$4^5 = (2^2)^5 = 2^{10}$

Exercise 29

Write these as a single power and then calculate the answer.

1 $2^2 \times 2^2$ **2** $2 \times 2 \times 2 \times 2 \times 2$ **3** $2^4 \div 2^2$

4 $5^5 \div 5^2$ **5** $\dfrac{3^8}{3^2}$ **6** $(2^2)^5$

7 $0.1 \times (0.1)^2$ **8** $2.1^{10} \div 2.1^8$ **9** $\dfrac{4^2 \times 4^5}{4^3}$

10 $20^2 \times 20^2$

Exercise 29*

For Questions 1–8, write as a single power and then calculate the answer.

1 $8^4 \times 8^5 \div 8^6$ **2** $(7^2)^3 \div 7^3$ **3** $5^5 \div 25$ **4** $216 \div 6^2$

5 $125^2 \div 5^3$ **6** $(10^3)^3 \div 1000$ **7** $8^4 \div 4^6$ **8** $\dfrac{125^3}{25^3}$

9 Given that $2^{20} = 1\,048\,576$, calculate 2^{21} and 2^{19}.

10 A large sheet of paper is 0.1 mm thick. It is cut in half, and one piece is placed on top of the other. These two pieces are then cut in half a second time to make the pile four sheets thick. Copy and complete this table.

No. of times done	No. of sheets in pile	Height of pile mm
2	4	0.4
3		
5		
10		
50		

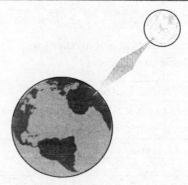

How many times would you have to do this for the pile to reach the Moon, approximately 3.84×10^5 km away?

Direct proportion

In mathematics, there are many ways to relate two quantities together. Here are a few examples.

Change of units: 1 mile = 1.609 km

Gradients: 1 in 5

Velocities: 30 miles travelled in 1 hour (30 miles/hour)

Scales: 1 : 50

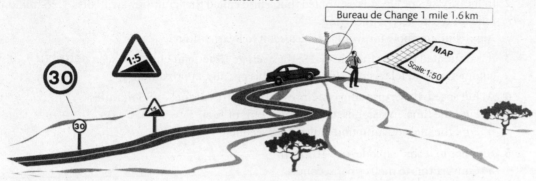

Exchange rates: £1 = $1.60

Ratio: 4 : 5

Densities: 13 g is the mass of 1 cm³ (13 g/cm³)

Problem solving: 3.4 m of timber costs $6.80 ($2/m)

Equations: $3x + 5 = 16$

Graphs (two axes)

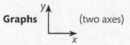

Two different ways of doing the same calculation are shown in Example 14.

Example 14

The committee that organised the Athens Olympics in 2004 recycled 108 tonnes of paper, saving 1836 trees and cutting energy consumption by 442 800 kWh. Calculate the amount of energy saved by recycling one tonne of paper.

Method 1 'unitary method'

108 tonnes   save   442 800 kWh

So   1 tonne   saves $\frac{442\,800}{108}$ kWh = 4100 kWh

Method 2 'per method'

Energy saved per tonne $= \dfrac{\text{Total energy saved}}{\text{Total tonnage}} = \dfrac{442\,800}{108} = 4100$ kWh per tonne

> **Remember**
>
> 'Per' means divide. For example, 'km per hour' means km divided by hours, or km/hour.

Exercise 30

For Questions 1–3, find the exchange rate in the form £1 = ...

1 £6 = ¥1080

2 Aus$62 = £26

3 NZ$22 = £8

4 4 m of timber costs $33.60.
   a What is the cost of 1 m?   **b** What is the cost of 9 m?

5 In one of the strongest hurricanes to sweep across South America, 63 cm of rain fell in 6 hours.
   a Find the amount of rain that fell in millimetres per hour.
   b Find the amount of rain that fell per minute.

6 A military jet uses 3000 litres of fuel on a 45-minute flight.
   a For how long would it travel using 1 litre?
   b How many litres does it use in 1 minute?

Exercise 30*

1 6 Hong Kong dollars can be exchanged for 80 Japanese yen.
   a How many dollars can be exchanged for 200 yen?
   b How many yen can be exchanged for 200 dollars?

2 In its 12 years of life, it is estimated that a bird called the 'chimney swift' flies 1.25 million miles and can sleep while flying.
   Approximately how far would you expect it to fly in 1 hour?

3 It is estimated that by the age of 18, the average American child has seen 350 000 commercials on television. How many is this per day, approximately?

4 At full speed, the cruise ship QM2 uses 40 000 litres of fuel in 55 minutes.
   a Find the time, in seconds, taken to use 1 m^3 of fuel.
   b Find the fuel consumption in litres per second.

5 Air bags in a car 'explode' at 340 km/h.
   a Convert this to metres per second.
   b How long would the air bag take to move 10 cm?
      Give your answer to the nearest thousandth of a second.

Converting measurements
Converting lengths

Remember

10 mm = 1 cm
1000 mm = 1 m
100 cm = 1 m
1000 m = 1 km

Example 15

Change 3 km to cm.

$$3\,km = 3 \times 1000\,m \qquad \text{(as 1 km = 1000 m)}$$
$$= 3 \times 1000 \times 100\,cm \qquad \text{(as 1 m = 100 cm)}$$
$$= 3 \times 10^5\,cm$$

Example 16

Change 5×10^6 mm to km.

$$5 \times 10^6\,mm = \frac{5 \times 10^6}{1000}\,m \qquad \text{(as 1000 mm = 1 m)}$$
$$= \frac{5 \times 10^6}{1000 \times 1000}\,km \qquad \text{(as 1000 m = 1 km)}$$
$$= 5\,km$$

Exercise 31

Fill in the gaps in the following table.

	km	m	cm	mm
1	5			
2		2000		
3			5000	
4				10^6

Exercise 31*

Fill in the gaps in the following table.

	km	m	cm	mm
1	2.5×10^4			
2		5×10^6		
3			50	
4				9×10^9

5 A nanometre is 10^{-9} metres.

   a How many nanometres are there in 200 km?

   b How many km are there in 10^{18} nanometres?

Converting areas

A diagram is useful, as shown in the following examples.

Example 17

A rectangle measures 1 m by 2 m. Find the area in mm².

1 m is 1000 mm.
2 m is 2000 mm.
So the diagram is as shown on the right.

So the area is 1000×2000 mm²
$$= 2\,000\,000 \text{ mm}^2$$
$$= 2 \times 10^6 \text{ mm}^2$$

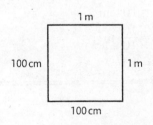

Example 18

Change 30 000 cm² to m².

$$1 \text{ m}^2 = 1 \text{ m} \times 1 \text{ m}$$
$$= 100 \text{ cm} \times 100 \text{ cm}$$
$$= 10\,000 \text{ cm}^2$$

So $30\,000 \text{ cm}^2 = \dfrac{30\,000}{10\,000} \text{ m}^2$
$$= 3 \text{ m}^2$$

Exercise 32

Fill in the gaps in the following table.

	km²	m²	cm²	mm²
1	2			
2		50		
3			6×10^6	
4				10^{13}

Exercise 32*

Fill in the gaps in the following table.

	km²	m²	cm²	mm²
1		6000		
2			6×10^{10}	
3				2×10^{21}
4	7×10^{-2}			

Converting volumes

Again diagrams are very helpful.

Remember

1 litre = 1000 cm³

Example 19

A cuboid measures 1 m by 2 m by 3 m.
Find the volume in mm³.

1 m is 1000 mm.
2 m is 2000 mm.
3 m is 3000 mm.

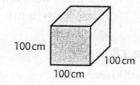
1000 mm   3000 mm
2000 mm

So the volume is $1000 \times 2000 \times 3000 \, \text{mm}^3 = 6 \times 10^9 \, \text{mm}^3$

Example 20

Change $10^7 \, \text{cm}^3$ to m³.

$$1 \, \text{m}^3 = 1 \, \text{m} \times 1 \, \text{m} \times 1 \, \text{m}$$
$$= 100 \, \text{cm} \times 100 \, \text{cm} \times 100 \, \text{cm}$$
$$= 10^6 \, \text{cm}^3$$

So $10^7 \, \text{cm}^3 = \dfrac{10^7}{10^6} \, \text{m}^3$

$$= 10 \, \text{m}^3$$

100 cm
100 cm
100 cm

Exercise 33

Fill in the gaps in the following table.

	km³	m³	cm³	mm³
1	1			
2		8		
3			4×10^3	
4				10^{15}

5 How many litres are there in 1 m³?

6 How many m³ are there in 10 000 litres?

Exercise 33*

Fill in the gaps in the following table.

	km³	m³	cm³	mm³
1		600		
2			3×10^8	
3				5×10^{25}
4	4×10^{-6}			

5 How many litres are there in $512\,\text{m}^3$?

6 How many m^3 are there in 10^6 litres?

7 A picometre is $10^{-12}\,\text{m}$. How many cubic picometres are there in $1\,\text{km}^3$?

Exercise 34 (Revision)

Calculate the following showing all of your working.

1 $\frac{3}{7} + \frac{2}{5}$

2 $\frac{3}{7} - \frac{2}{5}$ **3** $\frac{3}{7} \times \frac{2}{5}$

4 $\frac{3}{7} \div \frac{2}{5}$

5 $2\frac{3}{5} + 1\frac{1}{7}$ **6** $2\frac{3}{5} - 1\frac{1}{7}$

7 $2\frac{3}{5} \times 1\frac{1}{7}$

8 $2\frac{3}{5} \div 1\frac{1}{7}$

9 Divide $36\,\text{m}$ in the ratio of $1:2$.

10 Divide $105\,\text{kg}$ in the ratio of $3:4$.

11 Divide \$400 in the ratio of $2:3$.

12 Divide 360 minutes in the ratio of $4:5$.

13 Divide £133 in the ratio of $1:2:4$.

14 Divide $352\,\text{km}$ in the ratio of $2:3:6$.

Write the following in standard form correct to 3 significant figures.

15 0.012 345

16 0.012 355

17 0.000 159 5

18 0.008 888

19 $(1.25 \times 10^{-2}) \times (3.45 \times 10^5)$

20 $(7.58 \times 10^{-5}) \times (1.35 \times 10^{12})$

21 $(7.25 \times 10^{-3}) \times (3.45 \times 10^{-10})$

22 $(8.5 \times 10^{-2}) \times (3.45 \times 10^{-7})$

23 $(1.25 \times 10^{-2}) \div (3.45 \times 10^5)$

24 $(7.58 \times 10^{-5}) \div (1.35 \times 10^{12})$

25 $(7.25 \times 10^{-3}) \div (3.45 \times 10^{-10})$

26 $(8.5 \times 10^{-2}) \div (3.45 \times 10^{-7})$

27 Four tonnes of limestone blocks cost \$600. Find the cost of
  a 1 tonne   **b** 11 tonnes   **c** 500 kg.

28 Seven identical pens cost \$8.40. Find the cost of
  a one pen   **b** five pens.   **c** a dozen pens.

29 Stella Pajunas typed 216 words in 1 minute in Chicago, US to set a new world record. If she maintained this rate, how many words would you expect her to type in
  a 45 seconds   **b** 50 seconds   **c** an hour?

30 Avind Pandya of India ran backwards from Los Angeles to New York, US in 107 days covering 5000 km.
  a If he maintained this rate, how far would you expect him to travel in
    (i) 1 day   **(ii)** a year?
  b How long would it have taken him to travel
    (i) 10 km   **(ii)** 3500 m?
  c Calculate his speed in mm/s writing your answer in standard form to 3 significant figures.

Exercise 34* (Revision)

Calculate the following, showing all of your working.

1 $\frac{2}{5} \times \frac{5}{11} \times \frac{3}{8}$

2 $\frac{1}{3} + \frac{4}{7} - \frac{2}{15}$

3 $\frac{4}{5} \div \frac{2}{7} \times \frac{3}{14}$

4 $\frac{1}{7} \div \left(\frac{3}{5}\right)^2$

5 $1 \times 1\frac{1}{2} \times 1\frac{1}{3} \times 1\frac{1}{4} \times 1\frac{1}{5} \times 1\frac{1}{6} \times 1\frac{1}{7}$

6 $\left(2\frac{3}{7}\right)^2 \div \left(1\frac{3}{7}\right)^2$

7 The ratio of $5 : x$ is equal to the ratio of $x : 20$. Calculate the value of x.

8 A bed of roses consists of m roses. The ratio of pink roses to white roses is $2 : 3$.
Find the number of pink roses expressed in terms of m.

9 Xavier, Yi and Zazoo decide to share their lottery winnings of £11 000 such that Yi has three times as much as Zazoo and Xavier has a half of Yi's winnings.
How much should each receive?

10 The plan for an office block is produced to a scale of $1 : 50$.

 a Find the length, in mm, which represents the height of the building on the plan if the actual height is 25 m.

 b Find the area of the actual front door, in m², if the door on the plan has an area of 80 cm².

Write the following in standard form correct to 3 significant figures:

11 $(1.36 \times 10^{-3})^2$

12 $(3.75 \times 10^{-5})^2 \times (4.35 \times 10^{-7})^2$

13 $\sqrt{5.875 \times 10^{-12}}$

14 $\sqrt{\dfrac{3.85 \times 10^{-9}}{1.47 \times 10^{-3}}}$

15 If $p = 9.47 \times 10^{-5}$ and $q = 4.31 \times 10^{-3}$, find the following in standard form correct to 3 significant figures.

 a pq

 b pq^2

 c p^2q

 d $\left(\dfrac{p}{q}\right)^2$

16 The smallest mammal is the Kitti's hog-nosed bat in Thailand which has a body length of 29 mm. Find this length in km in standard form correct to 3 significant figures.

17 The biggest known star is the M-class supergiant Betelgeuse which has a diameter of 980 million km.

 a Assuming it to be a sphere, calculate its surface area in mm², giving your answer in standard form correct to 3 significant figures.

 b Given that the Earth has a radius of 6370 km, express its surface area as a percentage of Betelgeuse's. Give your answer in standard form correct to 3 significant figures.
 (The surface area of a sphere $= 4\pi r^2$, where r is the radius of the sphere.)

18 Kaylan Ramji Sain of India grew a moustache to a length of 339 mm from 1976 until 1993. Calculate the speed of his moustache growth in km/s. Give your answer in standard form correct to 3 significant figures.

Simplifying fractions

Algebraic fractions are simplified in the same way as arithmetic fractions.

Multiplication and division

Example 1

Simplify $\dfrac{4x}{6x}$. $\dfrac{\overset{2}{\cancel{4}}x}{\underset{3}{\cancel{6}}x} = \dfrac{2\cancel{x}}{3\cancel{x}} = \dfrac{2}{3}$

Example 2

Simplify $\dfrac{3x^2}{6x}$. $\dfrac{3x^2}{6x} = \dfrac{\overset{1}{\cancel{3}} \times x \times \overset{1}{\cancel{x}}}{\underset{2}{\cancel{6}} \times \underset{1}{\cancel{x}}} = \dfrac{x}{2}$

Example 3

Simplify $(27xy^2) \div (60x)$. $(27xy^2) \div (60x) = \dfrac{27xy^2}{60x} = \dfrac{\overset{9}{\cancel{27}} \times \overset{1}{\cancel{x}} \times y \times y}{\underset{20}{\cancel{60}} \times \underset{1}{\cancel{x}}} = \dfrac{9y^2}{20}$

Exercise 35

Simplify these.

1 $\dfrac{4x}{x}$ **2** $\dfrac{6y}{2}$ **3** $(6x) \div (3x)$ **4** $\dfrac{12a}{4b}$

5 $\dfrac{3ab}{6a}$ **6** $(9a) \div (3b)$ **7** $\dfrac{12c^2}{3c}$ **8** $\dfrac{4a^2}{8a}$

9 $\dfrac{12x}{3x^2}$ **10** $\dfrac{8ab^2}{4ab}$ **11** $\dfrac{3a}{15ab^2}$ **12** $(3a^2b^2) \div (12ab^2)$

Exercise 35*

Simplify these.

1 $\dfrac{5y}{10y}$ **2** $\dfrac{12a}{6ab}$ **3** $(3xy) \div (12y)$ **4** $\dfrac{3a^2}{6a}$

5 $\dfrac{10b}{5b^2}$ **6** $(18a) \div (3ab^2)$ **7** $\dfrac{12xy^2}{4xy}$ **8** $\dfrac{3a^2b^2}{6ab^3}$

9 $\dfrac{15abc}{5a^2b^2c^2}$ **10** $(3a^2) \div (12ab^2)$ **11** $\dfrac{abc^3}{(abc)^3}$ **12** $\dfrac{150a^3b^2}{400a^2b^3}$

Example 4

Simplify $\dfrac{3x^2}{y} \times \dfrac{y^3}{x}$. $\dfrac{3x^2}{y} \times \dfrac{y^3}{x} = \dfrac{3 \times x \times \overset{1}{\cancel{x}}}{\underset{1}{\cancel{y}}} \times \dfrac{\overset{1}{\cancel{y}} \times y \times y}{\underset{1}{\cancel{x}}} = 3xy^2$

Example 5

Simplify $\dfrac{2x^2}{y} \div \dfrac{2x}{5y^3}$. $\dfrac{2x^2}{y} \div \dfrac{2x}{5y^3} = \dfrac{\overset{1}{\cancel{2}} \times x \times \overset{1}{\cancel{x}}}{\underset{1}{\cancel{y}}} \times \dfrac{5 \times \overset{1}{\cancel{y}} \times y \times y}{\underset{1}{\cancel{2}} \times \underset{1}{\cancel{x}}} = 5xy^2$

Key Point

To divide by a fraction, turn the fraction upside down and multiply.

Exercise 36

Simplify these.

1. $\dfrac{3x}{4} \times \dfrac{5x}{3}$

2. $\dfrac{x^2 y}{z} \times \dfrac{xz^2}{y^2}$

3. $\dfrac{x^2}{y} \times \dfrac{z}{x^2} \times \dfrac{y}{z}$

4. $\dfrac{4c \times 7c^2}{7 \times 5c}$

5. $\dfrac{3x}{4} \div \dfrac{x}{8}$

6. $4 \div \dfrac{8}{ab}$

7. $\dfrac{2b}{3} \div 4$

8. $\dfrac{2x}{3} \div \dfrac{2x}{3}$

9. $\dfrac{2x}{y^2} \div \dfrac{x}{y}$

10. $\dfrac{5ab}{c^2} \div \dfrac{10a}{c}$

Exercise 36*

Simplify these.

1. $\dfrac{3x}{2} \times \dfrac{x}{9}$

2. $\dfrac{4a}{3} \times \dfrac{5a}{2} \times \dfrac{3a}{5}$

3. $\dfrac{3x^2 y}{z^3} \times \dfrac{z^2}{xy}$

4. $\dfrac{45}{50} \times \dfrac{p^2}{q} \times \dfrac{q^3}{p}$

5. $\dfrac{3x}{y} \div \dfrac{6x^2}{y}$

6. $\dfrac{15x^2 y}{z} \div \dfrac{3xz}{y^2}$

7. $\dfrac{2x}{y} \times \dfrac{3y}{4x} \times \dfrac{2y}{3}$

8. $\dfrac{x^2}{y^2} \div \dfrac{5z}{y^2}$

9. $\left(\dfrac{x}{2y}\right)^3 \times \dfrac{2x}{3} \div \dfrac{2}{9y^2}$

10. $\dfrac{\sqrt{a^3 b^2}}{6a^3} \times \dfrac{3a^5 b}{(a^3 b^2)^2} \div \dfrac{ab}{\sqrt{a^3 b^2}}$

Addition and subtraction

Example 6

Simplify $\dfrac{a}{4} + \dfrac{b}{5}$

$\dfrac{a}{4} + \dfrac{b}{5} = \dfrac{5a + 4b}{20}$.

Example 7

Simplify $\dfrac{3x}{5} - \dfrac{x}{3}$.

$\dfrac{3x}{5} - \dfrac{x}{3} = \dfrac{9x - 5x}{15} = \dfrac{4x}{15}$

Example 8

Simplify $\dfrac{2}{3b} + \dfrac{1}{2b}$. $\dfrac{2}{3b} + \dfrac{1}{2b} = \dfrac{4 + 3}{6b} = \dfrac{7}{6b}$

Example 9

Simplify $\dfrac{3 + x}{7} - \dfrac{x - 2}{3}$.   Remember to use brackets here. Note sign change

$$\dfrac{3 + x}{7} - \dfrac{x - 2}{3} = \dfrac{3(3 + x) - 7(x - 2)}{21} = \dfrac{9 + 3x - 7x + 14}{21} = \dfrac{23 - 4x}{21}$$

Exercise 37

Simplify these.

1. $\dfrac{x}{2} + \dfrac{x}{4}$

2. $\dfrac{x}{3} + \dfrac{x}{4}$

3. $\dfrac{a}{3} - \dfrac{a}{4}$

4. $\dfrac{a}{3} + \dfrac{b}{4}$

5. $\dfrac{2x}{3} - \dfrac{x}{4}$

6. $\dfrac{2a}{7} + \dfrac{3a}{14}$

7. $\dfrac{a}{4} + \dfrac{b}{3}$

8. $\dfrac{3x}{4} - \dfrac{x}{3}$

9. $\dfrac{2a}{3} - \dfrac{a}{2}$

10. $\dfrac{a}{4} + \dfrac{2b}{3}$

Exercise 37*

Simplify these.

1 $\frac{x}{6} + \frac{2x}{9}$

2 $\frac{2a}{3} - \frac{3a}{7}$

3 $\frac{2x}{5} + \frac{4y}{7}$

4 $\frac{3a}{4} + \frac{a}{3} - \frac{5a}{6}$

5 $\frac{3}{2b} + \frac{4}{3b}$

6 $\frac{2}{d} + \frac{3}{d^2}$

7 $\frac{2-x}{5} + \frac{3-x}{10}$

8 $\frac{y+3}{5} - \frac{y+4}{6}$

9 $\frac{x-3}{3} + \frac{x+5}{4} - \frac{2x-1}{6}$

10 $\frac{a}{a-1} - \frac{a-1}{a}$

Solving equations

Example 10

Solve $3x^2 + 4 = 52$.

$3x^2 + 4 = 52$ (Subtract 4 from both sides)

$3x^2 = 48$ (Divide both sides by 3)

$x^2 = 16$ (Square root both sides)

$x = \pm 4$

Check: $3 \times 16 + 4 = 52$

(Note that -4 is also an answer because $(-4) \times (-4) = 16$.)

Example 11

Solve $5\sqrt{x} = 50$.

$5\sqrt{x} = 50$ (Divide both sides by 5)

$\sqrt{x} = 10$ (Square both sides)

$x = 100$

Check: $5 \times \sqrt{100} = 50$

Example 12

Solve $\frac{\sqrt{x+5}}{3} = 1$.

$\frac{\sqrt{x+5}}{3} = 1$ (Multiply both sides by 3)

$\sqrt{x+5} = 3$ (Square both sides)

$x + 5 = 9$ (Subtract 5 from both sides)

$x = 4$

Check $\frac{\sqrt{4+5}}{3} = 1$

Exercise 38

Solve these equations.

1 $4x^2 = 36$

2 $\frac{x^2}{3} = 12$

3 $x^2 + 5 = 21$

4 $\frac{x^2}{2} + 5 = 37$

5 $2x^2 + 5 = 23$

6 $5x^2 - 7 = -2$

7 $\frac{x+12}{5} = 5$

8 $\frac{x^2+4}{5} = 4$

9 $\sqrt{x} + 27 = 31$

10 $4\sqrt{x} + 4 = 40$

Exercise 38*

Solve these equations.

1 $x^2 - 5 = 20$

2 $4x^2 + 26 = 126$

3 $\frac{x^2}{7} - 3 = 4$

4 $\frac{x^2-11}{7} = 10$

5 $4\sqrt{x} + 4 = 40$

6 $\sqrt{\frac{x-3}{4}} + 5 = 6$

$$7 \quad \frac{40 - 2x^2}{2} = 4 \qquad\qquad 8 \quad 22 = 32 - \frac{2x^2}{5} \qquad\qquad 9 \quad (3 + x)^2 = 169$$

$$10 \quad \sqrt{\frac{3x^2 + 5}{2}} + 4 = 8$$

Using formulae

A formula is a way of describing a relationship, using algebra. For example, the formula to calculate the volume of a cylindrical can is $V = \pi r^2 h$ where V is the volume, r is the radius and h is the height.

Remember

When using any formula:
Write down the **facts**
with the correct units.
Then write down the
equation, **substitute**
the facts, and do the
working.

Example 13

Find the volume of a cola can that has a radius of 3 cm and a height of 11 cm.

Facts	$r = 3$ cm, $h = 11$ cm, $V = ?$ cm^3
Equation	$V = \pi r^2 h$
Substitution	$V = \pi \times 3^2 \times 11$
Working	$\pi \times 3^2 \times 11 = \pi \times 9 \times 11 = 311$ cm^3 (3 s.f.)
	Volume $= 311$ cm^3 (3 s.f.)

Some commonly-used formulae

You will need the following formulae to complete Exercises 39 and 39*. The formulae are covered more fully later in this book.

In a right-angled triangle,
Pythagoras's theorem:
$a^2 = b^2 + c^2$

Area of parallelogram
$= bh$

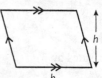

Area of trapezium
$= \dfrac{h}{2}(a + b)$

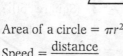

Area of a triangle $= \frac{1}{2}$ base $\times$ height

Circumference of a circle $= 2\pi r$

Area of a circle $= \pi r^2$

Speed $= \dfrac{\text{distance}}{\text{time}}$

Challenge

Prove the formulae for the area of a parallelogram and the area of a trapezium.

Exercise 39

1 The area of a parallelogram is 31.5 cm^2 and its base is 7 cm long. Find its height.

2 The radius of a circle is 7 cm. Find the circumference of the circle, and its area.

3 The area of this triangle is 72 cm^2.
Find its height h.

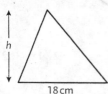

Unit 2 : Algebra

4 Find YZ.

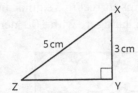

5 The Earth, which is 150 million km from the Sun, takes 365 days to complete one orbit, assumed circular.
  a Find the length of one orbit, giving your answer in standard form correct to 3 significant figures.
  b Find the speed of the Earth around the Sun in km per hour correct to 2 significant figures.

6 The area of this trapezium is 21.62 cm². Find its height h.

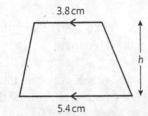

7 It takes light $8\frac{1}{3}$ minutes to reach the Earth from the Sun. Calculate the distance between the Sun and the Earth if light travels at 300 000 km/s.

8 A bullet from a high-velocity military machine gun travels at 108 000 km/h. Assuming that the bullet does not slow down, find the time, in seconds, that it takes to travel 3 km.

Exercise 39*

1 The circumference of a circle is 88 cm. Find its radius.

2 The area of triangle ACD is 25.2 cm².

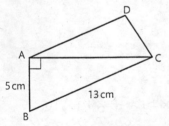

Find AC, and the perpendicular height of D above AC.

3 The area of this trapezium is 30.8 cm².

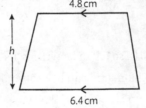

Find its height h.

4 Can a brick which measures 9 cm by 12 cm by 25 cm pass down a cylindrical pipe of diameter 14 cm?

5 The diagram shows the cross-section of a metal pipe. Find the area of the shaded part correct to 3 significant figures.

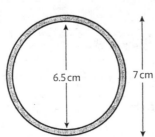

6 Find the area of the shaded region in each of the diagrams correct to 3 significant figures. Use your answers to work out the shaded region of a similar figure with 100 identical circles.

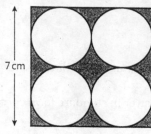

7 cm

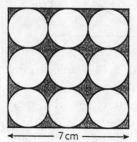

7 cm

7 How long would the minute hand of a clock have to be if its end were to move at 100 km/hour?

8 A family goes on a cruise around the world. They have a pet hamster which remains in their cabin, which is at sea level, throughout the voyage. The captain has a parrot on the bridge. If the bridge is 35 m above sea level, calculate how much further the parrot travels than the hamster. (Assume that the voyage is a circular path.) Give your answer correct to 3 significant figures.

9 The area of the shaded region is 20 cm². Find the value of x, correct to 3 significant figures.

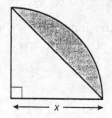

x

Positive integer indices

Remember

$1000 = 10 \times 10 \times 10$, so it can be written as 10^3.

$1\,000\,000\,000$ is a billion, and can be written as 10^9.

This is because a billion is equal to $10 \times 10 \times 10 \times 10 \times 10 \times 10 \times 10 \times 10 \times 10$.

Similarly, $3 \times 3 \times 3 \times 3 \times 3$ can be written as 3^5, and, in algebra, $a \times a \times a \times a \times a \times a$ can be written as a^6.

To help you to understand how the rules of indices work, look carefully at these examples.

Operation	Example	Rule
Multiplying	$a^4 \times a^2 = (a \times a \times a \times a) \times (a \times a)$ $= a^6 = a^{4+2}$	**Add** the indices $(a^m \times a^n = a^{m+n})$
Dividing.	$a^4 \div a^2 = \dfrac{a \times a \times a \times a}{a \times a} = a^2 = a^{4-2}$	**Subtract** the indices $(a^m \div a^n = a^{m-n})$
Raising to a power	$(a^4)^2 = (a \times a \times a \times a) \times (a \times a \times a \times a)$ $= a^8 = a^{4 \times 2}$	**Multiply** the indices $(a^m)^n = a^{mn}$

Example 14

Use the rules of indices first to simplify $6^3 \times 6^4$.
Then use your calculator to calculate the answer.

$6^3 \times 6^4 = 6^7 = 279\,936$ (Add the indices)

Example 15

$9^5 \div 9^2 = 9^3 = 729$ (Subtract the indices)

Example 16

$(4^2)^5 = 4^{10} = 1\,048\,576$ (Multiply the indices)

Notice that some answers become very large even though the index is quite small.

2nd term.

Exercise 40

Use the rules of indices to simplify these. Then use your calculator to calculate the answer.

1 $2^4 \times 2^6$	**2** $4^3 \times 4^4$	**3** $2^{10} \div 2^4$
4 $\dfrac{7^{13}}{7^{10}}$	**5** $(2^3)^4$	**6** $(6^2)^4$

Use the rules of indices to simplify these.

7 $a^3 \times a^2$	**8** $c^6 \div c^2$	**9** $(e^2)^3$	**10** $a^2 \times a^3 \times a^4$
11 $\dfrac{c^8}{c^3}$	**12** $2 \times 6 \times a^4 \times a^2$	**13** $2a^3 \times 3a^2$	**14** $2(e^4)^2$

Exercise 40*

Use the rules of indices to simplify these. Then use your calculator to calculate the answer.
Give your answers correct to 3 significant figures and in standard form.

1 $6^6 \times 6^6$	**2** $7^{12} \div 7^6$	**3** $(8^3)^4$	**4** $4(4^4)^4$

Use the rules of indices to simplify these.

5 $a^5 \times a^3 \times a^4$	**6** $(12c^9) \div (4c^3)$	**7** $2(e^4)^2$	**8** $(2g^4)^3$
9 $3(2j^3)^4$	**10** $3m(2m^2)^3$	**11** $3a^2(3a^2)^2$	**12** $\dfrac{2a^8 + 2a^8}{2a^8}$

13 $\dfrac{12b^8}{6b^4} + 6b^4$ **14** $\dfrac{b^4 + b^4 + b^4 + b^4 + b^4 + b^4}{b^4}$

Inequalities

Number lines

Remember

These are examples of how to show inequalities on a number line.

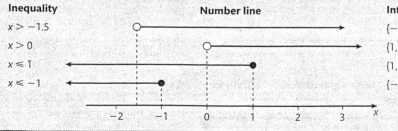

Inequality	Number line	Integer solutions
$x > -1.5$		$\{-1, 0, 1, 2, \dots\}$
$x > 0$		$\{1, 2, 3, 4, \dots\}$
$x \leqslant 1$		$\{1, 0, -1, -2, \dots\}$
$x \leqslant -1$		$\{-1, -2, -3, -4, \dots\}$

Inequalities are solved in the same way as algebraic equations, **except** that when multiplying or dividing by a negative number the inequality sign is reversed.

Example 17

Solve the inequality $4 < x \leqslant 10$. Show the result on a number line.

$4 < x \leqslant 10$ (Split the inequality into two parts)

$4 < x$ and $x \leqslant 10$

$x > 4$ and $x \leqslant 10$

Note that x cannot be equal to 4.

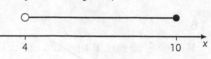

Example 18

Solve the inequality $4 \geqslant 13 - 3x$. Show the result on a number line.

$4 \geqslant 13 - 3x$ (Add $3x$ to both sides)

$3x + 4 \geqslant 13$ (Subtract 4 from both sides)

$3x \geqslant 9$ (Divide both sides by 3)

$x \geqslant 3$

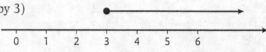

Example 19

Solve the inequality $5 - 3x < 1$. List the four smallest integers in the solution set.

$5 - 3x < 1$ (Subtract 5 from both sides)

$-3x < -4$ (Divide both sides by -3, so **reverse** the inequality sign)

$x > \dfrac{-4}{-3}$

$x > 1\frac{1}{3}$

Thus the four smallest integers are 2, 3, 4 and 5.

Example 20

Solve the inequality $x \leqslant 5x + 1 < 4x + 5$. Show the inequality on a number line.

$x \leqslant 5x + 1 < 4x + 5$ (Split the inequality into two parts)

a $x \leqslant 5x + 1$ (Subtract $5x$ from both sides)

$-4x \leqslant 1$ (Divide both sides by -4, so **reverse** the inequality sign)

$x \geqslant -\frac{1}{4}$

b $5x + 1 < 4x + 5$ (Subtract $4x$ from both sides)

$x + 1 < 5$ (Subtract 1 from both sides)

$x < 4$

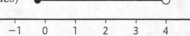

Remember

$x > 4$ means that x cannot be equal to 4, while $x \geqslant 4$ means that x can be equal to 4 or greater than 4.

When finding the solution set of an inequality:

Collect up the algebraic terms on one side.

When dividing or multiplying both sides by a negative number, **reverse** the inequality sign.

For Questions 1–4, insert the correct symbol, $<$, $>$ or $=$.

1 $-3 \;\square\; 3$ **2** $30\% \;\square\; \frac{1}{3}$ **3** $-3 \;\square\; -4$ **4** $0.3 \;\square\; \frac{1}{3}$

5 Write down the inequalities represented by this number line.

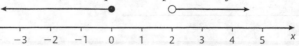

6 Write down the single inequality represented by this number line.

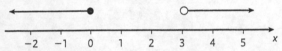

For Questions 7–14, solve the inequality, and show the result on a number line.

7 $x - 3 > 2$

8 $x - 3 \leqslant 1$

9 $4 < 7 - x$

10 $10 \geqslant 13 - x$

11 $4x \geqslant 3x + 9$

12 $6x + 3 < 2x + 19$

13 $2(x + 3) < x + 6$

14 $5(x - 1) > 2(x + 2)$

Solve these inequalities. List the integers in each solution set.

15 $4 < x \leqslant 6$

16 $2 < x \leqslant 4.5$

17 $-1 < x \leqslant 1.5$

18 $2 \leqslant 2x < x + 5$

19 $4 < 2x + 1 \leqslant 7$

Exercise 41*

1 Write down the inequalities represented by this number line.

Explain why your two answers *cannot* be combined into a single inequality.

For Questions 2–7, solve the inequality and show the result on a number line.

2 $3x \leqslant x + 5$

3 $5x + 3 < 2x + 19$

4 $3(x + 3) < x + 12$

5 $2(x - 1) > 7(x + 2)$

6 $\frac{x}{2} - 3 \geqslant 3x - 8$

7 $x < 2x + 1 \leqslant 7$

8 Solve the inequality, and then list the four largest integers in the solution set.

$$\frac{x + 1}{4} \geqslant \frac{x - 1}{3}$$

9 Find the largest prime number y that satisfies $4y \leqslant 103$.

10 List the integers that satisfy both these inequalities.

$$-3 \leqslant x < 4 \quad \text{and} \quad x > 0$$

Exercise 42 (Revision)

Simplify these.

1 $\dfrac{3y}{y}$

2 $\dfrac{4x}{4}$

3 $\dfrac{9x^2}{3x}$

4 $\dfrac{2a}{3} \times \dfrac{6}{a}$

5 $\dfrac{6b}{4} \div \dfrac{3b}{2a}$

6 $\dfrac{10x^2}{3} \times \dfrac{9}{5x}$

7 $\dfrac{y}{4} + \dfrac{y}{5}$

8 $\dfrac{x}{3} - \dfrac{x}{5}$

9 $\dfrac{2a}{5} + \dfrac{b}{10}$

Solve these.

10 $\dfrac{x^2}{2} + 2 = 10$

11 $\dfrac{x^2 + 2}{2} = 19$

12 $\sqrt{\dfrac{4 + x}{6}} = 2$

Use the rules of indices to simplify these.

13 $a^4 \times a^6$

14 $b^7 \div b^5$

15 $(c^4)^3$

For Questions 16–19, rewrite each expression, replacing ☐ by the correct symbol, $<$, $>$ or $=$.

16 $-2 \ \square \ -3$

17 $\frac{1}{8} \ \square \ \frac{1}{7}$

18 $0.009 \ \square \ 0.01$

19 $0.1 \ \square \ 10\%$

20 Write down the single inequality represented by this number line.

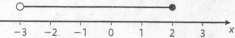

What is the smallest integer that x can be?

For Questions 21–23, solve the inequality and show each result on a number line.

21 $x - 4 > 1$ **22** $5x \leqslant 3x + 9$ **23** $5(x - 2) \geqslant 4(x - 2)$

24 Solve the inequality $x + 5 \leqslant 6x$.

25 List the integers in the solution set $3 \leqslant x < 5$.

26 The area of a circle is 33 cm^2. Taking the area of the circle to be πr^2, find its radius correct to 3 significant figures.

27 The Niagara Falls, one of the world's most spectacular waterfalls, is eroding the rockface over which the water cascades at the rate of 0.9 m/year.
Find the length of the erosion since the Falls were formed at the end of the Ice Age, 12 600 years ago.

Exercise 42* (Revision)

Simplify these.

1 $\dfrac{20a}{5b}$ **2** $\dfrac{35x^2}{7xy}$ **3** $\dfrac{12ab^2}{48a^2b}$

4 $\dfrac{2a}{b} \times \dfrac{b^2}{4a}$ **5** $\dfrac{30}{xy^2} \div \dfrac{6x^2}{x^2y}$ **6** $\dfrac{(3a)^2}{7b} \div \dfrac{a^3}{14b^2}$

7 $\dfrac{3a}{2} + \dfrac{a}{10}$ **8** $\dfrac{2}{3b} + \dfrac{3}{4b} - \dfrac{5}{6b}$ **9** $\dfrac{x+1}{7} - \dfrac{x-3}{21}$

Solve these.

10 $3x^2 + 5 = 32$ **11** $2 = \dfrac{\sqrt{2x} + 2}{2}$ **12** $\sqrt{100 - 4x^2} = 6$

For Questions 13–15, use the rules of indices to simplify each expression.

13 $a^5 \times a^6 \div a^7$ **14** $(2b^3)^2$ **15** $3c(3c^2)^3$

16 Write down the single inequality represented by the number line.
What is the smallest integer that satisfies the inequality?

For Questions 17–19, solve the inequality and show each result on a number line.

17 $7x + 3 < 2x - 19$ **18** $2(x - 1) < 5(x + 2)$ **19** $\dfrac{x-2}{5} \geqslant \dfrac{x-3}{3}$

20 Find the largest prime number y which satisfies $3y - 11 \leqslant 103$.

21 List the integers which satisfy both these inequalities simultaneously.
$-3.5 < x < 3$ and $4x + 1 \leqslant x + 2$

22 Find the circumference of a circle of area 200 cm^2.

23 The fastest speed of a ball that has been served in tennis is 222 km/hour.
How long would it have taken the ball to travel 24 m, the length of a tennis court?
Give your answer in seconds correct to 2 significant figures.

24 A pulsar (an imploding star) rotates at an incredible rate of 30 times a second.
If its diameter is 12 km, find the speed of a point on its equator.
Give your answer correct to 3 significant figures.

Simultaneous equations

Activity 9

Lorna is trying to decide between two Internet service providers, Pineapple and Banana. Pineapple charges $9.99/month plus 1.1 cents/minute online, while Banana charges $4.95/month plus 1.8 cents/minute online.

If C is the cost in cents and t is the time (in minutes) online per month then the cost of using Pineapple is $C = 999 + 1.1t$, and the cost of using Banana is $C = 495 + 1.8t$.

- Copy and complete this table to give the charges for Pineapple.

Time online t (minutes)	0	500	1000
Cost C (cents)			

- Draw a graph of this data with t along the horizontal axis and C along the vertical axis.
- Make a similar table for the Banana charges. Add the graph of this data to your previous graph.
- How many minutes online per month will result in both companies charging the same amount?

In Activity 10, you solved the simultaneous equations $C = 999 + 1.1t$ and $C = 495 + 1.8t$ graphically. From the graph you can also tell which is the cheaper option for any number of minutes online.

Example 1

Solve the simultaneous equations $y = \frac{1}{2}x + 2$ and $y = 4 - x$ graphically.
First, make a table of values for each equation.

x	0	2	4
$y = \frac{1}{2}x + 2$	2	3	4

x	0	2	4
$y = 4 - x$	4	2	0

Next, draw accurate graphs for both equations on one set of axes.

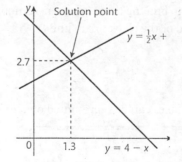

The solution point is approximately $x = 1.3$, $y = 2.7$.

Exercise 43

1 Copy and complete these tables, and then draw both graphs on one set of axes.

x	0	2	4
$y = x + 1$			

x	0	2	4
$y = 2x - 2$			

Solve the simultaneous equations $y = x + 1$, $y = 2x - 2$ using your graph.

2 On one set of axes, draw the graphs of $y = 3x - 1$ and $y = 2x + 1$ for $0 \leqslant x \leqslant 6$.
Then, solve the simultaneous equations $y = 3x - 1$ and $y = 2x + 1$ using your graph.

For Questions 3–6, solve the simultaneous equations graphically, using $0 \leqslant x \leqslant 6$ in each question.

3 $y = 2x + 2$ $\qquad$ $y = 3x - 1$ $\qquad\qquad$ **4** $y = 4x + 3$ $\quad$ $y = 2x + 6$

5 $y = \frac{1}{2}x + 2$ $\qquad$ $y = x$ $\qquad\qquad$ **6** $y = \frac{1}{2}x + 1$ $\quad$ $y = 4 - x$

Exercise 43*

1 On one set of axes, draw the graphs of $y = 2x + 1$ and $y = 3x - 5$ for $0 \leqslant x \leqslant 6$.
Then, solve the simultaneous equations $y = 2x + 1$ and $y = 3x - 5$ using your graphs.

For Questions 2–4, solve the simultaneous equations graphically, using $0 \leqslant x \leqslant 6$ for each pair.

2 $x + y = 6$ $\qquad$ $3x - y = 1$ $\qquad\qquad$ **3** $2x + 3y = 6$ $\qquad$ $2y = x - 2$

4 $6x - 5 = 2y$ $\qquad$ $3x - 7 = 6y$

5 The Purple Mobile Phone Company offers the following two pricing plans to customers. Plan A costs \$15/month, with calls at 25c/minute, while Plan B costs \$100/month with calls at 14c/minute.
  a Find an equation that gives C, the cost in dollars, in terms of t, the call time in minutes per month for each plan.
  b Plot the graphs of these equations on one set of axes.
  c What call time per month costs the same under both plans?

6 Copy and complete this table to show the angle that the minute hand of a clock makes with the number 12 for various times after 12 noon.

Time after 12 noon (hours)	0	$\frac{1}{4}$	$\frac{1}{2}$	$\frac{3}{4}$	1	$1\frac{1}{4}$	$1\frac{1}{2}$	$1\frac{3}{4}$	2	$2\frac{1}{4}$	$2\frac{1}{2}$
Angle (degrees)	0		180			90					

  a Use the table to draw a graph of angle against time. Show the time from 0 hours to 6 hours along the x-axis, and the angle from 0° to 360° along the y-axis.
  b Draw another line on your graph to show the angle that the hour hand makes with the number 12 for various times after 12 noon.
  c Use your graph to find the times between 12 noon and 6pm when the hour hand and the minute hand of the clock are in line.

Inequalities

Inequalities in two variables can be represented on a graph.

Example 2

Find the region representing $x + y < 4$.

First draw the line $x + y = 4$.
This line divides the graph into two regions.
One of these regions satisfies $x + y < 4$, and the other satisfies $x + y > 4$.

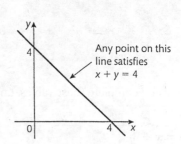

Any point on this line satisfies $x + y = 4$

To decide which region satisfies $x + y < 4$, take any point in one of the regions, for example $(1, 1)$.

Substitute $x = 1$ and $y = 1$ into $x + y$ to see if the result is less than 4.

$1 + 1 < 4$

So $(1, 1)$ is in the required region, because it satisfies $x + y < 4$.

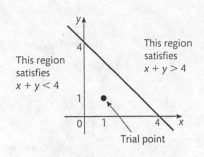

Therefore the required region is **below** the line $x + y = 4$.

Therefore this is the solution.

The line $x + y = 4$ is drawn as a broken line, to show that points on the line are *not* required. (Draw a solid line if points on the line *are* required.)

Notice that the **unwanted** region is always shaded.

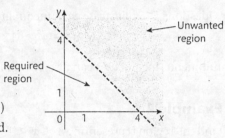

Inequalities in one variable can also be represented on a graph.

Example 3

Find the region that represents $y \leqslant 3$.

Draw the line $y = 3$ as a solid line because points on the line satisfy $y \leqslant 3$.

Points below the line $y = 3$ have y values less than 3, so the required region is below the line, and the unwanted region, above the line, is shaded.

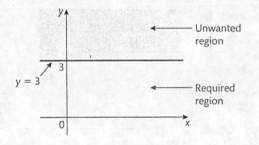

Find the line representing the equality.

If points on the line are required, draw a solid line. Otherwise, draw a broken line.

Find the required region by using a trial point that is not on the line.

Shade the **unwanted** region.

Exercise 44

For Questions 1–4, describe the **unshaded** region in each graph.

1
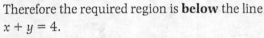

2

3
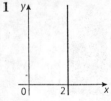

4
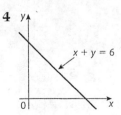

For Questions 5–8, illustrate each inequality on a graph.

5 $x < 1$ **6** $y \geqslant -2$ **7** $y \geqslant 8 - x$ **8** $y < 6 - 2x$

Exercise 44*

For Questions 1–4, describe the **unshaded** region in each graph.

1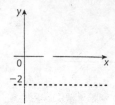

2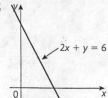

3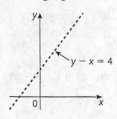

4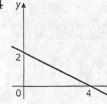

For Questions 5–8, illustrate each inequality on a graph.

5 $x \leqslant -5$ **6** $3x + 4y > 12$ **7** $y - 3x > 4$ **8** $2y - x \geqslant 4$

Simultaneous inequalities can also be represented on a graph.

Example 4

Find the region representing $1 \leqslant x < 4$.

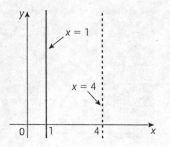

Example 5

Find the region representing $x + y \leqslant 4$ and $y - x < 2$.

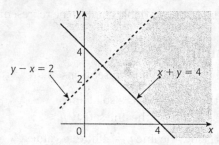

Investigate

What would several inequalities drawn on one graph look like if the **wanted** region was shaded?

Exercise 45

For Questions 1–6, describe the **unshaded** region in the graph.

1   **2**   **3**

4   **5** 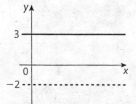   **6**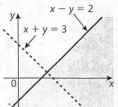

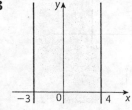

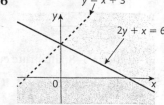

For Questions 7–10, illustrate each inequality on a graph.

7 $2 \leqslant y < 5$

8 $x < -1$ or $x \geqslant 4$

9 $y > 5 - x$ and $y \geqslant 2x - 2$

10 $x \geqslant 0, y > 2x - 3$ and $y \leqslant 2 - \dfrac{x}{2}$

Exercise 45*

For Questions 1–4, describe the **unshaded** region in the graph.

1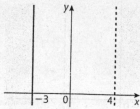

2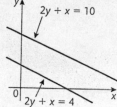

3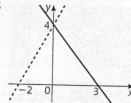

4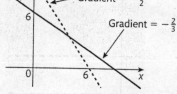

For Questions 5 and 6, illustrate each set of inequalities on a graph.

5 $y \geqslant 2x, x + 2y \leqslant 4$ and $y + 2x > 1$

6 $x \geqslant 0, y > 0, y < \dfrac{x}{2} + 4$ and $y \leqslant 6 - 2x$

7　**a** On a graph, draw the triangle with vertices $(-2, 0), (0, 2)$ and $(2, -2)$.
　　b Find the three inequalities that define the region inside the triangle.
　　c What is the smallest integer value of y that satisfies all three inequalities?

8 Illustrate on a graph the region that satisfies $y > x^2 - 4$ and $y \leqslant 0$.

Exercise 46 (Revision)

1 Copy and complete these tables, and then draw both graphs on one set of axes.

x	0	2	4
$y = x + 1$			

x	0	2	4
$y = 5 - x$			

Solve the simultaneous equations $y = x + 1$ and $y = 5 - x$ using your graphs.

2 Copy and complete these tables, and then draw both graphs on one set of axes.

x	0	2	4
$y = 3 - x$			

x	0	2	4
$y = x - 1$			

Solve the simultaneous equations $y = 3 - x$ and $y = x - 1$ using your graphs.

3 Solve $y = 3 - x$ and $y = 2x - 3$ graphically using $0 \leqslant x \leqslant 4$.

4 Solve $y = x - 1$ and $y = \frac{1}{2}x + 1$ graphically using $0 \leqslant x \leqslant 6$.

5 Solve $y = x + 4$ and $y = 1 - 2x$ graphically using $-2 \leqslant x \leqslant 2$.

6 Solve $y = 2x + 2$ and $y = -x - 4$ graphically using $-4 \leqslant x \leqslant 2$.

7 Solve $y = \frac{1}{2}x + 1$ and $y = 4 - x$ graphically using $0 \leqslant x \leqslant 6$.

8 Solve $y = 1 - \frac{1}{2}x$ and $y = 2x - 2$ graphically using $0 \leqslant x \leqslant 6$.

9 Solve $x + y = 6$ and $y = \frac{1}{2}x - 1$ graphically using $0 \leqslant x \leqslant 6$.

10 Solve $y = 3 - 2x$ and $y = x - 4$ graphically using $0 \leqslant x \leqslant 4$.

For Questions 11–24 describe the **unshaded** region in each graph.

11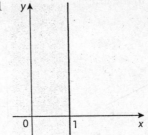

12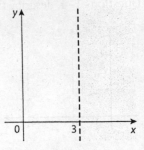

13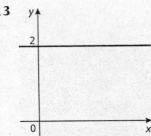

14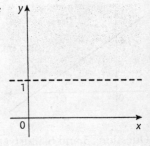

15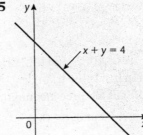

16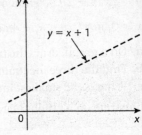

Illustrate each inequality on a graph, shading the unwanted region.

17 $x > 2$ **18** $y \leqslant 5$ **19** $x \leqslant 4$ **20** $y \geqslant 3$

Describe the **unshaded** region in each graph.

21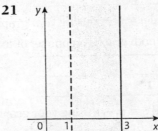

22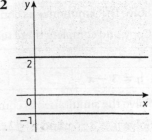

Illustrate each inequality on a graph, shading the unwanted region.

23 $y < 1$ or $y > 2$ **24** $3 \leqslant x \leqslant 6$

Exercise 46* (Revision)

1 On one set of axes, draw the graphs of $y = x + 3$ and $x + y = 6$ for $0 \le x \le 6$.
Then solve the simultaneous equations $y = x + 3$ and $x + y = 6$ using your graphs.

2 On one set of axes, draw the graphs of $y = 3 - x$ and $y = 2x - 4$ for $0 \le x \le 6$.
Then solve the simultaneous equations $y = 3 - x$ and $y = 2x - 4$ using your graphs.

Solve the following simultaneous equations graphically, using $0 \le x \le 6$ in each question.

3 $y = 3x - 5$ and $y = x - 1$

4 $y = 4 - \frac{1}{2}x$ and $y = 6 - x$

5 $y = 6 - \frac{1}{2}x$ and $y = x - 2$

6 $y = \frac{1}{2}x - 2$ and $y = 6 - 2x$

7 Solve $3x + 8y = 24$ and $2y = x + 2$ graphically using $0 \le x \le 4$.

8 Solve $6x + 5y = 30$ and $3y = 12 - x$ graphically using $0 \le x \le 4$.

9 Solve $y - 3x - 6 = 0$ and $y + 2x + 2 = 0$ graphically using $-4 \le x \le 2$.

10 Solve $2y + x + 3 = 0$ and $3y - x + 1 = 0$ graphically using $-4 \le x \le 2$.

For Questions 11–24 describe the **unshaded** region in each graph.

11

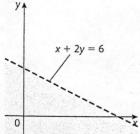

12

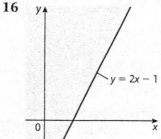

13

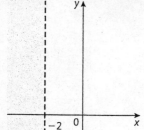

14

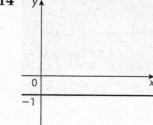

15

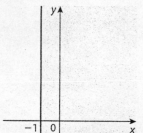

16

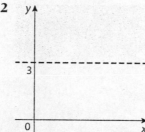

Illustrate each inequality on a graph, shading the unwanted region.

17 $x < -2$

18 $y \ge -4$

19 $x + y \le 8$

20 $2x + y \ge 4$

Describe the **unshaded** region in each graph.

21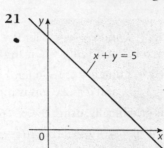

22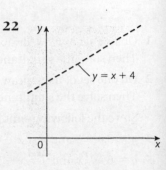

Illustrate each inequality on a graph, shading the unwanted region.

23 $x + y \geqslant 2$ and $x < 2$

24 $y \leqslant x + 1$ and $y > -1$

Tangent ratio

The history of mankind is full of examples of how we used features of **similar triangles** to build structures (for example Egyptian pyramids), survey the sky (as Greek astronomers did), and survey the land (for instance to produce Eratosthenes' map of the world).

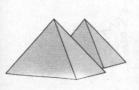
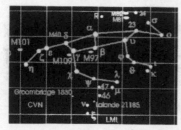

Trigonometry (triangle measurement) is used today by architects, engineers, surveyors and scientists. It allows us to solve **right-angled triangles** without the use of scale drawings.

The sides of a right-angled triangle are given special names which must be easily recognised.

Remember

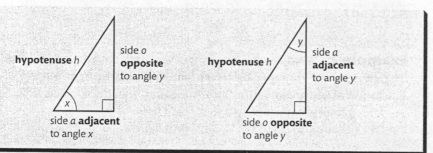

Activity 10

Triangles X, Y and Z are similar.

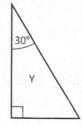
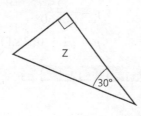

- For each triangle, measure the sides opposite (*o*) and adjacent (*a*) to the 30° angle in millimetres.
- Calculate the ratio of $\frac{o}{a}$ to 2 decimal places for X, Y and Z.
- What do you notice?

You should have found that the ratio $o:a$ for the 30° angle is the same for all three triangles. This is the case for *any* similar right-angled triangle with a 30° angle, and should not surprise you because you were calculating the **gradient** of the same slope each time.

The actual value of $\frac{\text{opposite}}{\text{adjacent}}$ for 30° is 0.577 350 (to 6 d.p.).

The ratio $\frac{\text{opposite}}{\text{adjacent}}$ for a given angle x is a fixed number. It is called the **tangent of x**, or **tan x**.

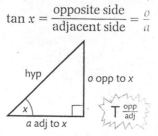

Calculating sides

If you know an angle – e.g. an angle of elevation or angle of depression – and one side, you can find the length of another side by using the tangent ratio.

Example 1

Find the length of the side p correct to 3 significant figures.

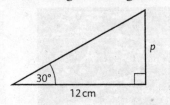

$$\tan 30° = \frac{p}{12 \text{ cm}}$$

$$12 \text{ cm} \times \tan 30° = p$$

$$p = 6.93 \text{ cm} \quad (\text{to 3 s.f.})$$

`1 2 × tan 3 0 =`   **6.92820**   (to 6 s.f.)

Example 2

PQ represents a 25 m tower, and R is a surveyor's mark p m away from Q.
The **angle of elevation** of the top of the tower from the surveyor's
mark R on level ground is 60°.
Find the distance RQ correct to 3 significant figures.

$$\tan 60° = \frac{25}{p}$$

$$p \times \tan 60° = 25$$

$$p = \frac{25}{\tan 60°}$$

$$p = 14.4 \text{ (to 3 s.f.)}$$

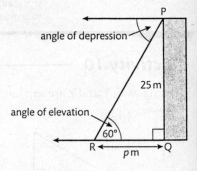

`2 5 ÷ tan 6 0 =`   **14.4338**   (to 6 s.f.)

Exercise 47

In this exercise, give answers correct to 3 significant figures.

In Questions 1–2, which sides are the hypotenuse, opposite and adjacent to the given angle a?

1

2

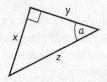

3 What is the value of $\tan b$ for this triangle?

For Questions 4–9, find the length of side x.

4

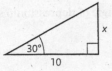

5

6

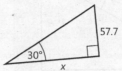

7

8

9

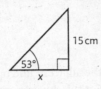

10 The angle of elevation of the top of a cliff from a boat 125 m away from its foot is 35°. Find the height of the cliff.

11 Find the cross-sectional area of the pitched roof WXY if WX = WY.

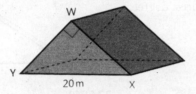

Exercise 47*

Give answers correct to 3 significant figures.

For Questions 1–4, find the length of the side marked x.

1

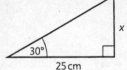

2

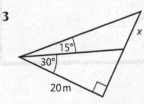

3

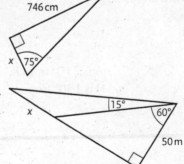

4

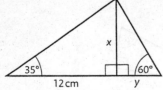

5 The gnomon (central pin) of the giant equatorial sundial in Jaipur, India, is an enormous right-angled triangle. The angle the hypotenuse makes with the base is 27°. The base is 44 m. Find the height of the gnomon.

For Questions 6 and 7, find the lengths of x and y.

6

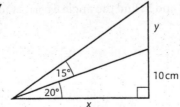

7

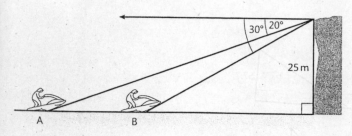

8 From the top of a 25 m high cliff, the angle of depression of a jet ski at point A is 20°. The jet ski moves in a straight line towards the base of the cliff so that its angle of depression 5 s later at point B is 30°.

a Calculate the distance AB.

b Find the average speed of the jet ski in kilometres/hour.

Calculating angles

If you know the adjacent and opposite sides of a right-angled triangle, you can find the angles in the triangle. For this 'inverse' operation, you need to use the [INV] [tan] buttons on your calculator.

Example 3

The diagram shows Luther on a slide.
Find angles x and y to the nearest degree.

$$\tan x = \frac{3}{4.5} \text{ and } \tan y = \frac{4.5}{3}$$

Key Points

To calculate an angle from a tangent ratio, use the [INV] [tan] or [SHIFT] [tan] buttons.

 33.6901 (to 6 s.f.)

 56.3099 (to 6 s.f.)

So $x = 34°$ and $y = 56°$ (to the nearest degree).

$$T \frac{\text{opp}}{\text{adj}}$$

Remember

Bearings are measured
• clockwise
• from North.

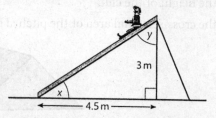

A is 310° from B.
B is 130° from A.

Exercise 48

Find the angles that have these tangents, giving your answers correct to 2 significant figures.

1 1.000

2 0.268

3 Find the angle with a tangent of 2.747, giving your answer correct to 3 significant figures.

For Questions 4 and 5, find the angle a correct to 1 decimal place.

4

14

14

5

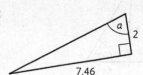

a

2

7.46

For Questions 6 and 7, find the angle x correct to 1 decimal place.

6

7

8 A bell tower is 65 m high. Find the angle of elevation, to 1 decimal place, of its top from a point 150 m away on level ground.

9 Ollie is going to walk in the rain forest. He plans to go up a slope along a straight footpath from point P to point Q. The hill is 134 m high, and distance PQ *on the map* is 500 m. Find the angle of the hill.

Exercise 48*

1 ABCD is a rectangle.
Find angles a and b to the nearest degree.

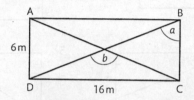

For Questions 2 and 3, find angle a.

2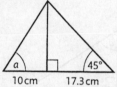

3 Rectangle

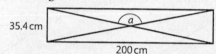

4 The grid represents a map on which villages X, Y and Z are shown. The sides of the grid squares represent 5 km.
Find, to 1 decimal place, the bearing of
  a Y from X   **b** X from Y
  c Z from X   **d** Z from Y

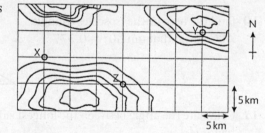

5 A bird watcher spots an eagle perched at the top of a tree of height h metres. The angle of elevation from the ornithologist to the eagle is 20° and she stands 100 m from the base of the tree. Find h.

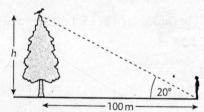

6 Find the area of an equilateral triangle with total perimeter 60 cm.

7 A 5 m flagpole is secured by two ropes PQ and PR. Point R is the mid-point of SQ. Find angle a to 1 decimal place.

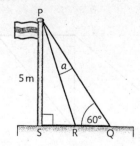

8 Given that $\tan 30° = \dfrac{1}{\sqrt{3}}$ and $\tan 60° = \sqrt{3}$,
show that the exact value of the height of the
tree in metres is given by $25\sqrt{3}$.

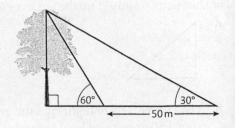

Exercise 49 (Revision)

In this exercise, give answers correct to 3 significant figures.
For Questions 1–6, find the value of x.

1
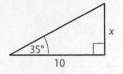

2

3
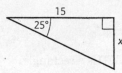

4
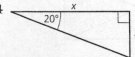

5
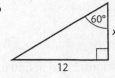

6
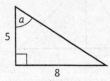

For Questions 7–9, find angle a.

7

8
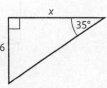

9
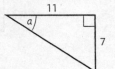

10 Find the area of the isosceles triangle
ABC, given that $\tan 30° = 0.577$.

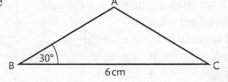

11 Calculate the angle between the longest side and the diagonal of a 577 mm by 1000 mm
rectangle.

For Questions 12–15, find the value of angle θ.

12
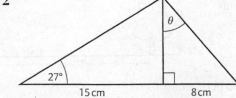

13
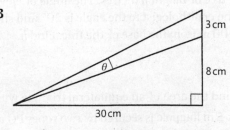

14
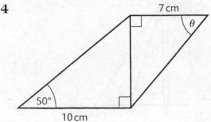

15
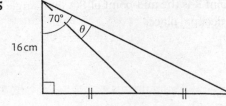

Unit 2 : Trigonometry

Exercise 49* (Revision)

Give answers to 3 significant figures.
For Questions 1–4, find the lengths of sides x and y.

1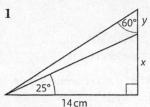

2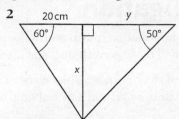

3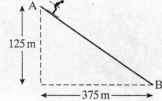

4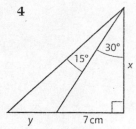

5 The angle of elevation to the top of the CN Tower, Toronto, Canada, from a point on the ground 50 m away from the tower is 84.8°. Find the height of the CN Tower.

6 A harbour H is 25 km due North of an airport A. A town T is 50 km due East of H.
 a Calculate the bearing of T from A. **b** Calculate the bearing of A from T.

7 The diagram shows the lines of sight of a car-driver.

 a Calculate the driver's 'blind' distance D in metres.

 b Why is this distance important in the car design?

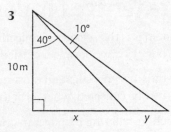

8 Ezola skis down a straight slope from point A to point B. The difference in height between these two points is 125 m, and the actual distance AB as viewed on a map is 375 m. Calculate the angle of this ski slope.

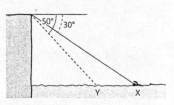

9 Find the angle that the line $y = 3x - 7$ makes with the x-axis.

10 Find the angle that the line $2x + 3y - 4 = 0$ makes with the y-axis.

11 An area of an equilateral triangle is 1000 cm². Calculate the perimeter of the triangle.

12 An equilateral triangle has sides of 10 m. Its area is equal to that of a circle. Calculate the circumference of the circle.

13 From the top of a cliff of height y m the angle of depression of a Channel swimmer at X is 30°. She swims directly towards the base of the cliff such that, 1 minute later, at Y, the angle of depression from the top of the cliff to the swimmer is 50°. Given that her speed between X and Y is 0.75 m/s, find the height of the cliff.

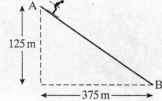

14 Show that the area of an equilateral triangle of side $2x$ is $x^2\sqrt{3}$, given that $\tan 60° = \sqrt{3}$.

15 A lighthouse is 25 m high. From its top, the angles of depression of two buoys due North of it are 45° and 30°.

 Given that $\tan 30° = \dfrac{1}{\sqrt{3}}$, show that the distance x, in metres, between the buoys is $25(\sqrt{3} - 1)$.

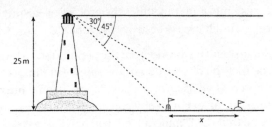

Statistical investigation

Statistics is the science of collecting information (**data**) and analysing it. People do this to gather evidence, form conclusions (and make predictions), and then to take decisions.

A statistical investigation must have a clear purpose. This determines what data is needed, how it must be collected, what form it needs to be in, and how it should be displayed.

A school catering manager might survey pupils about their favourite foods before creating new menus. A government minister of transport must consider projected traffic flows when a new major road is being planned.

Data can be **qualitative** (such as opinions, colours, or clothes size) or **quantitative** (such as a length measurement, a mass, or a frequency).

Primary data is data that has been collected directly by a **researcher**. A researcher may also refer to **secondary data** – data that has been collected and made available by other organisations. Many **databases** are made available to the public by government departments and large companies, either by direct request or through the Internet.

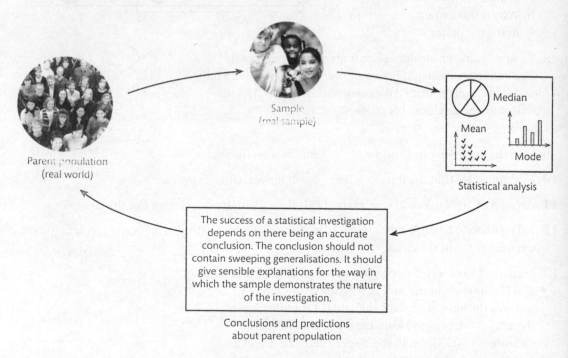

Parent population
(real world)

Sample
(real sample)

Median
Mean
Mode
Statistical analysis

The success of a statistical investigation depends on there being an accurate conclusion. The conclusion should not contain sweeping generalisations. It should give sensible explanations for the way in which the sample demonstrates the nature of the investigation.

Conclusions and predictions
about parent population

In a perfect world, the researcher would take a **census** (in which everything or everybody is surveyed). However, this is rarely practical, and is often impossible. A **survey** costs money and takes time, and sometimes the process of 'testing' can destroy the 'objects'. For example, testing rope until it breaks destroys the rope.

So, instead, a **sample** from the population to be examined is tested. The size of the sample and the method of selection of the sample are important but no matter how carefully a sample is chosen, there is always a possibility of some **bias**. This is often allowed for by the inclusion of a **margin of error** in any conclusions. For example, the results of an election opinion poll might state that the Democrat vote will be 37% ± 3%.

Collecting data

There are many ways of collecting data.

Questionnaires

Data logging

Databases

Simulation

It is important to be methodical and accurate when writing down the data, and when transcribing it into another form, for instance into a computer file, table, chart or diagram.

Frequency tables

Before attempting to calculate any statistics, or to recognise any trends or patterns, it is sensible to sort the data and group it together.

These marks were obtained by a set of pupils in an IGCSE mock examination.

42	54	60	48	73	50	59	45	84	49
67	47	70	78	77	67	55	68	42	59
54	41	69	65	41	56	80	59	44	68
82	41	71	42	55	64	51	69	89	72
72	46	85	40	78	67	66	52	42	89
46	41	62	51	73	50	41	58	44	69

Tally charts

Tally charts are used to count the numbers within each group. There are 60 numbers in the data set. The smallest is 40 and the largest is 89. It is most convenient to choose five groups of numbers: 40–49, 50–59, and so on.

Working through the figures in order, a *neat* tally mark is made alongside the appropriate group. Tally marks are grouped into fives for easier counting, and a 'slash' across ||||| denotes '5'.

The frequencies can be added to check that the total is the same as the total number of items in the data set.

Group	Tally	Frequency
40–49	ɪɪɪɪ ɪɪɪɪ ɪɪɪɪ |||	18
50–59	ɪɪɪɪ ɪɪɪɪ ||||	14
60–69	ɪɪɪɪ ɪɪɪɪ |||	13
70–79	ɪɪɪɪ ||||	9
80–89	ɪɪɪɪ		6
		Total = 60

Averages

The **mean** of the data is the $\dfrac{\text{total of all values}}{\text{total number of values}}$.

The **mode** of the data is the value that occurs the most often.

The **median** of the data is the value in the middle when the data is arranged in ascending order. (For an even number of values, the median is the mean of the middle pair of values.)

Example set of data: 8, 9, 13, 19, 8, 15, 8

$$\textbf{Mean} = \frac{8 + 9 + 13 + 16 + 8 + 15 + 8}{7} = \frac{77}{7} = 11$$

Rearranging the data: 8, 8, 8, 9, 13, 15, 16: **Mode** = 8 (3 times). **Median** = 9 (middle value).

Comparing the mean, median and mode

The mean, median and mode are all different averages of a set of data. An average is a single value that should give you some idea about all the data. For example, if you are told one batsman has a mean score of 12 and another has a mean score of 78 then you immediately have some idea about the batsmen without knowing all their individual scores.

When the data is plotted as a bar chart, if the result is roughly 'bell shaped' or 'normal' then the mean, median and mode all give approximately the same answers.

If there are a few very high or very low values (called 'outliers') then these can distort the averages, and you have to choose which average is the best to use.

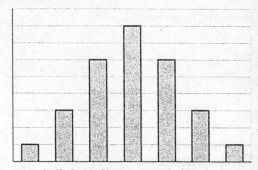

A 'bell shaped' or 'normal' distribution

Example 1

The wages per week in a small firm are

$500,	$500,	$650,	$660,
$670,	$680,	$3000	

The mean is £951 (to the nearest dollar)
The median is $660
The mode is $500

The mode comes from the two lowest values and is not representative.

The mean is not that useful as it is distorted by the managing director's large salary.
The median is the best to use as it tells us most about the data.

However, it all depends on your point of view! In wage negotiations, the managing director can claim that the 'average' salary is good at $951, while the shop steward can claim that the 'average' salary is poor at $500. Both are telling the truth, it all depends which 'average' you use.

	Mean	**Median**	**Mode**
Advantages	Uses all the data	Easy to calculate Not affected by extreme values	Easy to calculate Not affected by extreme values
Disadvantages	Distorted by extreme values	Does not use all the data	Does not use all the data Less representative than the mean or median
When to use	When the data is distributed reasonably symmetrically	When there are some extreme values	For non-numerical data When the most typical value is needed

Exercise 50

Use your calculator where appropriate.

1 A biased dice has six faces numbered 1, 2, 3, 4, 5 and 6. It has been thrown 40 times, and
these are the scores.

1	6	3	2	3	2	1	2	4	1
4	2	4	1	6	1	3	5	2	6
3	2	3	1	2	5	4	2	5	1
2	1	6	2	4	1	3	6	3	5

Construct a tally chart for the data. Comment on the bias.

2 These are the weights, in kilograms, of 50 newborn babies in a hospital during August.

1.35	2.05	2.71	3.00	2.34	3.36	2.44	2.70	3.48	1.68
2.66	2.59	2.03	3.76	3.11	3.03	2.23	4.18	2.95	2.50
3.10	2.09	4.65	2.68	1.28	3.77	3.88	3.60	3.88	2.34
1.58	2.84	1.64	4.22	2.88	1.86	4.00	2.41	3.25	2.89
3.92	3.05	3.60	1.97	3.54	3.45	2.85	4.06	2.12	2.85

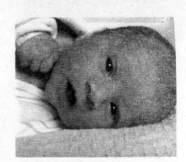

Construct a tally chart using the groups $1.0 \leqslant w < 1.5$, $1.5 \leqslant w < 2.0$, and so on.
What proportion of these babies weighs less than 3 kg?

For Questions 3 and 4, find the mean, the median and the mode for each data set.

3 7, 12, 14, 14, 3

4 21, 31, 11, 16, 18, 4, 4

Exercise 50*

1 These are the times taken by a group of pupils to solve a puzzle.
48 s   52 s   88 s   34 s   37 s   38 s   45 s
Calculate the mean and median times.

2 A keen golfer plays four rounds of golf. His scores are 88, 98, 91 and 91.
Which type of average (mean, median or mode) will he prefer to call his 'average'?

3 A group of 12 boys has a mean age of 10.75 years. One of them, aged 13.5, leaves the
group. What is the mean age of the remaining boys?

4 Find six numbers, five of which are smaller than the mean of the group.

Displaying data

Discrete data

Discrete data are measurements collected from a source that can be listed and counted. The number of absentees from a school is an example of discrete data.

The table and the bar chart show the numbers of driving tests taken by a group of students before they passed the test.

Number of driving tests	Frequency
1	23
2	35
3	20
4	12
5	6
6	3
7	0
8	1

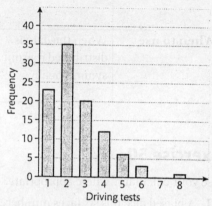

The labelling of the horizontal axis and the spaces between the bars indicate that the results are discrete.

Continuous data

Continuous data are measurements collected from a source that *cannot* be listed and counted.

The table and bar chart show the lives of a sample of 75 Everglo batteries tested in continuous use in a model train until they were exhausted.

Lifetime (hours)	Frequency
$6 \leqslant h < 7$	8
$7 \leqslant h < 8$	12
$8 \leqslant h < 9$	14
$9 \leqslant h < 10$	22
$10 \leqslant h < 11$	16
$11 \leqslant h < 12$	3

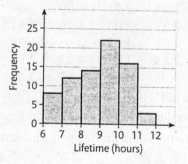

The labelling of the horizontal axis and the 'touching' bars indicate continuous data.

Frequency polygon

A frequency polygon is made by joining the mid-points of the tops of the bars in a bar chart or histogram. It is useful for indicating trends, and for comparing two or more sets of results.

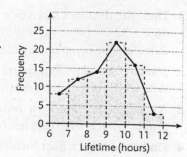

Pie chart

These are the annual costs of running a typical family car.

	Cost ($)
Petrol	1000
Car tax	150
Insurance	350
Maintenance	900
Depreciation	3600
Total = $6000	

Cost as a proportion of the total	Angle of sector
$\frac{1000}{6000}$	$\frac{1000}{6000} \times 360° = 60°$
$\frac{150}{6000}$	$\frac{150}{6000} \times 360° = 9°$
$\frac{350}{6000}$	$\frac{350}{6000} \times 360° = 21°$
$\frac{900}{6000}$	$\frac{900}{6000} \times 360° = 54°$
$\frac{3600}{6000}$	$\frac{3600}{6000} \times 360° = 216°$
	Total = 360°

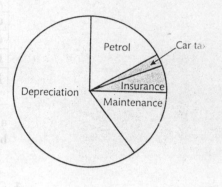

A pie chart is most suitable for displaying proportions.

Pictogram

A pictogram is like a bar chart, except that the frequencies are shown by pictures instead of bars.

Month	Car thefts (one car = 100 thefts)
Mar	🚗 🚗 🚗 🚗 🚗
Apr	🚗 🚗 🚗 🚗 🚗 🚗
May	🚗 🚗 🚗 🚗 🚗 🚗 🚗
Jun	🚗 🚗 🚗 🚗
Jul	🚗 🚗 🚗
Aug	🚗 🚗 🚗

Operation Clampdown was introduced by a police authority on 1 June to combat car crime

Exercise 51

Use your calculator where appropriate.

1 The table shows how 900 students travel to school.

Mode of travel	Walk	Cycle	Car	Bus	Train
Number travelling	510	80	50	140	120

Display the data as a pie chart. What percentage of pupils travel to school by public transport?

2 This frequency table shows the heights of a class of schoolgirls.

Height (cm)	Frequency
$130 \leqslant h < 135$	3
$135 \leqslant h < 140$	5
$140 \leqslant h < 145$	8
$145 \leqslant h < 150$	5
$150 \leqslant h < 155$	4
$155 \leqslant h < 160$	5

Construct a frequency polygon to represent this information. Choose a scale such that each group is 1 cm wide.

3 A sports centre runs several evening clubs for junior members.
The table shows the membership numbers.

	Girls	Boys
Aerobics	12	6
Climbing	6	10
Gymnastics	16	4
Hockey	14	18
Judo	6	16
Tennis	12	8

a Display the details separately for the boys and the girls on one frequency polygon diagram.

b Investigate other ways of representing this information.

4 Calculate the mean of this data set: 5, 8, 3, 11, 9, 6, 8, 6.
The mean of another set of seven figures is 5. Calculate the mean of the combined set of 15 figures.

Exercise 51*

1 The bar chart shows the numbers of children in each of 30 families. *The information was gathered by asking a class of pupils to indicate their family details on a questionnaire.*

a Calculate the mean and median number of children per family in this sample.

b Comment on the bias in the sample and the effect that it will have on the statistics. Suggest how an unbiased sample could be gathered.

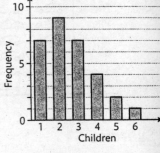

2 The pie chart displays the distribution of mobile phones among the pupils at a school.
If 318 pupils have a mobile, how many pupils are there in the school?

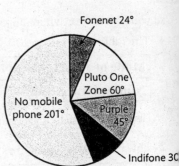

3 In a cross-country race, these times were recorded by a team of eight runners.

| 13 min 53 s | 14 min 5 s | 14 min 28 s | 14 min 30 s |
| 14 min 56 s | 15 min 11 s | 15 min 12 s | 15 min 13 s |

Calculate the mean time.

4 Explain briefly how the random number generator on your calculator can be used to simulate the experiment of throwing three coins and recording the number of heads.

a Do this simulation and record the results for 40 throws in a tally chart.

b Display the results in a bar chart.

c Comment briefly on your results.

Exercise 52 (Revision)

1 These are the maximum temperatures in °C that a gardener recorded one week.

21, 20, 23, 20, 22, 20, 21

Find the mean, median and mode of these temperatures.

2 In a primary school competition to grow the tallest sunflower, the following heights (in metres) were obtained.

1.32, 1.87, 2.03, 1.56, 1.95, 1.48, 1.12, 2.15

a Calculate the mean and median heights of these sunflowers.
b The 2.15 m height had been recorded incorrectly and was actually 2.75 m. Calculate the new mean height.

3 The day after a school disco a sample of pupils were asked to give it a rating from A to E, A being excellent and E being dreadful. The following results were obtained.

D, B, E, E, C, A, E, B, A, B, E, A, B, D, A, D, D, A, D, B, A, D, E, C, E, A, E, E, A, A

a Construct a tally chart and then draw a bar chart to display this information.
b Can you draw any conclusions about the disco?

4 The following ingredients are needed to make a strawberry smoothie.

250 g of strawberries, 150 g of banana, 200 g yoghurt and 120 g of iced water.

Display this information in a pie chart, marking the size of the angles clearly.

5 Tara suspects that a dice is biased. She throws it 50 times with the following results.

6 1 4 4 2 5 4 1 2 3 4 3 2 4 1 2 6 3 2 6 6 6 6 1 4
5 6 4 6 4 2 6 3 1 2 2 6 5 4 3 1 6 6 2 5 5 2 1 4 5

a Construct a tally chart for the data.
b Construct a bar chart to represent this information.
c Comment on Tara's suspicions.

6 These are the times, t, in seconds, that it took a group of students to solve a puzzle.

42, 67, 54, 79, 72, 56, 54, 47, 41, 41, 46, 42, 78, 71, 69, 70, 49,
62, 48, 78, 65, 42, 40, 51, 73, 77, 41, 65, 78, 73, 50, 67, 66, 44,
67, 50, 59, 55, 50, 51, 66, 41, 45, 69, 52, 58, 42, 44, 43, 68

a Construct a tally chart using the groups $40 \leqslant t < 45$, $45 \leqslant t < 50$ and so on.
b Draw a bar chart to represent this information.
c What percentage of the class took less than 1 minute to solve the puzzle?

7 This frequency table shows the times taken by competitors on a charity fun run.

Time (min)	$60 \leqslant t < 70$	$70 \leqslant t < 80$	$80 \leqslant t < 90$	$90 \leqslant t < 100$	$100 \leqslant t < 110$	$110 \leqslant t < 120$
Frequency	8	14	26	17	13	7

a How many competitors were there?
b Draw a frequency polygon to display this information.
c What percentage of the competitors took at least 100 minutes?

8 A group of ten girls has a mean height of 1.42 m.

a What is the total height of all ten girls? One girl with height 1.57 m leaves the group.
b What is the mean height of the remaining girls?

9 These are the number of goals Mark scored each match **last** season. 0, 1, 0, 0, 2, 0, 1, 1, 3, 0

a Find the mean, median and mode of this data.
b This season Mark has played in 12 games. The mean number of goals he has scored is 1.
What is the mean number of goals he has scored over both seasons?

Exercise 52* (Revision)

1 The lengths of tracks in minutes on a CD were as follows

6.2, 3.6, 10.8, 2.4, 2.3, 8.7, 14.7, 10.9

a Calculate the mean and median length of tracks on the CD.

b The time of 10.9 had been misread and was actually 16.9. Calculate the correct mean.

2 These are the number of times that each pupil in 7R has been late this term.

0, 4, 5, 1, 3, 4, 6, 0, 1, 2, 3, 0, 0, 2, 6, 5, 4, 0, 1, 3, 2, 0, 4, 3, 2, 1, 1, 4, 6, 0

a Find the mean, median and mode of this data.

b Which average would the head teacher prefer to use in his report to the Governors?

c Draw a bar chart to display the information.

3 The pie chart shows the results of a school mock election
If Socialist received 183 votes, how many pupils voted altogether?

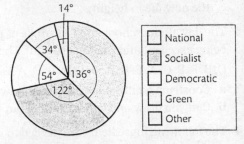

	National
	Socialist
	Democratic
	Green
	Other

4 The weight marked on a cereal packet is 500 g. These are the weights in grams of fifty packets of cereal.

506, 507, 513, 502, 504, 507, 509, 505, 506, 515, 503, 508, 510, 500, 511, 505, 493, 514, 499, 501, 510, 506, 503, 516, 495, 519, 503, 507, 523, 508, 511, 517, 506, 497, 513, 516, 512, 503, 511, 504, 512, 501, 508, 505, 506, 509, 507, 513, 508, 521

a Construct a tally chart using groups $490 \leqslant w < 495$, $495 \leqslant w < 500$ and so on up to $520 \leqslant w < 525$.

b Draw a bar chart to display the information.

c Does the evidence suggest that 500 g is a mean weight or a minimum weight?

5 This frequency table shows the speeds of some cars on a road.

Speed s (km/h)	$30 \leqslant s < 35$	$35 \leqslant s < 40$	$40 \leqslant s < 45$	$45 \leqslant s < 50$	$50 \leqslant s < 55$	$55 \leqslant s < 6$
Frequency	4	18	43	54	5	2

a How many cars had their speeds recorded?

b Draw a frequency polygon to display this information.

c What do you think the speed limit is on this road? Give a reason for your answer.

6 This frequency table shows how much time per week 120 teenagers spent watching television.

Time t (hrs)	$0 \leqslant t < 2$	$2 \leqslant t < 4$	$4 \leqslant t < 6$	$6 \leqslant t < 8$	$8 \leqslant t < 10$	$10 \leqslant t < 12$	$12 \leqslant t \leqslant 14$
Frequency	18	12	10		21	25	19

a What is the frequency for the group $6 \leqslant t < 8$?

b Draw a bar chart to display the information.

c What percentage watched between 4 and 10 hours of television per week?

7 A football team of eleven players has a mean height of 1.83 m. One player is injured and is replaced by a player of height 1.85 m. The new mean height of the team is now 1.84 m. What is the height of the injured player?

8 Sharonda is doing a biology project using two groups of worms. The mean length of the first group of ten worms is 8.3 cm.

a What is the total length of all the worms in the first group?

The mean of the second group of 8 worms is 10.7 cm.

b Calculate the mean length of all eighteen worms.

1   $a = 1.2 \times 10^3$, $b = 3.5 \times 10^{-2}$ and $c = 4.9 \times 10^2$.

If $p = \dfrac{2abc}{(a-c)^2}$, the value of p to 3 s.f. is

   A 8.17×10^{-2} **B** 8.16×10^{-2} **C** 5.80×10 **D** 4.96×10^2

2   How many mm^3 are there in $1\,km^3$?

   A 10^6 **B** 10^9 **C** 10^{12} **D** 10^{18}

3   The smallest possible solution of $3(x-1) \geqslant 2(x-5)$ is

   A -7 **B** 7 **C** -4 **D** -13

4   The unshaded area is described by the inequalities

   A $x + y > 4$ **B** $x + y \leqslant 4$
       $2y - x > 2$ $2y - x < 2$
   C $x + y < 4$ **D** $x + y \leqslant 4$
       $2y - x < 2$ $2y - x > 2$

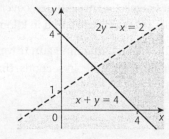

5   The solutions to the simultaneous equations $x + y = 14$, $2x - y = -8$ is

   A $(7, 7)$ **B** $(12, 2)$ **C** $(-2, 12)$ **D** $(2, 12)$

6   If the solution to the simultaneous equations $x + y = 3$, $3x - y = 13$ is (p, q), the value of $(p - q)^3$ is

   A 27 **B** -27 **C** 125 **D** 15

7   The area of an equilateral triangle of side 5 m to the nearest integer is

   A $10\,m^2$ **B** $11\,m^2$ **C** $12\,m^2$ **D** $13\,m^2$

8   The angle of depression from an 80 m high lighthouse to a ship 100 m away to the nearest degree is

   A $51°$ **B** $39°$ **C** $30°$ **D** $72°$

9   Ten girls have a mean age of 15 years 6 months. If one is 11 years old, the mean age of the others is

   A 14 years   **B** 15 years   **C** 16 years   **D** 17 years

10   A football team's record of goals scored over the first 5 games of the season is shown in the bar chart. The mean goals scored per game is x. In their next match the team win $4-3$ which changes the mean goals scored per games to y. The value of $y - x$ is:

   A 0.5   **B** 1
   C 1.5   **D** 2

1 Write each of these numbers as an ordinary number.

  a $(5 \times 10^{-2}) + (4 \times 10^{-4})$ **b** $(5 \times 10^{-2}) \times (4 \times 10^{-4})$

2 Show that

  a $\frac{3}{5} \times \frac{15}{21} = \frac{3}{7}$ **b** $3\frac{1}{4} + 2\frac{1}{3} = 5\frac{7}{12}$ **c** $\frac{14}{15} \div \frac{7}{25} = 3\frac{1}{3}$

3 Share these amounts in the ratios given.

  a $767\,\text{kg}$ in the ratio $5:8$ **b** $\$4.48$ in the ratio $2:7:5$

4 $\$30$ can be exchanged for 170 Egyptian pounds.
  How many Egyptian pounds would you get for $\$12$?

5 A car travels for three hours, covering a distance of 147 miles.
  Work out its average speed in kilometres per hour, assuming 1 mile = 1.6 km.

6 A bullet from a machine gun travels at 108 m/s. Assuming that the bullet does not slow
  down, find the time, in seconds, that it takes to travel 3 km.

7 Simplify these.

  a $\frac{45a^5}{15a^3}$ **b** $\frac{7x}{y} \div \frac{14x^2}{y}$ **c** $\frac{8a^3}{3b} \times \frac{9b^2}{22a} \times \frac{11b}{4a^2}$

8 Write these as single fractions.

  a $\frac{a}{2} + \frac{b}{7}$ **b** $\frac{3y}{5} - \frac{2y}{4}$ **c** $\frac{7(x-3)}{4} + \frac{3(2-y)}{5}$

9 Solve these equations for x.

  a $2x^2 - 7 = 43$ **b** $\frac{11 + \sqrt{x}}{3} = 5$ **c** $\sqrt{\frac{3x^2 + 20}{2}} = 8$

10 The formula for converting a temperature, C, in degrees Celsius (°C) to a temperature, F, in
  degrees Fahrenheit (°F) is $F = \frac{9C + 160}{5}$.

  a Use the formula to convert

      i 20 °C to degrees Fahrenheit.   **ii** 100 °F to degrees Celsius.

  b What temperature is the same on both scales?

11 Simplify these.

  a $y^4 \times y^9$ **b** $b^{16} \div b^7$ **c** $8a^2 \times 2b^8$

12 Simplify these.

  a $b^2 \times b^4 \times b^3$ **b** $(7a^4)^2$ **c** $p^9 \div p$

13 Simplify these.

  a $\frac{21a^2bd}{48(ab)^2}$ **b** $\frac{49a^5b^3c^7}{7a^3b^4c^4}$ **c** $\frac{(3f)^2}{24f^5}(2f)^3$

14 Solve these inequalities and illustrate the answers on a number line.

  a $x - 8 > 4$ **b** $7x + 2 \leqslant 4x + 23$

15 n is an integer such that $-6 < n \leqslant 2$. List all the possible values of n.

16 **a** On graph paper draw x- and y-axes such that $-2 \leqslant x \leqslant 12$ and $-2 \leqslant y \leqslant 12$.

  b On these axes draw both the lines $y = 2x - 1$ and $x + y = 11$.

  c Use your graph to solve the simultaneous equations $y = 2x - 1$ and $x + y = 11$.

  d Illustrate the region $y \geqslant 2x - 1$ and $x + y \leqslant 11$ by shading the **unwanted** regions.

17 Calculate the lengths x and y and the angle z in the diagrams below.

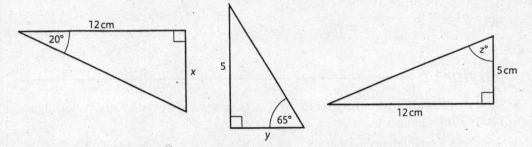

18 On a map of a Fun Park, the roller coaster (R) is 10 cm East of the cafe (C) and the ghost-train (G) is 7 cm South of the roller coaster.
Draw a suitable diagram and **calculate** the bearing of the ghost-train from the cafe.

19 The amounts of money, in pence, placed in the charity box over 20 days by a class of pupils is given in the table.

20	19	33	18	13	25	21	31	15	18
27	27	26	38	16	19	27	25	30	20

a Find the mean, median and mode of these amounts.
The mean amount collected per day over the next 10 days was 22.2 pence.
b Calculate the mean amount collected per day over the total period of 30 days.

20

Grade	Number of pupils
A	p
B	q
C	r
D	s
E	40

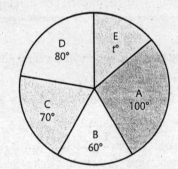

The table and the pie-chart show the number of grades that were obtained in a Russian examination by the pupils at Moscow Grammar School.
Find the values of p, q, r, s and t.

Compound percentages

Activity 11

- Show that if $120 **increases** in value by 8%, its new value will be $129.60. Calculate 120×1.08. Comment.
- Show that if $120 **decreases** in value by 8%, its new value will be $110.40. Calculate 120×0.92. Comment.
- Copy and complete these two tables.

To increase by (%)	Multiply by
15	
70	
	1.56
	1.02

To decrease by (%)	Multiply by
15	
70	
	0.8
	0.98

Compound percentages are used when one percentage is followed by another in a calculation.

Example 1

$120 is invested at 15% interest for 3 years. Find the value of the investment after 1 year, 2 years, and 3 years.

To increase by 15%, multiply by $\left(1 + \dfrac{15}{100}\right) = 1.15$.

After 1 year, the investment is $120 \times 1.15 = $138.

After 2 years, the investment is $138 \times 1.15 = $158.70.
(Notice that, after 2 years, you could write ($120 \times 1.15) \times 1.15 = $120 \times 1.15^2 = $158.70.)

After 3 years, the investment is $= $120 \times 1.15 \times 1.15 \times 1.15$
$$= $120 \times (1.15)^3$$
$$= $182.51 \text{ (to 2 d.p.)}$$

In the second and third years, the interest earned by the investment has *not* been withdrawn. It has been added to the investment. This is called **compound interest**.

Key Points

- To **increase** a number by $R\%$, multiply it by $\left(1 + \dfrac{R}{100}\right)$.
- To **decrease** a number by $R\%$, multiply it by $\left(1 - \dfrac{R}{100}\right)$.

Exercise 53

1. What would you multiply by to increase an amount by
   a 10%   b 20%   c 30%   d 1%   e 15%   f 25%

2. A house increases in value by 3% in the first year and by 4% in the second year. Copy and complete 'to work out the answer, multiply the original value by $1.03 \times ...$'.

3. There are 500 pupils in a school. In the next year, the numbers increase by 10%, and, in the following year, they increase by 20%. Find the number of pupils at the end of these two years.

4 A sum of money is invested at 4% compound interest for three years.
Copy and complete 'to calculate the answer, multiply by ...'.

5 $450 is invested at 7% compound interest. Find the value after two years.

6 In Snowland, inflation pushes up prices by 35% per year. A car costs $40 000.
Find the price of the same model of car three years later.

Exercise 53*

Where appropriate, give your answers correct to 3 significant figures.

1 What would you multiply by to increase an amount by
   a 12.5%   **b** 4%   **c** 5%

2 Increase 80 km by 12.5%.

3 Increase $400 by 4%, and then by another 5%.

4 What is the result of increasing a quantity by 20%, followed by decreasing it by 20%?

5 $550 is invested at 8% compound interest. Find the value after three years.

6 An antique vase appreciates (gains) in value at the rate of 15% per year.
If the vase is bought for $5000, how much will it be worth in 5 years, 10 years and 15 years?

7 A car depreciates (loses) in value at the rate of 15% per year.
If the car is bought for $20 000, how much will it be worth in 5 years, 10 years and 15 years?

8 If Julius Caesar had invested the equivalent of 1 cent at 1% per annum compound interest,
what would it have been worth in $ after 2000 years?

Multiples, factors and primes

The **multiples** of 4 are 4, 8, 12, 16 ...

The only numbers that divide exactly into 10 are called **factors of 10**, and these are 1, 2, 5 and 10.

If a number has no factors apart from 1 and itself, it is called a **prime number**.

Remember

Prime numbers are divisible only by 1 and themselves.

They are 2, 3, 5, 7, 11, 13, 17, 19, 23, 29, 31, 37, 41, 43, 47, 53, ..., ..., 67, 71, 73, ...

Notice that 1 is *not* a prime number.

Example 2

Express 84 as the product of prime factors.

(Divide repeatedly by prime numbers)

$$84 = 2 \times 42$$
$$\quad\;\; = 2 \times 2 \times 21$$
$$\quad\;\; = 2 \times 2 \times 3 \times 7$$

2	84
2	42
3	21
7	7
	1

Therefore $84 = 2^2 \times 3 \times 7$

Exercise 54

For Questions 1 and 2, list the first five multiples.

1 7   **2** 6

For Questions 3 and 4, list the factors.

3 12   **4** 30

For Questions 5 and 6, express the number as a product of prime factors.

5 28   **6** 60

7 Is 7 a multiple of 161?

For Questions 8 and 9, express the number as a product of prime factors.

8 210   **9** 88

10 Why is 511 not a prime?

Exercise 54*

1 Is 13 a prime factor of 181?

For Questions 2 and 3, list the prime factors.

2 399   **3** 231

4 What is the highest prime factor of 385?

For Questions 5 and 6, write down, in numerical order, all the factors.

5 75   **6** 54

For Questions 7 and 8, express the number as a product of prime factors.

7 165   **8** 399

9 What are the two prime numbers missing in this sequence
43, 47, 53, ..., ..., 67, 71, 73?

For Questions 10 and 11, express the number as a product of prime factors, using indices where necessary.

10 504   **11** 1008

Highest common factor (HCF) and lowest common multiple (LCM)

HCF: The highest factor that is common to a group of numbers.

LCM: The lowest multiple that is common to a group of numbers.

Example 3

Find the HCF of 12 and 42.

Express 12 and 42 as the product of prime factors: $12 = 2 \times 2 \times 3$; $42 = 2 \times 3 \times 7$.

The *common* prime factors are 2, 3.

The *highest* common factor (HCF) is 6 (which is 2×3).

Example 4

Find the LCM of 12 and 42.

Express 12 and 42 as the product of prime factors: $12 = 2 \times 2 \times 3$; $42 = 2 \times 3 \times 7$.

Multiples of 12 must contain the factors 2, 2 and 3.

Multiples of 42 must contain the factors 2, 3 and 7.

Common multiples must contain the factors 2, 2, 3 and 7.

Common multiples are 84, 168, 252, ...

The *lowest* common multiple (LCM) is 84.

N.B. $12 \times 42 = 504$ is a common multiple, but not the lowest one.

The factors in Examples 3 and 4 can be illustrated by a Venn diagram.

Can you see how this diagram can help you to calculate the highest common factor and lowest common multiple of the numbers 42 and 12?

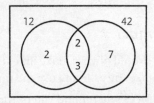

Key Points

- **HCFs** are used when cancelling fractions.
  For example, divide the top and bottom of $\frac{12}{42}$ by 6 to get $\frac{2}{7}$.
  They are also used in algebra when factorising, for example, $12x + 42y = 6(2x + 7y)$.

- **LCMs** are used in adding and subtracting fractions when the lowest common denominator has to be found.
  For example, $\frac{1}{12} + \frac{5}{42} = \frac{7}{84} + \frac{10}{84} = \frac{17}{84}$.

Exercise 55

You may find it helpful to draw a Venn diagram for these questions.

For Questions 1–3, find the highest common factor.

1 6 and 8   **2** 20 and 35   **3** 22 and 44

For Questions 4–6, find the lowest common multiple.

4 2 and 3   **5** 5 and 6   **6** 6 and 15

For Questions 7–9, find the highest common factor.

7 $2x$ and $4xy$ **8** $6a^2$ and $9a$ **9** $12y^2$ and $8xy^2$

For Questions 10–12, find the lowest common multiple.

10 x and y **11** $2a$ and $3b$ **12** $4y$ and $3x$

Simplify

13 $\frac{6}{8}$ **14** $\frac{20}{35}$ **15** $\frac{22}{44}$ **16** $\frac{6}{32}$

17 $\frac{15}{75}$

Calculate

18 $\frac{1}{6} + \frac{1}{8}$ **19** $\frac{3}{20} - \frac{1}{35}$ **20** $\frac{1}{5} + \frac{1}{6}$ **21** $\frac{5}{6} - \frac{1}{4}$

Exercise 55*

For Questions 1–7, find the HCF and LCM.

1 12 and 18

2 30 and 105

3 $3xy$ and $2yzc$

4 $4xy$ and $6xy$

5 x^2y and xyz

6 x^3y and xy^4

7 $6x^2yz$ and $9xy^2z^2$

Calculate

8 $\frac{1}{6} + \frac{3}{8}$

9 $\frac{7}{9} - \frac{5}{12}$

10 $\frac{11}{20} + \frac{4}{35}$

11 $\frac{11}{18} - \frac{5}{24}$

Calculator work

Remember

- You should always make an **estimate** to check your calculator answer. To make the estimate use the numbers corrected to 1 significant figure.

- **BIDMAS** will remind you of the correct order of operations. First calculate the **B**rackets, and then the **I**ndices, followed by the **D**ivision and the **M**ultiplication, and, lastly, the **A**ddition and the **S**ubtraction.

- For long calculations you may have to add brackets. It helps to write out the sum before starting to use the calculator.

- Be familiar with your calculator. Read and keep the instruction booklet.

- A scientific calculator applies BIDMAS automatically.

Some calculator functions

- To calculate 4^5, press ☐ ☐ ☐ (the answer is 1024)

- To enter 3.4×10^{-2} (written in standard form), press ☐ ☐ ☐ ☐ ☐ ☐

- To enter $\frac{6}{7}$, press ☐ ☐ ☐

- ▸To enter $2\frac{4}{5}$, press ☐ ☐ ☐ ☐ ☐

Example 5

Calculate these correct to 3 significant figures.

a $\sqrt{23.8^2 - 18.4^2}$

☐ ☐ 23.8 ☐ ☐ 18.4 ☐ ☐ ☐ The answer is 15.1 (to 3 s.f.).

b $2\frac{5}{7} - 3.8 \times 10^{-2}$

2 ☐ 5 ☐ 7 ☐ 3.8 ☐ ☐ 2 ☐ The answer is 2.68 (to 3 s.f.).

c $\sqrt{\dfrac{7.98^3}{5.91 + 1.09}}$

This is a long calculation, so write it out first: $\sqrt{(7.98^3 \div (5.91 + 1.09))} =$

☐ ☐ 7.98 ☐ 3 ☐ ☐ 5.91 ☐ 1.09 ☐ ☐ ☐

The answer is 8.52 (to 3 s.f.).

Activity 12

- Key any three-digit number into your calculator.
- Multiply your number by 7, then multiply the answer by 11, then multiply this answer by 13.
- Repeat this with other three-digit numbers.

What do you notice? Can you explain this?

Activity 13

Suppose that petrol costs $5 per litre. The petrol tank of a truck holds 142.15469 litres.

How much does a tank of petrol cost?

(Now turn your calculator upside-down.) Where should the driver buy his petrol?

Where should he buy his petrol if he buys 236.6851 litres at $3 per litre?

Try and make up some puzzles yourself.

Exercise 56

For each question, calculate the answer correct to 3 significant figures.

1 $4.7 + 9.7 \div 4.3$

2 $(4.7 + 9.7) \div 4.3$

3 $4.7 + \dfrac{9.7}{4.3}$

4 $\dfrac{4.7 + 9.7}{4.3}$

5 $(7.6 + 3.8)^2 + 4.2$

6 $\frac{1}{9} + \frac{3}{11}$

7 $3\frac{2}{9} + 9.7$

8 $11.8 + 2.08^2$

9 $\dfrac{9.73 - 1.08}{1.83^2}$

10 $\sqrt{12.3^2 + 7.8^2}$

11 $\sqrt{\dfrac{38}{12} + 4.08}$

12 $\dfrac{3.8^2 + 4.9}{2.7}$

13 $43.2 \div (4.7 - 0.87)$

14 $\dfrac{4.98}{2.09 \times 1.96}$

15 5^4

16 $(2.4)^6$

17 $(2.4)^5 \times 2.4$

18 $4680 + (2.4 \times 10^5)$

19 $\dfrac{2.8 \times 10^8}{1.6 \times 10^{-2}}$

Exercise 56*

For each question, calculate the answer correct to 3 significant figures.

1 $\dfrac{45}{2.3 \times 5.7}$

2 $\dfrac{0.059}{12.08 - 1.9}$

3 $\dfrac{12.8 - 55.9}{38.7 + 3.98}$

4 $13\frac{1}{7} - \frac{7}{9}$

5 $\dfrac{4.89}{1.8 \times 4.3} - \dfrac{3}{1.89}$

6 $\dfrac{1}{(0.678)^7}$

7 $(0.468)^3 + (0.0987)^5$

8 $(3.4)^5 \div 10^4$

9 $\left(\dfrac{1}{0.95} + \dfrac{1}{9.98}\right)^9$

10 Does $(0.5)^2 + (0.6)^2 + (0.7)^2 + (0.8)^2$ equal $(0.5 + 0.6 + 0.7 + 0.8)^2$?
   Explain your answer.

Exercise 57 (Revision)

Copy and complete these tables:

1

% increase	multiply by
5	1.05
12	
25	
	1.75
	1.99

2

% decrease	multiply by
5	0.95
12	
25	
	0.25
	0.01

3 $500 is invested at 7% compound interest per year. Find the value of the investment after
   a 1 year   **b** 2 years   **c** 3 years

4 £100 is invested at 4% compound interest per year. Find the value of the investment after
   a 1 year   **b** 2 years   **c** 5 years

5 A valuable oil painting is worth $25 000 and appreciates by 5% per year.
How much is it worth after
   a 1 year   **b** 2 years   **c** 5 years

6 A brand new car is purchased for $45 000 and depreciates by 8% per year.
How much is it worth after
   a 1 year   **b** 2 years   **c** 5 years

7 Find the highest common factor of
   a 4 and 10   **b** 5 and 30   **c** 16 and 24   **d** 14 and 30.

8 Find the lowest common multiple of
   a 3 and 4   **b** 6 and 7   **c.** 15 and 40   **d** 18 and 70.

Use your calculator to find the answer to the following, correct to 3 significant figures

9 $1.2 + 1.2^2$

10 $1.2 + 1.2^2 + 1.2^3$

11 $\dfrac{251.7 + 3.6 \times 10^2}{2.5 \times ^1}$

12 $4\frac{2}{7} + 1.35^3$

13 $\dfrac{11.2 + 3.7}{\sqrt{6.3}}$

14 $\dfrac{\sqrt{21.3}}{17.3 - 2.6}$

15 $\left(\dfrac{3.5 \times 10^3}{6.7 \times 10^2}\right)^2$

16 $\sqrt{5.75 \times 10^{12}}$

17 $(2^3 + 3^4 + 4^3 + 3^2)^2$

18 $\sqrt{3^4 + 3^4 + 3^4}$

19 $\sqrt{3.82 \times 10^7} \div \sqrt{6.7 \times 10^{-5}}$

20 $\sqrt{9.5^3 - 5.9^3}$

Exercise 57* (Revision)

1 $2500 is invested at 6% compound interest. Calculate
   a its value after 5 years   **b** after how many years its value has doubled.

2 $100 000 is invested at 3% compound interest. Calculate
   a its value after 10 years   **b** after how many years its value is increased by 50%.

3 An antique silk rug is worth $12 000 after it has appreciated by 5% for ten consecutive years. How much was it worth ten years ago?

4 A valuable wooden carving is valued at $750 000 after it has appreciated by 12% for five consecutive years. How much was it worth
   a five years ago?   **b** two years ago?

5 A television is worth $525 after it has depreciated by 7% for three consecutive years. How much was it worth three years ago?

6 A farm tractor is worth \$11 500 after it has depreciated by 9% for four consecutive years. How much was it worth
  a four years ago   **b** two years ago?

7 Find the HCF and LCM of
  a 60 and 70   **b** 140 and 84   **c** 42 and 70

8 Find the HCF and LCM of
  a $2x^2yz$ and $12xy^2z^2$ **b** $20pq$ and $35pq$ **c** $12a^2b^3c^4$ and $18a^4b^3c^2$

For Questions 9–14, use your calculator to find the answer correct to 3 significant figures.

9 $\left(\dfrac{1}{2.5 \times 10^{-3}} - \dfrac{1}{2.6 \times 10^{-3}}\right)^5$

10 $\sqrt{5 + \sqrt{5 + \sqrt{5 + \sqrt{5 + \sqrt{5}}}}}$

11 $\sqrt[3]{10 + \sqrt[3]{10 + \sqrt[3]{10 + \sqrt[3]{10 + \sqrt[3]{10}}}}}$

12 $\dfrac{1}{\sqrt{\pi^3 \times 10^{-3}}}$

13 $\sqrt[3]{5.5 \times 10^3 \times \tan(30)^\circ}$

14 $\tan\left(\sqrt{\dfrac{10\pi^3}{\pi^2 - 1}}\right)^\circ$

Simple factorising

Expanding $2a^2(a + 3b)$ gives $2a^3 + 6a^2b$. The reverse of this process is called **factorising**.

If the common factors are not obvious, first write out the expression to be factorised in full, writing numbers in prime-factor form. Identify each term that is common to all parts, and use these terms as common factors to be placed outside the bracket.

Example 1

Factorise $x^2 + 4x$.

$$x^2 + 4x = x \times x + 4 \times x$$
$$= x(x + 4)$$

red terms   black terms

Example 2

Factorise $6x^2 + 2x$.

$$6x^2 + 2x = 2 \times 3 \times x \times x + 2 \times x$$
$$= 2x(3x + 1)$$

red terms   black terms

Example 3

Factorise $14a^3b - 6a^2b^2$.

$$14a^3b - 6a^2b^2 = 2 \times 7 \times a \times a \times a \times b - 2 \times 3 \times a \times a \times b \times b$$
$$= 2a^2b(7a - 3b)$$

red terms   black terms

Exercise 5

Factorise these completely.

1 $x^2 + 3x$

2 $x^2 - 4x$

3 $5a - 10b$

4 $xy - xz$

5 $2x^2 + 4x$

6 $3x^2 - 18x$

7 $ax^2 - a^2x$

8 $6x^2y - 21xy$

9 $9p^2q + 6pq$

10 $ap + aq - ar$

Exercise 5❋

Factorise these completely.

1 $5x^3 + 15x^4$

2 $3x^3 - 18x^2$

3 $9x^3y^2 - 12x^2y^4$

4 $x^3 - 3x^2 - 3x$

5 $abc^2 - ab^2 + a^2bc$

6 $4p^2q^2r^2 - 12pqr + 16pq^2$

7 $30x^3 + 12xy - 21xz$

8 $0.2h^2 + 0.1gh - 0.3g^2h^2$

9 $\frac{1}{8}x^3y - \frac{1}{4}xy^2 + \frac{1}{16}x^2y^2$

10 $\pi r^2 + 2\pi rh$

11 $16p^2qr^3 - 28pqr - 20p^3q^2r$

12 $ax + bx + ay + by$

13 $(x - y)^2 - (x - y)^3$

Further simplifying of fractions

To simplify $\frac{234}{195}$ it is easiest to factorise it first.

$$\frac{234}{195} = \frac{2 \times 3^{\cancel{2}^{3}} \times \cancel{13}^{1}}{\cancel{3}_{1} \times 5 \times \cancel{13}_{1}} = \frac{6}{5}$$

Algebraic fractions are also best simplified by factorising first.

Example 4

Simplify $\frac{x^2 + 5x}{x}$.

$$\frac{x^2 + 5x}{x} = \frac{{}^{1}\cancel{x}(x + 5)}{\cancel{x}_{1}} = x + 5$$

Example 5

Simplify $\frac{2a^3 - 4a^2b}{2ab - 4b^2}$.

$$\frac{2a^3 - 4a^2b}{2ab - 4b^2} = \frac{{}^{1}\cancel{2}a^2(a - 2b)}{{}_{1}\cancel{2}b(a - 2b)} = \frac{a^2}{b}$$

Exercise 59

Simplify these.

1 $\dfrac{x^2 + x}{x}$

2 $\dfrac{2x + 2y}{2z}$

3 $\dfrac{2r + 2s}{r + s}$

4 $\dfrac{a^2 - ab}{ab}$

5 $\dfrac{at - bt}{ar - br}$

6 $\dfrac{x - xy}{z - zy}$

Exercise 59*

Simplify these.

1 $\dfrac{ax + ay}{a}$

2 $\dfrac{z}{z^2 + z}$

3 $\dfrac{6x^2 + 9x^4}{3x^2}$

4 $\dfrac{8x^3y^2 - 24x^2y^4}{12x^2y^2}$

5 $\dfrac{y^2 + y}{y + 1}$

6 $\dfrac{6x^2 - 12x^2y}{3xz - 6xyz}$

7 $\dfrac{(a^2 + 2ac) - (ab^2 + 2ac)}{a^2 - ab^2}$

8 $\dfrac{2a^2 - ab}{3a^2 - ab} \times \dfrac{3ab - b^2}{2a^2 - ab}$

9 $\dfrac{1}{3x - y} \div \dfrac{1}{15x - 5y}$

10 $\dfrac{5x^3 - 10x^2}{10x - 5x^2}$

Equations with fractions

Equations with fractions are easy to deal with because both sides of the equation can be multiplied by the lowest common denominator to clear the fractions.

Equations with numbers in the denominator

Example 6

$\frac{2x}{3} - 1 = \frac{x}{2}$ (Multiply both sides by 6)

$4x - 6 = 3x$

$\quad x = 6$ Check: $4 - 1 = 3$

Example 7

$\frac{3}{4}(x - 1) = \frac{1}{3}(2x - 1)$ (Multiply both sides by 12)

$9x - 9 = 8x - 4$

$\quad x = 5$ Check: $\frac{3}{4}(5 - 1) = \frac{1}{3}(10 - 1)$

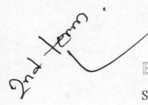

Exercise 60

Solve for x the equations in Questions 1–11.

1 $\dfrac{3x}{4} = 6$ **2** $\dfrac{x}{5} = -2$

3 $\dfrac{x}{4} = \dfrac{1}{2}$ **4** $\dfrac{3x}{8} = 0$

5 $\dfrac{2x}{3} = -4$ **6** $\dfrac{1}{3}(x + 7) = 4$

7 $\dfrac{3(x - 10)}{7} = -6$ **8** $\dfrac{x}{2} - \dfrac{x}{3} = 1$

9 $x - \dfrac{2x}{7} = 10$ **10** $\dfrac{1}{4}(x + 1) = \dfrac{1}{5}(8 - x)$

11 $\dfrac{3 - x}{3} = \dfrac{2 + x}{2}$

12 Ruby does one-third of her journey to school by car, and one-half by bus. Then she walks the final kilometre. How long is her journey to school?

13 Sam was boasting about a fish he had caught. When he was asked how long it was, he said he could not remember, but he did remember that the tail was 30 cm long, the head was one-sixth of the total length, and the body was equal to half of the tail plus two heads. How long was Sam's fish?

Exercise 60*

Solve for x the equations in Questions 1–8.

1 $\dfrac{2x - 3}{5} = 3$ **2** $\dfrac{3}{8}(5x - 3) = 0$

3 $\dfrac{x + 1}{5} = \dfrac{x + 3}{6}$ **4** $\dfrac{x + 1}{7} - \dfrac{3(x - 2)}{14} = 1$

5 $\dfrac{6 - 3x}{3} - \dfrac{5x + 12}{4} = -1$ **6** $\dfrac{2(x + 1)}{5} - \dfrac{3(x + 1)}{10} = x$

7 $\dfrac{2x - 3}{2} - \dfrac{x - 2}{3} = \dfrac{7}{6}$ **8** $\dfrac{2x + 1}{4} - x = \dfrac{3x + 1}{8} + 1$

9 $\left(\dfrac{x}{14} + 3\dfrac{1}{2}\right)$ is twice $\left(\dfrac{x}{21} + 1\dfrac{2}{3}\right)$. Find the value of x.

10 Diophantus was a famous ancient Greek mathematician. This was carved on his tomb.

> Here lie the remains of Diophantus. He was a child for one-sixth of his life. After one-twelfth more, he became a man. After one-seventh more, he married, and five years later his son was born. His son lived half as long as his father and died four years before his father.

How old was Diophantus when he died?

11 Ethan goes on a charity journey. He walks one-tenth of the way at 6 km/hour, runs one-sixth at 12 km/hour, cycles one-fifth at 24 km/hour, and completes the remaining 32 km by car at 48 km/hour. How many kilometres does he travel, and how long does it take?

Equations with x in the denominator

When the denominator contains x, the same principle of clearing fractions applies.

Example 8

$$\frac{3}{x} = \frac{1}{2}$$ (Multiply both sides by $2x$)

$$\frac{3}{x} \times 2x = \frac{1}{2} \times 2x$$

$$6 = x$$ Check: $\frac{3}{6} = \frac{1}{2}$

Example 9

$$\frac{4}{x} - x = 0$$ (Multiply both sides by x)

$$\frac{4}{x} \times x - x \times x = 0 \times x$$ (Remember to multiply everything by x)

$$4 - x^2 = 0$$

$$x^2 = 4$$

$$x = \pm 2$$ Check: $\frac{4}{2} - 2 = 0$ and $\frac{4}{-2} - (-2) = 0$

Exercise 61

Solve these for x.

1 $\frac{10}{x} = 5$ **2** $\frac{12}{x} = -4$ **3** $\frac{3}{x} = 5$

4 $\frac{4}{x} = -\frac{1}{2}$ **5** $\frac{3}{5} = \frac{6}{x}$ **6** $\frac{8}{x} = -\frac{10}{3}$

7 $\frac{35}{x} = 0.7$ **8** $0.3 = -\frac{15}{2x}$ **9** $\frac{5}{3x} = 1$

10 $\frac{9}{x} - x = 0$

Exercise 61*

Solve these for x.

1 $\frac{52}{x} = 13$ **2** $2.5 = -\frac{20}{x}$ **3** $\frac{15}{2x} = 45$

4 $\frac{8}{x} = -\frac{1}{8}$ **5** $\frac{2.8}{x} = 0.7$ **6** $\frac{16}{x} = \frac{x}{4}$

7 $\frac{12}{x} - 3x = 0$ **8** $\frac{3.2}{x} - 4.3 = 5.7$ **9** $\frac{1}{2x} + \frac{1}{3x} = 1$

10 $\frac{1}{ax} + \frac{1}{bx} = 1$

Simultaneous equations

Elimination method

Solving simultaneous equations graphically is time-consuming. It can also be inaccurate, as the solutions are read from a graph. Algebraic methods are often preferable.

Example 10

Solve the simultaneous equations $2x - y = 35$, $x + y = 118$.

$$2x - y = 35 \quad (1)$$
$$x + y = 118 \quad (2)$$
$$3x = 153 \quad \text{(Adding equations (1) and (2))}$$
$$x = 51$$

Substituting $x = 51$ into (1) gives $102 - y = 35 \implies y = 67$

The solution is $x = 51$, $y = 67$.

Check: Substituting $x = 51$, $y = 67$ into (2) gives $51 + 67 = 118$.

The method of Example 10 only works if the numbers before either x or y are of opposite sign and equal value. The equations may have to be multiplied by suitable numbers to achieve this.

Example 11

Solve the simultaneous equations $x + y = 5$, $6x - 3y = 3$.

$$x + y = 5 \quad (1)$$
$$6x - 3y = 3 \quad (2)$$

Multiply *both* sides of equation (1) by 3.

$$3x + 3y = 15 \quad (3)$$
$$6x - 3y = 3 \quad (2)$$
$$9x = 18 \quad \text{(Adding equations (3) and (2))}$$
$$x = 2$$

Substituting $x = 2$ into (1) gives $2 + y = 5 \implies y = 3$

The solution is $x = 2$, $y = 3$.

Check: Substituting $x = 2$ and $y = 3$ into (2) gives $12 - 9 = 3$.

Exercise 62

Solve these simultaneous equations.

1 $x + y = 8$, $x - y = 2$ **2** $x - y = 1$, $2x + y = 8$ **3** $x + y = 3$, $-x + y = 1$

4 $x + 3y = 4$, $2y - x = 1$ **5** $x + y = 0$, $3x - 2y = 5$

Exercise 62*

Solve these simultaneous equations.

1 $x + y = 11$, $x - y = 5$ **2** $2x - y = 3$, $x + y = 9$ **3** $3x + y = 8$, $3x - y = -2$

4 $-2x + y = -2$, $2x - 3y = 6$ **5** $x - y = -6$, $y + 2x = 3$

If the numbers in front of x or y are not of opposite sign, multiply by a negative number, as shown in Example 12.

Example 12

Solve the simultaneous equations $x + 2y = 8$, $2x + y = 7$.

$$x + 2y = 8 \qquad (1)$$
$$2x + y = 7 \qquad (2)$$

Multiply *both* sides of equation (2) by -2.

$$x + 2y = 8 \qquad (3)$$
$$\underline{-4x - 2y = -14} \qquad (2)$$
$$-3x = -6 \qquad \text{(Adding equations (3) and (2))}$$
$$x = 2$$

Substituting $x = 2$ into (1) gives $2 + 2y = 8 \Rightarrow y = 3$

The solution is $x = 2, y = 3$.

Check: Substituting $x = 2$ and $y = 3$ into equation (2) gives $4 + 3 = 7$.

Sometimes *both* equations have to be multiplied by suitable numbers, as in Example 13.

Example 13

Solve the simultaneous equations $2x + 3y = 5$, $5x - 2y = -16$.

$$2x + 3y = 5 \qquad (1)$$
$$5x - 2y = -16 \qquad (2)$$

Multiply (1) by 2. $\qquad 4x + 6y = 10 \qquad (3)$

Multiply (2) by 3. $\qquad \underline{15x - 6y = -48} \qquad (4)$

$$19x = -38 \qquad \text{(Adding equations (3) and (4))}$$
$$x = -2$$

Substituting $x = -2$ into (1) gives $-4 + 3y = 5 \Rightarrow y = 3$

The solution is $x = -2, y = 3$.

Check: Substituting $x = -2$ and $y = 3$ into equation (2) gives $-10 - 6 = -16$.

Key Points

To solve two simultaneous equations by elimination:
- **Label** equations (1) and (2).
- **Multiply** one or both equations by suitable numbers so that the numbers in front of the terms to be eliminated are the same and the signs are different.
- **Eliminate** by adding the equations. Solve the resulting equation.
- **Substitute** your answer in one of the original equations to find the other answer.
- **Check** by substituting both answers into the other original equation.

Exercise 63

Solve these simultaneous equations.

1 $3x + y = 11$, $x + y = 7$

2 $x + 3y = 8$, $x - 2y = 3$

3 $2x + y = 5$, $3x - 2y = -3$

4 $2x + 3y = 7$, $3x + 2y = 13$

5 $2x + 5y = 9$, $3x + 4y = 10$

Exercise 63*

Solve these simultaneous equations.

1 $2x + y = 5$, $3x - 2y = 11$

2 $3x + 2y = 7$, $2x - 3y = -4$

3 $3x + 2y = 4$, $2x + 3y = 7$

4 $7x - 4y = 37$, $5x + 3y = 44$

5 $3x + 2y = 3, 7x - 3y = 1.25$

6 $\dfrac{x}{2} + \dfrac{y}{3} = 4, \dfrac{y}{4} - \dfrac{x}{3} = \dfrac{1}{6}$

7 $\dfrac{a+1}{b+1} = 2, \dfrac{2a+1}{2b+1} = \dfrac{1}{3}$

8 $\dfrac{2}{x} - \dfrac{1}{y} = 3, \dfrac{4}{x} + \dfrac{3}{y} = 16$ Hint: let $p = \dfrac{1}{x}, q = \dfrac{1}{y}$.

Substitution method

The puzzle in Example 14 is often used as a brain-teaser. See if you can 'guess' the solution first.

Example 14

A bottle and a cork together cost $1. The bottle costs 90c more than the cork. Find the cost of the bottle.

Let b be the cost of the bottle in cents, and c be the cost of the cork in cents. The total cost is 100c, and so

$$b + c = 100 \qquad (1)$$

The bottle costs 90c more than the cork, and so

$$b = c + 90 \qquad (2)$$

Substituting (2) into (1) gives

$$(c + 90) + c = 100$$
$$2c = 10 \Rightarrow c = 5$$

Substituting in (1) gives $b = 95$.
Therefore the bottle costs 95c, and the cork costs 5c.
Check: Equation (2) gives $95 = 5 + 90$.

Exercise 64

Solve the following simultaneous equations by substitution.

1 $x = y + 2$
$x + 4y = 7$

2 $y = x + 3$
$y + 2x = 6$

3 $y = 3x + 3$
$5x + y = 11$

4 $x = y - 3$
$x + 3y = 5$

5 $y = 2x - 7$
$3x - y = 10$

Exercise 64*

Solve the following simultaneous equations by substitution.

1 $3x + 4y = 11$
$x = 15 - 7y$

2 $x = 7 - 3y$
$2x - 2y = 6$

3 $y = 5 - 2x$
$3x - 2y = 4$

4 $3x - 2y = 7$
$4x + y = 2$

5 $x - 2y + 4 = 0$
$5x - 6y + 18 = 0$

Solving problems using simultaneous equations

Example 15

Tickets at a concert cost either $10 or $15. The total takings were $8750.
Twice as many $10 tickets were sold as $15 tickets. How many tickets were sold?

Let x be the number of $10 tickets sold, and y the number of $15 tickets sold.
The total takings were $8750, and so

$$10x + 15y = 8750$$

Divide by 5 to simplify.

$$2x + 3y = 1750 \qquad (1)$$

Twice as many $10 tickets were sold as $15 tickets, and so

$$x = 2y \qquad (2)$$

To check that equation (2) is correct, substitute simple numbers that obviously work,
such as $x = 10$, $y = 5$.

Substituting (2) into (1) gives
$$4y + 3y = 1750$$
$$7y = 1750$$
$$y = 250$$

and so $x = 500$, from (2).

750 tickets were sold altogether.

Check: In (1), $1000 + 750 = 1750$.

Key Points
- Define your variables.
- Turn each sentence from the problem into an equation.
- Solve the equations by the most appropriate method.

Exercise 65

1 Find two numbers with a sum of 112 and a difference of 54.

2 Find two numbers with a mean of 14 and a difference of 4.

3 Two times one number added to four times another gives 34. The sum of the two numbers is 13. Find the numbers.

4 For this rectangle, find x and y and the area.

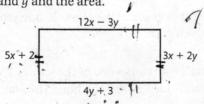

5 At McEaters, Phoebe bought two burgers and three colas, which cost her $3.45. Her friend Pete bought four burgers and two colas, and this cost him $4.94. How much did each item cost?

6 A telephone box accepts only 20c coins or 50c coins. On one day, 39 coins were collected with a total value of $11.40. Find how many 50c coins were collected.

7 At a shooting range, each shot costs 20c. If you hit the target, you receive 30c. Mira has 20 shots and makes a loss of 70c. How many hits did she get?

Example 16

Ahmed makes a camel journey of 20 km. The camel travels at 12 km/h for the first part of the journey, but then conditions worsen and the camel can only travel at 4 km/h for the second part of the journey. The journey takes 3 hours. Find the length of each part of the journey.

Let x be the length in km of the first part of the journey, and y be the length in km of the second part.

$$x + y = 20 \qquad (1) \qquad \text{(Total length is 20 km)}$$

Use the formula time $= \dfrac{\text{distance}}{\text{speed}}$.

$$\frac{x}{12} + \frac{y}{4} = 3 \qquad (2) \qquad \text{(Total time taken is 3 hours)}$$

Multiply equation (2) by 12.

$$x + 3y = 36 \qquad (3)$$
$$x + y = 20 \qquad (1) \qquad \text{(Subtract equation (1) from equation (3))}$$
$$2y = 16$$
$$y = 8$$

From equation (1), if $y = 8$ then $x = 12$, so the first part is 12 km and the second part is 8 km.

Check: These values work in equations (1) and (2).

Exercise 65*

1 Find the intersection of the lines $y = x + 1$ and $3y + 2x = 13$ without drawing the graphs.

2 The denominator of a fraction is 5 more than the numerator. If both the denominator and numerator are increased by 3, the fraction becomes $\frac{3}{4}$. Find the original fraction.

3 Imran can row at 3 m/s against the current and at 6 m/s with the current. Find the speed of the current.

4 One year ago, Joan was five times as old as her pony. In one year's time, the sum of their ages will be 22. How old is Joan now?

5 To cover a distance of 10 km, Jacob runs some of the way at 15 km/h, and walks the rest of the way at 5 km/h. His total journey time is 1 hour. How far did Jacob run?

6 A 2-digit number is increased by 36 when the digits are reversed. The sum of the digits is 10. Find the original number.

7 To go to school in the morning, I first walk to the garden shed at 6 km/h to collect my bicycle. I then cycle to school at 15 km/h. The total journey normally takes 18.5 minutes. One day, I am late and I run to the shed at 18 km/h and cycle at 27 km/h. The journey takes me 10 min 10 s. How far is it from my house to the garden shed?

Exercise 66 (Revision)

Factorise

1 $x^2 - 8x$

2 $3x^2 + 12x$

3 $6xy^2 - 30x^2y$

4 $12x^3 + 9x^2 - 15x$

Simplify

5 $\dfrac{x^2 - x}{x}$

6 $\dfrac{x^2 + xy}{x^2 - xy}$

Solve these equations.

7 $\dfrac{3x - 4}{4} = 2$

8 $\frac{1}{4}(x - 2) = \frac{1}{7}(x + 1)$

9 $\dfrac{2x + 7}{4} - \dfrac{x + 1}{3} = \dfrac{3}{4}$

10 $\dfrac{4}{n} - 1 = \dfrac{2}{n}$

11 Sarah shares out some sweets with her friends. She gives one-eighth of the sweets to Anna, one-sixth to Mattie and one-third to Rosie. She then has nine sweets left over for herself. How many sweets did she have to start with?

For Questions 12–15, solve the pairs of simultaneous equations.

12 $y - x = 4$ and $y + 2x = 1$

13 $y + x = 3$ and $y - 2x = 3$

14 $3x + 2y = 10$ and $5x - 4y = 2$

15 $5x - 2y = -1$ and $10x - 3y = 1$

16 At a sale, Andre buys two CDs and three tapes for $25.50.
His friend Charlie buys four CDs and five tapes for $47.50.
What is the cost of each item if all the CDs cost the same and all the tapes cost the same?

17 Husna is collecting 10c and 20c pieces. When she has 30 coins, the value of them is $4.10. How many of each coin does she have?

Exercise 66* (Revision)

Factorise

1 $3x^4 - 12x^3$

2 $\frac{4}{3}\pi r^3 + \frac{2}{3}\pi r^2$

3 $24x^3y^2 - 18x^2y$

4 $15a^2b^3c^2 - 9a^3b^2c^2 + 21a^2b^2c^3$

Simplify

5 $\dfrac{x^2 - xy}{xy - y^2}$

6 $\dfrac{ax - bx}{x^2 + xy} \div \dfrac{2a - 2b}{2x^2 + 2xy}$

For Questions 7–10, solve the equations.

7 $\frac{2}{7}(3x - 1) = 0$

8 $\dfrac{3x + 2}{5} - \dfrac{2x + 5}{3} = x + 3$

9 $1\frac{2}{3}(x + 1) = x + 5\frac{2}{3}$

10 $\dfrac{1}{x} + \dfrac{1}{2x} - \dfrac{1}{3x} = 2\frac{1}{3}$

11 Mrs Taylor has lived in many countries. She spent the first third of her life in Barbados, the next sixth in India, one-quarter in Malaysia, $3\frac{1}{2}$ years in Thailand, and one-fifth in Burma, where she is now living. How old is Mrs Taylor?

For Questions 12–15, solve the pairs of simultaneous equations.

12 $5x + 4y = 22$ and $3x + 5y = 21$

13 $5x + 3y = 23$ and $x + 2y = 6$

14 $3x + 8y = 24$ and $x - 2y = 1$

15 $6x + 5y = 30$ and $3x + 4y = 18$

16 The straight line $ax + by = 1$ passes through the points $(1, 4)$ and $(3, 1)$. Find the values of a and b.

17 In ten years' time, Azim will be twice as old as his son Chen. Ten years ago, Azim was seven times as old as Chen. How old are Azim and Chen now?

Unit 3 : Graphs

Travel graphs

This section covers distance–time graphs and speed–time graphs.

Distance–time graphs

Example 1

A veteran car takes part in the annual London to Brighton motor rally, and then returns to London. Here is its distance–time graph.

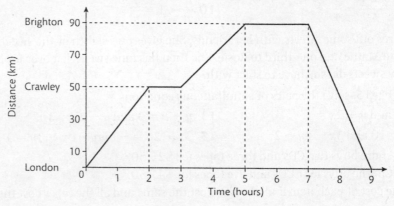

a What is the speed of the car from London to Crawley?

The speed from London to Crawley is $\dfrac{50\,\text{km}}{2\,\text{h}} = 25\,\text{km/h}$.

b The car breaks down at Crawley. For how long does the car break down?

The car is at Crawley for 1 hour.

c What is the speed of the car from Crawley to Brighton?

The speed from Crawley to Brighton is $\dfrac{40\,\text{km}}{2\,\text{h}} = 20\,\text{km/h}$.

d The car is towed on a trailer back to London from Brighton. At what speed is the car towed?

The speed from Brighton to London is $\dfrac{90\,\text{km}}{2\,\text{h}} = 45\,\text{km/h}$.

Exercise 67

1 Isaac travels southbound on a motorway from Hamburg, while Lester travels northbound on the same road from Hannover. This distance–time graph (where the distance is from Hannover) shows the journeys of both travellers.

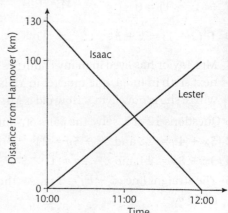

a What is Isaac's speed in kilometres per hour?

b What is Lester's speed in kilometres per hour?

c At what time does Isaac reach Hannover?

d How far apart are Isaac and Lester at 10:30?

e At what time do Isaac and Lester pass each other?

2 This distance–time graph shows the journeys of a car and a motorcycle between Manchester (M) and Birmingham (B).

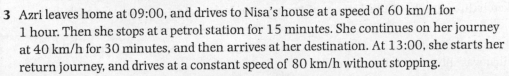

a When did the car stop, and for how long?

b When did the car and the motorcycle pass each other?

c How far apart were the car and the motorcycle at 09:30?

d After the motorcycle's first stop, it increased its speed until it arrived in Birmingham. The speed limit on the road was 70 miles/hour. Was the motorcyclist breaking the speed limit?

e Over the whole journey (excluding stops), what was the mean speed of the car?

f What was the mean speed of the motorcycle (excluding stops)?

3 Azri leaves home at 09:00, and drives to Nisa's house at a speed of 60 km/h for 1 hour. Then she stops at a petrol station for 15 minutes. She continues on her journey at 40 km/h for 30 minutes, and then arrives at her destination. At 13:00, she starts her return journey, and drives at a constant speed of 80 km/h without stopping.

a Draw a distance–time graph to illustrate Azri's journey.

b Use this graph to estimate at what time Azri returns home.

4 Fatima and Sue train for a triathlon by swimming 1 km along the coast, cycling 9 km in the same direction along the straight coast road, and then running directly back to their starting point via the same road. The times of this training session are shown in the table.

Activity	Fatima's time (minutes)	Sue's time (minutes)
Swimming	20	15
Rest	5	5
Cycling	10	15
Rest	10	5
Running	35	50

a Draw a distance–time graph (in kilometres and hours) to illustrate this information, given that Fatima and Sue both start at 09:00. Let the time axis range from 09:00 to 10:30.

b Use your graph to estimate when Fatima and Sue finish.

c Use your graph to estimate when Fatima and Sue are level.

d Calculate the mean speed of both athletes over the whole session, excluding stops.

Exercise 67*

1 A goat is tethered to a pole at A in the corner of a square field ABCD. The rope is the same length as the side of the field. The goat starts at B and trots at a constant speed to corner D keeping the rope taut. Sketch graphs for this journey:

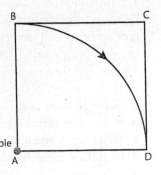

a Distance from A against time

b Distance from B against time

c Distance from C against time

d Distance from D against time

2 Ismael and Jack are two footballers who are put through an extra training session of running at *identical constant speeds*. For all three exercise drills, Ismael and Jack always start simultaneously from A and C, respectively.

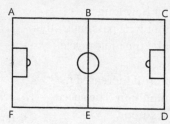

Drill 1   is that Ismael and Jack both run one clockwise circuit.

Drill 2   is that Ismael runs a circuit clockwise, and Jack runs a circuit anticlockwise.

Drill 3   is that Ismael and Jack run directly towards D and F, respectively.

Sketch three graphs of the distance of Ismael from Jack against time, one for each drill.

3 Three motorcyclists A, B and C set out on a journey along the same road. Part of their journey is shown in the travel graph.

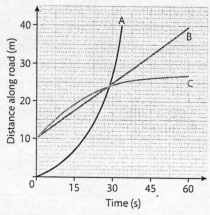

   a   Place the riders in order (first, second and third) after
       (i) 0 s   **(ii)** 15 s   **(iii)** 30 s
   b   When are all the riders the same distance along the road?
   c   Which rider travels at a constant speed?
   d   Which rider's speed is gradually
       (i) increasing?   **(ii)** decreasing?

4 The diagram shows the distances, in kilometres, between some junctions on a motorway.

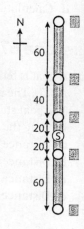

   The junctions are numbered as ▯, ▯ ... and ⑤ is the service area.
   Driver A (northbound) joins ▯ at 08:00, arrives at ⑤ at 09:00, rests for half an hour, and then continues his journey, passing ▯ at 12:00.
   Driver B (southbound) joins ▯ at 08:00, arrives at ⑤ at 10:00, rests for 1 hour, and then continues her journey, passing ▯ at 12:00.
   a   Draw a graph of the distance in kilometres from ▯ against the time in hours to show both journeys.
   b   When does driver A pass driver B?
   c   What are A and B's final speeds?
   d   Find their mean speeds, excluding stops.

Speed–time graphs

Travel graphs of speed against time can be used to find out more about speed changes and distances travelled.

Example 2

A train changes speed as shown in the speed–time graph.

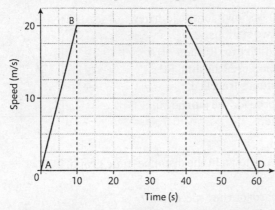

The train's speed is *increasing* between A and B, so it is *accelerating*.

The train's speed is *decreasing* between C and D, so it is *decelerating* (*retarding*).

The train's speed is constant at 20 m/s (and therefore the acceleration is zero) between B and C for 30 s. It has travelled 600 m (20 × 30 m).

This is the area under the graph between B and C.

a Find the *total* distance travelled by the train, and thus find the mean speed for the whole journey.

Total distance travelled = area under graph
$$= (\tfrac{1}{2} \times 10 \times 20) + (30 \times 20) + (\tfrac{1}{2} \times 20 \times 20) = 900$$

Therefore, mean speed $= \dfrac{900\,\text{m}}{60\,\text{s}} = 15$ m/s

b Find the train's acceleration between A and B, B and C, and C and D.
Acceleration between A and B = gradient of line AB
$$= \frac{20\,\text{m/s}}{10\,\text{s}} = 2\ \text{m/s}^2$$

Between B and C the speed is constant, so the acceleration is zero.
Acceleration between C and D = gradient of line CD
$$= \frac{-20\,\text{m/s}}{20\,\text{s}} = -1\ \text{m/s}^2$$
(The − sign indicates retardation.)

> **Key Point**
>
> In a speed–time graph,
>
> **acceleration**
> = gradient of line
> $= \dfrac{\text{change in speed}}{\text{time}}$
>
> **distance travelled**
> = area under the graph

Exercise 68

1 A speed–time graph for a journey of 15 s is shown.
  a Find the acceleration over the first 10 s.
  b Find the retardation over the last 5 s.
  c Find the total distance travelled.
  d Find the mean speed for the journey.

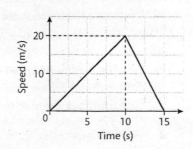

2 A speed–time graph for a journey of 3 hours is shown.
   a Find the acceleration over the first 2 hours.
   b Find the retardation over the last hour.
   c Find the total distance travelled.
   d Find the mean speed for the journey in kilometres per hour.

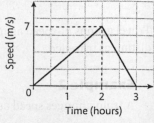

3 A speed–time graph is shown for the journey
   of a train between two stations.
   a Find the acceleration over the first 40 s.
   b Find the retardation over the final 80 s.
   c Find the total distance travelled.
   d Find the mean speed for the journey in
      metres per second.

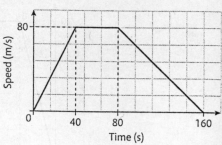

4 An insect's journey is shown in the distance–time graph.
   Find
   a the insect's outward journey speed in m/s
   b how long the insect remained stationary
   c the insect's return journey speed in m/s.

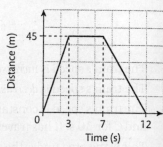

5 Luke leaves home at 08:00, and cycles to Martha's house at a speed of 20 km/h for one
   hour. He stays there for two hours before returning home at a speed of 30 km/h.
   a Draw a distance–time graph to illustrate Luke's journey.
   b Use this graph to find the time Luke returns home.

6 Find the acceleration of the boat's journey shown
   in the speed–time graph.

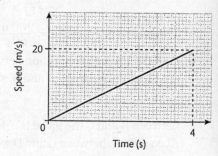

Exercise 68*

1 A cycle-taxi accelerates from rest to 6 m/s in 10 s, remains at that speed for 20 s, and then
   slows steadily to rest in 12 s.
   a Draw the graph of speed, in metres per second, against time in seconds for this journey.
   b Use your graph to find the cycle-taxi's initial acceleration.
   c What was the final acceleration?
   d What was the mean speed over the 42 s journey?

2 The acceleration of the first part of the journey shown
   is 3 m/s².
   a Find the maximum speed S metres per second.
   b Find the total distance travelled.
   c Find the mean speed of the whole journey.

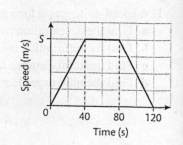

3 The speed–time graph shows an initial constant retardation of 2 m/s² for t seconds.

  a Find the total distance travelled.

  b Find the deceleration at $3t$ seconds.

  c Find the mean speed of the whole journey.

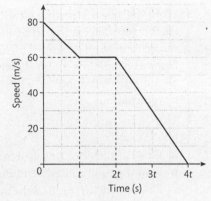

4 Sasha and Kim race over d metres. Sasha accelerates from rest for 6 s to a speed of 8 m/s, which she maintains for the next 40 s before she tires and uniformly decelerates at $\frac{4}{7}$ m/s² until she stops.

Kim accelerates from rest for 4 s to a speed of 8 m/s, which she maintains until 44 s have elapsed before she also tires and uniformly decelerates to a stop at $\frac{1}{2}$ m/s².

  a Draw the speed–time graph in metres per second and seconds for both girls on the same axes.

  b Use your graph to find who wins the race.

  c What was the mean speed for each runner?

  d Over what distance, in m, do the girls race?

  e Who is in the lead after **(i)** 100 m?   **(ii)** 300 m?

5 Sketch a distance–time graph for a flying duck such that its entire journey is described as:

  An initial constant speed, followed by a gradual reduction in speed until the duck is stationary, after which it gradually accelerates to reach a constant speed faster than its initial speed.

6 Mrs Lam leaves home for work at 07:00 driving at a constant speed of 60 km/h. After 45 minutes she increases her speed to 80 km/h for a further 45 minutes. She stays at work for 4 hours before she returns home at 70 km/h to meet Mr Lam who gets home at 2 pm.

  Draw a distance–time graph and use it to find out if Mrs Lam is late to meet her husband.

Exercise 69 (Revision)

1 The graph shows the journeys of Cheri and Felix, who went on a cycling trip.

  a How long did Cheri stop?

  b At what time did Felix start?

  c Find Cheri's mean speed.

  d How far apart were they at 10:20?

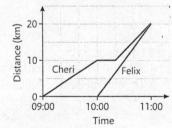

2 A bumble-bee flies out from its hive to some flowers, and returns to its hive some time later. Its journey is shown on the distance–time graph.

  a Find the bumble-bee's outward journey speed in metres per second.

  b How long does the bee stay at the flowers?

  c Find the bumble-bee's return journey speed in metres per second.

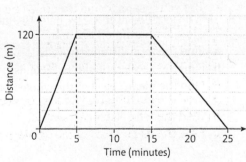

3 Wong sails from a resting position in his boat to a uniform speed of 4 m/s in 30 s. He then remains at this speed for a further 60 s, before he slows down at a constant retardation until he stops 15 s later.

Draw a speed–time graph showing this journey, and use it to find Wong's

   a initial acceleration   **b** acceleration at 60 s

   c retardation   **d** mean speed for the whole journey

4 The speed–time graph illustrates the journey of a cyclist.

   a Find the distance travelled in the first 50 s.

   b Find the total distance travelled.

   c Find the mean speed of the cyclist.

   d Find the acceleration when $t = 80$ s.

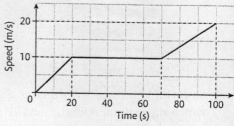

5 Izzat's journey is shown in the speed–time graph. Find Izzat's

   a initial acceleration

   b acceleration at 30 seconds

   c final acceleration

   d mean speed for the whole journey.

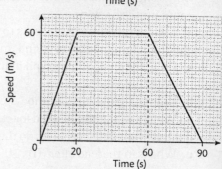

Exercise 69* (Revision)

1 A squash ball is hit against a wall by Ray. He remains stationary throughout the ball's flight, as shown in the distance–time graph.

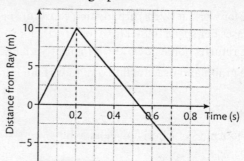

   a Find the speed with which the ball approaches the wall.

   b When does the ball pass Ray, and at what speed?

2 This speed–time graph is for a toy racing car. The initial retardation is 2 m/s².

   a Find the total distance travelled.

   b Find the deceleration at $6t$ seconds.

   c Find the mean speed of the whole journey.

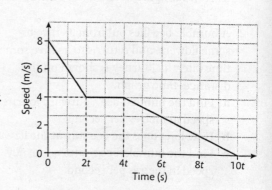

3 A hawk accelerates from rest to 12 m/s in 6 seconds followed by a further acceleration to 18 m/s in 3 seconds. It then remains at that speed for 10 seconds before retarding at x m/s^2 to rest.

  a Given that the hawk's total flight is 288 m, find x.

  b Draw the speed–time graph for the hawk's journey.

  c Use this graph to find the hawk's

    (i) initial acceleration

    (ii) mean speed for the whole length of the hawk's flight.

4 In the school sports day, Kimi wins the 400 m in 62.5 seconds.

  a Calculate his average speed in

    (i) m/s   **(ii)** km/h.

The speed–time graph of Kimi's race is shown.

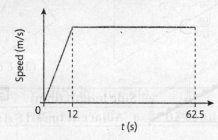

  b Calculate his maximum speed in m/s.

  c Find his initial acceleration in m/s^2.

5 This speed–time graph illustrates the speed of a lorry in metres per second.

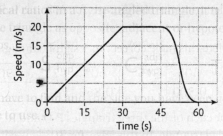

Which of these statements are true? Which are false?

Show working to justify your answers.

  a The initial acceleration over the first 30 s changes.

  b The braking distance is 150 m.

  c The mean speed for the whole journey is 12.5 m/s.

  d The greatest speed is 60 km/hour.

Sine and cosine ratios

Calculating sides

Activity 14

Triangles P, Q and R are all similar.

- For each triangle, measure the three sides in millimetres, and then copy and complete the table.

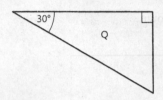

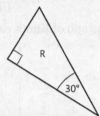

Triangle	Side length (mm)		
	o Opposite to 30°	*a* Adjacent to 30°	*h* Hypotenuse
P			
Q			
R			

- Calculate the *o* : *h* ratio for each of the three triangles P, Q and R. This ratio is the same for *any* similar right-angled triangle. In this case, for 30°, it is 0.5. It is called the **sine ratio** of 30°, or **sin 30°**.

- Use a calculator to complete this table correct to 3 significant figures.

d	0°	15°	30°	45°	60°	75°	90°
sin *d*			0.500		0.866		

- Calculate the *a* : *h* ratio for each of the three triangles P, Q and R. This ratio is the same for any similar right-angled triangle. In this case, for 30°, it is equal to 0.866 (to 3 decimal places) and is called the **cosine ratio** of 30°, or **cos 30°**.

- Use a calculator to complete this table correct to 3 significant figures.

d	0°	15°	30°	45°	60°	75°	90°
sin *d*			0.866		0.500		

Key Points

$$\sin f = \frac{\text{opposite side}}{\text{hypotenuse}} = \frac{o}{h} \qquad \cos f = \frac{\text{adjacent side}}{\text{hypotenuse}} = \frac{a}{h}$$

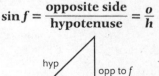

Example 1

Find the value of p correct to 3 significant figures.

$$\sin 32° = \frac{p}{10}$$

$$10 \times \sin 32° = p$$

$$p = 5.30 \text{ (to 3 s.f.)}$$

 $S \frac{\text{opp}}{\text{hyp}}$

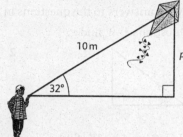

10 m

p m

32°

$\boxed{1\ 0\ \times\ \sin\ 3\ 2\ =}$ $\quad$ **5.29919** (to 6 s.f.)

So the height of the kite is 5.30 metres.

Example 2

Find the length of side q correct to 3 significant figures.

$$\cos 26° = \frac{q}{35}$$

$$35 \times \sin 26° = p$$

$$q = 31.5 \text{ (to 3 s.f.)}$$

 $C \frac{\text{adj}}{\text{hyp}}$

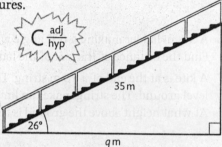

35 m

26°

q m

$\boxed{3\ 5\ \times\ \cos\ 2\ 6\ =}$ $\quad$ **31.4578** (to 6 s.f.)

So the length of the side is 31.5 m.

Remember

When using trigonometrical ratios in a right-angled triangle, it is important to choose the correct one.
- Identify the sides of the triangle as opposite, adjacent or hypotenuse to the angle you are interested in.
- Write down these ratios.

$S \frac{\text{opp}}{\text{hyp}}$ $\qquad$ $C \frac{\text{adj}}{\text{hyp}}$ $\qquad$ $T \frac{\text{opp}}{\text{adj}}$

- Mark off the side you have to find and the side you have been given. The ratio with the two marked sides is the correct one to use.

Example 3

Find the length y cm of the minute hand correct to 3 significant figures.

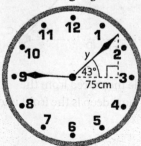

$$\cos 43° = \frac{75}{y}$$

$$y \times \cos 43° = 75$$

$$y = \frac{75}{\cos 43°}$$

$$y = 103 \text{ (to 3 s.f.)}$$

 $S \frac{\text{opp}}{\text{hyp}}$

 $C \frac{\text{adj}}{\text{hyp}}$ ✓

 $T \frac{\text{opp}}{\text{adj}}$

$\boxed{7\ 5\ =\ \cos\ 4\ 3\ =}$ $\quad$ **102.550** (to 6 s.f.)

So the length of the minute hand is 103 cm.

Exercise 70

Give your answers to the questions in this exercise correct to 3 significant figures.

For Questions 1–3, find x.

1

2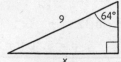

3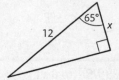

For Questions 4 and 5, find y.

4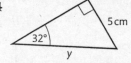

5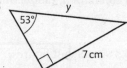

6 A 3.8 m ladder making a 65° angle with the ground rests against a vertical wall.
Find the distance of the foot of the ladder from the wall.

7 A kite is at the end of a 70 m string. The other end of the string is attached to a point on
level ground. The string makes an angle of 75° with the ground.
At what height above the ground is the kite flying?

Exercise 70*

Give your answers to the questions in this exercise correct to 3 significant figures.

1 Find BC in this isosceles triangle.

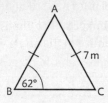

2 Calculate
  a AD
  b BD
  c the area ABC
  d the angle C

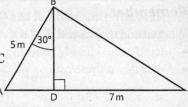

3 The cable car climbs at 48° to the horizontal
up the mountainside. BC = 52 m and DE = 37 m.
Calculate
  a the total length of the cable AC
  b the vertical height gained from C to A

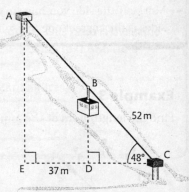

4 A submarine dives at a constant speed of 10 m/s at a diving angle measured from the
vertical of 75°. If the submarine starts its dive from the surface, how deep is the front end
of the submarine at the end of a 1-minute dive?

5 Jamila runs from point A on a constant bearing of 300° at 5m/s for 15 minutes at which point she stops at point B. Calculate how far Jamila has run
  a North from A   **b** West from A   **c** from A

6 The Petronas Tower II in Kuala Lumpur, Malaysia, is one of the tallest buildings in the world. From a position X on level ground, the angle of elevation to the top is 80°. Position Y lies 84.81 m further back from X in a direct line with the building. The angle of elevation from Y to the top is 70°. How high is the Petronas Tower II?

Calculating angles

> ### Activity 15
>
> If you are given the **opposite** side to angle θ and the **hypotenuse** in a triangle, you can find **sin θ**.
>
> You can use the 〔INV〕 〔sin〕 buttons on a calculator to find the angle θ.
>
> If you are given the **adjacent** side to angle θ and the **hypotenuse**, you can find **cos θ**.
>
> You can use the 〔INV〕 〔cos〕 buttons on a calculator to find the angle θ.
>
> - Check that the calculator is in degree mode.
> - Copy and complete the table for d and b to the nearest degree.
>
> | sin α | 0 | 0.259 | 0.500 | 0.866 | 0.966 | 1 |
> |---|---|---|---|---|---|---|
> | α | | | 30° | | | |
> | cos β | 1 | 0.966 | 0.866 | 0.500 | 0.259 | 0 |
> | β | | | | 60° | | |

Example 4

Find the angle d correct to 3 significant figures.

$$\cos d = \frac{1.5}{2}$$
$$d = 41.4° \text{ (to 3 s.f.)}$$

〔INV〕〔COS〕〔(〕〔1〕〔.〕〔5〕〔÷〕〔2〕〔)〕〔=〕 `41.4096` (to 6 s.f.)

Example 5

Find angle b correct to 3 significant figures.

$$\sin b = \frac{3.5}{4.3}$$
$$b = 54.5° \text{ (to 3 s.f.)}$$

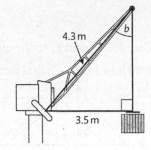

〔INV〕〔sin〕〔(〕〔3〕〔.〕〔5〕〔÷〕〔4〕〔.〕〔3〕〔)〕〔=〕 `54.4840` (to 6 s.f.)

Remember

To find an angle in a right-angled triangle.

• Write down these ratios.

 $S\dfrac{\text{opp}}{\text{hyp}}$
 $C\dfrac{\text{adj}}{\text{hyp}}$
 $T\dfrac{\text{opp}}{\text{adj}}$

• Mark off the sides of the triangle you have been given.

• The correct ratio to use is the one with two sides marked.

• Use the INV and sin, cos or tan buttons on a calculator to find the angle, having made sure that the calculator is in degree mode.

Exercise 71

Give your answers to the questions in this exercise correct to 3 significant figures.

For Questions 1 and 2, find each angle marked d.

1

2

3 A small coin is thrown off the Eiffel Tower in Paris. It lands 62.5 m away from the centre of the base of the 320 m-high structure. Find the angle of elevation from the coin to the top of the tower.

4 A steam train travels at 20 km/h for 15 minutes along a straight track inclined at $f°$ to the horizontal. During this time it has risen vertically by 150 m. Calculate the angle $f°$.

5 A control-line toy aircraft at the end of a 15 m wire is flying in a horizontal circle of radius 5 m.

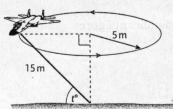

What angle does the wire make with the ground?

Exercise 71*

Give your answers to the questions in this exercise correct to 3 significant figures.

For Questions 1–3, find each angle marked d.

1

2

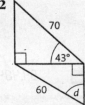

3

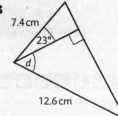

4 Mercedes starts from point M. She cycles 15 km North, and then 4 km East. She finally stops at point P. Find the bearing of P from M, and then the bearing of M from P.

5 A camera tripod has three equally spaced legs, each of length 1.75 m, around a circle of radius 52 cm on horizontal ground. Find the angle that the legs make with the horizontal.

Investigate

- Investigate the value of $(\sin x)^2 + (\cos x)^2$ for all angles x. Show this result algebraically.
- Compare the value of $\dfrac{\sin x}{\cos x}$ for all angles x with $\tan x$. Show this result algebraically.
- For an equilateral triangle with sides of 2 units, express the sine, cosine and tangent ratios of angles of 30° and 60° as exact fractions.

Exercise 72

Give your answers to the questions in this exercise correct to 3 significant figures.

For Questions 1–3, find each side marked x.

1   **2**   **3**

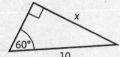

For Questions 4 and 5, find each angle marked a.

4 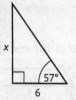   **5**

For Questions 6–11, find each marked side or angle.

6   **7** 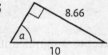   **8**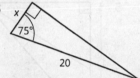

9 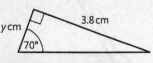   **10**    **11**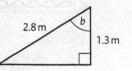

12 Shah runs from position A on a constant bearing of 060° for 500 m. How far North of A is he at the end of his run?

13 A straight 20 m wheelchair ramp rises 348 cm. Find the angle that the slope makes with the horizontal.

Exercise 72*

Give your answers to Questions 1–10 correct to 2 significant figures.

For Questions 1 and 2, find x.

1

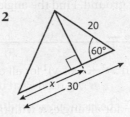

9, x, 30°

2

20, 60°, x, 30

For Questions 3 and 4, find the angles marked a and y.

3

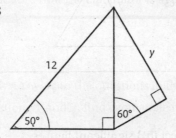

12, y, 50°, 60°

4

60°, 30°, x, 20, y

For Questions 5 and 6, find the value of θ.

5

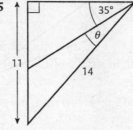

35°, θ, 11, 14

6

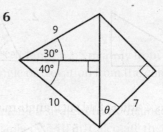

9, 30°, 40°, 10, 7, θ

Give your answers to Questions 7–12 correct to 3 significant figures.

7 A hot-air balloon drifts for 90 minutes at a constant height on a bearing of 285° at a steady speed of 12 km/h. How far is it then from its starting position

  a North or South?

  b West or East?

8 For this cross-section of a railway bridge, calculate the depth of the valley, and the length of the bridge.

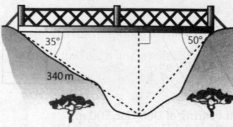

35°, 50°, 340 m

9 Calculate the height H of these stairs.

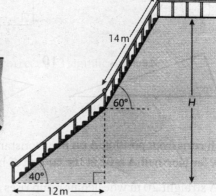

14 m, 60°, H, 40°, 12 m

10 A helicopter hovers in a fixed position 150 m above the ground. The angle of depression from the helicopter to church A (due West of the helicopter) is 32°. The angle of elevation from church B (due East of the helicopter) to the helicopter is 22°. Calculate the distance d between the two churches.

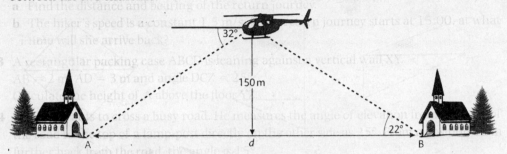

11 A motorboat is 10 km South of a lighthouse and is on a course of 053°. What is the shortest distance between the motorboat and the lighthouse?

12 Karen sits on her garden swing, and swings. At the highest point, the angle that the 3 m chain makes with the vertical is 60°. Find the difference in height between the highest and lowest points of her ride. Give your answer correct to 2 s.f.

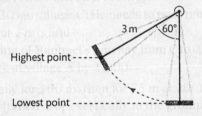

Investigate

Draw on the same axes, the graphs of $y = \tan a$, $y = \sin a$ and $y = \cos a$ for $0° \leqslant a \leqslant 90°$. Comment.

Investigate

A swing is on a chain of length x, measured in metres. Show that the height (in metres) travelled by the swing above its lowest position, when the chain makes an angle of a (degrees) to the vertical, is given by $h = x(1 - \cos a)$.

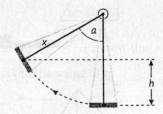

Investigate how h changes with a for swings of length 2 m, 3 m and 10 m.

What limits should you impose on angle a?

Exercise 73 (Revision)

Give your answers to the questions in this exercise correct to 3 significant figures.
For Questions 1–6, find the length of the side marked x and the size of angle d.

1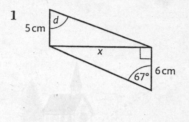

2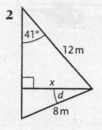

3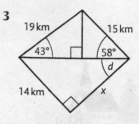

4

5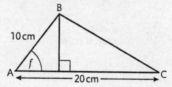

6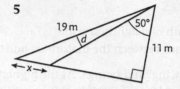

7 Calculate the area of an equilateral triangle of sides 10 cm.

8 The coordinates of triangle ABC are A(1, 1), B(7, 1) and C(7, 5). Calculate the value of angle CAB.

9 The area of triangle ABC is 50 cm².
   a Find the value of sin f.
   b What is the size of angle f?

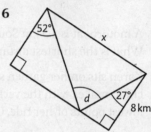

10 The centre of the clock face is in a tower 20 m above the ground. The hour hand is 1 m long.

   a How far above the ground is the end of the hour hand at 2am (h in the diagram)?
   b At 7am?
   c At 10.30am?

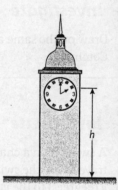

Exercise 73* (Revision)

1 A scuba diver dives directly from A to B, and then to C, and then to the seabed floor at D. She then realises that she has only 4 minutes' worth of air left in her tank. She can ascend vertically at 10 cm/s.
AB = 8 m, BC = 12 m and CD = 16 m.
Can she reach the surface before her air supply runs out?

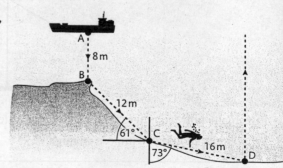

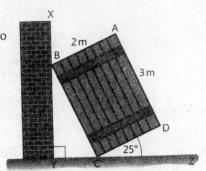

2 A hiker walks from her base camp for 10 km on a bearing of 050°, and then walks a further 14 km on a new bearing of 140°. There is then a thunderstorm, and she decides to return directly back to base camp.

  a Find the distance and bearing of the return journey.

  b The hiker's speed is a constant 1.5 m/s. If her return journey starts at 15:00, at what time will she arrive back?

3 A rectangular packing case ABCD is leaning against a vertical wall XY. AB = 2 m, AD = 3 m and angle DCZ = 25°. Calculate the height of A above the floor YZ.

4 A rabbit wants to cross a busy road. He measures the angle of elevation from the edge of the road to the top of a lamp-post directly on the other side as 25°. From a position 12 m further back from the road, the angle is 15°.

  a Calculate the width of the road.

  b The rabbit can scamper across the road at 1 m/s. Find the time he could take to cross the road.

  c The traffic on the road travels at 60 miles/hour. Find how far apart the vehicles must be for him to cross safely. (1 mile ≈ 1600 m.)

5 Raphael hikes from village A for 7 km on a bearing of 040° to village B. He then hikes 12 km on a bearing of 130° to village C. He needs to return back to A by 18:00. If he departs from C at 16:00 at 2 m/s find

  a the distance and bearing of Raphael's journey from C to A

  b if Raphael arrives back at village A by 18:00.

6 A firework travels vertically for 100 m, then for 75 m at 20° to the vertical, then 50 m at 20° to the horizontal. It then drops vertically to the ground in 1 minute. Find the average speed of its descent in m/s.

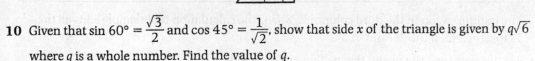

7 A ship 5 km North of an oil-rig travels on a bearing of 150°. Calculate the closest distance that the ship passes by the oil-rig.

8 A large fun-fair swing consists of a 5 m chain at the end of which is the seat. Find how far above the lowest point the seat is when the chain makes an angle of 50° with the vertical.

9 Given that sin 60° = $\frac{\sqrt{3}}{2}$, show that side x of the triangle is given by $p\sqrt{3}$, where p is a whole number. Find the value of p.

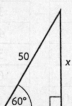

10 Given that sin 60° = $\frac{\sqrt{3}}{2}$ and cos 45° = $\frac{1}{\sqrt{2}}$, show that side x of the triangle is given by $q\sqrt{6}$ where q is a whole number. Find the value of q.

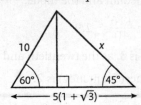

Data calculation tables

A table can be used to organise large sets of data and this reduces the chance of errors being made.

Remember

- Discrete data are measurements, collected from a source, that can be listed and counted. For example, the number of absentees from school will be 1 or 2 or 3, and so on.
- Continuous data are measurements, collected from a source, that cannot be listed and counted. For example, the weight of an average portion of cake could be anything from 100 g to 150 g.
- The **mean** $= \dfrac{\text{total of all values}}{\text{total number of values}}$.
- The **mode** is the value with the highest frequency.
- The **median** is the value (or mean of the pair of values) in the middle, after the results have been sorted into ascending order.
- The **range** = largest value − smallest value.
- **Sigma** (Σ) means 'the sum of'.
- The **mid-point** is the mean of the *exact* group boundaries.

Averages from discrete data

Example 1

These tables show the results of rolling a dice 40 times.

Data				
1	3	5	3	5
1	6	3	2	3
4	2	4	1	6
3	2	3	1	2
2	1	6	2	4
2	1	2	4	1
1	3	5	2	3
5	4	2	5	1

$\rightarrow$

Frequency table

Score x	Frequency f
1	9
2	10
3	8
4	5
5	5
6	3
	Total = 40

$\rightarrow$

Calculation table

Score x	Frequency f	$x \times f$
1	9	$1 \times 9 = 9$
2	10	$2 \times 10 = 20$
3	8	$3 \times 8 = 24$
4	5	$4 \times 5 = 20$
5	5	$5 \times 5 = 25$
6	3	$6 \times 3 = 18$
	Total = 40	Total = 116

The multiplication in the third column, $x \times f$, of the calculation table simply adds all the 1s together, then the 2s, and so on.

For this data, find the mean, the median and the mode.

Mean $= \dfrac{\text{total of all values}}{\text{total number of values}} = \dfrac{116}{40} = 2.9$.

The median (or middlemost value) is 3, as the twentieth and twenty-first values are both 3.

The mode (the score with the highest frequency) is 2.

Mean from grouped data (discrete or continuous)

Example 2

This frequency table gives the times of the first 60 runners in a cross-country race.

The exact values of the times have not been recorded, but the *boundaries* of the groups are defined *exactly* in the calculation table.

For example, the group '17–18' includes all the times from 17.00 to 17.99. (17.00 is in this group. 18.00 is in the next group.) So the group is defined exactly as $17 \leqslant t < 18$.

The mid-points of the groups are used to *estimate* the sum of the figures in each group.

Frequency table

Time (min)	Frequency
17–18	4
18–19	7
19–20	8
20–21	13
21–22	12
22–23	9
23–24	7
	$\Sigma = 60$

$\rightarrow$

Calculation table

Time (min)	Mid-point t	Frequency f	$t \times f$
$17 \leqslant t < 18$	17.5	4	$17.5 \times 4 = 70$
$18 \leqslant t < 19$	18.5	7	$18.5 \times 7 = 129.5$
$19 \leqslant t < 20$	19.5	8	$19.5 \times 8 = 156$
$20 \leqslant t < 21$	20.5	13	$20.5 \times 13 = 266.5$
$21 \leqslant t < 22$	21.5	12	$21.5 \times 12 = 258$
$22 \leqslant t < 23$	22.5	9	$22.5 \times 9 = 202.5$
$23 \leqslant t < 24$	23.5	7	$23.5 \times 7 = 164.5$
		$\Sigma = 60$	$\Sigma = 1247$

For example, we estimate each of the twelve times in the group '21–22' as 21.5. So the sum of these twelve times is estimated as $12 \times 21.5 = 258$. (In fact, some will be less than 21.5, and some will be more then 21.5, but 'on average' 258 will be a good estimate of the total.) The **modal class** is 20–21 min as it has the highest frequency.

Estimate of the mean $= \dfrac{1247}{60} = 20.8$ (to 3 s.f.)

Group boundaries

Take care to find the *exact* boundaries. Data tables can be misleading.

- Is the data discrete or continuous?
- Has the data been 'rounded' from continuous to discrete as it was collected?
- How is age rounded differently from other quantities?

Activity 16

Define the exact group boundaries, and find the class widths and mid-points for these sets of data.

- The times taken by 25 swimmers to breaststroke 50 m
- The lengths of 50 earthworms to the nearest centimetre

Time (s)	55–60	60–65	65–70	70–75
Frequency	3	8	12	2

Length (cm)	3–5	6–8	9–11	12–14
Frequency	16	17	12	5

- The noon temperature in degrees Celsius in London over 20 days in May
- The weights of 20 babies born in a hospital ward on 1 January
- The ages of children at a nursery/primary school

Temperature (°C)	0–8	8–16	16–24	24–32
Frequency	3	5	11	1

Weight (kg)	1–2	2–3	3–4	4–5
Frequency	4	6	7	3

Age (years)	2–3	4–5	6–7	8–9
Frequency	8	12	11	9

Exercise 74

1 A dice has six faces numbered 1, 2, 3, 4, 5 and 6. It is thrown 30 times and the scores are recorded here.

1	6	3	2	3	2	1	2	4	1
4	2	4	1	6	1	3	5	2	6
3	2	3	1	2	5	4	2	5	1

Construct a calculation table with three columns.
   a Work out the frequencies of each score and add them to your table.
   b Calculate the mean score.
   c Write down the median score.

2 The sizes of each family living in a block of flats are shown in the table.
   a How many families are there?
   b Construct a calculation table to work out the mean number of children per family.
   c What is the median number of children per family?

No. of children x	Frequency f
0	12
1	14
2	16
3	6
4	2

3 A questionnaire filled in by all the students at a college who had passed their driving test yielded these results.
   a How many students completed the questionnaire?
   b Construct a calculation table to work out the mean number of attempts required to pass the test.
   c What percentage of this group needed more than two attempts to pass?

No. of driving tests t	No. of pupils p
1	19
2	29
3	18
4	9
5	5

4 The UK National Lottery draws six numbered balls and a bonus ball. Balls are coloured white, blue, pink, green or yellow. Note that the white group has nine, and not ten, balls.

The data in this table shows the bonus ball numbers drawn in the first 415 UK National Lottery draws.

Draw up a calculation table, and calculate the mean value of the bonus ball number.

Colour	Number x	Frequency f
White	1–9	73
Blue	10–19	70
Pink	20–29	92
Green	30–39	80
Yellow	40–49	100
		Total = 415

Exercise 74*

1 Conservationists monitor crocodiles in their natural habitat using aerial photographs. The lengths of 30 crocodiles are recorded in the table.

Construct a calculation table and estimate the mean length of the crocodiles.

Length l (cm)	Frequency f
$170 \leqslant l < 180$	4
$180 \leqslant l < 190$	3
$190 \leqslant l < 200$	11
$200 \leqslant l < 210$	7
$210 \leqslant l < 220$	5

2 The waiting times, to the nearest minute, of patients at a morning surgery are shown in the table.

Construct a calculation table giving the exact boundaries of each group and estimate the mean waiting time.

Waiting time t (min)	No. of patients n
5–9	7
10–14	8
15–19	5
20–24	5
25–29	4
30–34	1

3 This table shows the numbers drawn (excluding the bonus ball) in the first 415 UK National Lottery draws.

Draw up a calculation table, and calculate the mean value of the numbers.

Colour	Number x	Frequency f
White	1–9	471
Blue	10–19	480
Pink	20–29	520
Green	30–39	493
Yellow	40–49	526

Exercise 75 (Revision)

For each question in this exercise copy out the table of data.

1 At a fair, Tobias fires an air rifle at a target. The table shows the information about his scores.
  a How many times did he fire at the target?
  b Find his median score.
  c Find the mode of his scores.
  d Calculate his mean score.

Score	Frequency
0	8
1	4
2	5
4	2
6	1

2 The table shows the number of appointments patients made with a doctor last week.
  a How many patients made appointments?
  b Find the median number of appointments per patient.
  c Find the mode of the number of appointments per patient.
  d Calculate the mean number of appointments per patient.

Number of appointments	Frequency
1	16
2	12
3	7
4	4
5	1

3 Alicia is amusing herself by counting the number of times her Maths teacher says 'Um' during a lesson. The table shows her results for a term.
  a How many Maths lessons did Alicia have during the term?
  b Write down the modal class.
  c Calculate an estimate for the mean number of 'Ums' per lesson.

Number of 'Ums'	Frequency
1–5	3
6–10	7
11–15	14
16–20	16
21–25	10

4 As part of her routine examination a vet weighs every cat she sees in her surgery. The table shows her results for a week.

   a How many cats did she see that week?

   b Write down the modal class.

   c Calculate an estimate of the mean weight of these cats.

Weight (w kg)	Number of cats
$0 < w \leqslant 2$	2
$2 < w \leqslant 4$	6
$4 < w \leqslant 6$	10
$6 < w \leqslant 8$	5
$8 < w \leqslant 10$	1

5 The times of some cross-country runners in a race are given in the table.

   a How many runners took part in the race?

   b Write down the modal class.

   c Calculate an estimate of the mean time of these runners.

Time (t min)	Frequency
$11.5 < t \leqslant 14.5$	3
$14.5 < t \leqslant 17.5$	7
$17.5 < t \leqslant 20.5$	11
$20.5 < t \leqslant 23.5$	4

6 The table shows information about the ages of students who sing in the choir.

   a How many students sing in the choir?

   b Write down the modal class.

   c Calculate the mean of these ages.

   d Another student joins the choir on her 13th birthday. Will the mean increase or decrease? Give a reason for your answer.

Age (a years)	Frequency
$12 \leqslant a < 13$	8
$13 \leqslant a < 14$	5
$14 \leqslant a < 15$	9
$15 \leqslant a < 16$	6
$16 \leqslant a < 17$	4

Exercise 75* (Revision)

For each question in this exercise copy out the table of data.

1 Fifty people took part in a golf tournament. The table shows the scores.

   a How many people scored 72?

   b Find the median score.

   c Find the mode of the scores.

   d Calculate the mean score.

Score	Frequency
70	2
71	5
72	
73	11
74	15
75	9

2 The table shows the number of children in some families. The mean number of children per family is 2.2.

   a Calculate the value of x.

   b Find the median number of children per family.

   c What percentage of families have less than two children?

Number of children	Number of families
0	2
1	6
2	11
3	x
4	3
5	1

3 Each day for a month Ricky keeps a record of the number of calls he makes on his mobile phone.
The table shows the results.
  a In which month did Ricky do his survey?
  b Write down the modal class interval.
  c Work out an estimate of the mean number of calls per day.

Number of calls	Frequency
1–5	2
6–10	4
11–15	7
16–20	9
21–25	6

4 The table shows the number of words in some essays written in an English exam.
  a How many students took the exam?
  b Write down the modal class interval.
  c Calculate an estimate of the mean number of words per essay.

Number of words	Number of essays
401–600	150
601–800	425
801–1000	350
1001–1200	75

5 A Monro is the name given to any mountain in Scotland over 3000 feet in height. The table shows the distribution of Monros by height.
  a How many Monros are there in Scotland?
  b Write down the modal class interval.
  c Calculate an estimate of the mean height of a Monro, giving your answer to the nearest ten feet.

Height (h feet)	Frequency
$3000 < h \leqslant 3300$	300
$3300 < h \leqslant 3600$	135
$3600 < h \leqslant 3900$	80
$3900 < h \leqslant 4200$	20
$4200 < h \leqslant 4500$	5

6 The table shows the maximum speed of serve of 50 players in a professional tennis tournament.
  a What value, in terms of x, should go in the blank space in the frequency column of the table?
The calculation of the estimate of the mean speed gave the result 107.8 mph.
  b Calculate the value of x.

Speed (s mph)	Frequency
$90 < s \leqslant 100$	x
$100 < s \leqslant 110$	23
$110 < s \leqslant 120$	
$120 < s \leqslant 130$	5

1 If $100 is invested for 10 years at 5% compound interest, the total amount at the end of this period to the nearest dollar is

   A $150   **B** $161   **C** $162   **D** $163

2 The value of $\sqrt{\dfrac{4.57 \times 10^4}{7.54 \times 10^{-4}}}$ in standard form to 3 s.f. is

   A 7.78×10^3 **B** 7.79×10^3 **C** 7.79×10^{-1} **D** 3.67×10^{15}

3 If the solution to the simultaneous equations $x + y = 3$, $3x - y = 13$ is (p, q), the value of $(p - q)^3$ is

   A 27   **B** −27   **C** 125   **D** 15

4 Usain Bolt ran the 100 m in a World Record time of 9.58 s. His speed in km/h to 3 s.f. is

   A 10.4 km/h   **B** 37.6 km/h   **C** 62.6 km/h   **D** 174 km/h

5 The speed-time graph shows Carl's 400 m race which he completes in 56 s. The value of his maximum speed v m/s is:

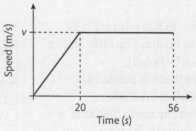

   A 2 m/s   **B** 4 m/s   **C** 8 m/s   **D** 10 m/s

6 Hans and Heidi ski down a mountain in a straight line from P.
Hans skis at 10 m/s. 10 s after Hans starts, Heidi leaves P at 12 m/s and catches Hans at Q.
The distance PQ is:

   A 60 m   **B** 600 m   **C** 720 m   **D** 800 m

7 A right-angled triangle has sides in the ratio 6 : 8 : 10.
If θ is the smallest angle, the value of $(\cos\theta)^2$ is

   A $\dfrac{9}{16}$ **B** $\dfrac{16}{25}$ **C** $\dfrac{9}{25}$ **D** $\dfrac{25}{16}$

8 A lighthouse L is on a bearing of 047° from a harbour H and 10 km away. A ship S leaves the harbour and sails on a bearing of 100°. The closest distance of S from L to 3 s.f. is

   A 10.0 km   **B** 7.31 km   **C** 6.02 km   **D** 7.99 km

9 The length of earthworms in a survey is shown in the table below measured to the nearest cm. The mean estimate is 8 cm:

Length (cm)	2–4	5–7	8–10	11–15
Frequency	3	9	x	5

The total number of earth worms in this survey is

   A 23   **B** 24   **C** 25   **D** 26

10 The mean difference of the successive squares of the first six prime numbers is

   A 24   **B** 24.2   **C** 33   **D** 33.6

1 a $1500 is invested for 4 years at 7% compound interest. Calculate the interest earned.

 b $1500 is invested for 7 years at 4% compound interest. Calculate the interest earned.

2 An antique glass bowl was valued at $10 000 at the end of 1998. This value increased by 25% in 1999 and by 22% in 2000. Its value fell by 30% in 2001. What was its value on January 1st 2002?

3 Light travels at 2.998×10^8 m/s. Calculate an estimate of how far light travels in a year. Give your answer, correct to 3 significant figures, in km, using standard form.

4 a Calculate the value of $\dfrac{2.94^2}{34.9 + 67.1} + \dfrac{0.0089}{1.2}$, writing down the full calculator display.

 b Give the answer to 3 significant figures, in standard form.

5 Find the highest common factor and lowest common mulitple of 12 and 30.

6 Simplify these algebraic expressions.

 a $\dfrac{30}{xy^2} \div \dfrac{6x^2}{x^2y}$

 b $\dfrac{x+1}{7} - \dfrac{x-3}{21}$

7 Solve the equation $\dfrac{2(x+1)}{5} - \dfrac{3(x+1)}{10} = x$

8 Solve these simultaneous equations. $4x - 3y = 26$
$2x + y + 10 = 0$

9 A model aeroplane accelerates from rest to 10 m/s in 20 s, remains at that speed for 30 s and then slows steadily to rest in 10 s.

 a Draw the speed–time graph for the aeroplane's journey.

 b Use this graph to find the aeroplane's
   i initial acceleration
   ii final acceleration
   iii mean speed over the whole 60 second journey.

10 This distance–time graph shows the journeys of Elisa and Albert, who cycle from their houses to meet at a lake. After staying at the lake for 90 minutes, Elisa cycles back home at 8 km/hour and Albert returns home at 12 km/hour.

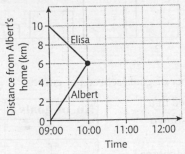

 a Copy the graph and represent these facts on your graph.
 b When did each person arrive home?
 c What is the mean speed in m/s for each cyclist?
   (Exclude the stop.)

11 XYZ is an obtuse-angled triangle. Point P is the foot of the perpendicular from Y onto XZ produced.
XY = 10 cm, XZ = 4 cm, Angle PXY = 40°
Calculate

a length YP

b length XP

c length ZP

d angle PZY

e angle XZY

f area of triangle XYZ.

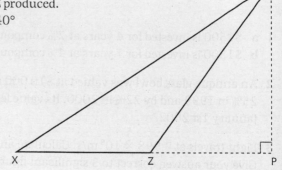

12 Find the sides x and z and the angle y.

a

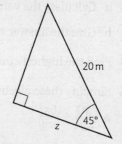

10 m

60°

x

b

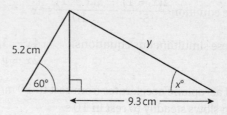

8.66 m

y

10 m

c

20 m

45°

z

13 Find the angle x and the side y.

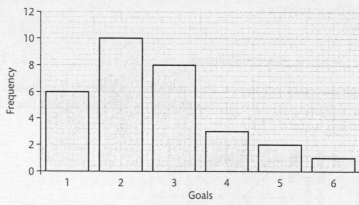

5.2 cm

60°

y

$x°$

9.3 cm

14 The bar chart shows the goals scored per game by Liverpool FC.

Frequency

12

10

8

6

4

2

0

1 2 3 4 5 6

Goals

Calculate the mean number of goals scored per game.

15 The number of goals scored in a hockey season by Gabi is shown in the table.

a Copy and complete the table to find the sum of all the goals scored by Gabi.

b Find the mean number of goals Gabi scored per game.

No. of goals x	Frequency f	fx
0	12	
1	14	
2	16	
3	6	
4	2	
	$\Sigma f =$	$\Sigma fx =$

Inverse percentages

In Unit 1 you calculated simple percentages, starting with the original quantity. Now you will work back to find the original quantity.

> **Remember**
>
> To increase a number by $R\%$, multiply it by $\left(1 + \dfrac{R}{100}\right)$.
>
> To decrease a number by $R\%$, multiply it by $\left(1 - \dfrac{R}{100}\right)$.

Example 1

A house in Italy is sold for €138 000, giving a profit of 15%.

Find the original price that the owner paid for the house.

Let €x be the original price.

$x \times 1.15 = 138\,000$ Selling price after a 15% increase

$x = \dfrac{138\,000}{1.15}$

$x = €120\,000$

Example 2

An ancient Japanese book is sold for ¥34 000 (yen), giving a loss of 15%.

Find the original price that the owner paid for the book.

Let ¥x be the original price.

$x \times 0.85 = 34\,000$ Selling price after a 15% decrease

$x = \dfrac{34000}{0.85}$

$x = ¥40\,000$

Exercise 76

Give all your answers correct to 3 significant figures.

1. Decrease $456 by 3%.
2. A price doubles. What is this, as a percentage increase?
3. A pair of shoes is sold for $48 at a profit of 20%. Find the original cost to the shop.
4. Find the price of this chest of drawers *before* the reduction.
5. The height of a tree increases from 12 m to 13.6 m. Find the percentage increase.
6. A garden chair is bought for $68 and sold at a 12% profit. Find its selling price.
7. The value of a carpet increases by 6% to $78.56. Find the original value.

Sale Price 10% OFF!
$54

Exercise 76*

Give all your answers correct to 3 significant figures.

1 A portable digital radio is sold for $52.80, giving the shop a profit of 20%. Find the cost to the shop of the radio.

2 The price of a house is reduced by 15% to $153 000. What was the original price of the house?

3 A rare stamp is worth $81 after an increase in value of 35%. What was its value before the increase?

4 In a chemistry laboratory, there are two beakers, each containing 500 ml of a liquid. 5% of beaker A is water and 4.5% of beaker B is water. As a percentage, how much more water is in beaker A?

5 A farm is sold for $457 000, which gives a profit of 19%. Find the profit.

6 Miss Meg's salary increases by $450 to $1760. Find her percentage salary increase.

7 A DVD recorder is reduced by 20% and then by a further 10% of the original price. What was the original price of the DVD recorder if it was sold for $324 after the second reduction?

Estimating

On many occasions it is acceptable and desirable not to calculate an exact answer, but to make a sensible estimate.

You have already been approximating numbers to a certain number of decimal places (2.47 = 2.5 to 1 d.p.), significant figures (34.683 = 34.7 to 3 s.f.) and rounding to a fixed suitable value (12 752 = 13 000 to the nearest 1000). Estimation is usually done without a calculator and is performed in a way that makes the working simple!

Activity 17

Write down the following estimates to the stated degree of accuracy.

- The length of your pen.   (nearest cm)
- The area of this page.   (nearest cm²)
- The time for you to walk a km.   (nearest 30 s)
- The weight of a dog.   (nearest kg)
- The population of your town.   (nearest 1000)
- The acute angle between these lines.   (nearest degree)

Check your estimates.

Example 3

Estimate the answers to these calculations.

a 19.7×3.1

$\quad 19.7 \times 3.1 \qquad \approx 20 \times 3 = 60 \qquad$ (exact answer is 61.07)

b 121.3×98.6

$\quad 121.3 \times 98.6 \qquad \approx 120 \times 100 = 12\,000 \qquad$ (exact answer is 11960.18)

c $252.03 \div 81.3$

$\quad 252.03 \div 81.3 \qquad \approx 240 \div 80 = 3 \qquad$ (exact answer is 3.1)

d $(11.1 \times (7.8 - 5.1))^2$

$\quad (11.1 \times (7.8 - 5.1))^2 \approx (10 \times (8-5))^2$

$\qquad\qquad\qquad\qquad = (30)^2 = 900 \qquad$ (exact answer is 898.2009)

Estimating using standard form

It is often useful to use standard form to work out an estimate. Make sure you can write a number in standard form. You can refer to the rules of indices on page 45.

Example 4

$$\text{Change to 1 s.f.} \quad \text{Write in standard form} \quad \text{Use rules of indices}$$

$$0.067\,68 \times 38\,750 \simeq 0.07 \times 40\,000 = 7 \times 10^{-2} \times 4 \times 10^4 \simeq 30 \times 10^{-2+4} = 30 \times 10^2 = 3000$$

$$0.0753 \div 0.003\,68 \simeq \frac{0.08}{0.004} = \frac{8 \times 10^{-2}}{4 \times 10^{-3}} = 2 \times 10^{(-2)-(-3)} = 2 \times 10^1 = 20$$

Example 5

Use standard form to calculate an estimate of $\sqrt{3.3 \times 10^7}$.

$$\sqrt{3.3 \times 10^7} = \sqrt{33 \times 10^6} \qquad \text{(Change 33 to the nearest square whole number)}$$

$$\simeq \sqrt{36 \times 10^6}$$

$$= \sqrt{36} \times \sqrt{10^6} = 6 \times 1000 = 6000$$

Example 6

Use standard form to work out an estimate for $(4.5 \times 10^7) + (4.5 \times 10^6)$.

Write your answer, correct to 1 significant figure, in standard form.

$$(4.5 \times 10^7) + (4.5 \times 10^6) = (45 \times 10^6) + (4.5 \times 10^6) \quad \text{(Change the index numbers to be the same)}$$

$$= 49.5 \times 10^6 \qquad \text{(Add like terms)}$$

$$\simeq 5 \times 10^7$$

Exercise 77

Estimate the answers to these.

1 3.1×47.9

2 $23.2 \div 7.8$

3 $(7.3 + 12.1) \times 15.9$

4 79.868×0.101

5 $20.92 \div 0.11$

Use standard form to calculate these, giving each answer in standard form.

6 $(2 \times 10^4) \times (3 \times 10^3)$

7 $(8 \times 10^6) \div (2 \times 10^3)$

8 $(2 \times 10^4) + (3 \times 10^3)$

9 $(9 \times 10^4) - (3 \times 10^3)$

10 $\sqrt{2.5 \times 10^5}$

Use standard form to calculate an estimate for these, giving each answer in standard form correct to 1 significant figure.

11 2670×760

12 $8490 \div 56.9$

13 $(6.8 \times 10^6) + (2.3 \times 10^5)$

14 $(4.8 \times 10^6) - (3.2 \times 10^5)$

15 $\sqrt{6.3 \times 10^5}$

Exercise 77*

Estimate the answers to these.

1 $\dfrac{3.1 \times 19.7}{14.8}$

2 1.98^3

3 Estimate the area of the rectangle.

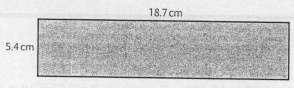

18.7 cm

5.4 cm

4 Estimate the green shaded area.

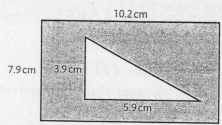

10.2 cm

7.9 cm 3.9 cm

5.9 cm

Use standard form to calculate these, giving each answer in standard form.

5 $(3 \times 10^3) \times (4 \times 10^5)$

6 $(6 \times 10^8) \div (3 \times 10^5)$

7 $(7 \times 10^8) + (6 \times 10^6)$

Use standard form to calculate an estimate for these, giving each answer as an ordinary number correct to 1 significant figure.

8 $(1.2 \times 10^2) \times (4.5 \times 10^2)$

9 $(7.98 \times 10^{-4}) \div (3.79 \times 10^{-3})$

10 A warehouse sells 8.7 million cups at an average price of $0.65 each. Use standard form to calculate an estimate of the total sales. Give your answer correct to 1 significant figure.

11 A human kidney has a volume of 120 cm^3. It contains 1.35 million nephrons (small tubes). Use standard form to calculate an estimate of the average number of nephrons per cubic centimetre. Give your answer correct to 1 significant figure.

Use standard form to calculate an estimate for these, giving each answer in standard form, correct to 1 significant figure.

12 5003×393

13 47.8×0.0059

14 $\dfrac{80\,920}{0.004\,18}$

15 $\dfrac{0.6597}{729.8}$

Exercise 78 (Revision)

1 Ali sells her goat for $150, giving her a profit of 25%. Find the original price she paid for the goat.

2 Liam sells a bag of rice for $24, giving him a profit of 20%. Find the original price he paid for the rice.

3 Amir sells an oil painting for $2125, giving him a loss of 15%. Find the original price he paid for the painting.

4 Guvinda ran 400 m in 77.6 seconds which was an improvement of 3%. Find her original time.

For Questions 5–16, make an estimate of the following correct to 1 significant figure in standard form:

5 $(1.9 \times 10^3) \times (5.1 \times 10^4)$

6 $(4.9 \times 10^7) \times (8.1 \times 10^5)$

7 $(6.8 \times 10^6) \times (7.9 \times 10^8)$

8 $(1.1 \times 10^6) \times (9.9 \times 10^4)$

9 $(7.9 \times 10^5) \div (4.1 \times 10^3)$

10 $(3.9 \times 10^7) \div (1.8 \times 10^4)$

11 $(8.9 \times 10^9) \div (1.9 \times 10^2)$

12 $(9.7 \times 10^{12}) \div (1.9 \times 10^4)$

13 $18\,000\,000 \times 19\,000\,000$

14 $49\,000 \times 61\,000\,000\,000$

15 $21\,800\,000 \div 19\,700$

16 $79\,950\,000\,000\,000 \div 3900$

Exercise 78* (Revision)

1 A valuable clock has an original price of x. It is sold for $5000 after its original price has been increased by 5% and then by 10%. Find x.

2 A stone garden statue has an original price of y. It is sold for $2500 after its original price has been increased by 15% and then by 20%. Find y.

3 A doll's house has an original price of x. It is sold for $350 after its original price has been decreased by 5% and then by 10%. Find x.

4 A mountain bike has an original price of y. It is sold for $1750 after its original price has been decreased by 15% and then by 20%. Find y.

5 An antique chair has an original price of p. It is sold for $1250 after its original price has been increased by 7% and then decreased by 9%. Find p.

6 An exclusive apartment in New York has a value of q. It is sold for $1 million after its original price decreases by 10% and then increases by 25%. Find q.

7 A pearl necklace has a value of $7500 after it has appreciated by 3% per year for 3 years. Find its original value 3 years ago.

8 A signed shirt by a world famous footballer has a value of $1800 after it has depreciated by 5% per year for 5 years. Find its original value 5 years ago.

For Questions 9–14, make an estimate of the following correct to 1 significant figure in standard form.

9 $0.005\,912 \times 290\,000\,000$

10 $0.000\,007\,987 \div 0.001\,967$

11 $(3.89 \times 10^{-7}) \times (5.91 \times 10^{-6})$

12 $4\,890\,000\,000 \times 0.000\,9$

13 $(5.81 \times 10^{-5}) \div (2.98 \times 10^{-6})$

14 $(7.71 \times 10^{-3}) + (3.98 \times 10^{-4})$

Change of subject

It is sometimes helpful to write an equation, or formula, in a different way. For example, to draw the graph of the equation $2y - 4 = 3x$ it is easier to compile a table of values if y is the subject, that is, on its own and on one side. Notice that the method used in Example 1 is the same as that used to solve a similar equation, as shown in Example 2.

To rearrange an equation or formula, apply the same rule that is used to solve equations.

Do the same operation to both sides.

Example 1

Make y the subject in $2y - 4 = 3x$.

$2y - 4 = 3x$ (Add 4 to both sides)

$2y = 3x + 4$ (Divide both sides by 2)

$y = \dfrac{3x + 4}{2}$

$y = \dfrac{3}{2}x + 2$

Example 2

Solve $2y - 4 = 2$.

$2y - 4 = 2$ (Add 4 to both sides)

$2y = 6$ (Divide both sides by 2)

$y = 3$

The rearranged equation in Example 1 also allows you to state the gradient of the line, that is, $\frac{3}{2}$, and the y intercept of $+2$. Both of these are required to *sketch* the graph of the line.

Exercise 79

Make x the subject of the equations.

1 $x + 2 = a$	**2** $x - p = 5$	**3** $c = x + a$	**4** $5x = b$
5 $3x + a = b$	**6** $t - 2x = s$	**7** $ax + b = 4$	**8** $f = ex - g$
9 $x(a + b) = c$	**10** $8b + cx = d$	**11** $3(x + b) = a$	**12** $a(x + b) = c$
13 $a = \dfrac{x}{b}$	**14** $\dfrac{px}{q} = r$	**15** $p + q = \dfrac{x}{r}$	**16** $d = \dfrac{x - b}{c}$

Exercise 79*

In Questions 1–11, make x the subject of the equations.

1 $ax + b = c$	**2** $d = \dfrac{x - b}{c}$	**3** $c = \dfrac{bx}{d}$	**4** $a(x + c) = e$
5 $P = \pi x + b^2$	**6** $\dfrac{bx}{d^2} = T$	**7** $\pi - x = b$	**8** $ab - dx = c$
9 $\dfrac{a}{x} = b$	**10** $\dfrac{a + b}{x} = c$	**11** $p = q + \dfrac{s}{x}$	

12 Make r the subject of the formula $A = 2\pi r$.

13 Make h the subject of the formula $V = \frac{1}{3}\pi r^2 h$.

14 Make x the subject of the formula $y = mx + c$.

15 Make s the subject of the formula $v^2 = u^2 + 2as$.

16 Make a the subject of the formula $m = \frac{1}{2}(a + b)$.

17 Make a the subject of the formula $S = \dfrac{a(1 - r^n)}{1 - r}$.

18 Make a the subject of the formula $S = \dfrac{n}{2}\{2a + (n - 1)d\}$.

Example 3

Make x the subject of the equation $ax^2 + b = c$.

$ax^2 + b = c$ (Subtract b from both sides)

$ax^2 = c - b$ (Divide both sides by a)

$x^2 = \dfrac{c - b}{a}$ (Square root both sides)

$x = \sqrt{\dfrac{c - b}{a}}$

Example 4

Make x the subject of the equation $ax + bx = c$.

$ax + bx = c$ (Factorise)

$x(a + b) = c$ (Divide both sides by $(a + b)$)

$x = \dfrac{c}{(a + b)}$

Exercise 80

In questions 1–8, make x the subject of the equation.

1 $ax^2 = b$
2 $\dfrac{x^2}{a} = b$
3 $x^2 + C = 2D$

4 $\dfrac{x^2}{a} + b = c$
5 $ax^2 + 2b = c$
6 $ax + dx = t$

7 $a(x - b) = x$
8 $a(x + 1) = b(x + 2)$

9 Make r the subject of the formula $A = 4\pi r^2$.

10 Make v the subject of the formula $a = \dfrac{v^2}{r}$.

11 Make r the subject of the formula $V = \frac{4}{3}\pi r^3$.

12 Make l the subject of the formula $T = 2\pi\sqrt{l}$.

Exercise 80*

In Questions 1–7, make x the subject of the equation.

1 $Rx^2 = S$
2 $g = cx^2 + a$
3 $ax = bx - c$

4 $c - dx = ex + f$
5 $\tan b + a(x + c) = x$
6 $p = \sqrt{s + \dfrac{x}{t}}$

7 $\dfrac{Ab - x^2}{D} = a$

8 Make r the subject of the formula $V = \frac{1}{3}\pi r^2 h$.

9 Make v the subject of the formula $mgh = \frac{1}{2}mv^2$.

10 Make x the subject of the formula $y = \dfrac{1}{a^2 + x^2}$.

11 Make a the subject of the formula $s = \frac{1}{12}(b - a)^2$.

12 Make Q the subject of the formula $r = \dfrac{S}{\sqrt{PQ}}$.

13 Make d the subject of the formula $F = \dfrac{k}{\sqrt[3]{d}}$.

14 Make x the subject of the formula $y = \dfrac{1}{\sqrt{1 - x^2}}$.

15 Make x the subject of the formula $y = \dfrac{x + p}{x - p}$.

Using formulae

Activity 18

The period T seconds taken for a pendulum of
length L metres to swing to and fro is given by
the formula

$$T = 2\pi\sqrt{\dfrac{L}{g}}$$

where g is the acceleration due to gravity = 9.81 m/s^2.

- What is the period of a pendulum 1 km long?
  (Give your answer correct to 1 decimal place.)
- Show that, when L is made the subject, the formula
  becomes
  $$L = g\left(\dfrac{T}{2\pi}\right)^2$$
- Use the rearranged formula to complete this table,
  giving your answers correct to 2 significant figures.

Period	1 s	10 s	1 minute	1 hour	1 day
Length					

Here are some useful formula that you can use.

Circles

The **area of a circle** with radius r is πr^2.

The **area of a semicircle** with radius r is $\dfrac{\pi r^2}{2}$.

The **perimeter of a shape** is the distance all the way around the shape.
For a circle, the perimeter is called its circumference.

The **circumference of a circle** with radius r is $2\pi r$.

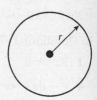

Activity 19

Take a rectangular piece of paper and make it into a hollow cylinder.

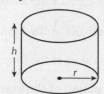

Use your paper to explain why the **curved surface area of your cylinder = $2\pi rh$**.

Example 5

The volume $V\,\text{m}^3$ of a pyramid with a square base of length a m and a height of h m is given by $V = \frac{1}{3}a^2h$.

a Find the volume of the Great Pyramid of Cheops, where $a = 232$ m and $h = 147$ m.

Substituting $a = 232$ m and $h = 147$ m gives
$V = \frac{1}{3} \times 232^2 \times 147 = 2\,640\,000\,\text{m}^3$ to 3 s.f.

b Another square-based pyramid has a volume of $853\,000\,\text{m}^3$ and a height of 100 m. Find the length of the side of the base.

First make a the subject of the formula: $a^2 = 3\dfrac{V}{h} \Rightarrow a = \sqrt{\dfrac{3V}{h}}$

Substituting $V = 853\,000$ and $h = 100$ gives $a = \sqrt{\dfrac{3 \times 853\,000}{100}} \Rightarrow a = 160$ m to 3 s.f.

Exercise 81

1 The time T minutes, to cook a joint of meat weighing w kg is $T = 45w + 20$.
  a Find the time to cook a joint weighing 3 kg.
  b Find the weight of a joint of meat that took 110 minutes to cook.

2 The area $A\,\text{cm}^2$ of a triangle with base b cm and height h cm is given by $A = \dfrac{bh}{2}$.
  a Find A when $b = 12$ and $h = 3$.
  b Find h when $A = 25$ and $b = 5$.

3 The increase in length e cm of an aluminium rod of length l cm when heated by T degrees Celsius is $e = 0.00003lT$.
  a Find e when $l = 100$ and $T = 50$.
  b Find T if $l = 150$ and $e = 0.9$.

4 A formula to calculate income tax I in \$ for a salary of \$$S$ is $I = 0.2(S - 8250)$.
  a Find I if $S = 20\,000$.
  b Find S if $I = 3500$.

5 Find the formulae for the area A and the perimeter P of the shape shown.
  a Find the area and perimeter when $r = 5$ cm.
  b Find r when $A = 100\,\text{cm}^2$.
  c Find r when $P = 50$ cm.
  d Find the value of r that makes $A = P$ numerically.

Exercise 81*

1 The cost C, in \$, of using a phone is $C = 0.02n + 18$ where n is the number of units used.
  a Find C when 200 units are used.
  b Find the number of units used if the cost is \$26.

2 The stopping distance d m of a car travelling at v km/h is given by $d = 0.0065v^2 + 0.75v$.
  a Find the stopping distance of a car travelling at 50 km/h.
  b Estimate the speed of a car which takes 100 m to stop.

3 The volume V cm³ of a cone with base radius r cm and height h cm is $V = \frac{1}{3}\pi r^2 h$.
   a Find V when $r = 5$ and $h = 8$.
   b Find h when $V = 113$ and $r = 6$.
   c Find r when $V = 670$ and $h = 10$.

4 A formula used in mechanics is $v = \sqrt{u^2 + 20s}$ where u and v are speeds in m/s and s is distance in metres.
   a Find v when $u = 12$ and $s = 30$.
   b Find s when $v = 16$ and $u = 4$.
   c Find u when $v = 12$ and $s = 5$.

5 Find the formulae for the area and the perimeter of the shape shown.
   a Find the area and perimeter when $r = 4$ cm.
   b Find r when $A = 30\,\text{cm}^2$.
   c Find r when $P = 70\,\text{cm}$.
   d Find the value of r that makes $A = P$ numerically.

Exercise 82 (Revision)

Make x the subject of these equations.

  1 $ax = b$ 2 $\dfrac{x}{c} = a$ 3 $bx + c = a$

Make y the subject of these equations.

  4 $by^2 = d$ 5 $\sqrt{ay} = b$ 6 $ay - cy = d$ 7 $c(y - b) = y$

8 The speed v m/s of a racing car t seconds after exiting a corner in a race is given by $v = 20 + 5t$.
   a Find the speed of the car after 5 seconds.
   b Make t the subject of the formula.
   c Find how long it takes the car to reach 30 m/s.

9 Eloise is organising a leavers' prom. The caterer tells her that the cost C dollars of each meal when n meals are supplied is given by $C = 20 + \dfrac{300}{n}$.
   a Find the cost of each meal if 50 people come to the prom.
   b How many people must come if the cost of each meal is to be less than \$23?
   c What must Eloise pay the caterer if 75 meals are supplied?

10 The diagram shows a pattern of shapes made from cubes of side 1 cm.

 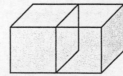 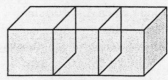

Shape number 1   Shape number 2   Shape number 3

A rule to find the surface area of a shape in this pattern is
   'Multiply the shape number by two, add one and then multiply your answer by two'
A is the surface area of shape number n.
   a Find and simplify a formula for A in terms of n.
   b Find A when $n = 100$.
   c Make n the subject of your formula in part **a**.
   d What shape number has a surface area of $214\,\text{cm}^2$?

Exercise 82* (Revision)

Make x the subject of these equations.

1 $c - ax = b$

2 $\dfrac{b}{x} + d = a$

3 $a(b - x) = \tan c$

Make y the subject of these equations.

4 $\dfrac{a}{y^2} + c = b$

5 $a(y - c) + d = by$

6 $c = a + \sqrt{\dfrac{b - y}{d}}$

7 A formula to find the number, N, of rolls of wallpaper needed to wallpaper a room with wall area $A\,\text{m}^2$ is given by $N = 2 + 0.4A$.

   a A room measures 5 m long by 4 m wide by 2.5 m high. Find the number of rolls required to wallpaper this room, ignoring any windows or doors.

   b Make A the subject of the formula.

   c Find A if $N = 30$.

8 Lisa is pushing her baby daughter on a swing. The time, t seconds, to complete one swing is given by $t = 2\pi\sqrt{\dfrac{l}{10}}$ where l m is the length of the swing.

   a Find t if $l = 2$ m.

   b Make l the subject of the formula.

   c Find l if $t = 2.5$ seconds.

9 The cost, $\$C$, of each ticket for a concert is given by $C = 10 + \dfrac{200}{n}$ where n is the number of people buying a ticket.

   a Find C when $n = 50$.

   b Make n the subject of the formula.

   c Find n if the cost of each ticket is $\$11.25$.

10 The surface area, $A\,\text{cm}^2$, of a cylindrical drinks can with height h cm and radius r cm is given by $A = 2\pi r(r + h)$.

   a Find the surface area of a coke can with $h = 11$ cm and $r = 3.25$ cm.

   b Make h the subject of the formula.

   c Another can has $A = 500\,\text{cm}^2$ and $r = 4$ cm. Find the value of h.

Quadratic graphs $y = ax^2 + bx + c$

You have seen how to plot straight lines of type $y = mx + c$; but, in reality, many graphs are curved.

Quadratic expressions are those in which the highest power of x is x^2, and they produce curves called **parabolas**.

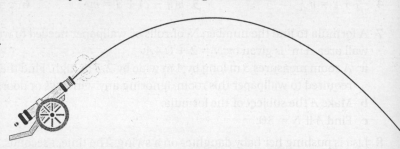

Activity 20

Mathematicians and scientists often try to find a formula to connect two quantities. The first step is usually to plot the graph. For these three graphs, which are parts of parabola, suggest two quantities from real life that might be plotted as x and y to produce the shapes of graphs shown here.

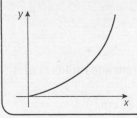 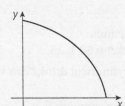 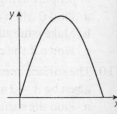

Example 1

Plot the curve $y = 2x^2 - 3x - 2$ in the range $-2 \leqslant x < 4$.
Construct a table of values and plot a graph from it.

x	-2	-1	0	1	2	3	4
$2x^2$	8	2	0	2	8	18	32
$-3x$	6	3	0	-3	-6	-9	-12
-2	-2	-2	-2	-2	-2	-2	-2
y	12	3	-2	-3	0	7	18

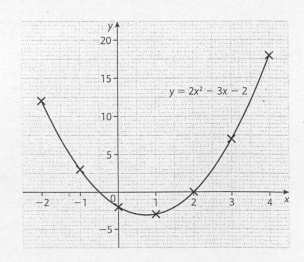

Key Points

- Expressions of the type $y = ax^2 + bx + c$ are called **quadratics**. When they are plotted, they produce **parabolas**.

- If $a > 0$, the curve is U-shaped.

- If $a < 0$, the curve is an inverted U shape.

- Plot enough points to enable a smooth curve to be drawn, especially where the curve turns.

- Do not join up the points with straight lines. Plotting intermediate points will show you that this is incorrect.

Example 2

Plot the curve $y = -3x^2 + 3x + 6$ in the range $-2 \leqslant x \leqslant 3$.

Construct a table of values and plot a graph from it.

x	-2	-1	0	1	2	3
$-3x^2$	-12	-3	0	-3	-12	-27
$+3x$	-6	-3	0	3	6	9
$+6$	6	6	6	6	6	6
y	-12	0	6	6	0	-12

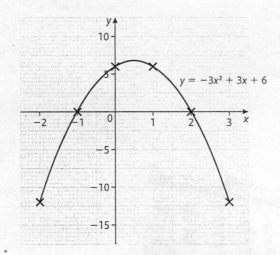

Example 3

Alexandra keeps goats, and she wants to use a piece of land beside a straight stone wall for grazing. This grazing land must be rectangular in shape, and it is to be fenced off by a fence of total length 50 m.

What plot dimensions will provide the goats with the largest grazing area?

What range of values can the rectangle width (x metres) take in order for the enclosed area to be at least 250 m²?

If the total fence length is 50 m, the dimensions of the rectangle are x by $(50 - 2x)$.

Let the area enclosed be A square metres. Then

$$A = x(50 - 2x)$$
$$\Rightarrow \quad A = 50x - 2x^2$$

147

Construct a table of values, and plot a graph from it.

x	0	5	10	15	20	25
$50x$	0	250	500	750	1000	1250
$-2x$	0	-50	-200	-450	$-800\cdot$	-1250
A	0	200	300	300	200	0

The solutions can be read from the graph.

The maximum enclosed area is when $x = 12.5$ m, giving dimensions of 12.5 m by 25 m, and an area of 313 m^2 (to 3 s.f.).

If $A \geqslant 250$ m^2, x must be in the range $7 \leqslant x \leqslant 18$.

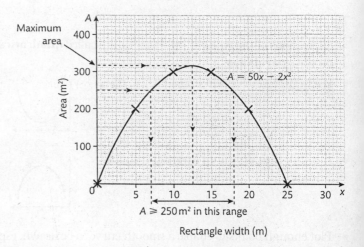

Exercise 83

For Questions 1−4, draw a graph for each equation after compiling a suitable table between the stated x values.

1 $y = x^2 + 2$ for $-3 \leqslant x \leqslant 3$

2 $y = x^2 + 2x$ for $-3 \leqslant x \leqslant 3$

3 Copy and complete this table for the equation $y = 2x^2 + 3x + 2$. Then plot the graph for x values in the range $-3 \leqslant x \leqslant 2$.

x	-3	-2	-1	0	1	2
y	11			2		16

4 A water tank has a square base of side length x metres, and a height of 2 m.

   a Show that the volume V, in cubic metres, of water in a full tank is given by the formula $V = 2x^2$.

   b Copy and complete this table, and use it to draw the graph of V against x.

x (m)	0	0.4	0.8	1.2	1.6	2.0
V (m^3)	0					8

   c Use your graph to estimate the dimensions of the base that give a volume of 4 m^3.

   d What volume of water could be held by the tank if its base area is 0.36 m^2?

   e A hotel needs a water tank to hold at least 3 m^3. If the tank is to fit into the loft, its side length cannot be more than 1.8 m. What range of x values enables the tank to fit into the roof space?

5 On a Big-Dipper ride at a funfair, the height y metres of a carriage above the ground t seconds after the start is given by the formula $y = 0.5t^2 - 3t + 5$ for $0 \leqslant t \leqslant 6$.

   a Copy and complete this table, and use it to draw the graph of y against t.

   b Use your graph to find the height above the ground at the start of the ride.

   c What is the minimum height above the ground and at what time does this occur?

   d What is the height above the starting point after 6 s?

   e Between what times is the carriage at least 3 m above the ground?

t (s)	0	1	2	3	4	5	6
y (m)		2.5		0.5		2.5	

Exercise 83*

1 Draw a graph for $y = -x^2 + 2$ after compiling a suitable table for x values in the range $-3 \leqslant x \leqslant 3$.

2 The population P (in millions) of bacteria on a piece of cheese after t days is given by the equation $P = kt^2 + t + 1$, where k is a constant that is valid for $2 \leqslant t \leqslant 12$.

   a Study this table carefully to find the value of k, and then copy and complete the table.

t (days)	2	4	6	8	10	12
P (millions)	10					265

   b Draw a graph of P against t.

   c Use your graph to estimate the bacteria population after 5 days.

   d How many days does it take for the bacteria population to exceed 10^8?

3 The depth of water, y m, at the entrance of a tidal harbour t hours after midday is given by the formula $y = 4 + 3t - t^2$ where $0 \leqslant t \leqslant 4$.

   a Copy and complete this table, and use it to draw a graph of y against t.

t (hours after 12:00)	0	1	1.5	2	3	4
y (m)		6			4	

   b Use your graph to find the depth of water at the harbour entrance at midday.

   c At what time is the harbour entrance dry?

   d What is the maximum depth of water at the entrance and at what time does this occur?

   e A large ferry requires at least 5 m of water if it is to be able to enter a harbour. Between what times of the day can it safely enter the harbour? Give your answers to the nearest minute.

4 An open box is made from a thin square metal sheet measuring 10 cm by 10 cm. Four squares of side x centimetres are cut away, and the remaining sides are folded upwards to make a box of depth x centimetres.

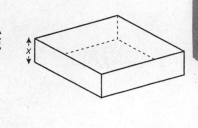

   a Show that the external surface area A cm^2 is given by the formula $A = 100 - 4x^2$ where $0 \leqslant x \leqslant 5$.

   b Draw the graph of A against x by first constructing a table of values.

   c Use your graph to find values of x which will produce a box with an external surface area of between 50 cm^2 and 75 cm^2 inclusive.

5 The total stopping distance y metres of a car in dry weather travelling at a speed of x m.p.h. is given by the formula $y = 0.015x^2 + 0.3x$ where $20 \leqslant x \leqslant 80$.

   a Copy and complete this table and use it to draw a graph of y against x.

x (miles/hour)	20	30	40	50	60	70	80
y (m)		22.5		52.5		94.5	

   b Use your graph to find the stopping distance for a car travelling at 55 m.p.h.

   c At what speed does a car have a stopping distance of 50 m?

   d The stopping distance is measured from when an obstacle is observed to when the car is stationary. Hence, a driver's reaction time before applying the brakes is an important factor.

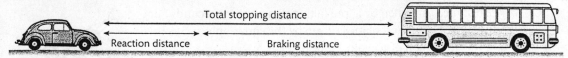

Total stopping distance

Reaction distance   Braking distance

Sam is driving at 75 m.p.h. when she sees a stationary school bus ahead of her. Given that she just manages to stop before hitting the bus, and her braking distance is 83.5 m, calculate her reaction time. (1 mile $\simeq$ 1600 m)

Investigate

A rescue helicopter has a searchlight.

- Find the radius of the illuminated circle when the light is 50 m above the ground.
- What is the radius when the light is 100 m above the ground?
- Calculate the illuminated area for heights of both 50 m and 100 m.
- Show that the relationship of the illuminated area A and the vertical height of the beam H is given by $A = \pi(H \tan 25°)^2$.
- Investigate their relationship by drawing a suitable graph.
- Consider beam angles other than 50°.

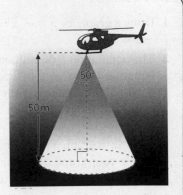

Exercise 84 (Revision)

Draw the graphs of these equations between the stated x values by first compiling suitable tables of values.

1 $y = x^2 + x - 2$ $-3 \leqslant x \leqslant 3$ **2** $y = x^2 + 2x - 3$ $-4 \leqslant x \leqslant 2$

For Questions 3–6, copy and complete the tables, then draw the graph.

3 $y = x^2 + x + 3$

x	-2	-1	0	1	2	3
y	5		3			

4 $y = x^2 + x - 5$

x	-2	-1	0	1	2	3
y				-5	-3	

5 $y = x^2 + 2x + 7$

x	-3	-2	-1	0	1	2	3	4
y			6		10			

6 $y = x^2 + 3x - 4$

x	-3	-2	-1	0	1	2	3	4
y			6		0			

7 The area A cm² of a semicircle formed from a circle of diameter d cm is given by the approximate formula $A \approx 0.4d^2$. Draw a graph of A against d for $0 \leqslant d \leqslant 6$ and use this graph to find

   a the area of a semicircle of diameter 3.5 cm

   b the diameter of a semicircle of area 8 cm²

   c the perimeter of the semicircle whose area is 10 cm²

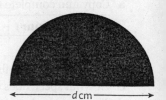

8 The distance y m fallen by a free-falling parachutist t seconds after jumping out of an aeroplane is given by the equation $y = 5t^2$.

   a Draw the graph of y against t for $0 \leqslant t \leqslant 5$.

   b Use your graph to estimate:

     (i) The distance fallen after 3.5 s.

     (ii) The number of seconds it takes the parachutist to fall 25 m.

Exercise 84* (Revision)

Draw the graphs of these equations between the stated x values by first compiling tables of values.

1 $y = 2x^2 + 3x - 4$ $-3 \leqslant x \leqslant 3$ **2** $y = -3x^2 + 2x + 4$ $-3 \leqslant x \leqslant 3$

3 $y = (x + 3)(2x - 5)$ $-4 \leqslant x \leqslant 4$ **4** $y = (2x - 3)^2$ $0 \leqslant x \leqslant 4$

5 Lee is designing a bridge to cross a jungle river.
She decides that it will be supported by a parabolic
arch. The equation $y = -\frac{1}{2}x^2$ is used as the
mathematical model to design the arch, where
the axes are as shown in the diagram.

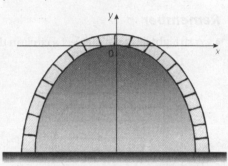

  a Draw the graph of the arch for $-4 \leqslant x \leqslant 4$.

  b Draw the line $y = -6$ to represent the water
     level of the river, and use your graph to
     estimate the width of the arch 10 m above
     the river, if the scale of the graph is 1 unit to 5 m.

6 The equation for the flight path of a golfer's shot is $y = 0.2x - 0.001x^2$, for $0 \leqslant x \leqslant 200$,
where y m is the ball's height, and x m is the horizontal distance moved by the ball.

  a Draw a graph of y against x by first compiling a suitable table of values between the
     stated x values.

  b Use your graph to estimate the maximum height of the ball.

  c Between what distances is the ball at least 5 m above the ground?

7 The profit p (£1000s) made by a new coffee shop Caffeine Rush, t months after opening,
is given by the equation $p = 10t - kt^2$, valid for $0 \leqslant t \leqslant 4$, where k is a constant.

  a Use this table to find the value of k, and then copy and complete the table.

t	0	1	2	3	4
p		7			

  b Draw the graph of p against t.

  c Use it to estimate

    (i) The greatest profit made by Caffeine Rush and when it occurred.

    (ii) When Caffeine Rush starts to make a loss.

8 Consider the graph of $y = px^2 + qx + r$ where p, q and r are integers.
Sketch the following graphs if

  a $p > 0$, $q = 0$ and $r > 0$ **b** $p > 0$, $q = 0$ and $r < 0$

  c $p = 0$, $q > 0$ and $r > 0$ **d** $p < 0$, $q = 0$ and $r > 0$

Circles

Angles can be calculated in shapes involving parallel lines, triangles, quadrilaterals and other straight-sided shapes called polygons. This section shows you how to calculate angles in circles.

Remember

- A straight line can intersect a circle in three ways. It can be a **diameter**, a **chord** or a **tangent**.

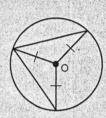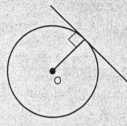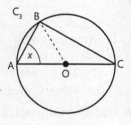

- A **tangent** 'touches' the circle. It is perpendicular to the radius at the point of contact.
- Always look for **isosceles triangles** in circle problems.

Angles in a semicircle and tangents

Key Point

An angle in a semicircle is always a right angle.

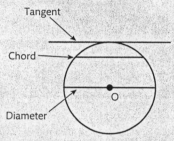

Activity 21

Proving the result

- Copy and complete the table one row at a time by calculating the angles.

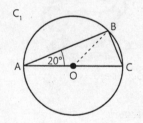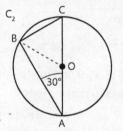

	∠BAO	∠ABO	∠AOB	∠BOC	∠OCB + ∠OBC	∠OBC	∠ABC
C_1	20°						
C_2	30°						
C_3	x						

Using x in the final case generalises the process used for C_1 and C_2. If you add reasons for each step, you have a formal proof to show $\angle ABC = 90°$ when AC is a diameter.

Let	$\angle BAC = x$	
Then	$\angle ABO = x$	($\triangle ABO$ is isosceles $\Rightarrow \angle BAO = \angle ABO$)
	$\angle BOC = 2x$	(Exterior angle of $\triangle ABO$)
	$\angle OCB + \angle OBC = 180° - 2x$	(Angle sum of $\triangle BCO$)
	$\angle OBC = 90° - x$	($\triangle BCO$ is isosceles $\Rightarrow \angle OBC = \angle OCB$)
	$\angle ABC = (90° - x) + x = 90°.$	

In Exercises 85 and 85*, you are asked to give reasons. This is an important part of any explanation or proof.

Key Points

Examples of reasons in explanations or proofs:

- Angle sum of △ ...
- Angles on straight line at ...
- Angles at the point ...
- Alternate angles, ... ∥ ...
- Radius ⊐ tangent at ...

- △ ... is isosceles. ⇒ ... = ...
- Vert. opp. angles at ...
- Exterior angle of △ ...
- Angle in a semicircle

The reason *follows* the statement or equation, and abbreviations can be used.

Key Point

To enable other people to read your work, you need to be consistent with your mathematical language. On diagrams, always label points with capital letters, lengths with lower-case letters, and angles with lower-case letters.

Exercise 85

For Questions 1–12, find the size of each lettered angle. Give your reasons.

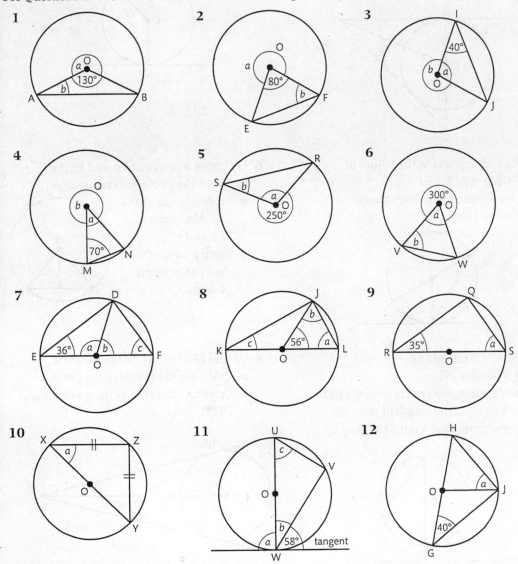

1 **2** **3**

4 **5** **6**

7 **8** **9**

10 **11** **12**

Exercise 85*

1

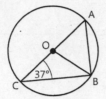

a Find ∠OBC.
b Find ∠ABO.
c Add a reason to your answers to parts **a** and **b**.

3 AB and BC are tangents to the circle, AB = AC, and ∠AOB = 67.5°.
a Find ∠ABO.
b From triangle CBA, find ∠ABC.
Give a reason with each answer.
Hence show that OB bisects ∠ABC.

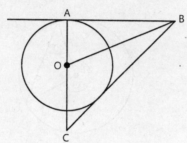

2 ∠OFG = x.

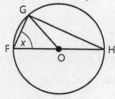

a Find ∠OGF in terms of x.
b Find ∠GOH in terms of x.
c Add a reason to your answers in parts **a** and **b**.

4 a Find, in terms of x, the values of ∠OUT, ∠TOU and ∠TUV.
b Add a reason to your answers in part **a**.

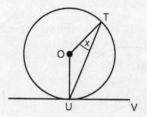

5 Find, in terms of x, the values of ∠OBA, ∠AOB and ∠COB.
Give a reason with each step of your working.

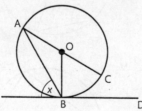

6 Let OM = x centimetres, and let the radius of the circle be r centimetres.
a From triangle OAM, find AM in terms of x and r.
b From triangle OBM, find BM in terms of x and r.

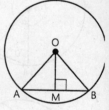

7 OR = 7.5 cm and SR = 9 cm.
a Calculate ST.
b Calculate the area of triangle RTS.
c Calculate the length of the perpendicular from S to RT.

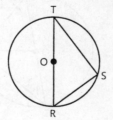

8 DE and EF are tangents to the circle.
a Calculate the radius of the circle.
b Calculate the area of the quadrilateral DEFO.

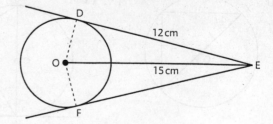

Angle at centre is twice angle at circumference

Activity 22

Proving the result.

- Copy and complete the table one row at a time by calculating the sizes of the angles on each diagram.

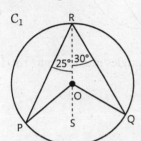   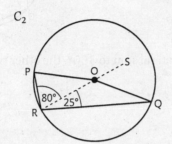

	∠ORP	∠ORQ	∠RPO	∠POS	∠RQO	∠QOS	∠PRQ	∠POQ
C₁	25°	30°						
C₂	80°	25°						
C₃	x	y						

- What is the relationship between ∠PRQ and ∠POQ?
  Using x, you have generalised. Adding reasons gives the formal proof of another 'theorem'.

Let   ∠ORP = x and ∠ORQ = y

Then   ∠RPO = x (△RPO is isosceles, so ∠RPO = ∠PRQ)

So   ∠POS = $2x$ (Exterior angle of △PRO)

            ∠RQO = y (△RQO is isosceles, so ∠RQS = ∠PRQ)

So   ∠QOS = $2y$ (Exterior angle of △QRO)

Thus   ∠PRQ = $x + y$ and ∠POQ = $2x + 2y$

that is,   ∠POQ = $2 × $ ∠PRQ

Example 1

Show that ∠APB = ∠AQB.

Draw lines AO and BO to make an angle at the centre.
The chord AB divides the circle into two segments.
P and Q are in the same segment.

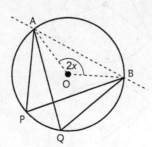

Let   ∠AOB = $2x$

Then   ∠APB = x

and   ∠AQB = x

So   ∠APB = ∠AQB

Unit 4 : Geometry

Example 2

Show that $\angle ABC + \angle CDA = 180°$.

Draw the radii AO and CO.

Let	$\angle ABC = x$
Then	$\angle AOC = 2x$
Let	$\angle CDA = y$
Then	$\angle AOC$ reflex $= 2y$

$$2x + 2y = 360°$$
$$x + y = 180°$$

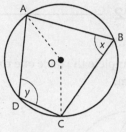

As the four angles of a quadrilateral sum to 360°, the other two angles must sum to 180° as well.

Key Point

- The angle subtended at the centre of a circle is twice the angle at the circumference.
- Angles in the same segment are equal.
- Opposite angles of a cyclic quadrilateral sum to 180°.

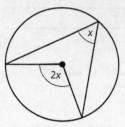
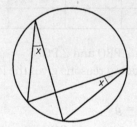
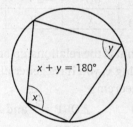

Exercise 86

Find the size of each lettered angle.

1

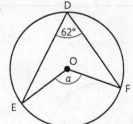

2

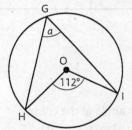

3

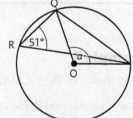

4

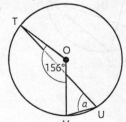

5

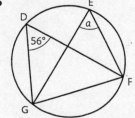

6

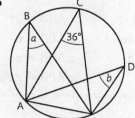

7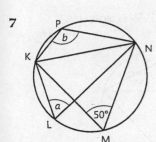

8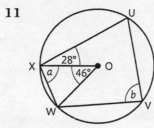

9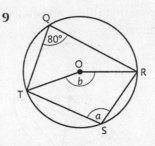

10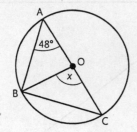

11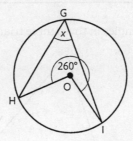

12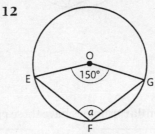

Exercise 86*

1 Show that $x = 96°$, giving a reason for each step of your working.

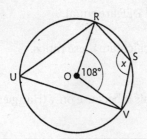

2 Show that $x = 50°$, giving a reason for each step of your working.

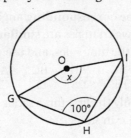

3 Find, giving reasons, the size of angle x.

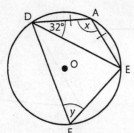

4 Show that $x = 160°$, giving a reason for each step of your working.

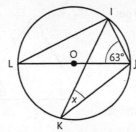

5 Find, giving reasons, the sizes of angles x and y.

6 Find, giving reasons, the size of angle x.

7 Prove that OBC is an equilateral triangle.

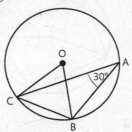

8 Prove that triangle UVW is isosceles.

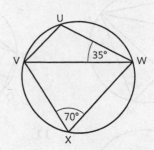

Similar triangles

Key Points

Similar triangles have these properties.

| Corresponding angles equal | ⟷ | Similar triangles | ⟷ | Ratios of the corresponding sides are equal |

If any one of these facts is true, then the other two must also be true.

Activity 23

- Measure each of the angles in these three triangles.

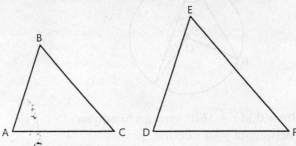

You should find that the **corresponding** angles in triangles ABC and DEF are equal.
This is because these two triangles are **similar** in shape.

- Now measure each of the nine sides, and use your measurements to calculate these ratios.

$$\frac{AC}{DF}, \quad \frac{AB}{DE}, \quad \frac{BC}{EF}, \quad \frac{AB}{GH}, \quad \frac{AC}{GI}, \quad \frac{EF}{HI}$$

You should find that only the first three ratios give the same result. This is because only triangles ABC and DEF are similar in shape.

Example 3

Which of these triangles are similar to each other?

The angle sum of a triangle is 180°.

Therefore in T_1 the angles are 55°, 60° and 65°, in T_2 the angles are 45°, 60° and 75°, and in T_3 the angles are 55°, 60° and 65°.

Thus the triangles in T_1 and T_3 are similar in shape.

T_1 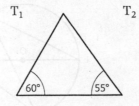 T_2 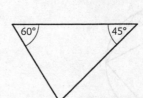 T_3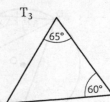

Example 4

The ancient Egyptians used similar triangles to work out the heights of their pyramids. The unit they used was the cubit, a measure based on the length from a man's elbow to his fingertips.

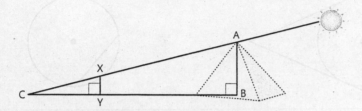

The shadow of a pyramid reached C, which was 500 cubits from B.
The Egyptian surveyor found that a pole of length 4 cubits had to be placed at Y, 20 cubits from C, for its shadow to reach C as well.
What was the height of the pyramid AB?

As AB and XY are both vertical, the triangles CAB and CXY are similar in shape.

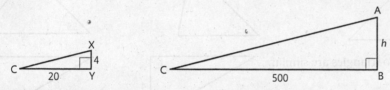

So the ratios of their corresponding sides are equal.

$$\frac{AB}{XY} = \frac{CB}{CY} \qquad \frac{AB}{4} = \frac{500}{20} = 25 \qquad AB = 4 \times 25 = 100$$

So the height of the pyramid is 100 cubits.

Example 5

Use the method of similar triangles to work out x and y.

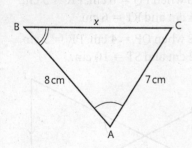

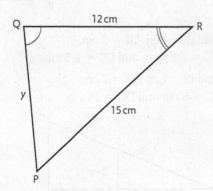

$\hat{A} = \hat{Q}$, $\hat{B} = \hat{R}$ ($\Rightarrow \hat{C} = \hat{P}$).   So △s ABC, QRP are similar.

$$\Rightarrow \frac{AB}{QR} = \frac{AC}{QP} = \frac{BC}{RP} \qquad\qquad \Rightarrow \frac{8}{12} = \frac{7}{y} = \frac{x}{15}$$

$$\frac{8}{12} = \frac{x}{15} \Rightarrow \frac{8 \times 15}{12} = x \quad \Rightarrow x = 10 \text{ cm}$$

$$\frac{8}{12} = \frac{7}{y} \Rightarrow \frac{12}{8} = \frac{y}{7} \qquad \Rightarrow \frac{12 \times 7}{8} = y \quad \Rightarrow y = 10.5 \text{ cm}$$

Investigate

• Investigate these two results.

Tangents from an external point are equal in length.

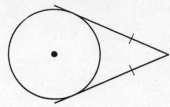

The perpendicular bisector of any chord passes through the centre of the circle.

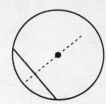

Exercise 87

1 Which two triangles are similar in shape?

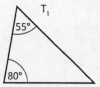

T₁ 55° 80°   T₂ 55° 45°   T₃ 85° 55°

2 These two triangles are similar. Find x.

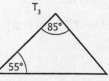

8   4

12   x

3 Triangles ABC and DEF are similar. Find x and y.

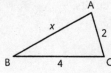

A   x   2   B   4   C

D   10   4   E   y   F

4 a Find BE when AB = 7 cm, AC = 10.5 cm and DC = 4.5 cm

b Find BE when AB = 7 cm, BC = 5 cm and DC = 24 cm.

5 a Find RS when PQ = 6 cm, PR = 5 cm, QR = 4 cm and RT = 6 cm.

b Find PS when QP = 4 cm, PR = 3 cm, QR = 2 cm and ST = 10 cm.

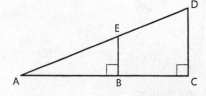

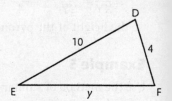

6 Badia and Caimile are sitting on a see-saw. Caimile is sitting 1 m from the pivot, and his end is on the ground. The pivot is 60 cm high. If Badia is sitting 1.5 m from the pivot, how high is she in the air?

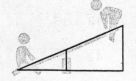

1 Triangles ABC and AEF are similar, and $\frac{AE}{AB} = \frac{1}{2}$. Calculate the coordinates of E and F.

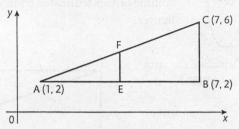

2 a Prove that △GHI and △FHJ are similar.
  b Calculate FJ.
  c Calculate GF.

3 a Prove that △ADE and △ABC are similar.
  b Calculate DE.
  c Calculate BD.

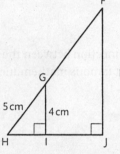

HG : HF = 1 : 4

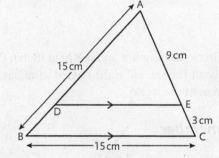

4 In this diagram, which three triangles are similar?
  a Calculate the values of PS and RS.
  b Hence calculate the area of the triangle PQS.

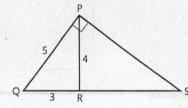

5 A crude method of estimating large heights
is to take a 'line of sight' over two poles.

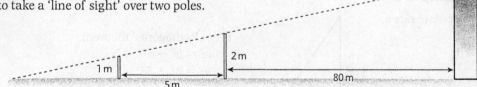

Poles of length 1 m and 2 m are placed 5 m apart, with the longer pole 80 m from the base
of a clock tower, and a line of sight is taken. Calculate the height of the tower.

6 You can estimate the width of a river
without getting your feet wet.
You need just one landmark on the far
bank.
Mark L, M, N and P by line of sight, and
then use similar triangles.
If LP = 10 m, calculate the width of the
river.

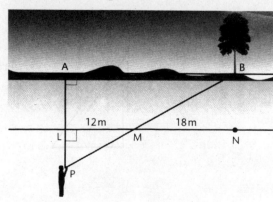

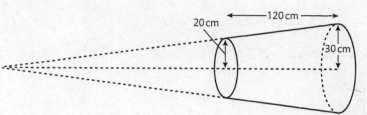

20 cm
←—— 120 cm ——→
30 cm

7 The volume of a cone is given by the formula $V = \frac{1}{3}\pi r^2 h$, where r is the radius of the cone and h is its height. Use this formula, and similar triangles, to calculate the volume of this truncated cone, correct to 3 significant figures.

8 ABCD a rectangle. Calculate the sum of the angles ∠AED and ∠BEF.

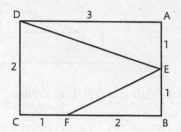

Pythagoras' theorem

The Greek philosopher and mathematician Pythagoras found a connection between the lengths of the sides of right-angled triangles. It is probably the most famous mathematical theorem in the world.

Remember

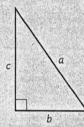

$$a^2 = b^2 + c^2$$

Side a is always the hypotenuse.

Example 6

Calculate side a.

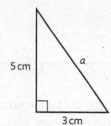

5 cm a 3 cm

From Pythagoras' theorem,
$$a^2 = b^2 + c^2$$
$$a^2 = 3^2 + 5^2$$
$$= 34$$
$$a = \sqrt{34}$$
$$a = 5.83 \text{ cm (3 s.f.)}$$

Example 7

Calculate side b.

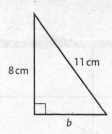

8 cm   11 cm b

From Pythagoras' theorem,
$$a^2 = b^2 + c^2$$
$$11^2 = b^2 + 8^2$$
$$b^2 = 11^2 - 8^2$$
$$= 57$$
$$b = \sqrt{57}$$
$$b = 7.55 \text{ cm (3 s.f.)}$$

Activity 24

Proof of Pythagoras' theorem

There are many elegant proofs of Pythagoras' theorem.

One of the easiest to understand involves a square of side $(a + b)$.

Inside the large square is a smaller one of side c.

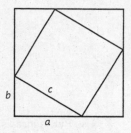

Given that the area of the large square is $(a + b)^2$ and is clearly equal to the area of the four identical triangles plus the area c^2 of the smaller square, form an equation.

Now simplify it to show that $c^2 = a^2 + b^2$, and hence prove that Pythagoras was correct!

Example 8

If point A(2, 3) and B(5, 7), find:

a Length AB

b Coordinates of point M if M is the mid-point of AB.

a

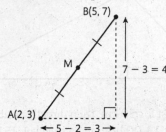

From Pythagoras' theorem.

$$a^2 = b^2 + c^2$$
$$AB^2 = 3^2 + 4^2$$
$$= 25$$
$$AB = \sqrt{25}$$
$$AB = 5$$

b The simplest method to find the coordinates of M is to find the mean of the x-values and y-values.

$$x\text{-value} = \frac{2 + 5}{2} = 3\tfrac{1}{2}$$

$$y\text{-value} = \frac{3 + 7}{2} = 5$$

M(3.5, 5)

Exercise 88

Find length a in these right-angled triangles to 3 s.f.

1

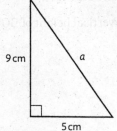

9 cm

a

5 cm

2

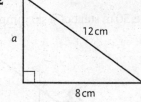

12 cm

a

8 cm

3

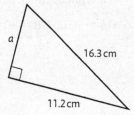

a

16.3 cm

11.2 cm

4 A fishing boat sails from Port Morant in Jamaica. It travels 15 km due East, then 25 km due South. How far is it from Port Morant at this position?

5 A 3.5 m ladder rests against a vertical wall such that its foot is 1.5 m away from the wall. How far up the wall is the top of the ladder?

6 Let OQ = y centimetres, and let the radius of the circle be r.

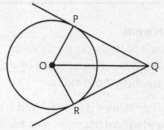

Use Pythagoras's theorem to answer parts **a**, **b** and **c**.

a From triangle OPQ, find PQ in terms of y and r.

b From triangle ORQ, find RQ in terms of y and r.

c Hence show that PQ = RQ.

Exercise 86*

Find length a in these right-angled triangles to 3 s.f.

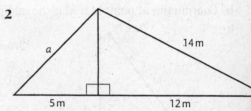

3 Thuso sails his boat from Mogadishu directly North-East for 50 km, then directly South-East for 100 km. He then sails directly back to Mogadishu at 13:00 hours at a speed of 25 km/h. What time does he arrive?

4 Find the length AB in this rectangular block.

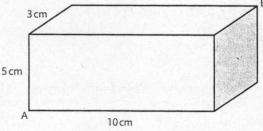

5 A fierce guard dog is tethered by a 15 m chain to a post that is 6 m from a straight path. For what distance along the path is a trespasser in danger from the dog?

6 OA = 50 m. OB = 80 m. The 50 m start of a ski jump at A has a vertical height of 30 m.

a Use similar triangles to calculate the vertical height of the 80 m start at B.

b Use Pythagoras' theorem to calculate the horizontal distance between the two starts.

Congruence

If two figures are the same size and shape (one can be placed exactly on top of the other) they are **congruent**.

For example, an isosceles triangle ABC can be cut from a folded piece of paper.

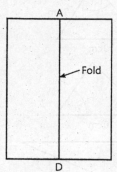 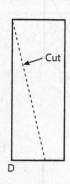 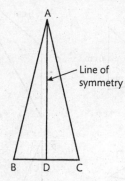 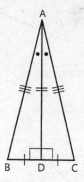

Because of the fold (line of symmetry), triangle ABD is exactly the same size and shape as triangle ACD. The triangles are congruent. Reversing this argument, $\triangle ABD \equiv \triangle ACD$ *because* of the line of symmetry.

Example 9

Why are triangles ABC and XYZ congruent?
AB = XY
AC = XZ
BC = YZ

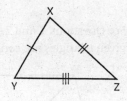 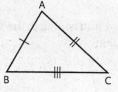

Triangles ABC and XYZ are congruent because the lengths of their corresponding sides are equal (SSS).

Remember

When trying to prove congruence, or calculate a length or angle using congruence:
- Only use the given facts.
- Draw a reasonably accurate and neat diagram to show all known facts.
- Give a reason for each statement in the proof.

Example 10

Triangles ABC and XYZ are congruent (SAS). Write down three deductions that follow from this fact.

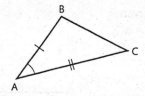 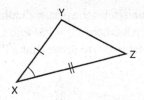

$\angle ACB = \angle XZY$, BC = YZ, $\angle ABC = \angle XYZ$

Exercise 89

For Questions 1–6, state whether pairs of triangles are congruent and give the condition, e.g. SSS, SAS.

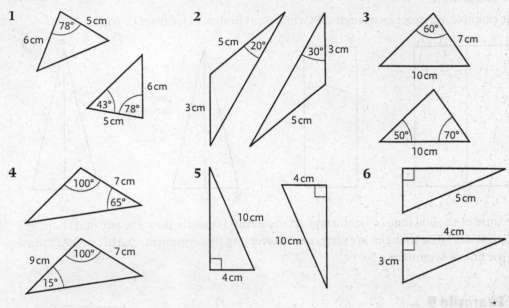

1 78° 5 cm 6 cm 43° 78° 5 cm 6 cm

2 5 cm 20° 30° 3 cm 3 cm 5 cm

3 60° 7 cm 10 cm 50° 70° 10 cm

4 100° 7 cm 65° 9 cm 100° 7 cm 15°

5 4 cm 10 cm 10 cm 4 cm

6 5 cm 4 cm 3 cm

Exercise 89*

For Questions 1 and 2, copy and complete the workings (with SSS, SAS, etc.) to show the two given triangles are congruent.

1 W X Z Y

2 A 10 cm B E C 10 cm D

3 Prove that AD = BC.

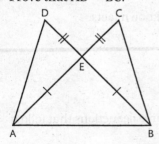

D C E A B

4 Prove that ∠XDC = ∠XCD

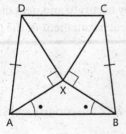

D C X A B

5 The construction shows the perpendicular from P on to the line AB where PL = PM and LX = MX. Use congruent triangles to prove that ∠PYB = 90°.

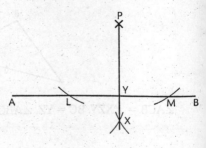

P Y A L M B X

6 ABCD is a square, and the points E, F, G and H shown are such that EB = FC = GD = HA.
Prove that EFGH is also a square.

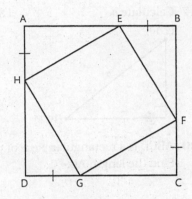

Exercise 90 (Revision)

For Questions 1–12 find the size of each lettered angle.

1
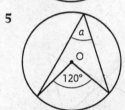

2
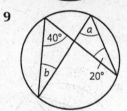

3
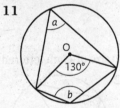

4
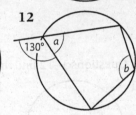

5
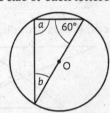

6
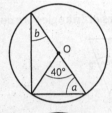

7
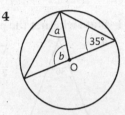

8
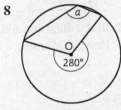

9
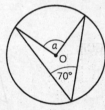

10
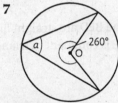

11
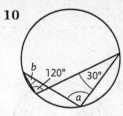

12
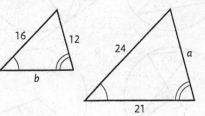

13 The two triangles are similar.
Calculate *a* and *b*.

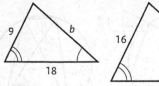

14 Calculate *a*.

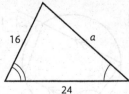

15 Calculate *a* and *b*.

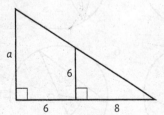

16 Calculate *a* and *b*.

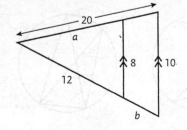

17 Calculate *a*.

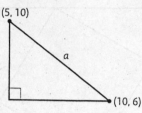

(5, 10)

(10, 6)

18 Calculate *b*.

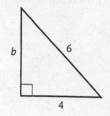

b   6

4

19 Calculate *c*.

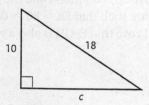

10   18

c

20 ABCD is a rectangular piece of paper.
Find the length AC.

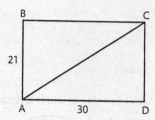

B   C

21

A   30   D

21 State whether the pairs of triangles are congruent. If they are congruent, give reasons.

a

3   30°   3

2   30°   2

b

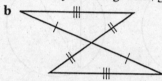

22 In the diagram, AEC and BED are straight lines.
Prove that triangle ABE is conguent to triangle CDE.

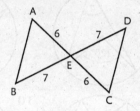

A   D

6   7

E

7   6

B   C

Exercise 90° (Revision)

For Questions 1–12 find the size of each lettered angle.

1

b O

a

2

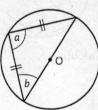

a

O

b

3

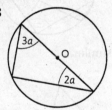

3*a*

O

2*a*

4

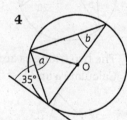

b

a

O

35°

5

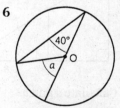

30°

O   *a*

20°

6

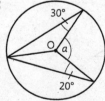

40°

a   O

7

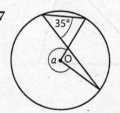

35°

a O

8

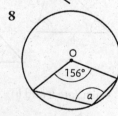

O

156°

a

9

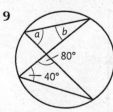

a   *b*

80°

40°

10

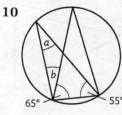

a

b

65°   55°

11

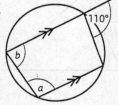

110°

b

a

12

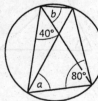

b

40°

80°

a

13 Calculate *a*.

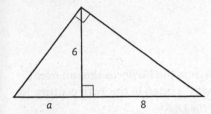

14 Calculate *a* and *b*.

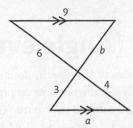

15 Calculate *a* and *b*.

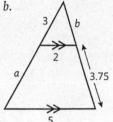

16 a Show that the two triangles are similar.

b Calculate *a* and *b*.

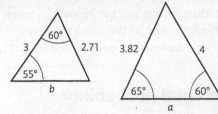

17 Calculate *a*.

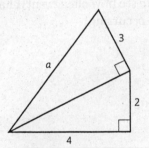

18 Calculate *a*.

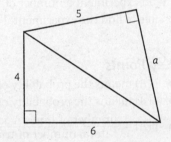

19 a Calculate the radius of the circle

b Calculate *a*.

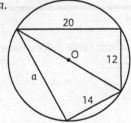

20 The diagram shows a circle with radius 5 cm.
AB = 8 cm and X is the
mid-point of AB
Find **a** XC
b AC

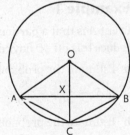

21 In the diagram, BCD and ACE are
straight lines. Use congruent triangles
to prove that AB is parallel to DE.

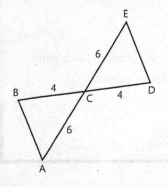

22 In the diagram, ABCD is a parallelogram.
Use congruent triangles to prove that
AX = XD.

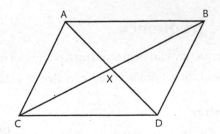

Unit 4 : Geometry

Probability (single event)

Probability theory enables us to analyse random events to assess the *likelihood* that an event will occur. One of the earliest known works on probability was written in the 16th century by Italian mathematician (and gambler) Cardano, '*On Casting the Die*'.

Consider these statements. They all involve a degree of uncertainty, which could be estimated through experiment or using previous knowledge.

- I doubt if I will ever win the lottery.
- My dog will probably not live beyond 15 years.
- It is unlikely to snow in the Sahara Desert.
- Roses will probably never grow at the South Pole.

Experimental probability

If in an experiment a number of trials are carried out to see how often event A happens, it is possible to find the experimental probability, $p(A)$, of A occurring.

Key Points

- $p(A)$ means the probability of event A happening.
- $p(\overline{A})$ means the probability of event A *not* happening.
- $p(A) = \dfrac{\text{number of times } A \text{ occurs}}{\text{total number of trials}}$

Example 1

Event A is that a particular parrot lands on Mrs Bableo's bird table before 9am each day. It does this on 40 days over a period of 1 year (365 days).

a Estimate the probability that tomorrow event A happens.

$p(A) = \dfrac{40}{365} = \dfrac{8}{73}$

b Estimate the probability that tomorrow event A does not happen.

$p(\overline{A}) = \dfrac{325}{365} = \dfrac{65}{73}$

Notice that $p(A) + p(\overline{A}) = 1$.

Relative frequency

Relative frequency can help build up a picture of a probability as an experiment takes place.

It *usually* leads to a more accurate conclusion as the number of trials increases.

Remember

Relative frequency $= \dfrac{\text{number of successes}}{\text{total number of trials}}$

Example 2

Bill is curious about the chances of a piece of toast landing buttered-side up.
He suspects that this event, A, is unlikely to happen.
He then conducts eight trials, with the results shown in this table.

Trial number	1	2	3	4	5	6	7	8
Butter lands upwards	✗	✓	✗	✗	✓	✓	✗	✗
Relative frequency	$\frac{0}{1} = 0$	$\frac{1}{2}$	$\frac{1}{3}$	$\frac{1}{4}$	$\frac{2}{5}$	$\frac{3}{6}$	$\frac{3}{7}$	$\frac{3}{8}$

He plots the results on a **relative frequency diagram**.

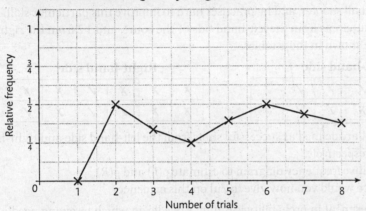

From these eight trials, Bill estimates that the probability of his toast landing buttered-side up is $\frac{3}{8}$.

How could he improve his estimation of p(A)? Comment on Bill's initial assumption.

Exercise 91

1 Molly is suspicious that a particular dice is biased towards the odd numbers so she carries out a number of trials. The results are given in this table.

Trial number	1	2	3	4	5	6	7	8	9	10
Odd number	✗	✓	✓	✗	✓	✗	✓	✗	✓	✓

  a Draw a relative frequency diagram to investigate Heidi's suspicion.
  b To what conclusion do these results lead? How could the experiment be improved?

2 October in Jamaica has a reputation for being a particularly wet month. The data in this table was collected by a weather station in Kingston for the first 20 days in one October.

Day number	1	2	3	4	5	6	7	8	9	10
Rain	✓	✗	✗	✓	✓	✓	✗	✗	✓	✓
Day number	11	12	13	14	15	16	17	18	19	20
Rain	✗	✓	✓	✓	✓	✗	✗	✓	✓	✗

  a Draw a relative frequency diagram to investigate the experimental probability of rain in Kingston in the first 20 days of October.
  b What conclusion can you draw from this data?

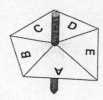

3 A spinner for a word game is an irregular pentagon that has sections divided into five parts denoted by letters A, B, C, D and E. Sanjeev is a keen player, and he experiments to see if he can calculate an estimate of the probability of a vowel being spun. The results are shown as ticks (a vowel) and crosses (no vowel).

✓ ✓ ✗ ✗ ✗ ✗ ✗ ✗ ✓ ✓

✓ ✓ ✗ ✗ ✓ ✗ ✗ ✓ ✗ ✓

Draw a relative frequency diagram to investigate the experimental probability of the spinner landing on a vowel.

Exercise 91*

1 Zul is a basketball shooter, and he practises hard to improve this particular skill.
He makes 12 attempts from the left-hand side of the court, and 12 from the right.
His results are shown in these tables.

Left-hand side

✗	✗	✓	✗	✓	✓
✓	✓	✗	✓	✓	✓

Right-hand side

✗	✗	✗	✓	✗	✗
✗	✗	✓	✓	✗	✓

Events L and R are defined as successful shots from the left-hand side and right-hand side of the court, respectively.
a Draw a relative frequency diagram to estimate p(L) and p(R).
b What advice would you now give to Zul on this evidence?

2 A famous experiment in probability is Buffon's needle, in which a sewing needle of length x centimetres is dropped onto a sheet of paper with parallel lines drawn on it that are x centimetres apart. Event A is defined as the needle landing across a line.
a Draw a relative frequency diagram to find an estimate of p(A) using at least 50 trials.
b Complicated probability theory predicts that p(A) = $\frac{2}{\pi}$. Compare your result with this one, and record more trials to see if this takes your result any closer to the expected probability of event A.

3 A bag contains 100 marbles of similar size and texture. The marbles are either white or purple, and the number of each is not known. A marble is randomly taken from the bag and replaced before another is withdrawn. 20 marbles are sampled in this way, with the results as shown in this table.

W	W	P	P	W	W	W	P	W	W
P	P	W	W	W	W	W	P	P	P

Events W and P are defined as the withdrawal of white and purple marbles, respectively.
a Use a relative frequency diagram to estimate the values of p(W) and p(P). Comment.
b Estimate how many marbles of each colour are in the bag.

Theoretical probability

If all possible outcomes are equally likely, it is possible to find out how many of these *ought* to be event A, that is, to calculate the theoretical probability, p(A).

> **Remember**
>
> $$p(A) = \frac{\text{number of desired outcomes}}{\text{total number of possible outcomes}}.$$

Example 3

A fair dice is rolled. Calculate the probability of a prime number being thrown.

The relevant prime numbers are 2, 3 and 5. Event A is the event of a prime being observed.

$p(A) = \dfrac{3}{6}$ (3 is the number of desired outcomes, and 6 is the total number of possible outcomes)

$\quad = \dfrac{1}{2}$

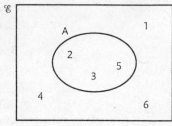

Example 4

A double-headed coin is tossed. If event A is that of a head being thrown, calculate $p(A)$ and $p(\overline{A})$.

$p(A) = \dfrac{2}{2} = 1$ (A certainty) $\qquad p(\overline{A}) = \dfrac{0}{2} = 0$ (An impossibility)

These results imply two important results in probability.

The first is that all probabilities can be measured on a scale from 0 to 1 inclusively.

Key Point

If A is an event,

$\qquad 0 \leqslant p(A) \leqslant 1$.

Investigate

Copy this scale across your page.

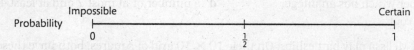

Label the scale, marking approximately where you think the probability of these five events A–E should be placed.

- A basketball captain wins the toss (A).
- A heart is drawn from a pack of cards (B).
- A heart is not drawn from a pack of cards (C).
- You will be abducted by aliens on your way home from school today (D).
- Your teacher will be wearing shoes for your next geography lesson (E).

The second result is that, if A is an event, it either occurs (A) or it does not ($\overline{A}$ or A').

It is certain that nothing else can happen.

Key Point

$\qquad p(A) + p(\overline{A}) = 1$

or, perhaps more usefully,

$\qquad p(\overline{A}) = 1 - p(A)$

Example 5

A card is randomly selected from a pack of 52 playing cards. Calculate the probability that a queen is *not* chosen.

Let event Q be that a queen is chosen.

$p(\overline{Q}) = 1 - p(Q) = 1 - \dfrac{4}{52} = \dfrac{48}{52} = \dfrac{12}{13}$

Exercise 92

1 Nelson is a keen collector of tropical fish. In his tank, there are four guppies, three angel fish, two cat fish, and one Siamese fighting fish. The tank has to be cleaned, so he randomly scoops one of these fish up in his net. Calculate the probability that it is

a a guppy

b an angel fish

c a tiger fish

d not a Siamese fighting fish

2 A letter is chosen randomly from a collection of Scrabble tiles that spell the word PERIODONTOLOGY. Calculate the probability that it is

  a an O   b a T

  c a vowel   d an N or another non-vowel

3 A card is randomly selected from a pack of 52 playing cards.
Calculate the probability that it is

  a a red card   b a king

  c a number card that is a multiple of 3   d an ace, jack, queen or king

4 The bar chart shows the sock colours worn by pupils in class 5C. If a pupil is chosen at random from 5C, calculate the probability that he or she will be wearing

  a grey (G) socks   b white (W) socks

  c red (R) or black (B) socks   d not red socks

5 A fair ten-sided dice with numbers from 1 to 10 on it is thrown. Calculate the probability of obtaining

  a a 1   b an even number

  c a number which has an integer square root   d a number of at most 7 and at least 4

6 David and Melissa play battleships. On their 10×10 grid of squares, both have these in their fleet.

Battleship   Submarine   Aircraft carrier

David shoots first, choosing a square randomly. Calculate the probability that he

  a hits Melissa's aircraft carrier   b hits nothing

  c hits her battleship or submarine   d does not hit her submarine

7 Gita wishes to estimate her probability of scoring a goal in hockey from a penalty. She does this by taking 10 penalties in succession, with the following results.

Score: S   Miss: M

S S M M M S S S M S

Use a relative frequency diagram to estimate the probability of scoring.

8 The probability of it snowing in New York on Dec 25th is 0.2.
What is the probability of New York not experiencing a 'White Christmas'?

9 A box contains twelve roses. Four are white, two are red and six are pink.
Sacha picks out one rose at random. What is the probability that it is

  a pink   b not red

  c white, red or pink   d yellow?

10 One letter is randomly chosen from this sentence.
'*I have hardly ever known a mathematician who was capable of reasoning*'.
What is the probability of the letter being

  a an 'e'   b an 'a'

  c a consonant (non-vowel)   d a 'z'?

Exercise 92*

1 A black dice and a white dice are thrown together, and their scores are added. Copy and complete the 'probability space' table showing all 36 possible outcomes.

  a Use your table to calculate the probability of obtaining
- **(i)** a total of 6
- **(ii)** a total of more than 10
- **(iii)** a total less than 4
- **(iv)** a prime number

  b What is the most likely total?

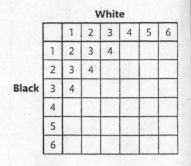

White

		1	2	3	4	5	6
	1	2	3	4			
	2	3	4				
Black	3	4					
	4						
	5						
	6						

2 Four marbles are in a red bag. They are numbered 2, 3, 5 and 7. A green bag contains four more marbles numbered 11, 13, 17 and 19. Two marbles, one from each bag, are randomly selected and the *difference* in the two scores is noted.

  a Construct a suitable 'probability space' table to calculate the probability of obtaining
- **(i)** a score of 6
- **(ii)** a score of at most 8
- **(iii)** a score of at least 12
- **(iv)** a square number

  b What are the least likely scores?

3 A regular five-sided spinner is spun twice, and the scores are multiplied.
Copy and complete the 'probability space' table.
Use the table to calculate the probability of scoring

  a an odd number

  b a number less than 9

  c a number of at least 15

  d a triangular number

First spin

Second spin		1	2	3	4	5
	1	1	2	3		
	2	2	4			
	3	3				
	4					
	5					

4 Three vets record the number of allergic reactions experienced by puppies given the same vaccination.

Vet	No. of puppies vaccinated	No. of allergic reactions
X	50	3
Y	60	7
Z	70	10

  a Calculate the probability that a puppy injected by vet X or Y will experience an allergic reaction.

  b Calculate the probability that a puppy injected by vet Y or Z will *not* experience any reaction.

  c If 7650 puppies are given this injection in a particular year, estimate how many of them will show signs of an allergy.

5 A regular three-sided spinner numbered 2, 4 and 6 is spun, and a six-sided dice is thrown. The highest number obtained is noted, and if the two numbers are equal, that number is taken. Using a probability space or other method, calculate the probability of obtaining

  a a multiple of 3   b a number less than 4

  c a non-prime number   d two consecutive numbers

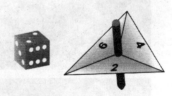

6 Five beads numbered 1, 2, 3, 4 and 5 are placed in bag X. Three beads numbered 1, 2 and 3 are placed in bag Y. One bead is withdrawn from X and one from Y. These represent the coordinates (for example (1, 3)) of a point on the positive x-axis and y-axis, respectively. Calculate the probability that after one selection from each bag, the selected point

  a lies on the line $y = x$ b lies on the line $x = 2$

  c lies on the line $y = 2x - 5$ d lies on the curve $y = x^2 - 6$

7 Two fair six-sided dice are thrown and their scores are multiplied.
  a Write down all the possible outcomes in a table.
  b Use this table to find the probability of the following scores being obtained.
   (i) 36   **(ii)** 11   **(iii)** a multiple of 5   **(iv)** at least 20.

8 Three discs are in a black box and are numbered 10, 30 and 50. Four discs are in a white box and these are numbered 9, 16, 25 and 36. Two discs, one from each box are randomly selected and the *highest* number of the two scores is noted.
  a Construct a suitable 'probability space' table.
  b Use this table to find the probability of the following scores being obtained.
   (i) 10   **(ii)** a prime number   **(iii)** an even number **(iv)** a square number

9 A pond contains 20 tadpoles, of which *f* are frog tadpoles and the others are toad tadpoles. If 10 more frog tadpoles are added to the pond, the probability of catching a frog tadpole is doubled. Find *f*.

10 A dartboard is in the shape of an equilateral triangle inside which is inscribed a circle.
A dart is randomly thrown at the board (assume that it hits the board).

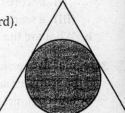

  a Given that $\tan 60° = \sqrt{3}$ and $\sin 60° = \dfrac{\sqrt{3}}{2}$, show that the probability of the dart hitting the board inside the circle is $\dfrac{\pi}{3\sqrt{3}}$.

  b If 100 darts are thrown at the board, and they all hit the board, how many would you expect to land outside the circle?

Investigate

The information in the table was compiled by the League of Dangerous Sports.
Rank the sports in terms of their safety.
Comment.

Activity	Deaths over 5-year period	Participation of adults (millions)
Air sports	51	1
Badminton	3	59
Boating/sailing	69	23
Cricket	2	20
Fishing	50	37
Football	14	128
Golf	1	110
Gymnastics	1	14
Hockey	2	9
Horse riding	62	39
Motor sports	65	11
Mountaineering	51	6
Running	9	200
Rugby	2	12
Swimming/diving	191	370
Tennis	1	45

Exercise 93 (Revision)

1 Alec wants to estimate his chances of scoring a goal from a penalty.
He does this by taking 12 penalties in succession, with these results.

✓	✗	✓	✓	✗	✓	✓	✓	✗	✗	✓	✓

Use a relative frequency diagram to estimate his chances of scoring. Comment.

2 The probability that a new truck gets a puncture in a tyre during its first 30 000 km is $\frac{2}{15}$.
What is the probability of a puncture-free first 30 000 km for this vehicle?

3 One letter is randomly chosen from this sentence: 'All the world's a stage and all the men
and women merely players.' What is the probability of the letter being
a an 'a'?   **b** a 't'?   **c** a vowel?   **d** an 'x'?

4 A $1 coin and a $2 coin are tossed.
Write down all the possible outcomes, and calculate the probability of obtaining
a two tails   **b** a head and a tail

5 Umar has $1, $10, $20 and $50 notes in his wallet. He has one of each type. He randomly
removes two notes together. Find the probability that these two notes total
a $11   **b** $70   **c** $80   **d** at least $11

6 Frances buys ten raffle tickets from 500 sold. If she does not win anything with any of the
first six tickets drawn, what is the probability that she will win with the seventh?

7 Jamal receives 50 emails. 32 are from England, 12 are from the US and the rest are from
China. He chooses one at random to read first. What is the probability that it is from
a France   **b** the US or China   **c** England   **d** not England?

8 A card is randomly selected from a pack of 52 playing cards.
Calculate the probability that it is
a a Queen   **b** a King or a Jack
c not a heart   **d** a red picture card.

9 A fair eight-sided dice has numbered faces from 1 to 8. After it is thrown, find the
probability of obtaining a
a 3   **b** prime number
c a number of at least 3   **d** a number of at most 5.

10 A box contains a red marble and three green marbles. Two are taken at random.
a Write down all the possible outcomes in a table.
b Use this table to find the probability of obtaining one marble of each colour.

Exercise 93* (Revision)

1 Germaine is a keen bird-watcher and spots an Australian magpie at the same place in a
rain forest from 1–10 January in three successive years. She keeps a record, shown below.

Jan Year	1	2	3	4	5	6	7	8	9	10
2002	1	0	1	0	1	1	1	0	1	1
2003	0	0	1	1	0	1	1	1	1	0
2004	0	0	1	0	0	1	0	0	1	1

a Use a relative-frequency graph to estimate the probability of seeing an Australian
magpie in this place for each year.
b Comment.

2 A black dice and a red dice are thrown at the same time. Their scores are multiplied together. Use a probability space diagram to calculate the probability of obtaining

   a a 4   **b** an even number   **c** at least 16

3 A region of Eastern China is called 'GUANGXI ZHUANGZU ZIZHIQU'. One letter is randomly chosen from this name.

   a Find the probability of the letter being

      (i) a A   **(ii)** a Z   **(iii)** a B

   b What is the most likely letter to be picked?

4 The sets A and B consist of the following numbers:

$$A = \{1, 3, 5, 7, 9, 11\} \qquad B = \{1, 5, 9, 13, 17, 21\}$$

A whole number from 1 to 25 inclusive is randomly chosen.
Find the probability that this number is in the set

   a A   **b** B′   **c** A∩B   **d** A∪B

5 A spinner is spun and a dice is thrown.
$5 is won when the score on the dice is at least the score on the spinner. How much would be won if this game were played 12 times in succession?

(Assume that there is no charge to play the game.)

6 Three coins are tossed simultaneously. List all the possible outcomes in a 'probability space' table and use it to calculate the probability of obtaining:

   a three heads

   b two heads and a tail

   c at least two heads

7 Three vets record the number of horses which are cured after being given a particular medicine.

Vet	Number of horses given medicine	Number of horses cured
Mr Stamp	18	16
Mrs Khan	14	11
Miss Abu	10	9

Calculate the probability that

   a a horse given the medicine by Mr Stamp or Miss Abu will be cured.

   b a horse given the medicine by Mrs Khan or Miss Abu will not be cured.

   c all three vets treat 714 horses in total with this medicine. How many would you expect not to be cured?

8 Baby Kiera has two toy boxes. The pink box contains a triangle, a circle and a star whilst the blue box contains a square, a rectangle and a star. She randomly picks one shape from each box to play with. She is happy as long as at least one of these shapes is a star.

   a Construct a suitable 'probability space' table.

   b Use this table to find the probability that:

      (i) Kiera is happy   **(ii)** Kiera is not happy.

9 A garden pond contains 40 Koi Carp, of which x are golden and the others are white. If 20 more golden Koi Carp are added to the pond, the probability of catching a golden Koi Carp is doubled. Find x.

10 Alfred buys eight raffle tickets from 250 sold. If he does not win anything from any of the first five tickets drawn, find the probability that he will win on the sixth draw.

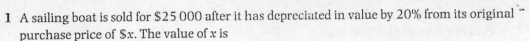

1 A sailing boat is sold for \$25 000 after it has depreciated in value by 20% from its original purchase price of \$$x$. The value of x is

   A 20 000   **B** 30 000   **C** 20 833   **D** 31 250

2 If $p(y + q) = r(s - y)$, the equation for y is given as

   A $y = \dfrac{rs - pq}{p + r}$ **B** $y = \dfrac{rs + pq}{p + r}$ **C** $y = \dfrac{p + r}{rs - pq}$ **D** $y = rs - ry - pq$

3 If $x = \dfrac{-b \pm \sqrt{b^2 - 4ac}}{2a}$, $a = 2$, $b = 3$ and $c = -4$, the greatest value of x to 3.s.f. is

   A -0.235 **B** 0.851 **C** 0.202 **D** 6.20

4 The quadratic curve $y = x^2 - 5x + 4$ cuts the y-axis at the point

   A $(1, 0)$ **B** $(4, 0)$ **C** $(0, 4)$ **D** $(0, 1)$

5 Two points on the curve $y = x^2 - 11x + 30$ are $(2, p)$ and $(4, q)$ The value of pq is:

   A 24   **B** 30   **C** -30 **D** 112

6 The value of angle ABO is

   A 24°   **B** 30°

   C 40°   **D** 60°

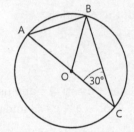

7 The length of diagonal AB in the unit cube is:

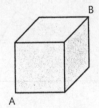

   A $\sqrt{3}$ **B** $\sqrt{2}$ **C** $\frac{2}{3}$ **D** 2

8 The probability that the letter t is randomly selected from this sentence is

   A $\frac{11}{64}$ **B** $\frac{11}{63}$ **C** $\frac{3}{16}$ **D** $\frac{5}{32}$

9 Three people toss a coin. If the first two outcomes were tails, the probability that the third person will throw a head is

   A $\frac{1}{4}$ **B** $\frac{1}{3}$ **C** $\frac{1}{2}$ **D** $\frac{1}{6}$

10 Which condition proves that the following two triangles are congruent?

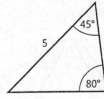

   A SAS   **B** SSS   **C** AAS   **D** ASA

1 A Mini motor car is sold for \$12 500 after it has depreciated by 12.5% from brand new. Find the original purchase price.

2 A valuable stone sculpture is purchased for \$25 000 after it has appreciated by 10%. Find the original price.

3 **Without a calculator** and showing your working, use standard form to calculate an **estimate** of

   a $\dfrac{497 + 33}{0.000802}$ b $\sqrt{2.62 \times 10^5}$

4 Make x the subject of this equation.

$$V = \frac{ab - x}{g}$$

5 Make x the subject of this equation.

$$a(x - b) = c(d - x)$$

6 Make x the subject of this equation.

$$P = \frac{\sqrt{x + Q}}{R}$$

7 Draw the graph of
   a $y = 2x^2 - 3x - 1$ for $-2 \leqslant x \leqslant 3$ and use it to solve the equation $2x^2 - 3x - 1 = 0$.
   b $y = -2x^2 + 3x + 1$ for $-2 \leqslant x \leqslant 3$ and use it to solve the equation $-2x^2 + 3x + 1 = 0$.

8 Find the value of the lettered sides of these right-angled triangles.
   a 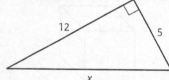   b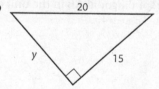

9 Find the distance between the points A(4, -3) and B(12, 6).

10 Find sides x and y.   11 Find sides x and y.

 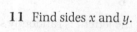

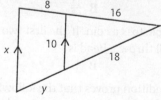

For Questions 12–15 find the size of each lettered angle.

12 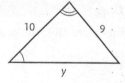   13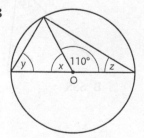

14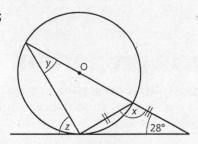

15

16 Two fair six-sided dice are rolled and the outcomes are added together.
   a Draw a probability space diagram showing all possible outcomes.
   b Calculate the probability that the sum will be
     (i) 10   **(ii)** a multiple of 5
     (iii) not 7   **(iv)** 15.

17 A letter is chosen randomly from a collection of Scrabble tiles that spell the expression
   TRIGONOMETRICAL CALCULATION. Calculate the probability that it is
   a an O   **b** an L
   c a vowel   **d** an E or a consonant.

18 Three fair coins are tossed at the same time.
   a Write down all the possible outcomes in a probability space table.
   b Find the probability of obtaining
     (i) three heads   **(ii)** two heads   **(iii)** at least one tail.

19 The formula for t is given by $t = \left(\dfrac{1 + e}{1 - e}\right)\sqrt{\dfrac{h}{5}}$

   a Make e the subject of the formula.
   b Calculate the value of e when $t = 2.5$ and $h = 1.25$.

20 Prove that the two triangles are congruent.

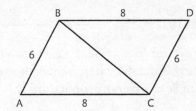

Inverse proportion

An example of **direct proportion** is: when one quantity is *multiplied* by two, the other quantity is also *multiplied* by two.

For example, if 2 kg of apples cost $3, then 4 kg of the same type of apple costs $6.

This relationship produces a straight line graph through the origin.

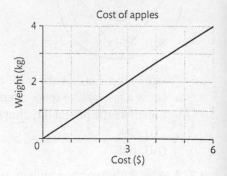

Cost of apples

However, with **inverse proportion**, when one quantity is *multiplied* by two, the other quantity is *divided* by two.

For example, if one machine produces 100 lamp shades in one hour, then it will take two machines half an hour to produce the same number of identical lamp shades.

This relationship produces a **hyperbola**.

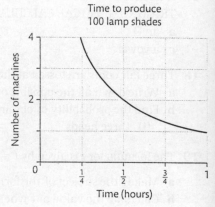

Time to produce 100 lamp shades

Activity 25

For a distance of 160 km, a train travels at a constant speed of 80 km/hour for 2 hours. Clearly, if the speed were halved, the same journey would take twice as long. This is an example of **inverse proportion**. This can be written as:

It takes 2 hours to travel 160 km at 80 km/hour

It takes 4 hours to travel 160 km at 40 km/hour

- Copy and complete this table for the train journey over x km:

Time (hours)	1	2	4	6	7	8
Speed (km/hour)		80	40			

To find the speed for 6 and 7 hours, look at the product of the speed and time for the other entries.

- Plot the points on a suitable graph of speed against time. Join the points with a *smooth curve*. Comment on the shape of the curve.

Example 1

Last year, a farmer used 3 ploughs to plough a field and it took 17 hours. This year the same job must be done in less than 5 hours. How many ploughs will be needed?

It took 17 hours to plough with 3 ploughs

So in 1 hour it will require 3×17 ploughs

So in 5 hours it will require $\dfrac{3 \times 17}{5}$ ploughs

$$= 10.2 \text{ ploughs}$$

Therefore to plough the field in *less* than 5 hours the farmer will need to use 11 ploughs.

Exercise 94

1 It takes 1 person 8 days to dig a trench. To dig a similar trench, how long would it take with
   a 2 people?   **b** 4 people?   **c** 3 people?

2 Using the facts in Question 1, how many people are required to dig a similar trench in
   a 1 day?   **b** half a day?   **c** a quarter of a day?

3 It has been estimated that it took 4000 men 30 years to build the largest pyramid, at Giza, in Egypt, over four and a half thousand years ago. How long would it have taken with
   a 2000 men?   **b** 8000 men?   **c** 100 men?

4 Over a given distance, a train travels at a constant speed of 120 km/hour for 3 hours.
   a Over the same distance, how long will it take a train travelling at 30 km/hour?
   b Find the speed of another train that takes 5 hours to travel the same distance.

5 A car's average fuel consumption is 40 km/litre (km per litre) and, over a certain distance, it uses 4 litres of fuel.
   a Over the same distance, another car uses 5 litres. Find its fuel consumption in km/litre.
   b The fuel consumption of another car is 8 km/litre. How much fuel did this car use to travel the same distance?

Exercise 94*

1 In an exam room, the total power of all the light bulbs has to be 3000 watts.
   a Copy and complete this table.

Number of light bulbs (N)	Power of each bulb (P)
6	
	600
2	
	100

   b Write down the relationship between N and P.

2 In a scientific experiment three different substances, A, B and C, *each of the same mass*, are used. Substance A has a density of 4 g/cm³ and a volume of 3 cm³.
   a Find the density of 1 cm³ of substance B.
   b Find the volume of substance C with a density of 8 g/cm³.

3 The construction of the English Channel Tunnel created about 10^5 man-years of employment in the U.K. Copy and complete this table.

Number of men	Number of tunnels	Time in years
100 000	4	
100 000		2
20 000	8	
	2	0.5

4 One cow belches 200 g of methane in a day. Copy and complete these statements.
 a 1 cow belches 1 kg of methane in ... days.
 b ... cows belch 1000 kg of methane in 5 days.
 c 3100 million cows belch ... tonnes of methane in 365 days.
   (It is estimated that there are about 3100 million cows in the world.)

5 During the Oxford v Cambridge Boat Race, a commentator said, 'Every member of the crew does the equivalent amount of work of someone who lifts a 25 kg sack of potatoes, from the floor to shoulder height, 36 times a minute for 18 minutes'.
 a What total mass is 'lifted' by one crew member during the 18-minute race?
   Give your answer to 2 significant figures.
 b A lorry is to be loaded with sacks of potatoes each of mass 25 kg.
   Work out, correct to 1 significant figure, how long it should take
   (i) 1 crew member to load 4 tonnes
   (ii) 8 crew members to load 4 tonnes
   (iii) 4 crew members to load 4 tonnes
   (iv) 8 crew members to load 1 tonne

Recurring decimals

Remember

- All fractions can be written as decimals which either *terminate* or produce a set of *recurring* digits.
- Fractions that produce terminating decimals have, in their simplest form, denominators with only 2 or 5 as factors. This is because 2 and 5 are the only factors of 10 (*decimal* system).
- The dot notation is used to indicate which digits recur.
  For example $0.232323 \ldots = 0.\dot{2}\dot{3}$, $0.056056056 \ldots = 0.0\dot{5}\dot{6}$.

Example 2

Change $0.\dot{5}$ to a fraction.

$$x = 0.5555555 \ldots \qquad \text{(Multiply both sides by 10)}$$
$$10x = 5.5555555 \ldots \qquad \text{(Subtract the top from the bottom)}$$
$$9x = 5 \qquad \text{(Divide both sides by 9)}$$
$$x = \tfrac{5}{9}$$

Example 3

Change $0.\dot{7}\dot{9}$ to a fraction.

$$x = \ \ 0.797979 \ldots \quad \text{(Multiply both sides by 100)}$$
$$100x = 79.797979 \ldots \quad \text{(Subtract the top from the bottom)}$$
$$99x = 79 \quad \text{(Divide both sides by 99)}$$
$$x = \tfrac{79}{99}$$

Example 4

Change $0.\dot{1}2\dot{3}$ to a fraction.

$$x = \ \ 0.123123 \ldots \quad \text{(Multiply both sides by 1000)}$$
$$1000x = 123.123123 \ldots \quad \text{(Subtract the top from the bottom)}$$
$$999x = 123 \quad \text{(Divide both sides by 999)}$$
$$x = \tfrac{123}{999}$$

Key Point

To change a simple recurring decimal to a fraction

No. of repeating digits	Divide by
1	9
2	99
3	999

Exercise 95

For Questions 1–3, change the fraction to a terminating decimal.

1 $\frac{3}{8}$ **2** $\frac{2}{25}$ **3** $\frac{9}{32}$

For Questions 4–7, change the fraction to a recurring decimal, writing your answers using the dot notation.

4 $\frac{2}{9}$ **5** $\frac{2}{11}$ **6** $\frac{4}{15}$ **7** $\frac{7}{18}$

For Questions 8 and 9, *without* doing any calculation, write down those fractions that produce terminating decimals.

8 $\frac{5}{11}, \frac{9}{16}, \frac{2}{3}, \frac{5}{6}, \frac{2}{15}$ **9** $\frac{2}{19}, \frac{3}{20}, \frac{5}{48}, \frac{5}{64}, \frac{13}{22}$

For Questions 10–15, change each of these recurring decimals to a fraction in its simplest form.

10 $0.\dot{3}$ **11** $0.\dot{5}$ **12** $0.\dot{7}$ **13** $0.\dot{0}\dot{7}$ **14** $0.\dot{0}\dot{3}$ **15** $0.\dot{0}\dot{5}$

Exercise 95*

For Questions 1–4, change each fraction to a recurring decimal, writing your answers using the dot notation.

1 $\frac{7}{15}$ **2** $\frac{7}{150}$ **3** $2\frac{10}{33}$ **4** $\frac{149}{495}$

For Questions 5–6, *without* doing any calculation, write down those fractions that produce terminating decimals.

5 $\frac{7}{17}, \frac{11}{16}, \frac{2}{3}, \frac{7}{40}, \frac{3}{15}$ **6** $\frac{3}{17}, \frac{19}{20}, \frac{3}{25}, \frac{5}{64}, \frac{13}{24}$

For Questions 7–12, change each of these recurring decimals to a fraction in its simplest form.

7 $0.\dot{2}\dot{4}$ **8** $0.3\dot{0}$ **9** $9.0\dot{1}\dot{9}$ **10** $0.0\dot{2}\dot{7}$ **11** $0.4\dot{1}\dot{2}$ **12** $0.38\dot{4}$

For Questions 13 and 14, change each recurring decimal to a fraction.

13 $0.1\dot{2}$ **14** $0.0\dot{5}\dot{6}$

15 Write $0.7\dot{3} \times 0.0\dot{5}$ as a recurring decimal.

Exercise 96 (Revision)

1 It takes one person four days to fully tile a bathroom. To tile a similar bathroom, how long would it take
   a two people   **b** four people   **c** eight people?

2 It takes one person five days to fully weed a garden. To weed a similar garden, how long would it take
   a two people   **b** four people   **c** ten people?

3 The Humber Bridge created about 12 000 man-years of employment in England. In theory it could have been built by 2000 workers in six years.
   Copy and complete the following table for the bridge's construction.

Number of years, n	Number of men, m
1	
	6000
3	
	3000
6	
	m
n	

4 It takes one person five days to lay a 10 m brick garden path.
   a How long would it take to lay a similar 10 m garden brick path using
   (i) two people   **(ii)** four people
   (iii) five people   **(iv)** p people?
   b How many people would it take to lay a similar 40 m garden path in
   (i) two days   **(ii)** four days
   (iii) five days   **(iv)** d days?

5 A helicopter travels between Grenada and Guadeloupe at a constant speed of 150 km/hr for 3 hours.
   a How long would the helicopter take to travel between Grenada and Guadeloupe travelling at 50 km/hr?
   b Calculate the speed of the helicopter travelling from Grenada to Guadeloupe if the journey took 5 hours.

6 Express the following fractions as recurring decimals using dot notation.
   a $\frac{1}{3}$ **b** $\frac{1}{9}$
   c $\frac{1}{12}$ **d** $\frac{1}{15}$

7 Change these recurring decimals into fractions.
   a $0.\dot{4}$ **b** $0.\dot{7}$
   c $0.5\dot{3}$ **d** $0.8\dot{2}$

8 Change these recurring decimals into fractions.
   a $0.\dot{3}0\dot{1}$ **b** $0.\dot{7}0\dot{7}$
   c $0.0\dot{3}\dot{5}$ **d** $0.00\dot{4}0\dot{9}$

Exercise 96* (Revision)

1 It takes two women four days to build an 8 m dry-stone wall.
Use this information to copy and complete the following table.

Number of women w	1	3	4	6		w	w	
Length of dry-stone wall x (m)	8		24		32	x		x
Time of construction t (days)		6		2	12		t	t

2 A single bumble bee travels 150 km to produce 1 g of honey.
Use this information to copy and complete the following table
for a similar colony of bumble bees.

Number of bees b	10	20		500	10^6	b	b	
Length of bee's journey x (km)	150		750		1000	x		x
Mass of honey m (g)		50	500	1000			m	m

3 A rugby pitch can be cut with two grass-cutters in 30 minutes.
Use this information to copy and complete the following table.

Number of grass-cutters n	2		4		n	n
Number of rugby pitches r	1	2	4	r		r
Time t (hours)	0.5	1.5		t	t	

4 A house in the Caribbean produces on average 20 kg of waste per week. Use this
information to copy and complete the following table:

Number of houses h	1	100		h		h
Mass of waste h (kg)	20		10^6	w	w	
Time t (weeks)	1	10	52		t	t

5 Convert the following recurring decimals into fractions.

  a $0.32\dot{1}$ **b** $0.7\dot{5}$ **c** $0.3\dot{7}\dot{8}$ **d** $1.0\dot{2}\dot{5}$

6 Express $0.2\dot{5}\dot{4} \div 0.7\dot{1}\dot{3}$ as a fraction.

Multiplying brackets

Activity 26

Finding the area of a rectangle entails multiplying two numbers together.
Multiplying $(x + 2)$ by $(x + 4)$ can also be done by finding the area of a rectangle.

This rectangular poster has sides $(x + 2)$ and $(x + 4)$.
Notice that the diagram shows the area of each part.

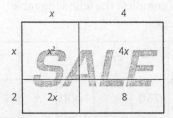

The total area is
$(x + 2)(x + 4) = x^2 + 4x + 2x + 8 = x^2 + 6x + 8$.

- Draw similar diagrams to calculate these:
  $(x + 5)(x + 2)$ $(10 + x)(x + 7)$ $(x + 1)(x + 1)$

- A very common mistake is to say that $(x + 2)^2 = x^2 + 2^2$.
  Show that $(x + 2)^2 \neq x^2 + 2^2$ by substituting various numbers for x.
  Are there any values of x for which $(x + 2)^2 = x^2 + 2^2$?
  What does $(x + 2)^2$ equal?
  Remember that $(x + 2)^2 = (x + 2)(x + 2)$.

- With imagination, this method can be extended to deal with negative numbers.

	x	-5
x	x^2	$-5x$
2	$2x$	-10

$(x + 2)(x - 5) = x^2 - 5x + 2x - 10 = x^2 - 3x - 10$

- Use diagrams to calculate these:
  $(x + 4)(x - 3)$, $(x - 1)(x - 6)$, $(x - 3)^2$, $(x + 2)(x - 2)$

First – Outside – Inside – Last

Brackets can be multiplied without drawing diagrams:

$$(x + 2) \times a = xa + 2a$$

$\Rightarrow \qquad (x + 2) \times (x + 4) = x(x + 4) + 2(x + 4)$

giving $\qquad (x + 2)(x + 4) = x^2 + 4x + 2x + 8 = x^2 + 6x + 8$

Example 1

Multiply out and simplify $(x + 1)(x + 2)$.

Multiply the **F**irst terms x^2
Multiply the **O**utside terms $2x$
Multiply the **I**nside terms x
Multiply the **L**ast terms 2

Add these terms to give $(x + 1)(x + 2) = x^2 + 2x + x + 2 = x^2 + 3x + 2$

Example 2

Multiply out and simplify $(x + 3)(x - 2)$.

Multiply the **F**irst terms x^2
Multiply the **O**utside terms $-2x$ (Note how the negative signs are dealt with)
Multiply the **I**nside terms $3x$
Multiply the **L**ast terms -6

Add these terms to give $(x + 3)(x - 2) = x^2 + (-2x) + 3x + (-6)$
$$= x^2 + x - 6$$

Example 3

Multiply out and simplify $(2x - 3)(3x - 5)$.

Multiply the **F**irst terms $6x^2$
Multiply the **O**utside terms $-10x$
Multiply the **I**nside terms $-9x$
Multiply the **L**ast terms $+15$

Add these terms to give $(2x - 3)(3x - 5) = 6x^2 + (-10x) + (-9x) + 15$
$$= 6x^2 - 19x + 15$$

Key Points

The mnemonic **FOIL** will remind you of what to do when multiplying out brackets.

FOIL stands for First, Outside, Inside, Last.

From each bracket,
- multiply the First terms
- multiply the Outside terms
- multiply the Inside terms
- multiply the Last terms.

Then add the four terms and simplify.

Unit 5 : Algebra

Exercise 97

Multiply out and simplify these expressions.

1 $(x + 4)(x + 1)$ **2** $(x - 7)(x + 3)$ **3** $(x + 2)(x - 6)$

4 $(x - 3)(x - 5)$ **5** $(x + 3)^2$ **6** $(x - 4)^2$

7 $(x + 5)(x - 5)$ **8** $(x + 2)(x - 8)$ **9** $(3x - 2)(5x + 1)$

10 $(x^2 - 5)(x + 2)$

Exercise 97*

For Questions 1–11, multiply out and simplify the expression.

1 $(x + 7)(x - 3)$ **2** $(x - 3)(x + 3)$ **3** $(x + 12)^2$

4 $(3x - 4)(4x - 3)$ **5** $(x - a)(x + b)$ **6** $(4x - 5)^2$

7 $(3x^2 + 1)(5x + 7)$ **8** $(x + 3)^2 - (x - 1)^2$ **9** $\left(\dfrac{a}{2} - \dfrac{b}{5}\right)^2$

10 $x(5x^3 + 3x^2)(2x + 1)$ **11** $\left(\dfrac{a}{b} + \dfrac{b}{a}\right)^2 - \left(\dfrac{a}{b} - \dfrac{b}{a}\right)^2$

12 Solve $2x^2 + (x + 4)^2 = (3x + 2)(x - 2)$.

13 If $(x + a)^2 + b = x^2 + 6x + 10$, find the values of a and b.

Activity 27

- Draw diagrams to show how to multiply out these expressions.
  $(x^2 + 2x + 3)(x + 1)$ $(x + y + 3)(x - 2y)$ $(x^2 + 2x - 3)(x^2 - 2x + 3)$

  Work out how to do the multiplication *without* using diagrams.

Exercise 98

1 A rectangle with length $(x + 2)$ cm and width $(x + 1)$ cm
has a square of length x cm cut out of it.
  a Find and simplify an expression for the area of the
    original rectangle.
  b Hence find an expression for the shaded area.
  c The shaded area is 11 cm². Find the value of x.

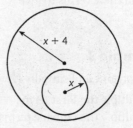

2 A circle of radius $(x + 4)$ m has a circle of radius x m cut
out of it.
  a Find and simplify an expression for the area of the
    original circle.
  b Hence find an expression for the shaded area.
  c The shaded area is 32π m². Find the value of x.

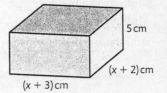

3 A concrete block is in the shape of a cuboid with length
$(x + 3)$ cm, width $(x + 2)$ cm and height 5 cm.
  a Find and simplify an expression for the volume of the
    block.
  b Find and simplify an expression for the surface area of
    the block.

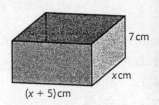

4 An open box has a rectangular base with dimensions
x cm and $(x + 5)$ cm.
  The height of the box is 7 cm.
  a Find and simplify an expression for the volume of the box.
  b Find and simplify an expression for the total surface area
    (inside and outside) of the box.

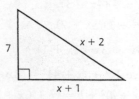

5 These two pictures have the same area. Find x.

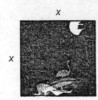 

6 A right-angled triangle has lengths as shown.
  a Use Pythagoras' theorem to form an equation in x.
  b Solve your equation to find x.

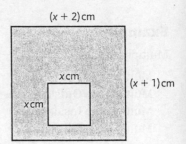

1 A circle of radius $(x + 6)$ cm has a circle of radius (x) cm cut out of it.

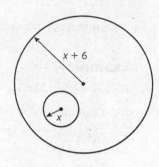

   a Find and simplify an expression for the area of the original circle.

   b The area that remains is 45π cm². Find x.

2 A rectangle with length $(4x + 5)$ cm and width $(x + 8)$ cm has four squares of side x cm cut out of its corners.

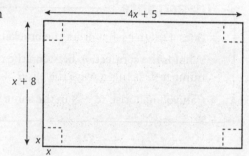

   a Find and simplify an expression for the area of the original rectangle.

   b The area that remains is 95.5 cm². Find the value of x.

3 A right-angled triangle has lengths as shown. Find x.

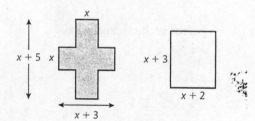

4 The diagram shows two flower beds which have the same area (all dimensions are in metres). What is the value of x?

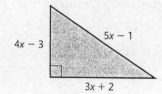

5 **a** If n is an integer, explain why $2n + 1$ must be an odd number.

   b Show that when two odd numbers are multiplied together, the answer is always odd.

   c Show that if you add 1 to the product of two consecutive odd numbers, the answer is always a perfect square.

6 A metal sheet of width 20 cm is to be bent into a chute with width 10 cm and height 6 cm with symmetrical cross-section as shown.

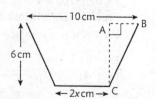

   a Find an expression in terms of x for the length CB.

   b Find an expression in terms of x for the length AB.

   c Use Pythagoras' theorem to find the value of x.

Unit 5 : Algebra

Factorising quadratic expressions

Factorising expressions such as $x^2 + 2x$ to give $x(x + 2)$ is easy, because x is a common factor.

Example 4

Factorise $x^2 - 12x$.

x is a common factor, and so

$x^2 - 12x = x(x - 12)$

Example 5

Expand $(x - 3)(x + 3)$ using **FOIL**.

$x^2 + 3x - 3x - 9 = x^2 - 9$

So factorising $x^2 - 9$ gives $(x - 3)(x + 3)$

Investigate

- Does it matter what order the brackets are in when you are factorising $x^2 - 9$?
- What is the connection between the numbers in the brackets (-3 and $+3$) and the number 9? Is this always the case?
- Can you factorise $x^2 + 9$ in the same way?

Exercise 99

Factorise these expressions.

1 $x^2 - 3x$

2 $x^2 + 2x$

3 $x^2 - 31x$

4 $x^2 + 42x$

5 $x^2 - 16$

6 $x^2 - 49$

Exercise 99*

Factorise these expressions.

1 $x^2 - 312x$

2 $x^2 + 51x$

3 $x^2 - 64$

4 $x^2 - 121$

5 $x^2 - 225$

6 $4x^2 - 16$

Expanding $(x + 2)(x - 5)$ using **FOIL** gives $x^2 - 3x - 10$.

Factorising is the reverse process. $x^2 - 3x - 10$ factorises to $(x + 2)(x - 5)$.

Example 6

Find a if $x^2 + 5x + 6 = (x + 3)(x + a)$.
Using FOIL, the last terms in each
bracket are multiplied to give 6.
So $3 \times a = 6$, and $a = 2$.

Check:

$(x + 3)(x + 2) = x^2 + 2x + 3x + 6$
$= x^2 + 5x + 6$

Example 7

Find a if $x^2 + x - 12 = (x + 4)(x + a)$.
Using FOIL, the last terms in each
bracket are multiplied to give -12.
So $4 \times a = -12$ and $a = -3$.

Check:

$(x + 4)(x - 3) = x^2 + 4x - 3x - 12$
$= x^2 + x - 12$

Exercise 100

Find a.

1 $x^2 + 3x + 2 = (x + 2)(x + a)$

2 $x^2 + 7x + 12 = (x + 3)(x + a)$

3 $x^2 + 3x - 4 = (x + 4)(x + a)$

4 $x^2 - 7x + 10 = (x - 5)(x + a)$

5 $x^2 + 4x + 4 = (x + 2)(x + a)$

6 $x^2 - 1 = (x + 1)(x + a)$

Exercise 100*

Find a.

1 $x^2 + 4x + 3 = (x + 1)(x + a)$

2 $x^2 + x - 12 = (x - 3)(x + a)$

3 $x^2 - 12x + 35 = (x - 5)(x + a)$

4 $x^2 + 2x - 15 = (x + 5)(x + a)$

5 $x^2 - 64 = (x + 8)(x + a)$

6 $x^2 + 4\frac{1}{2}x + 2 = (x + 4)(x + a)$

When the number in the first bracket is not given, then try all the factors of the last number in the expression. The two factors chosen must add to the number in front of the x term.

> **Key Point**
>
> If the last sign in the expression is $+$, then the numbers in both brackets will have the same sign as the middle term.

Example 8

Factorise $x^2 + 5x + 6$.

The last sign is $+$, and so both brackets will have the same sign as $+5x$, giving $(x + \ )(x + \ )$.

The missing numbers are both positive, multiply to give $+6$, and add to $+5$. The two numbers are $+3$ and $+2$.

Thus $x^2 + 5x + 6 = (x + 3)(x + 2)$.

Example 9

Factorise $x^2 - 7x + 6$.

The last sign is $+$, and so both brackets will have the same sign as $-7x$, giving $(x - \ )(x - \ )$.

The missing numbers are both negative, multiply to give $+6$, and add to -7. The two numbers are -1 and -6.

Thus $x^2 - 7x + 6 = (x - 1)(x - 6)$.

Exercise 101

Factorise these. (Notice that the last sign is always $+$.)

1 $x^2 - 3x + 2$

2 $x^2 - 4x + 3$

3 $x^2 - 7x + 12$

4 $x^2 + 8x + 16$

5 $x^2 - 9x + 8$

6 $x^2 - 2x + 1$

Exercise 101*

Factorise these. (Notice that the last sign is always $+$.)

1 $x^2 + 10x + 21$

2 $x^2 - 8x + 12$

3 $x^2 - 16x + 64$

4 $x^2 - 18x + 72$

5 $x^2 + 14x + 45$

6 $x^2 + 24x + 144$

> **Key Point**
>
> If the last sign in the expression is $-$, then the numbers in the brackets will have opposite signs.

Example 10

Factorise $x^2 - 5x - 6$.

The last sign is $-$, and so both brackets will have opposite signs, giving $(x + \ )(x - \ )$.
The missing numbers multiply to give -6 and add to -5.
The two numbers are $+1$ and -6.

Thus $x^2 - 5x - 6 = (x + 1)(x - 6)$.

Exercise 102

Factorise these. (Notice that the last sign is always $-$.)

1 $x^2 + x - 6$ **2** $x^2 - 3x - 10$ **3** $x^2 - 4x - 12$

4 $x^2 - 9x - 10$ **5** $x^2 + 5x - 14$ **6** $x^2 + 7x - 8$

Exercise 102*

Factorise these. (Notice that the last sign is always $-$.)

1 $x^2 + x - 30$ **2** $x^2 - 2x - 24$ **3** $x^2 + 7x - 60$

4 $x^2 - 9x - 70$ **5** $x^2 - 7x - 120$ **6** $x^2 + 10x - 75$

Exercise 103

Factorise these. (Notice that the last signs are mixed.)

1 $x^2 - 3x + 2$ **2** $x^2 + 2x - 3$ **3** $x^2 + 13x + 12$

4 $x^2 - 8x + 12$ **5** $x^2 - 8x + 16$ **6** $x^2 + x - 20$

Exercise 103*

Factorise these. (Notice that the last signs are mixed.)

1 $x^2 + 8x - 20$ **2** $x^2 - 7x - 18$ **3** $x^2 + 13x + 36$

4 $x^2 - 12x + 32$ **5** $x^2 + 8x - 48$ **6** $3 + 2x - x^2$

Solving quadratic equations (factorisation)

If $a \times b = 0$, what can be said about either a or b?

A little thought should convince you that either $a = 0$ or $b = 0$ (or both are zero).

Example 11

Solve $(x + 2)(x - 3) = 0$.

Either $(x + 2) = 0$ or $(x - 3) = 0$.

If $(x + 2) = 0$, then $x = -2$.

If $(x - 3) = 0$, then $x = 3$.

There are *two* solutions:

$x = -2$ or $x = 3$

Example 12

Solve $(x - 5)^2 = 0$.

$(x - 5)^2 = 0$ is the same as $(x - 5)(x - 5) = 0$.

If the first bracket $(x - 5) = 0$, then $x = 5$.

If the second bracket $(x - 5) = 0$, then $x = 5$.

There is *one* solution: $x = 5$.

Exercise 104

Solve these equations.

1 $(x + 1)(x + 2) = 0$ **2** $(x + 4)(x - 1) = 0$ **3** $0 = (x - 7)(x - 2)$

4 $(x + 8)^2 = 0$ **5** $x(x - 10) = 0$

Exercise 104*

Solve these equations.

1 $(x + 8)(x - 4) = 0$ **2** $0 = (x + 21)(x - 5)$ **3** $x(x - 8) = 0$

4 $(2x + 3)(4x - 3) = 0$ **5** $(x + 1)(x - 1)(2x + 5) = 0$

Example 13

Solve $x^2 + 5x + 6 = 0$.

$x^2 + 5x + 6 = 0$ factorises to
$(x + 3)(x + 2) = 0$. (See Example 8)
$\Rightarrow x = -3$ or $x = -2$.

Example 14

Solve $x^2 - 7x + 6 = 0$.

$x^2 - 7x + 6 = 0$ factorises to
$(x - 1)(x - 6) = 0$. (See Example 9)
$\Rightarrow x = 1$ or $x = 6$.

Example 15

Solve $x^2 - 5x = 6$.

$x^2 - 5x = 6$ must first be rearranged to $x^2 - 5x - 6 = 0$.
Then $x^2 - 5x - 6 = 0$ factorises to $(x + 1)(x - 6) = 0$. (See Example 10)
$\Rightarrow x = -1$ or $x = 6$.

Exercise 105

Factorise and solve these for x.

1 $x^2 - 3x + 2 = 0$ **2** $x^2 + x - 2 = 0$ **3** $x^2 + 6x + 8 = 0$

4 $x^2 - x - 12 = 0$ **5** $x^2 - 8x + 15 = 0$ **6** $x^2 + 8x + 16 = 0$

Exercise 105*

Factorise and solve these for x.

1 $x^2 - 9x + 20 = 0$ **2** $x^2 - 5x - 24 = 0$ **3** $x^2 + 21x + 108 = 0$

4 $x^2 - 18x + 56 = 0$ **5** $x^2 + 22x + 96 = 0$ **6** $24x^2 - 48x - 72 = 0$

7 $x^2 + 7x - 78 = 42$

If the quadratic expression has only two terms, the working is easier.

Example 16

Solve $x^2 - 12x = 0$.

$x^2 - 12x$ factorises to $x(x - 12)$. (See Example 4)
So $x^2 - 12x = 0 \Rightarrow x(x - 12) = 0$
Now, either $x = 0$ or $x - 12 = 0$, giving the two solutions $x = 0$ or $x = 12$.

Example 17

Solve $x^2 - 9 = 0$.

$x^2 - 9$ could be factorised and the working continued, but the following is easier.
$x^2 - 9 = 0 \Rightarrow x^2 = 9$
Square-rooting both sides gives $x = \pm\sqrt{9}$, so $x = 3$ or $x = -3$.
(Don't forget the negative square root!)

Key Point

To solve a quadratic equation, rearrange it so that the right-hand side is zero.

Then factorise the left-hand side to solve the equation.

Exercise 106

Solve for x.

1 $x^2 - 2x = 0$ **2** $x^2 + 7x = 0$ **3** $x^2 - 25x = 0$

4 $x^2 + 23x = 0$ **5** $x^2 - 4 = 0$ **6** $x^2 - 25 = 0$

Exercise 106*

Solve for x.

1 $x^2 - 125x = 0$ **2** $x^2 + 231x = 0$ **3** $x^2 - 64 = 0$

4 $x^2 - 169 = 0$ **5** $x^2 - 7 = 0$ **6** $x^2 + 9 = 0$

Problems leading to quadratic equations

Example 18

The product of two consecutive even numbers is 120. What are the numbers?

Let the first number be x. The second even number is two more than x and can be written as $x + 2$.

Then $\quad\quad x \times (x + 2) = 120 \quad\quad$ (Multiply out the bracket)

$\quad\quad\quad\quad\quad x^2 + 2x = 120 \quad\quad$ (Rearrange to equal zero)

$\quad\quad\quad x^2 + 2x - 120 = 0 \quad\quad$ (Factorise)

$\quad\quad (x - 10)(x + 12) = 0$

$\quad\quad\quad\quad\quad\quad x = 10 \text{ or } -12$

So, the numbers are 10 and 12 or -12 and -10.

There are two possible answers. Answering the question by 'trial and improvement' would find the positive answer, but probably not the negative answer.

Example 19

The length of a rectangular patio is 3 m more than the width. If the area is 28 m^2, find the length and width of the patio.

Let x be the width in metres. Then the length is $(x + 3)$ metres.

The area is $\quad x \times (x + 3) = 28 \quad\quad$ (Multiply out the bracket)

$\quad\quad\quad\quad\quad x^2 + 3x = 28 \quad\quad$ (Rearrange to equal zero)

$\quad\quad\quad x^2 + 3x - 28 = 0 \quad\quad$ (Factorise)

$\quad\quad (x + 7)(x - 4) = 0$

$\quad\quad\quad\quad\quad\quad x = -7 \text{ or } x = 4$

As the answer cannot be negative, the width is 4 m and the length is 7 m.

x

$x + 3$

Exercise 107

1 When x is added to its square, x^2, the answer is 12. Find the values of x.

2 When x is added to its square, x^2, the answer is 30. Find the values of x.

3 When x is subtracted from its square, the answer is 20. Find the values of x.

4 When x is subtracted from its square, the answer is 42. Find the values of x.

5 I think of a number. I then square it and add twice the original number. The answer is 35. What was the original number?

6 I think of a number. I then square it and subtract twice the original number. The answer is 24. What was the original number?

7 A rectangle has a length of $(x + 3)$ cm and a width of x cm.
  a Write down an expression for the area of the rectangle.
  b If the area is 18 cm², find the value of x.

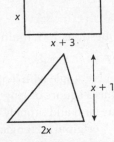

8 A triangle has a base of $2x$ cm and a height of $(x + 1)$ cm.
  a Write down an expression for the area of the triangle.
  b If the area is 42 cm², find the value of x.

9 The length of a mobile phone is 6 cm more than its width. The area of the face is 40 cm².
Find the length and width.

10 The length of a credit card is 3 cm more than its width. The area is 40 cm².
Find the length and width.

11 The rectangles shown have the same area.
Find x.

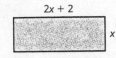

12 The right-angled triangles shown have
the same area. Find x.

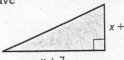
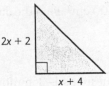

Exercise 107*

1 The product of two consecutive odd numbers is 143. Find the two numbers.

2 The product of two numbers is 96. One number is 4 more than the other number.
Find the two numbers.

3 The length of a picture is 10 cm more than the width. The area is 1200 cm².
Find the dimensions of the picture.

4 The length of a swimming pool is 80 m more than its width. The area is 2000 m².
Find the dimensions of the pool.

5 A ball is thrown vertically upwards so that its height above the ground after t seconds is
$(15t - 5t^2)$ m. At what times is it 10 m above the ground?

6 On a roller coaster, a carriage rolls down a sloping track and travels $(5t^2 + 5t)$ m in t seconds.
How long does it take to travel 30 m?

7 The sum of the squares of two consecutive integers is 145. Find the two integers.

8 The sum of the squares of two consecutive odd integers is 130. Find the two integers.

9 The sum of the first n integers 1, 2, 3, 4, ..., n is given by the formula $\frac{1}{2}n(n + 1)$.
How many integers must be taken to add up to 210?

10 An n-sided convex polygon has $\frac{1}{2}n(n - 3)$ diagonals. How many sides has a polygon with
135 diagonals?

11 The sides of two cubes differ by 2 cm and their volumes differ by 152 cm³.
Find the length of the side of the smaller cube.

12 Sammy spends $10 on some cans of drink. In another shop, she sees that the same cans
are each 10c cheaper, and she calculates that she could have bought five more cans for the
same money. How many cans did she buy?

Exercise 108 (Revision)

For Questions 1–3, multiply out and simplify.

1 $(x - 7)(x - 3)$ **2** $(x + 2)^2$ **3** $(2x + 3)(x - 5)$

4 A rectangle with length $(x + 3)$ cm and width $(x + 2)$ cm has a square of length x cm cut out of it.

    a Find and simplify an expression for the area of the original rectangle.

    b Hence find an expression for the shaded area.

    c The shaded area is 26 cm². Find the value of x.

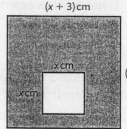

$(x + 3)$ cm

x cm

x cm

$(x + 2)$ cm

Factorise.

5 $x^2 - 36$ **6** $x^2 + 4x + 3$ **7** $x^2 + 2x - 8$

Solve for x.

8 $x^2 - 4x - 12 = 0$ **9** $x^2 - 5x = 0$ **10** $x^2 - 36 = 0$ **11** $x^2 - x = 20$

12 The width of a television screen is 10 cm more than the height. The area of the screen is 600 cm². Find the dimensions of the screen.

Exercise 108* (Revision)

For Questions 1–3, multiply out and simplify.

1 $(x + 9)(x - 12)$ **2** $(2x - 3)^2$ **3** $(3x - 1)(2x + 3)$

4 The foot of a ladder is 2 m away from a vertical wall. The height of the top of the ladder is $\frac{1}{2}$ m less than the length of the ladder. Find the length of the ladder.

5 An abstract painting is 96 cm square, and shows three circles that touch each other and the sides of the square.

The top two circles have the same radius.

What is the radius of the third circle?

Solve for x.

6 $x^2 - 121 = 0$ **7** $x^2 - 7x = 0$ **8** $x^2 - x = 56$ **9** $x^2 - 15x + 54 = 0$

10 Two integers differ by 6. The sum of the squares of these integers is 116. Find the two integers.

11 The stopping distance of a car travelling at x km/h is $\dfrac{x^2 + 30x}{150}$ metres.

    a A car takes 12 m to stop. Show that $x^2 + 30x = 1800$.

    b Find the value of x.

12 A square picture has a border 5 cm wide.

The picture area is $\frac{4}{9}$ of the total area.

Find the area of the picture.

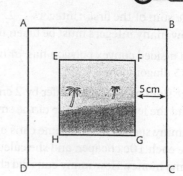

Solving $ax^2 + bx + c = 0$ using graphs

Solving linear simultaneous equations by graphs is a useful technique. The solutions are the intersection point of two straight lines.

The same method can be applied for the intersection points of a line (the x-axis where $y = 0$) and a curve $y = ax^2 + bx + c$ to give the solutions (roots) to $ax^2 + bx + c = 0$.

Example 1

Use the graph of $y = x^2 - 5x + 4$ to solve the equation $0 = x^2 - 5x + 4$.

The curve $y = x^2 - 5x + 4$ cuts the x-axis $(y = 0)$ at $x = 1$ and $x = 4$. Thus at these points, $0 = x^2 - 5x + 4$, and the solutions are $x = 1$ or $x = 4$.

Check: If $x = 1$, $y = 1^2 - 5 \times 1 + 4$
$$= 0$$
If $x = 4$, $y = 4^2 - 5 \times 4 + 4$
$$= 0$$

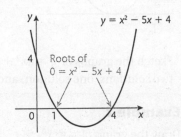

Exercise 109

Draw the graphs between the suggested x-values, and use them to solve the equations.

	Graph	x-values	Equation
1	$y = x^2 - 5x + 6$	$0 \leqslant x \leqslant 5$	$0 = x^2 - 5x + 6$
2	$y = x^2 - 6x + 5$	$0 \leqslant x \leqslant 6$	$0 = x^2 - 6x + 5$
3	$y = x^2 - 2x - 3$	$-2 \leqslant x \leqslant 4$	$0 = x^2 - 2x - 3$
4	$y = x^2 - 7x + 10$	$0 \leqslant x \leqslant 6$	$0 = x^2 - 7x + 10$
5	$y = x^2 - 3x$	$-2 \leqslant x \leqslant 4$	$0 = x^2 - 3x$
6	$y = x^2 - 5x$	$-1 \leqslant x \leqslant 6$	$0 = x^2 - 5x$

Exercise 109*

Draw the graphs between the suggested x-values, and use them to solve the equations. Check your solutions.

	Graph	x-values	Equation
1	$y = 2x^2 - 5x + 2$	$0 \leqslant x \leqslant 4$	$0 = 2x^2 - 5x + 2$
2	$y = 2x^2 - 3x - 5$	$-2 \leqslant x \leqslant 3$	$0 = 2x^2 - 3x - 5$
3	$y = 3x^2 - 8x + 4$	$0 \leqslant x \leqslant 4$	$0 = 3x^2 - 8x + 4$
4	$y = 4x^2 - 1$	$-2 \leqslant x \leqslant 2$	$0 = 4x^2 - 1$

Find the equations of these curves in the form $y = ax^2 + bx + c$.

5
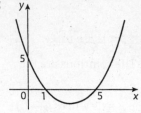

6
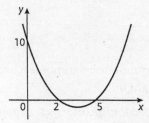

7
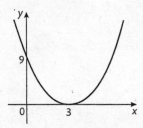

8
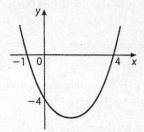

9 Sketch the graphs of $y = ax^2 + bx + c$ (if $a > 0$) when the equation $0 = ax^2 + bx + c$ has two solutions, one solution, and no solutions.

Example 2

Draw the graph of $y = x^2$.
Use this graph to solve the equation $2x^2 + x - 8 = 0$.

Rearrange $2x^2 + x - 8 = 0$ as $x^2 = 4 - \frac{1}{2}x$. This can be solved by finding the x-coordinates of the intersection points of the graphs $y = x^2$ and $y = 4 - \frac{1}{2}x$.

The graph on the right shows the solutions are approximately $x = -2.3$ or $x = 1.8$.

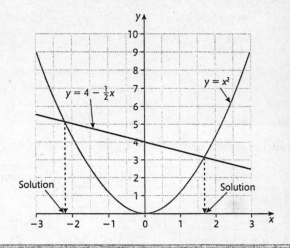

Exercise 110

Draw an accurate graph of $y = x^2$ for $-4 \leqslant x \leqslant 4$. Use this graph to solve these equations.

1 $x^2 - 5 = 0$

2 $x^2 - 3 = 0$

3 $x^2 - x - 2 = 0$

4 $x^2 + x - 3 = 0$

5 $x^2 + 2x - 7 = 0$

6 $x^2 - 2x - 6 = 0$

7 $x^2 - 4x + 2 = 0$

8 $x^2 + 4x + 1 = 0$

9 $2x^2 - x - 20 = 0$

10 $2x^2 + x - 16 = 0$

Exercise 110*

Draw an accurate graph of $y = x^2$ for $-4 \leqslant x \leqslant 4$. Use this graph to solve these equations.

1 $x^2 - x - 3 = 0$

2 $x^2 + x - 4 = 0$

3 $x^2 + 3x + 1 = 0$

4 $x^2 - 2x - 2 = 0$

5 $x^2 - 4x + 4 = 0$

6 $x^2 + 2x + 1 = 0$

7 $2x^2 + x - 12 = 0$

8 $2x^2 - x - 10 = 0$

9 $3x^2 - x - 27 = 0$

10 $3x^2 + x - 21 = 0$

Example 3

Draw the graph of $y = x^2 - 5x + 5$ for $0 \leqslant x \leqslant 5$. Use this graph to solve these three equations:

$$0 = x^2 - 5x + 5 \qquad 0 = x^2 - 5x + 3 \qquad 0 = x^2 - 4x + 4$$

To solve: $0 = x^2 - 5x + 5$

Find where the graph $y = x^2 - 5x + 5$ cuts the line $y = 0$ (the x-axis).

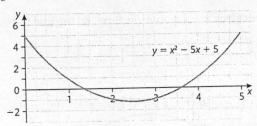

The graph cuts the x-axis at approximately $x = 1.4$ and $x = 3.6$.

So the approximate solutions to $0 = x^2 - 5x + 5$ are $x = 1.4$ or $x = 3.6$.

To solve: $0 = x^2 - 5x + 3$

$$0 = x^2 - 5x + 3 \qquad \text{(Add 2 to both sides)}$$
$$2 = x^2 - 5x + 5$$

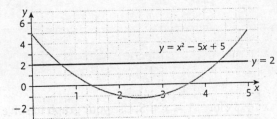

The graph of $y = x^2 - 5x + 5$ cuts the line $y = 2$ at $x = 0.7$ and $x = 4.3$ approximately.

So the approximate solutions to $0 = x^2 - 5x + 3$ are $x = 0.7$ or $x = 4.3$.

To solve: $0 = x^2 - 4x + 4$

$$0 = x^2 - 4x + 4 \qquad \text{(Add } 1 - x \text{ to both sides)}$$
$$1 - x = x^2 - 5x + 5$$

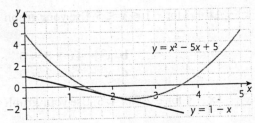

The graph of $y = x^2 - 5x + 5$ cuts the line $y = 1 - x$ at $x = 2$ approximately.

So the approximate solution to $0 = x^2 - 4x + 4$ is $x = 2$.

In this case, it looks as if this is an *exact* solution, but this would have to be checked by substitution.

Note: If the line had not cut the graph, there would be *no* real solutions.

Example 4

If the graph of $y = 6 + 2x - x^2$ has been drawn, find the equation of the line that should be drawn to solve

a $0 = 2 + 2x - x^2$ **b** $0 = 7 + x - x^2$

a $0 = 2 + 2x - x^2$ must be rearranged so that $6 + 2x - x^2$ is on the right-hand side.

Adding 4 to both sides gives $4 = 6 + 2x - x^2$, so the line to be drawn is $y = 4$.

b $0 = 7 + x - x^2$ must be rearranged so that $6 + 2x - x^2$ is on the right-hand side.

Adding $x - 1$ to both sides gives $x - 1 = 6 + 2x - x^2$, so the line to be drawn is $y = x - 1$.

The graphs are shown.

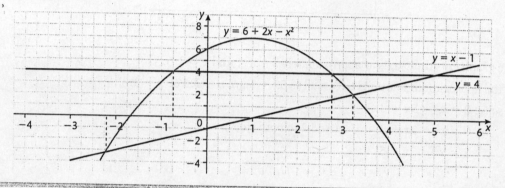

Exercise 111

1 Draw the graph of $y = x^2 - 3x$ for $-1 \leqslant x \leqslant 5$. Use your graph to solve these equations.

   a $x^2 - 3x = 0$ **b** $x^2 - 3x = 2$ **c** $x^2 - 3x = -1$

   d $x^2 - 3x = x + 1$ **e** $x^2 - 3x - 3 = 0$ **f** $x^2 - 5x + 1 = 0$

2 Draw the graph of $y = x^2 - 2x$ for $-2 \leqslant x \leqslant 4$. Use your graph to solve these equations.

   a $x^2 - 2x = 0$ **b** $x^2 - 2x = 5$ **c** $x^2 - 2x = -\frac{1}{2}$

   d $x^2 - 2x = 1 - x$ **e** $x^2 - 2x - 2 = 0$ **f** $x^2 - 4x + 2 = 0$

3 Draw the graph of $y = x^2 - 4x + 3$ for $-1 \leqslant x \leqslant 5$. Use your graph to solve these equations.

   a $x^2 - 4x + 3 = 0$ **b** $x^2 - 4x - 3 = 0$

   c $x^2 - 5x + 3 = 0$ **d** $x^2 - 3x - 2 = 0$

4 Draw the graph of $y = x^2 - 3x - 4$ for $-2 \leqslant x \leqslant 5$. Use your graph to solve these equations.

   a $x^2 - 3x - 4 = 0$ **b** $x^2 - 3x + 1 = 0$

   c $x^2 - 2x - 4 = 0$ **d** $x^2 - 4x + 2 = 0$

5 Romeo is throwing a rose up to Juliet's balcony. The balcony is 2 m away from him and 3.5 m above him. The equation of the path of the rose is $y = 4x - x^2$, where the origin is at Romeo's feet. Find by a graphical method where the rose lands. The balcony has a 1 m high railing. Does the rose pass over the railing?

6 A cat is sitting on a 2 m high fence when it spots a mouse 1.5 m away from the foot of the fence. The cat leaps along the path $y = -0.6x - x^2$, where the origin is where the cat was sitting and x is measured in metres. Find, by a graphical method, whether the cat lands on the mouse.

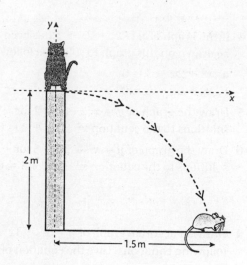

2 m

1.5 m

Exercise 111*

1 Draw the graph of $y = 5x - x^2$ for $-1 \leqslant x \leqslant 6$. Use your graph to solve these equations.

  a $5x - x^2 = 0$ **b** $5x - x^2 = 3$ **c** $5x - x^2 = x + 1$ **d** $x^2 - 6x + 4 = 0$

2 Draw the graph of $y = x - 2x^2$ for $-2 \leqslant x \leqslant 3$. Use your graph to solve these equations.

  a $x^2 - 2x^2 = 0$ **b** $x - 2x^2 = -4$ **c** $x - 2x^2 = -x - 3$ **d** $2x^2 - 2x - 2 = 0$

3 Draw the graph of $y = 2x^2 + 3x - 1$ for $-3 \leqslant x \leqslant 2$. Use your graph to solve these equations.

  a $2x^2 + 3x - 1 = 0$ **b** $2x^2 + 3x - 4 = 0$ **c** $2x^2 + 5x + 1 = 0$

4 Draw the graph of $y = 3x^2 - x - 2$ for $-2 \leqslant x \leqslant 3$. Use your graph to solve these equations.

  a $3x^2 - x - 2 = 0$ **b** $3x^2 - x - 4 = 0$ **c** $3x^2 - 3x - 1 = 0$

5 If the graph of $y = 5x^2 - 9x - 6$ has been drawn, find the equations of the lines that should be drawn to solve these equations.

  a $5x^2 - 10x - 8 = 0$ **b** $5x^2 - 7x - 5 = 0$

6 If the graph of $y = 4x^2 + 7x - 8$ has been drawn, find the equations of the lines that should be drawn to solve these equations.

  a $4x^2 + 8x - 5 = 0$ **b** $4x^2 + 4x - 3 = 0$

Exercise 112 (Revision)

1 Draw an accurate graph of $y = x^2$ for $-3 \leqslant x \leqslant 3$. Use this graph to solve the following equations. Give the equation of the lines you use to do this.

  a $x^2 - 7 = 0$ **b** $x^2 - x = 0$ **c** $x^2 - x - 3 = 0$
  d $x^2 + x - 4 = 0$ **e** $x^2 - 2x + 1 = 0$ **f** $x^2 + 2x - 1 = 0$

2 Draw an accurate graph of $y = x^2 + x - 1$ for $-4 \leqslant x \leqslant 3$. Use this graph to solve the following equations. Give the equation of the lines you use to do this.

  a $x^2 + x - 1 = 0$ **b** $x^2 + x = 0$ **c** $x^2 + x - 2 = 0$
  d $x^2 + x - 4 = 0$ **e** $x^2 + 3x - 1 = 0$ **f** $x^2 - 1 = 0$

3 If the graph of $y = x^2 - x$ has been drawn, find the equation of the lines that should be drawn on this graph to solve the following equations.

  a $x^2 - x - 2 = 0$ **b** $x^2 - 2x = 0$
  c $x^2 - 1 = 0$ **d** $x^2 - 3x - 3 = 0$

4 If the graph of $y = x^2 + 2x - 3$ has been drawn, find the equation of the lines that should be drawn on this graph to solve the following equations.

   a $x^2 + 2x - 3 = 0$ **b** $x^2 + 2x = 0$

   c $x^2 + 2x - 6 = 0$ **d** $x^2 + x - 3 = 0$

5 Draw the graph of $y = x^2 - 4x + 2$ for $-1 \leqslant x \leqslant 5$ and use this graph to find approximate solutions to the equation $x^2 - 4x + 2 = 0$. Check your answers.

6 Draw the graph of $y = x^2 - 2x - 5$ for $-2 \leqslant x \leqslant 4$ and use this graph to find approximate solutions to the equation $x^2 - 2x - 5 = 0$. Check your answers.

Exercise 112* (Revision)

1 Draw an accurate graph of $y = x^2 - 3x + 1$ for $-2 \leqslant x \leqslant 5$. Use this graph to solve the following equations. Give the equation of the lines you use to do this.

   a $x^2 - 3x + 1 = 0$ **b** $x^2 - 3x - 3 = 0$ **c** $x^2 - 4x + 1 = 0$

   d $x^2 - 4x - 1 = 0$ **e** $x^2 - 2x - 5 = 0$ **f** $x^2 - x - 3 = 0$

   g The equation $x^2 - 3x = k$ has only one solution. Find the value of k.

2 If the graph of $y = 2x^2 + 3x - 1$ has been drawn, find the equation of the lines that should be drawn on this graph to solve the following equations.

   a $2x^2 + 3x = 0$ **b** $2x^2 + 3x - 7 = 0$ **c** $2x^2 + 2x - 1 = 0$

   d $2x^2 + x - 3 = 0$ **e** $2x^2 + 7x - 4 = 0$ **f** $4x^2 + 6x - 1 = 0$

3 Thomas has just caught a heavy fish. The shape of his fishing rod is given by $y = 3.5x - x^2$ and the fishing line is given by $x + 2y = 8$. Use a graphical method to find the coordinates of the point where the line joins the rod. (Use $0 \leqslant x \leqslant 10$)

4 A food parcel has just been dropped by a low flying aeroplane flying over sloping ground. The path of the food parcel is given by $y = 40 - 0.005x^2$ and the slope of the ground is given by $y = 0.2x$. Use a graphical method to find the coordinates of the point where the food parcel will land. (Use $0 \leqslant x \leqslant 100$)

5 Draw the graph of $y = 2x^2 + 3x - 6$ for $-3 \leqslant x \leqslant 3$ and use this graph to find approximate solutions to the equation $2x^2 + 3x - 6 = 0$. Check your answers.

6 Draw the graph of $y = -2x^2 - 3x + 6$ for $-3 \leqslant x \leqslant 3$ and use this graph to find approximate solutions to the equation $-2x^2 - 3x + 6 = 0$. Check your answers.

Basic transformations

Reflection, rotation and translation are three basic transformations. They change the position of an object, but not its size or shape.

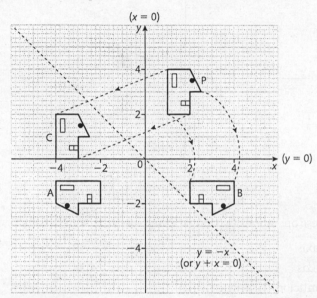

Transformation	Image of P	Definition	Notes
Reflection	A	• In the line $y = -x$	Line of reflection is also called the 'mirror line'.
Rotation	B	• 90° clockwise • About (0, 0)	Rotations are measured anticlockwise. A 90° clockwise rotation is actually a rotation of $-90°$.
Translation	C	• 5 left, 2 down • Or $\begin{pmatrix} -5 \\ -2 \end{pmatrix}$ This is called a vector.	The minus sign indicates a shift in the negative direction, which is: • for the top number: to the left parallel to the x-axis • for the bottom number: down parallel to the y-axis.

Remember

Transformation	Defined by ...	Notes
Translation	Translation vector	$\begin{pmatrix} 3 \\ -4 \end{pmatrix}$ means 3 right, 4 down
Reflection	Mirror line	State equation if requested e.g. $x = 2$, $y = x$, ... Line should be dotted.
Rotation	Angle, direction, centre	Direction is clockwise ($-$) or anticlockwise ($+$). Centre is stated as coordinates.

1 On graph paper, draw x- and y-axes from -6 to 6, and plot triangle P with vertices $(4, 1)$, $(4, 2)$, $(2, 1)$ as shown.

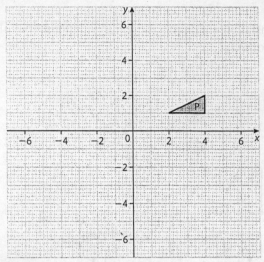

Then do each of these operations to triangle P, and label the images A, B, C, ..., F.

A	Reflect in y-axis
B	Reflect in x-axis
C	Rotate 90° anticlockwise about $(0, 0)$
D	Rotate 180° about $(0, 0)$
E	Translate by the vector $\begin{pmatrix} 2 \\ 4 \end{pmatrix}$
F	Translate by the vector $\begin{pmatrix} 2 \\ -6 \end{pmatrix}$

2 The tables below the figure give details of 12 reflections. Use this figure to copy and complete the tables.

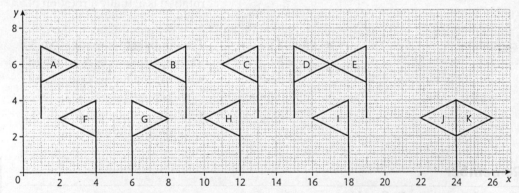

Object	Reflection in line	Image
A	$x = 5$	B
F	$x = 5$	
G		F
A	$x = 7$	
D		B
	$x = 9$	H

Object	Reflection in line	Image
K		J
	$x = 12$	G
	$x = 18$	
	$x = 10$	A
J		G
	$x = 17$	

3 Each triangle in this diagram can be transformed onto another one by rotation or translation.

Using the diagram, copy and complete this table.

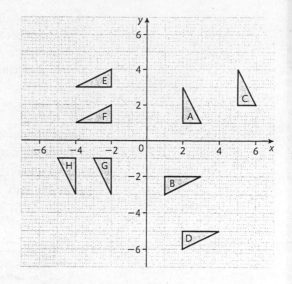

Object	Transformation			Image
B	Rotation	Centre (0, 0)	angle 90°	A
A	Translation	Vector $\begin{pmatrix} \\ \end{pmatrix}$		C
B	Rotation	Centre	angle	C
A	Rotation	Centre	angle 90°	F
F	Translation	Vector $\begin{pmatrix} 0 \\ 2 \end{pmatrix}$		
A	Rotation	Centre (−1, 0)	angle 90°	
A	Rotation	Centre (0, 0)	angle 180°	
G	Translation	Vector $\begin{pmatrix} \\ \end{pmatrix}$		H
A	Rotation	Centre	angle	H
B	Rotation	Centre	angle −90°	H
H	Translation	Vector $\begin{pmatrix} \\ \end{pmatrix}$		G
B	Rotation	Centre	angle	G

4 Draw x- and y-axes from −6 to 6. Draw the line $y = x$.
Plot triangle Q with vertices at (3, 1), (4, 2), (6, 1).

  a Reflect Q in the x-axis, and label the image A.

  b Reflect A in $y = x$, and label the image B.

  c Describe fully the single transformation which takes Q to B.

Exercise 113*

1 P(1, 2) is one corner of the coloured triangle in the figure. Point A is the image of P after reflection in the line $y = 3$. Without drawing, find the image of P after a reflection in the line

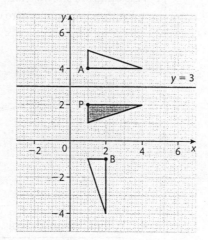

  a $y = 0$ **b** $y = 2$

  c $y = 6$ **d** $y = 10$

Point B is the image of P after a rotation of −90° about (0, 0). Without drawing, find the image of P after a rotation of −90° about the point

  e (2, 0)   **f** (3, 0)

  g (4, 0)   **h** (10, 0)

2 Draw x- and y-axes from −6 to 6. Draw the line $y = x$.
Plot triangle P with vertices at (1, 5), (3, 5), (3, 6).

  a Reflect P in $y = x$, and label the image A.

  b Reflect A in the x-axis, and label the image B.

  c Describe fully the single transformation which takes P to B.

3 Draw x- and y-axes from −6 to 6. Plot triangle R with vertices at (2, 1), (5, 1), (5, 3).

  a Rotate R by 90° about (0, 0), and label the image A.

  b Reflect A in the y-axis, and label the image B.

  c Describe fully the single transformation which takes R to B.

4 Draw x- and y-axes from -6 to 6. Plot triangle S with vertices at $(-2, 1)$, $(-5, 1)$, $(-5, 3)$.

 a Reflect S in the x-axis, and label the image A.

 b Rotate A by $-90°$ about O, and label the image B.

 c Describe fully the single transformation which takes S to B.

Enlargements

There are many real-world applications of enlargement, from photographic enlargement to microscopes and telescopes.

Photographs can be enlarged for displaying in frames. They can also be reduced in size for identity cards.

Remember

Transformation	Defined by	Notes
Enlargement	Scale factor k Centre	$k > 1$: Enlargement $k < 1$: Reduction Centre is stated as coordinates

Activity 28

This diagram shows triangle T transformed onto triangle T_1 and onto triangle T_2 by enlargements from centre O.

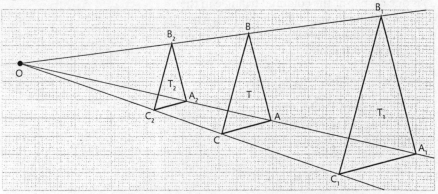

- Copy and complete this table, and comment on both sets of ratios.

T	AB =	AC =	BC =
T_1	$A_1B_1 =$	$A_1C_1 =$	$B_1C_1 =$
Ratios	$A_1B_1/AB =$	$A_1C_1/AC =$	$B_1C_1/BC =$
T_2	$A_2B_2 =$	$A_2C_2 =$	$B_2C_2 =$
Ratios	$A_2B_2/AB =$	$A_2C_2/AC =$	$B_2C_2/BC =$

- What factor changes the lengths of the triangle T to produce T_1 and to produce T_2? Your answers are called the 'scale factor of enlargement'.

Example 1

a Enlarge triangle ABC, with scale factor = 3 from the point O = (0, 0). Label the image $A_1B_1C_1$.

b Enlarge triangle $A_1B_1C_1$, with scale factor = $\frac{1}{2}$ from the point P = (−16, 12). Label the image $A_2B_2C_2$.

a
- Draw 'ray lines' from the point (0, 0) through the points A, B and C.
- On the ray line OA, mark the point A_1 such that $OA_1 = 3OA$.
- Repeat, marking B_1 on ray line OB and C_1 on ray line OC.
- Draw triangle $A_1B_1C_1$.

b
- Draw 'ray lines' from the point P (−16, 12) through the points A_1, B_1 and C_1.
- On the ray line PA_1, mark the point A_2 such that $PA_2 = \frac{1}{2}PA_1$.
- Repeat, marking B_2 on ray line PB_1 and C_2 on ray line PC_1.
- Draw triangle $A_2B_2C_2$.

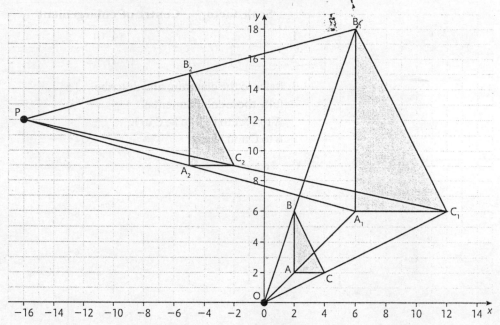

Exercise 114

Draw x- and y-axes from −4 to 4. Plot the triangle ABC, for points A(1, 0), B(1, −1) and C(0, −1). Draw the image for each of these enlargements.

	Scale factor	Centre of enlargement	Image
1	4	(0, 0)	P
2	$1\frac{1}{2}$	(4, 2)	R

3 Refer to the diagram and describe fully these enlargements: B to A; A to E; D to B; E to D and D to C.

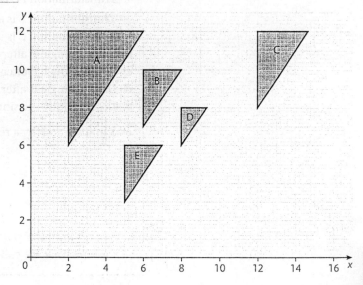

Exercise 114*

Draw x- and y-axes from -4 to 4. Plot the triangle DEF, for points D(2, 0), E(1, 0) and F(0, 2). Draw the image for each of these enlargements.

	Scale factor	Centre of enlargement	Image
1	2	(4, 0)	P
2	$1\frac{1}{2}$	$(-2, -2)$	R

3 Refer to the diagram, and describe fully these enlargements:
A to C; A to D; C to D; C to E;
B to C and E to D.

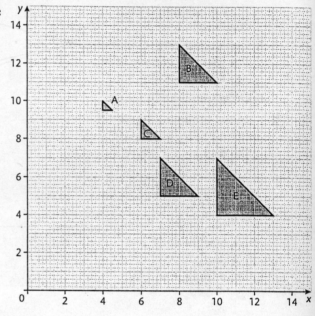

Combining transformations

We combine transformations when more than one operation is applied to a shape.

Example 2

a Plot points (1, 2), (1, 4) and (2, 4) to form triangle P.
Transformation A is a translation of $\begin{pmatrix} -5 \\ -4 \end{pmatrix}$
Transformation B is a reflection in $y = x$.
Transformation C is a clockwise rotation of 90° about centre (1, 1)

b Draw triangle P after it has been transformed by A and label this image Q.

c Draw triangle P after it has been transformed by B and label this image R.

d Draw triangle R after it has been transformed by C and label this image S.

e Describe fully the single transformation that maps P onto S.

e S maps onto P by a reflection in $y = 1$.

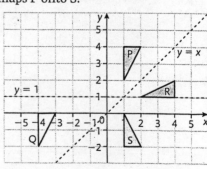

Example 3

Triangle T is shown in the diagram.

a T is reflected in line M to form image A.
Draw triangle A.

b A is reflected in $y = -1$ to form image B.
Draw triangle B.

c T undergoes a translation with vector $\begin{pmatrix} -2 \\ -3 \end{pmatrix}$
to form image C. Draw triangle C.

d T is enlarged by a scale factor of 2 with centre
$(3, 2)$ to form image D. Draw triangle D.

e Describe fully the transformation which maps T to B.

f Describe fully the transformation which maps C to D.

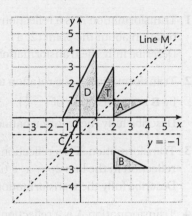

e T maps onto B by a clockwise rotation of 90° about centre $(0, -1)$.

f C maps onto D by an enlargement of scale factor 2 about centre $(-1, -4)$.

Exercise 115

Questions 1–8 refer to this diagram. For each one, describe the three transformations, and
comment on what you notice.

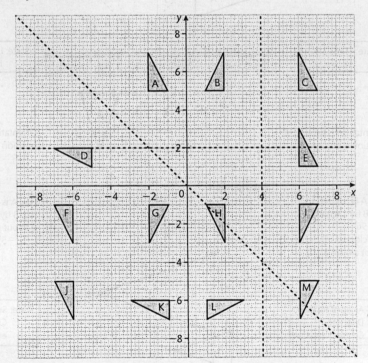

1 C to I, I to E and C to E

2 I to H, H to G and I to G

3 B to A, A to G and B to G

4 I to H, H to B and I to B

5 K to L, L to I and K to I

6 B to A, A to D and B to D

7 F to G, G to A and F to A

8 E to M, M to J and E to J

1 Transformation **A** is a reflection in the line $x = 2$. Transformation **A**, followed by **B** is a 90° rotation about the point $(2, 4)$.
Describe fully the transformation **B**. What single transformation is produced by **B**, then **A**?

2 Transformation **A** is a rotation about the origin through $-90°$. Transformation **A**, followed by **B** is a translation of $\begin{pmatrix} -4 \\ 2 \end{pmatrix}$.
Describe fully the transformation **B**. What single transformation is produced by **B**, then **A**?

3 Transformation **A** is a reflection in the line $x + y = 0$. Transformation **A**, followed by **B** is a reflection in the line $2y = x + 6$.
Describe fully the transformation **B**. What single transformation is produced by **B**, then **A**?

4 Transformation **A** is a reflection in the line $x + y = 0$. Transformation **A**, followed by **B** is a translation of $\begin{pmatrix} 8 \\ 8 \end{pmatrix}$.
Describe fully the transformation **B**. What single transformation is produced by **B**, then **A**?

Remember

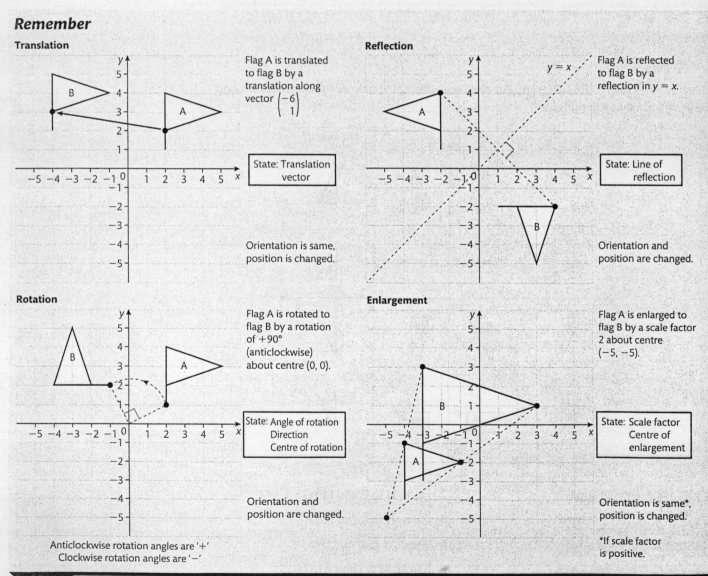

Translation

Flag A is translated to flag B by a translation along vector $\begin{pmatrix} -6 \\ 1 \end{pmatrix}$

State: Translation vector

Orientation is same, position is changed.

Reflection

Flag A is reflected to flag B by a reflection in $y = x$.

State: Line of reflection

Orientation and position are changed.

Rotation

Flag A is rotated to flag B by a rotation of $+90°$ (anticlockwise) about centre $(0, 0)$.

State: Angle of rotation
Direction
Centre of rotation

Orientation and position are changed.

Anticlockwise rotation angles are '+'
Clockwise rotation angles are '−'

Enlargement

Flag A is enlarged to flag B by a scale factor 2 about centre $(-5, -5)$.

State: Scale factor
Centre of enlargement

Orientation is same*, position is changed.

*If scale factor is positive.

Exercise 116 (Revision)

1 Find the image of point P(3, 4) after it has been
   a reflected in the x-axis
   b reflected in the y-axis
   c rotated about (0, 0) by 90° in a clockwise direction
   d translated along vector $\begin{pmatrix} 7 \\ -6 \end{pmatrix}$.

2 Find the image of point Q(−3, 5) after it has been
   a reflected in y = 0
   b reflected in x = 0
   c rotated about (0, 0) by +90°
   d translated along vector $\begin{pmatrix} -5 \\ 3 \end{pmatrix}$.

3 Draw the triangle ABC, where A is the point (1, 2), B is the point (1, 6) and C is the point (8, 2) on a set of axes where −8 ⩽ x ⩽ 16 and −8 ⩽ y ⩽ 10. Find the image of ABC after it has been
   a reflected in the x-axis
   b rotated about (0, 0) by 90° in an anticlockwise direction
   c translated along vector $\begin{pmatrix} -5 \\ 4 \end{pmatrix}$
   d enlarged by a scale factor of 2 about a centre of enlargement (0, 4).

4

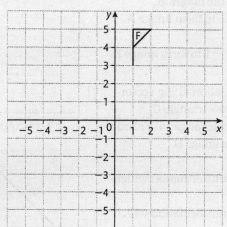

   a Reflect the flag F in the y-axis and label the image A.
   b Reflect A in the line y = x and label its image B.
   c Describe fully the single transformation which takes F to B.
   d Reflect A in the line x = 2; label this image C.
   e Describe the single transformation which takes F to C.

5 The image of a point P(x, y) is at point Q(4, 8) **after** P has undergone the following transformations in the order:
   Reflection in the x-axis.
   Rotation by 90° in a clockwise direction about (0, 0).
   Translation along vector $\begin{pmatrix} 3 \\ -3 \end{pmatrix}$.
   Find the value of x and y.

Exercise 116* (Revision)

1 The image of triangle ABC is at J(2, 3), K(2, 5), L(6, 3) **after** it has undergone the following transformations in the order:
   Reflection in the y-axis.
   Rotation by 90° in an anticlockwise direction about (0, 0).
   Translation along vector $\begin{pmatrix} -3 \\ 4 \end{pmatrix}$.
   Find the coordinates of triangle ABC.

2 a Triangle T undergoes the following transformations:

   (i) Reflection in the line $y = x$ after which the image is called A. Draw A.

   (ii) A 180° rotation about (1, 1) after which the image is called B. Draw B.

b C is the image of B after a 180° rotation about (0, 2) after which the image is called C. Draw C.

c Describe the single transformation that maps triangle A to triangle C.

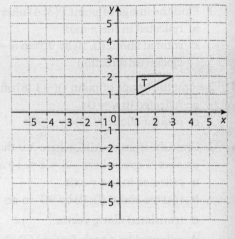

3 Triangle T is shown in the diagram.

a T is reflected in line L to form image A. Draw triangle A.

b A is reflected in $y = -1$ to form image B. Draw triangle B.

c T undergoes a translation along vector $\begin{pmatrix} -2 \\ -3 \end{pmatrix}$ to form image C. Draw triangle C.

d T is enlarged by a scale factor of 2 about centre (3, 2) to form image D. Draw triangle D.

e Describe fully the transformation which maps T to B.

f Describe fully the transformation which maps C to D.

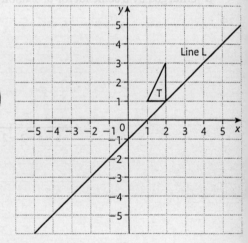

4 Transformation **A** is a reflection in the line $y = x$.

Transformation **B** is a rotation of 90° about the origin.

Transformation **C** is a translation along the vector $\begin{pmatrix} -2 \\ 2 \end{pmatrix}$.

Draw x- and y-axes, and plot an object near the origin.

Describe fully the single transformation which results from these combined transformations.

a A followed by **B**   **b B** followed by **A**   **c A** followed by **C**

d C followed by **A**   **e B** followed by **C**   **f C** followed by **B**

5 Draw x- and y-axes from -8 to 8.

Plot triangle P with vertices at (1, 2), (1, 4) and (2, 4).

Do these transformations, and label your images A–E, respectively.

a Rotate P by $-90°$ about (2, 0).

b Translate P along the vector $\begin{pmatrix} 4 \\ 2 \end{pmatrix}$.

c Reflect P in the line $x + y = 0$.

d Enlarge P from the point (5, 2) with scale factor 2.

e Enlarge P from the point (2, 0) with scale factor $\frac{1}{2}$.

Histograms

Histograms look similar to bar charts, but ...

♦ A bar chart measures *frequency* on the vertical axis.

♦ A histogram measures *frequency density* on the vertical axis.

♦ In a bar chart frequency is proportional to the *height* of the bar.

♦ **In a histogram frequency is proportional to the *area* of the bar.**

When data are divided into groups of *different* sizes, a histogram, rather than a bar chart, should be used to display the distribution.

A bar chart is only suitable when continuous data are divided into groups of the **same** size.

In a histogram, the **area** of each 'bar' is proportional to frequency.

Length × breadth
= (frequency density
  × width of group)
= frequency

$$\text{Frequency density} = \frac{\text{frequency}}{\text{width of group}}$$

Example 1

Ella records the length of 40 phone calls.
The results are shown in this table.

Length, t (min)	Frequency, f
$0 \leqslant t < 1$	2
$1 \leqslant t < 2$	5
$2 \leqslant t < 3$	8
$3 \leqslant t < 4$	9
$4 \leqslant t < 5$	7
$5 \leqslant t < 6$	4
$6 \leqslant t < 7$	3
$7 \leqslant t < 8$	2

Show the results on a bar chart.

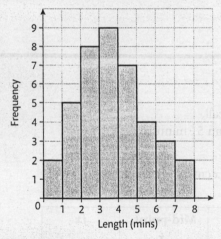

Ella decides to group the same results as shown in the following table.
Show the results on a bar chart and on a histogram.

The frequency densities are worked out in a calculation table. Four columns are needed.

Length, t (min)	Frequency, f	Width	Frequency density
$0 \leqslant t < 3$	15	3	$15 \div 3 = 5$
$3 \leqslant t < 4$	9	1	$9 \div 1 = 9$
$4 \leqslant t < 8$	16	4	$16 \div 4 = 4$

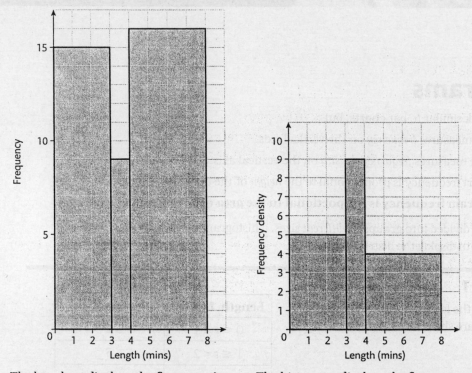

The bar chart displays the **frequencies**.   The histogram displays the **frequency densities**.

The bar chart with groups of different widths gives a very misleading impression of the original data.

The histogram gives a good impression of the original data though some of the fine detail has been lost by the grouping.

Calculation tables

Example 2

Use Ella's *grouped* data from Example 1 to

a Calculate an estimate of the mean length of a phone call.

b Estimate the number of phone calls that are longer than $5\frac{1}{2}$ minutes

c Calculate an estimate of the median.

a Extend the calculation table to include the mid-point (x cm) of each group and fx for each group.

Length, t mins	Frequency, f	Width	Frequency density	Mid-point, x	fx
$0 \leqslant t < 3$	15	3	$15 \div 3 = 5$	1.5	22.5
$3 \leqslant t < 4$	9	1	$9 \div 1 = 9$	3.5	31.5
$4 \leqslant t < 8$	16	4	$16 \div 4 = 4$	6	96
	$n = 40$				$\sum fx = 150$

Estimate of the mean $= \dfrac{\sum fx}{n} = \dfrac{150}{40} = 3.75$ mins

b The shaded area on the histogram represents the calls that are longer than $5\frac{1}{2}$ minutes. This area is $4 \times 2.5 = 10$. As the area represents frequency, an estimate of the number of phone calls over $5\frac{1}{2}$ minutes is 10.

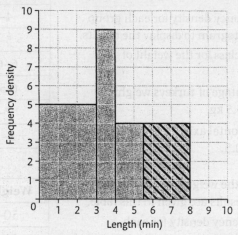

Note the original ungrouped data suggests that only seven calls were longer than $5\frac{1}{2}$ minutes, showing that accuracy is lost when data is grouped.

c The median cuts the total area of the histogram in half.
This is at $3\frac{5}{9}$ mins so an estimate of the median is $3\frac{5}{9}$ mins.

Exercise 117

1 A survey revealed these results for the time spent on homework for a Friday night by a group of 60 school children.

Calculate the frequency density for each group and construct a histogram to display the results.

Use these scales: horizontal axis, 1 cm = 10 min; vertical axis, 1 cm = 0.1.

Time, t (min)	Frequency of children, f
$0 \leqslant t < 30$	6
$30 \leqslant t < 60$	12
$60 \leqslant t < 80$	18
$80 \leqslant t < 100$	12
$100 \leqslant t < 120$	9
$120 \leqslant t < 180$	3

2 The ages (in completed years) of the teaching staff at a school are given in this table.

Calculate the frequency density for each group and construct a histogram to display the results.

Age, x (years)	Number of teachers, f
$22 \leqslant x < 25$	12
$25 \leqslant x < 30$	10
$30 \leqslant x < 35$	14
$35 \leqslant x < 40$	15
$40 \leqslant x < 45$	13
$45 \leqslant x < 50$	9
$50 \leqslant x < 60$	17

Use these scales: horizontal axis, 2 cm = 5 years; vertical axis, 1 cm = 0.5.

3 A turkey farmer produced 89 turkeys for the
Christmas market. Their weights are given in
this table.

a Calculate the frequency density for each group
and construct a histogram to display the results.

b What is the modal class for the weight of
the turkeys?

c Estimate the percentage of turkeys weighing
between 5 kg and 7.5 kg.

Use these scales: horizontal axis, 1 cm = 1 kg;
vertical axis, 1 cm = 0.5.

Weight (kg)	Frequency, f
2–4	7
4–5	7
5–6	10
6–6.5	12
6.5–7	19
7–8	16
8–10	18

4 A fruit farmer checks the weights of 100 apples for
quality control. The results are given in this table.

a Calculate the frequency density for each
group and construct a histogram to
display the results.

b What is the modal class for the
weight of the apples?

c Estimate the percentage of
apples weighing between 75 g
and 100 g.

d Calculate an estimate of the median.

Use these scales: horizontal axis, 1 cm = 10 g;
vertical axis, 1 cm = 0.2.

Weight (g)	Frequency, f
50–80	24
80–90	12
90–100	17
100–105	14
105–110	11
110–120	13
120–140	9

5 The ages (in completed years) of women giving
birth in a local hospital during January 2000
are given in this table.

a Calculate the frequency density for each
group in the table.

b Draw a histogram to illustrate the results.

c Calculate an estimate of the mean age of
the mothers.

Use these scales: horizontal axis, 1 cm = 2 years;
vertical axis, 1 cm = 1.

Age (years)	Frequency, f
14–16	7
16–20	38
20–25	60
25–30	68
30–35	52
35–45	25

6 The race times of cross-country runners are shown
in this table.

a Calculate the frequency density for each
group in the table.

b Draw a histogram to illustrate the results.

c Calculate an estimate of the mean time,
correct to the nearest second.

d Calculate an estimate of the median.

Use these scales: horizontal axis, 1 cm = 1 min;
vertical axis, 1 cm = 1.

Time, t (min)	Frequency, f
11–12	3
12–14	9
14–16	22
16–18	25
18–21	15
21–24	6

Exercise 117*

1 Fifty responses were received from a survey of French camp sites close to the Atlantic Coast. The 'size' of a camp site was defined by the number of mobile homes on it, and the results are shown in this table.

Size of camp site	Number of sites
0–100	4
100–200	7
200–350	13
350–500	17
500–750	6
750–1000	3

  a Draw a histogram of the data.
  b Estimate the percentage of sites with between 250 and 500 mobile homes.
  c Calculate an estimate of the mean number of mobile homes per site.
  d Calculate an estimate of the median.

2 This frequency table shows the distribution of ages in completed years for a cinema audience.

Age (years)	Frequency
18–20	24
20–25	42
25–30	24
30–40	16
40–50	30
50–60	30
60–80	34

  a Draw a histogram of the data.
  b Estimate the percentage of the audience aged between 35 and 55.
  c Calculate an estimate of the mean age of the audience.

3 The ages of children in completed years attending a summer camp are given in this table.

Age (years)	Number
3–5	30
6–7	26
8–9	30
10	15
11	13
12–14	21
15–17	21

  a Calculate the frequency density for each group in the table.
  b Calculate an estimate of the mean age of the children.
  c Given that the height of the first bar of the histogram is 5 cm, calculate the heights of the other bars in the histogram.

4 The ages (in completed years) of the members of a health and fitness club are shown in this table.

Age (years)	Number
18–19	30
20–24	57
25–29	69
30–34	42
35–39	36
40–49	36
50–59	30

  a Calculate the frequency density for each group in the table.
  b Calculate an estimate of the mean age of the membership.
  c Given that the height of the first bar is 5 cm, calculate the heights of the other bars.
  d Calculate an estimate of the median.

5 This table and the unfinished histogram represent the playing times of a sample of video films.

Playing time (min)	Frequency, f
60–80	
80–90	
90–95	13
95–100	18
100–110	17
110–120	6
120–150	3

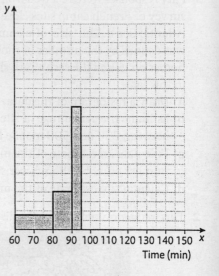

a Use the histogram to find the missing frequencies.

b Copy and complete the histogram, using the same scales and clearly labelling the vertical axis.

c Calculate the mean playing time of the videos.

6 A sample of batteries were tested by being continuously used to power a toy train. This table and the unfinished histogram represent the times it took for the train to stop moving.

Time (hours)	Frequency, f
4–5	10
5–5.5	9
5.5–6	16
6–6.5	18
6.5–7	7
7–8	
8–10	

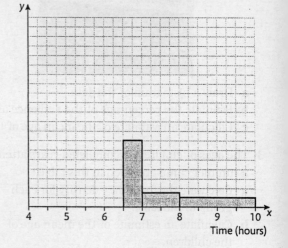

a Use the histogram to find the missing frequencies.

b Copy and complete the histogram, using the same scales and clearly labelling the vertical axis.

c Calculate the mean lifetime of the batteries.

Exercise 118 (Revision)

1 In an experiment the lengths of 80 daisy stalks were measured. The results are shown in the table.

a Construct a histogram for these results.

b Calculate an estimate of the number of daisies in this group that have a stalk length between 4 and 8 mm.

c Calculate an estimate of the mean length of the daisy stalks.

Length, l (mm)	Frequency, f
$0 < l \leqslant 5$	5
$5 < l \leqslant 10$	10
$10 < l \leqslant 20$	22
$20 < l \leqslant 30$	25
$30 < l \leqslant 50$	18

2 40 people were asked how long it takes them to travel to work. The results are shown in the table.

Time, t (mins)	Frequency, f
$15 < t \leqslant 20$	12
$20 < t \leqslant 30$	12
$30 < t \leqslant 40$	10
$40 < t \leqslant 70$	6

a Construct a histogram for these results

b Calculate an estimate of the number of people who took more than 60 minutes to travel to work.

c Calculate an estimate of the number of people who took less than 24 minutes to travel to work.

d Calculate an estimate of the median.

3 A farmer checks the masses of a sample of apples for quality control. The unfinished table and histogram show the results.

Mass, m (g)	Frequency, f
$60 < m \leqslant 70$	
$70 < m \leqslant 80$	22
$80 < m \leqslant 100$	40
$100 < m \leqslant 120$	20
$120 < m \leqslant 160$	

a Use the histogram to complete the table.

b Use the table to complete the histogram.

c Calculate an estimate of the number of apples with mass between 75 and 95 grams.

4 The unfinished table and histogram give the waiting times of 100 patients at a doctor's surgery. The number of patients waiting 10–12 minutes is 8 more than the number waiting 0–6 minutes.

Time t mins	Frequency
$0 < t \leqslant 6$	
$6 < t \leqslant 10$	26
$10 < t \leqslant 12$	
$12 < t \leqslant 14$	18
$14 < t \leqslant 22$	24

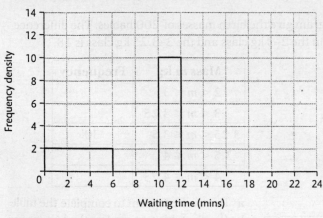

a Use the histogram to complete the table.

b Use the table to complete the histogram.

c Calculate an estimate of the percentage of patients who wait between 8 and 16 minutes.

d Calculate an estimate of the median waiting time.

Exercise 118* (Revision)

1 In an experiment the lengths of 100 cats' whiskers were measured. The results are shown in the table.

   a Construct a histogram for these results.

   b Calculate an estimate of the number of whiskers in this group that have a length between 5.5 and 8.5 mm.

   c Calculate an estimate of the mean length of the whiskers.

Length, l (mm)	Frequency, f
$5 < l \leqslant 6$	18
$6 < l \leqslant 6.5$	15
$6.5 < l \leqslant 7$	21
$7 < l \leqslant 8$	26
$8 < l \leqslant 10$	20

2 The table gives the times of 50 cyclists in a race.

   a Construct a histogram of these results.

   b Calculate an estimate of the percentage of cyclists in this group that had a time of less than 48 minutes.

   c Calculate an estimate for the median time.

   d Within which group does the median lie?

Time, t (mins)	Frequency, f
$20 < t \leqslant 30$	5
$30 < t \leqslant 35$	4
$35 < t \leqslant 40$	9
$40 < t \leqslant 50$	22
$50 < t \leqslant 70$	7
$70 < t \leqslant 100$	3

3 A retailer checks the lifetimes of a sample of 200 Christmas tree lights for quality control. The unfinished table and histogram show the results.

Life, t (hrs)	Frequency, f
$60 < t \leqslant 80$	
$80 < t \leqslant 90$	26
$90 < t \leqslant 95$	20
$95 < t \leqslant 100$	25
$100 < t \leqslant 115$	
$115 < t \leqslant x$	51

   a Use the histogram to complete the table.

   b Calculate the value of x.

   c Use the table to complete the histogram.

4 The unfinished table and histogram give the birth masses of 200 babies. The difference between the number of babies in the 2–3 kg class and the 3–3.25 kg class is 18.

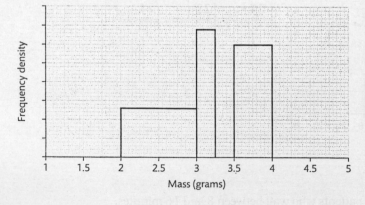

Mass m kg	Frequency
$2 < m \leqslant 3$	
$3 < m \leqslant 3.25$	
$3.25 < m \leqslant 3.5$	30
$3.5 < m \leqslant 4$	
$4 < m \leqslant 4.75$	

   a Use the histogram to complete the table.

   b Use the table to complete the histogram.

   c Calculate the probability that the birth mass is between 2.5 and 4.5 kg.

   d Calculate an estimate of the median mass.

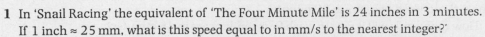

1 In 'Snail Racing' the equivalent of 'The Four Minute Mile' is 24 inches in 3 minutes.
   If 1 inch ≈ 25 mm, what is this speed equal to in mm/s to the nearest integer?
   A 3   **B** 4   **C** 8   **D** 200

2 The factors of $2x^2 + 5x - 3$ are
   A $(2x + 3)(x - 1)$ **B** $(2x - 3)(x + 1)$ **C** $(2x + 1)(x - 3)$ **D** $(2x - 1)(x + 3)$

3 The solutions to the equation $(x - 10)(2x + 1) = 0$ occur where the curve
   $y = (x - 10)(2x + 1)$ cuts the x-axis at
   A $x = -10, x = 1$ **B** $x = -10, x = \frac{1}{2}$ **C** $x = 10, x = -\frac{1}{2}$ **D** $x = -10, x = \frac{1}{2}$

4 Point P(1, 10) is transformed onto point Q by a reflection in the x-axis followed by a
   translation along vector $\binom{1}{10}$. Point Q is

   A (0, 20)   **B** (11, 9)   **C** (2, 0)   **D** (0, 2)

5 The value which divides a histogram in half is called the
   A mean   **B** mid-point   **C** mode   **D** median

6 The exact fraction which equal to $0.2\dot{3}$ is
   A $\frac{23}{100}$ **B** $\frac{23}{10}$ **C** $\frac{23}{99}$ **D** $\frac{7}{30}$

7 The volume of the box is

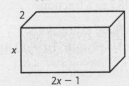

   A $2x^2 - 2x$ **B** $4x^2 - 2x$ **C** $4x - 2x^2$ **D** $4x^2 + 2x$

8 The graph of $y = x^2 - 2x + 1$ has been drawn. To solve the equation $x^2 - 3x + 4 = 0$,
   which line should be drawn?
   A $y = -3x + 4$ **B** $y = 3 - x$ **C** $y = x + 3$ **D** $y = x - 3$

9 Point P(a, b) is transformed on to point Q by a reflection in the x-axis. Point Q is
   A (a, $-b$)   **B** ($-a$, b)   **C** ($-a$, $-b$)   **D** (a, b)

10 An estimate of the mean of the following data is:

Time (s)	$0 < s \leqslant 10$	$10 < s \leqslant 20$	$20 < s \leqslant 40$
Frequency	4	10	6

   A 34   **B** 350   **C** 11   **D** 17.5

1 One girl can wash two cars in 90 minutes.
   a How many girls will it take to wash four similar cars in 45 minutes?
   b How many similar cars can ten girls wash in three hours?

2 Two men in a food factory can check 120 tubs of humous for quality control in 5 minutes. Copy and complete the following table.

Number of men m	Number of humous tubs h	Time t (min)
2	120	5
4		10
	1080	15
m	h	
	h	t
m		t

3 Convert the following into fractions.
   a $0.\dot{5}$ **b** $0.0\dot{3}\dot{7}$ **c** $0.5\dot{8}2\dot{1}$ **d** $1.05\dot{4}5\dot{6}$

4 Express $0.\dot{7}8\dot{9} \div 0.\dot{9}8\dot{7}$ as a fraction.

5 Expand and simplify these expressions.
   a $(x + 2)(x + 5)$ **b** $(y + 11)(y - 9)$ **c** $(3 - p)^2$

6 Expand and simplify these expressions.
   a $(5x - 3)(3x + 2)$ **b** $(a - 3b)^2$

7 Factorise these expressions.
   a $x^2 - 3x$ **b** $x^2 - 49$ **c** $x^2 + 3x + 2$ **d** $x^2 - 7x - 8$

8 Factorise these completely.
   a $x^2 - 10x + 21$ **b** $3p^4 - 12p^2$

9 Solve these quadratic equations by factorising.
   a $x^2 + 7x + 12 = 0$ **b** $x^2 - x - 6 = 0$
   c $x^2 - 3x = 0$ **d** $x^2 = 12x - 20$

10 **a** Use Pythagoras' theorem to form an equation in x for this right-angled triangle.

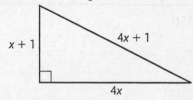

   b Solve your equation to find x.

11 The graph of $y = x^2 + 1$ has been drawn. What lines should be drawn on this graph to solve the following equations?
   a $x^2 = 5$ **b** $x^2 - 2x = 0$ **c** $x^2 + x - 1 = 0$

12 a On graph paper draw x- and y-axes such that $0 \leqslant x \leqslant 16$ and $0 \leqslant y \leqslant 16$.

b Draw and label the square A with corners at $(0, 2)$, $(2, 2)$, $(2, 4)$ and $(0, 4)$.

c Draw and label the square B with corners at $(10, 12)$, $(12, 12)$, $(12, 14)$ and $(10, 14)$.

d What translation transforms A to B?

A reflection in the line L transforms A to B.

e Draw and label the line L on your diagram. Give the equation of the line L.

A rotation with centre $(6, 8)$ transforms A to B.

f State the angle of this rotation.

A 90° clockwise rotation with centre C transforms A to B.

g What are the coordinates of C? Label the point C on your diagram.

An enlargement with centre $(0, 0)$ and scale factor 3 transforms A to E, then another enlargement transforms E to B.

h Draw the square E on your diagram.

i Find the coordinates of the centre, and the scale factor, of the second enlargement.

13 A health farm recorded the milk consumption, m ml, of its 180 clients one day. The table and histogram show the results.

Consumption, m (ml)	Frequency, f
$0 < m \leqslant 50$	9
$50 < m \leqslant 100$	42
$100 < m \leqslant 125$	30
$125 < m \leqslant 150$	
$150 < m \leqslant 200$	
$200 < m \leqslant 300$	22
$300 < m \leqslant 500$	

a Use the histogram to complete the table.

b Use the table to complete the histogram.

c Calculate an estimate of the median amount of milk consumed that day.

d Calculate an estimate of the percentage of clients who consumed between 70 and 170 ml of milk that day.

e Calculate an estimate of the mean amount of milk consumed that day.

Financial arithmetic

Wages and salaries

A **wage** is money paid weekly at a certain rate for a set number of hours per week. Overtime is hours worked above the normal weekly hours and is usually paid at a higher rate.

A **salary** is a fixed annual sum of money, usually paid each month.

(Note: 'per annum' means per year and is often written as p.a.)

Example 1

Freya's job as a waitress pays:

Normal time at $8.50 per hour.

Overtime at $1.5 \times \$8.50$ per hour.

Time (h)	Mon	Tue	Wed	Thu	Fri	Sat
Normal	8	8	4	8	8	
Overtime			1			8

Freya's wage = Normal time pay + Overtime pay

$= 36 \times \$8.50 \qquad + 1.5 \times 9 \times \$8.50$

$= \$420.75$

Example 2

Amin was offered two jobs as a mechanic in a local garage.

'Exhausts 4U' paid $26 208 p.a.

'Tyres R Us' paid $27 456 p.a.

How much more does he earn per week at 'Tyres R Us'?

'Exhausts 4U' pay per week $= \dfrac{\$26\,208}{52} = \$504$

'Tyres R Us' pay per week $= \dfrac{\$27\,456}{52} = \$528$

Logan earns $24 per week ($528 − $504) more at 'Tyres R Us'.

Exercise 119

1 Calculate the weekly wage earned by Delonn as a chef at $7.50 per hour if he works a 36 hour week.

2 Calculate the weekly wage earned by Tandi as a gardener at $6.80 per hour if she works a 42 hour week.

3 A cinema pays Patrice $5.20 per hour and $1.5 \times$ normal rate for overtime.
   Calculate Patrice's weekly wage for his time sheet below:

Time (h)	Mon	Tue	Wed	Thu	Fri	Sat	Sun
Normal	6	6	4	6	6	2	
Overtime			4				6

4 A flower shop pays Lela $4.80 per hour and 1.25 × normal rate for overtime. Calculate Lela's weekly wage for her time sheet below:

Time (h)	Mon	Tue	Wed	Thu	Fri	Sat	Sun
Normal		8	8	8	4	8	
Overtime							6

5 Calculate the salary earned by Jes as a life guard working a 38 hour week at $8.40 per hour.

6 Calculate the salary earned by Nur as a secretary working a 36 hour week at $9.40 per hour.

7 Isa earns a salary of $32 500 p.a. If he works an average of 34 hours a week calculate his hourly rate of pay

8 Viera is the manageress of a fast-food café and earns a salary of $36 400 p.a. If she works on average 50 hours a week calculate her hourly rate of pay

Exercise 119*

1 Shan works in a factory and is paid $12.20 per hour. If he works more than 36 hours in a week he is paid overtime at 1.5 times the basic rate.
   a If he works a 42 hour week calculate his wage for that week.
   b The week after he earned $585.60. How many hours overtime did Shan work?

2 Asma works as a tour guide and is paid $14.50 per hour. If she works more than 32 hours in a week she is paid 1.25 times the basic rate.
   a If she works a 40 hour week calculate her wage for that week.
   b The week after she earned $681.50. How many hours overtime did Asma work?

3 Andre's time sheet for a week as a train-driver is below. If he earns a normal rate of x per hour and an overtime rate of $2x$ per hour, find x if Andre earns $432 that week.

Time (h)	Mon	Tue	Wed	Thu	Fri	Sat
Normal	7	7		7	3	
Overtime			4		3	5

4 'Saksuni' cars pays its workers normal time of x per hour and an overtime rate of $1.5x$ per hour. If Ruby earns $525 for the timesheet below, find how much Ruby is paid at normal rate.

Time (h)	Mon	Tue	Wed	Thu	Fri	Sat
Normal	8	8	4	8	8	
Overtime			2		4	5

5 A Premiership footballer earns a wage of $120 000 per week. Calculate
   a his annual salary
   b how much he earns per minute

6 A doctor earns a salary of $120 000 p.a. Calculate
   a her weekly wage
   b how much she earns per minute.

7 Which salary would you prefer?
   A $28 000 p.a. with annual increases of $500.
   B $24 000 p.a. with annual increases of 10% of the previous year's salary.

8 Which salary would you prefer?
   A $40 000 p.a. with annual increases of $1000.
   B $32 000 p.a. with annual increases of 12% of the previous year's salary.

Foreign currency

Inflation occurs when prices are rising and is usually given as an annual rate e.g. 3.5%.

Deflation occurs when prices fall. Both of these situations will cause the foreign exchange rate in a country to change. Typical exchange rates are shown below for US $1.

Country	Currency	Exchange rate to US $1
Australia	Dollar	AUD$1.11
Brazil	Real	R$1.83
China	Yuan	CNY6.83
France	Euro	€0.74
Jamaica	Dollar	JMD$26.50
Mexico	Peso	MXN12.86
UK	Pound	£0.64

Example 3

How many Mexican pesos will $150 buy?

$1 = MXN12.86

$150 = 150 × MXN12.86 = MXN1929

Example 4

How many US dollars will JMD$9275 buy?

$1 = JMD$26.50 (Divide both sides by 26.50)

$\dfrac{1}{26.50}$ = JMD$1

JMD$9275 = 9275 × $\dfrac{1}{26.50}$ = $350

Exercise 120

All questions in this exercise refer to the exchange rate table above.

1 Convert $250 into
   a Australian dollars   b Brazilian reais
   c Chinese yuan   d euros

2 Convert $1500 into
   a Jamaican dollars   b Mexican pesos
   c UK pounds   d euros

3 Convert the following into US dollars
   a AUD$333   b CNY6147

4 Convert the following into US dollars
   a JMD$2120   b £1600

5 How many US dollars is a Jamaican dollar millionaire worth?

6 How many US dollars is a Mexican peso millionaire worth?

7 An American house is advertised in Jamaica for JMD$13 250 000 and in Mexico for MXN6 558 600. Which is cheaper in US dollars and by how much?

8 An American house is advertised in the UK for £192 000 and in France for €214 600. Which is cheaper in US dollars and by how much?

Example 5

Convert 1500 Jamaican dollars into UK pounds.

$$\$1 \qquad = \text{JMD}\$26.50 \quad \text{(Divide both sides by 26.50)}$$

$$\$\frac{1}{26.50} = \text{JMD}\$1$$

$$\text{JMD}\$1500 = 1500 \times \$\frac{1}{26.50} = \$56.60$$

$$\$1 \qquad = £0.64$$

$$\$56.60 \qquad = 56.60 \times £0.64 \quad = £36.22$$

$$\text{JMD}\$1500 = £36.22$$

Exercise 120*

All questions in this exercise refer to the exchange rate table on page 228.

1 Convert JMD$1000 into
   a UK pounds   b euros
   c Australian dollars   d Brazilian reais

2 Convert CNY1000 into
   a Jamaican dollars   b UK pounds
   c Mexican pesos   d euros

3 How many Jamaican dollars is a UK pound millionaire worth?

4 How many Brazilian reais is an Australian dollar millionaire worth?

5 Deon has JMD$10 000 and travels through Mexico and Brazil. He wishes to convert all his money to the country's currency through which he is travelling in the ratio of 2 : 3 respectively. How much of each currency will he have?

6 Shakina has AUD$6000 and travels through China, France and the UK. She wishes to convert all of her money to the country's currency through which she is travelling in the ratio of 1 : 2 : 3 respectively. How much of each currency will she have?

7 Nelson bought a house for x two years ago. Due to an inflation rate of 5% p.a. this house is now worth JMD$4 674 600. Calculate Nelson's original purchase price x.

8 Daisy bought an antique sculpture for y three years ago. Due to an inflation rate of 4% p.a. this sculpture is now worth €12 486. Calculate the sculpture's original purchase price y.

Exercise 121 (Revision)

For Questions 1–6 use the following currency conversion table.

Country	Currency	Rate/US $1
Australia	Dollar	1.11
Brazil	Reais	1.83
China	Yuan	6.83
France	Euro	0.74
India	Rupee	46.11
Jamaica	Dollar	26.50
Malaysia	Ringitt	3.99
Mexico	Peso	12.86
Russia	Rouble	30.00
UK	Pound	0.63

1 Convert $500 into Australian dollars.

2 Convert 500 Australian dollars into British pounds.

3 Convert 1000 Jamiaican dollars into euros.

4 Convert 100 pesos into Jamaican dollars.

5 Bud has 10 000 American dollars and travels through Jamaica, Mexico and Brazil. He wishes to convert all his money to the country's currency through which he is travelling in the ratio of $1:1:3$ respectively. How much of each currency will he have?

6 How many British pounds is a Rupee millionnaire worth?

7 Hector employs a bricklayer who charges $20/h, a plumber who charges $30/h and a labourer who charges $15/h. He uses each of them for 40 hours. How much will they charge him given that sales tax will be added to his bill at 15%?

10 Gerard earns a salary of $45 000 p.a. Marcelle earns a basic rate of $20/h and works a 40 hour week with 5% of this time added at an overtime rate of $1.5 \times$ basic rate. Who earns more per week and by how much?

Exercise 121* (Revision)

For Questions 1–4 use the currency conversion table in Exercise 229.

1 Tim has £1000 and wishes to convert this amount into Malaysian Ringitts and Chinese Yuan in the ratio of $2:3$ respectively. How much of each currency will he have?

2 Winnie has 1200 Malaysian Ringitts and wishes to convert 30% into Indian Rupees and the rest into Chinese Yuan. How much of each currency will she have?

3 Karen has x Rubles which she converts into £1000. Find x.

4 Tom has y Yuan which he converts into £1000 000. Find x.

5 Ringo earns $15/hr as a Car Salesman for a 35 hour week. If he works overtime he gets $x/hr. He takes home $625 for a 40 hour week. Find y.

6 Trang earns $y/hr as a Dentist's nurse for a 36 hour week. If she works overtime she gets $20/hr. She takes home $584 for a 40 hour week. Find y.

7 Park earns a salary of $65 000 p.a. as an Accountant. This is increased after one year to $70 000 p.a. If he works 40 hours per week, how much does his salary increase by per hour after his increase?

8 Suyin earns a salary of $120 000 p.a. as a Doctor. This is increased after one year by 8%. If she works 50 hours per week, how much does her salary increase by per hour after her increase?

Proportion

If two quantities are related to each other, given enough information, it is possible to write a formula describing this relationship.

Activity 29

Copy and complete this table to show which paired items are related.

Variables	Related? Y/N
Area of a circle (A) and its radius (r)	
Circumference of a circle (C) and its diameter (d)	
Volume of water in a tank (V) and its weight (w)	
Distance travelled (D) at constant speed and time taken (t)	
Number of pages in a book (N) and its thickness (t)	
Mathematical ability (M) and a person's height (h)	
Wave height in the sea (W) and wind speed (s)	
Grill temperature (T) and time to toast bread (t)	

Direct proportion

Linear relationships

When water is poured into an empty fish tank, each litre poured in increases the depth by a fixed amount.

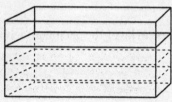

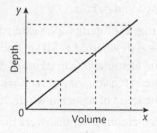

A graph of depth, y, against volume, x, is a straight line through the origin, showing a linear relationship.

In this case, y is **directly proportional** to x. If y is doubled, so is x. If y is halved, so is x, etc. This relationship can be expressed in *any* of these ways.

♦ y is directly proportional to x.
♦ y varies directly with x.
♦ y varies as x.

All these statements mean the same.

In symbols, direct proportion relationships can be written as $y \propto x$. The $\propto$ sign can then be replaced by '$= k$' to give $y = kx$, where k is the **constant of proportionality**. The graph of $y = kx$ is the equation of a straight line through the origin, with gradient k.

> **Key Point**
> y is directly proportional to x is written as $y \propto x$ and this means $y = kx$, for some fixed value k.
>
>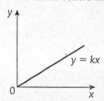

Example 1

The extension, y cm, of a spring is directly proportional to the mass, x kg, hanging from it.

If $y = 12$ cm when $x = 3$ kg, find

a the formula for y in terms of x

b the extension y cm when a 7 kg mass is attached

c the mass x kg that produces a 20 cm extension

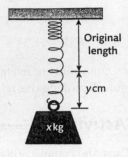

Original length

y cm

x kg

a y is proportional to x, so $y \propto x$ $y = kx$

     $y = 12$ when $x = 3$ $12 = k \times 3$

                                   $k = 4$

The formula is therefore $y = 4x$.

b When $x = 7$ $y = 4 \times 7$

The extension produced from a 7 kg mass is 28 cm.

c When $y = 20$ cm $20 = 4x$

                                   $x = 5$

The extension produced from a 5 kg mass is 20 cm.

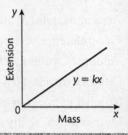

Extension | Mass | $y = kx$

Exercise 122

1 y is directly proportional to x. If $y = 10$ when $x = 2$, find

  a the formula for y in terms of x

  b y when $x = 6$

  c x when $y = 25$

2 d is directly proportional to t. If $d = 100$ when $t = 25$, find

  a the formula for d in terms of t

  b d when $t = 15$

  c t when $d = 180$

3 A bungee jumping rope's extension e m varies as the mass M kg of the person attached to it. If $e = 4$ m when $M = 80$ kg, find

  a the formula for e in terms of M

  b the extension for a person with a mass of 100 kg

  c the mass of a person when the extension is 6 m

4 An ice-cream seller discovers that, on any particular day, the number of sales (I) is directly proportional to the temperature ($t\,°C$). 1500 sales are made when the temperature is 20 °C. How many sales might be expected on a day with a temperature forecast of 26 °C?

5 The number of people in a swimming pool (N) varies as the daily temperature ($t\,°C$). 175 people swim when the temperature is 25 °C. The pool's capacity is 200 people. Will people have to queue and wait if the temperature reaches 30 °C?

Exercise 122*

1 The speed of a stone, v m/s, falling off a cliff is directly proportional to the time, t seconds, after release. Its speed is 4.9 m/s after 0.5 s.

  a Find the formula for v in terms of t.

  b What is the speed after 5 s?

  c At what time is the speed 24.5 m/s?

2 The cost, c cents, of a tin of salmon varies directly with its mass, m g. The cost of a 450 g tin is 150 cents.
   a Find the formula for c in terms of m.
   b How much does a 750 g tin cost?
   c What is the mass of a tin costing $2?

3 The distance a honey bee travels, d km, is directly proportional to the mass of honey, m g, it produces. A bee travels 150 000 km to produce 1 kg of honey.
   a Find the formula for d in terms of m.
   b What distance is travelled by a bee to produce 10 g of honey?
   c What mass of honey is produced by a bee travelling once around the world, a distance of 40 000 km?

4 The mass of sugar, m g, used in making oat-meal cookies varies directly as the number of cookies, n. 3.25 kg are used to make 500 cookies.
   a Find the formula for m in terms of n.
   b What mass of sugar is needed for 150 cookies?
   c How many cookies can be made using 10 kg of sugar?

5 The height of a tree, h m, varies directly with its age, y years. A 9 m tree is 6 years old.
   a Find the formula for h in terms of y.
   b What height is a tree that is 6 months old?
   c What is the age of a tree that is 50 cm tall?

Nonlinear relationships

Water is poured into an empty inverted cone. Each litre poured in will result in a different depth increase.

A graph of volume, y, against depth, x, will illustrate a direct **nonlinear relationship**.

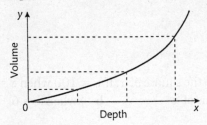

This relationship can be expressed in *either* of these ways.

♦ y is directly proportional to x cubed.
♦ y varies as x cubed.

Both these statements mean the same.

In symbols, this relationship is written as $y \propto x^3$. The $\propto$ sign can then be replaced by '$= k$' to give $y = kx^3$, where k is the constant of proportionality.

> **Key Point**
> 'y varies as x squared' is written as $y \propto x^2$ and this means $y = kx^2$, for some fixed value k.

Example 2

Express these relationships as equations with constants of proportionality.

a y is directly proportional to x squared. $y \propto x^2 \;\Rightarrow\; y = kx^2$

b m varies directly with the cube of n. $m \propto n^3 \;\Rightarrow\; m = kn^3$

c s is directly proportional to the square root of t. $s \propto \sqrt{t} \;\Rightarrow\; s = k\sqrt{t}$

d v squared varies as the cube of w. $v^2 \propto w^3 \;\Rightarrow\; v^2 = kw^3$

Example 3

The cost of Luciano's take-away pizzas in New York (C cents) is directly proportional to the square of the diameter (d cm) of the pizza. A 30 cm pizza costs 675 cents.

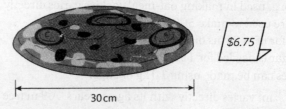

$\$6.75$

30 cm

a What is the price of a 20 cm pizza?

b What size of pizza should you expect for $4.50?

a C is proportional to d^2, so $C \propto d^2$ $C = kd^2$

$C = 675$ when $d = 30$ $675 = k(30)^2$

$k = 0.75$

The formula is therefore $C = 0.75d^2$.

When $d = 20$ $C = 0.75(20)^2$

$C = 300$

The cost of a 20 cm pizza is $3.

b When $C = 450$ $450 = 0.75d^2$

$d^2 = 600$

$d = \sqrt{600} = 24.5$ (3 s.f.)

A $4.50 pizza should be 24.5 cm in diameter.

Exercise 123

1 y is directly proportional to the square of x. If $y = 100$ when $x = 5$, find
  a the formula for y in terms of x **b** y when $x = 6$
  c x when $y = 64$

2 p varies directly as the square of q. If $p = 72$ when $q = 6$, find
  a the formula for p in terms of q **b** p when $q = 3$
  c q when $p = 98$

3 v is directly proportional to the cube of w. If $v = 16$ when $w = 2$, find
  a the formula for v in terms of w **b** v when $w = 3$
  c w when $v = 128$

4 m varies directly as the square root of n. If $m = 10$ when $n = 1$, find
  a the formula for m in terms of n **b** m when $n = 4$
  c n when $m = 50$

5 The distance fallen by a parachutist, y m, is directly proportional to the square of the time taken, t s. If 20 m are fallen in 2 s, find

   a the formula expressing y in terms of t **b** the distance fallen through in 3 s

   c the time taken to fall 100 m

6 'Espirit' perfume is available in bottles of different volumes of similar shapes. The price, $P, is directly proportional to the cube of the bottle height, h cm. A 10 cm high bottle is $50. Find

   a the formula for P in terms of h **b** the price of a 12 cm high bottle

   c the height of a bottle of 'Espirit' costing $25.60

Exercise 123*

1 If f is directly proportional to g^2, copy and complete this table.

g	2	4	
f	12		108

2 If m is directly proportional to n^3, copy and complete this table.

n	1		5
m	4	32	

3 The resistance to motion, R newtons, of the 'Storm' racing car is directly proportional to the square of its speed, s km/hour. When the car travels at 160 km/hour it experiences a 500 newton resistance.

   a Find the formula for R in terms of s.

   b What is the car's speed when it experiences a resistance of 250 newtons?

4 The height of giants, H metres, is directly proportional to the cube root of their age, y years. An 8-year-old giant is 3 m tall.

   a Find the formula for H in terms of y.   **b** What age is a 12 m tall giant?

5 The surface area of a sphere is directly proportional to the square of its radius. A sphere of radius 10 cm must be increased to a radius of x cm if its surface area is to be doubled. Find x.

6 The mass of spherical cannon balls is directly proportional to the cube of their diameter. A cannon ball of diameter 100 cm must be decreased to a diameter of y cm if its mass is to be halved. Find y.

Activity 30

The German astronomer Kepler (1571–1630) devised three astronomical laws.

Kepler's third law gives the relationship between the orbital period, t days, of a planet around the Sun, and its mean distance, d km, from the Sun.

In simple terms, this law states that t^2 is directly proportional to d^3.

- Find a formula relating t and d, given that the Earth is 150 million km from the Sun.
- Copy and complete this table.

Planet	d (million km)	Orbital period around Sun t (Earth days)
Mercury	57.9	
Jupiter		4315

- Try to find the values of d and t for other planets in the Solar System, and see if they fit the same relationship.

Key Point

If y is inversely proportional to x, the graph of y plotted against x looks like this.

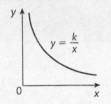

$$y = \frac{k}{x}$$

Inverse proportion

The temperature of a cup of coffee decreases as time increases.

A graph of temperature (T) against time (t) shows an **inverse relationship**.

This can be expressed as: 'T is inversely proportional to t.'
In symbols, this is written as

$$T \propto \frac{1}{t}$$

The $\propto$ sign can then be replaced by '$= k$', so

$$T = \frac{k}{t}$$

where k is the constant of proportionality.

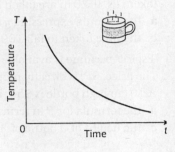

Example 4

Express these equations as relationships with constants of proportionality.

a y is inversely proportional to x squared. $\qquad y \propto \frac{1}{x^2} \quad \Rightarrow \quad y = \frac{k}{x^2}$

b m varies inversely as the cube of n. $\qquad m \propto \frac{1}{n^3} \quad \Rightarrow \quad m = \frac{k}{n^3}$

c s is inversely proportional to the square root of t. $\qquad s \propto \frac{1}{\sqrt{t}} \quad \Rightarrow \quad s = \frac{k}{\sqrt{t}}$

d v squared varies inversely as the cube of w. $\qquad v^2 \propto \frac{1}{w^3} \quad \Rightarrow \quad v^2 = \frac{k}{w^3}$

Example 5

Sound intensity, I dB (decibels), is inversely proportional to the square of the distance, d m, from the source. At a music festival, it is 110 dB, 3 m away from a loudspeaker.

a Find the formula relating I and d.

I is inversely proportional to d^2 $\qquad\qquad I = \frac{k}{d^2}$

$I = 110$ when $d = 3$ $\qquad\qquad 110 = \frac{k}{3^2}$

$\qquad\qquad\qquad\qquad\qquad\qquad\qquad k = 990$

The formula is therefore $I = \frac{990}{d^2}$.

b Calculate the sound intensity 2 m away form the speaker.

When $d = 2$ $\qquad\qquad\qquad\qquad\qquad I = \frac{990}{d^2} = 247.5$

The sound intensity is 247.5 dB, 2 m away (enough to cause deafness).

c At what distance away from the speakers is the sound intensity 50 dB?

When $I = 50$ $\qquad\qquad\qquad\qquad\qquad 50 = \frac{990}{d^2}$

$\qquad\qquad\qquad\qquad\qquad\qquad\qquad d^2 = 19.8$

$\qquad\qquad\qquad\qquad\qquad\qquad\qquad d = 4.45 \text{ (3 s.f.)}$

The sound intensity is 50 dB, 4.45 m away from the speakers.

Exercise 124

1 y is inversely proportional to x. If $y = 4$ when $x = 3$, find
  a the formula for y in terms of x
  b y when $x = 2$ **c** x when $y = 3$

2 d varies inversely with t. If $d = 10$ when $t = 25$, find
  a the formula for d in terms of t
  b d when $t = 2$ **c** t when $d = 50$

3 m varies inversely with the square of n. If $m = 4$ when $n = 3$, find
  a the formula for m in terms of n
  b m when $n = 2$ **c** n when $m = 1$

4 V varies inversely with the cube of w. If $V = 12.5$ when $w = 2$, find
  a the formula for V in terms of w
  b V when $w = 1$ **c** w when $V = 0.8$

5 Light intensity, I candle-power, from a lighthouse is inversely proportional to the square of the distance, d m, of an object from this light source. If $I = 10^5$ when $d = 2$ m, find
  a the formula for I in terms of d
  b the light intensity at 2 km

6 The life-expectancy, L days, of a cockroach varies inversely with the square of the density, d people/m^2, of the human population near its habitat. If $L = 100$ when $d = 0.05$, find
  a the formula for L in terms of d
  b the life-expectancy of a cockroach in an area where the human population density is 0.1 people/m^2

Exercise 124*

1 If a is inversely proportional to b^2, copy and complete this table.

b	2	5	
a	50		2

2 A scientist gathers this data.

t	1	4		10
r	20		4	2

  a Which of these relationships describes the collected data?

$$r \propto \frac{1}{\sqrt{t}} \qquad r \propto \frac{1}{t} \qquad r \propto \frac{1}{t^2}$$

  b Copy and complete the table.

3 The electrical resistance, R ohm, of a fixed length of wire is inversely proportional to the square of its radius, r mm. If $R = 0.5$ when $r = 2$, find
  a the formula for R in terms of r
  b the resistance of a wire of 3 mm radius

4 The cost of Mrs Janus's electricity bill, $\$C$, varies inversely with the average temperature, $t\,°C$, over the period of the bill. If the bill is $\$200$ when the temperature is $25\,°C$, find
  a the formula expressing C in terms of t
  b the bill when the temperature is $18\,°C$
  c the temperature generating a bill of $\$400$

5 The number of people shopping at Tang's Cornershop per day, N, varies inversely with the square root of the average outside temperature, $t\,°C$.

a Copy and complete this table.

Day	N	t
Mon	400	25
Tues		20
Wed	500	

b The remainder of the week (Thurs to Sat) has a hot spell with a constant daily average temperature of $30\,°C$. What is the average number of people per day who shop at Tang's for that week? (The shop is closed on Sundays.)

6 The time for a pendulum to swing, T s, is inversely proportional to the square root of the acceleration due to gravity, $g\,m/s^2$. On Earth $g = 9.8$, but on the Moon $g = 1.9$. Find the time of swing on the Moon of a pendulum whose time taken to swing on Earth is 2 s.

Activity 31

This graph shows an inverse relationship between the body mass, M kg, of mammals and their average heart pulse, P beats/min.

- Use the graph to complete this table.

	P (beats/min)	M (kg)
Hare		5
Dog	135	
Man		70
Horse	65	

- An unproven theory in biology states that the hearts of all mammals beat the same number of beats in an average life-span.

  - If man lives on average for 75 years, calculate the total number of heart beats in an average life-span.

  - Test out this theory by calculating the expected life-span of the creatures in the table above.

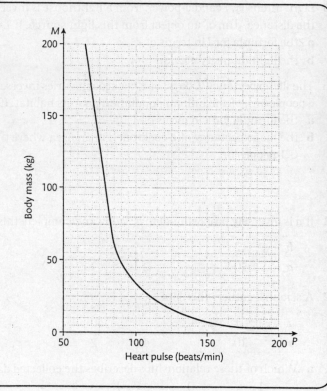

Exercise 125 (Revision)

1 y is directly proportional to x. If $y = 12$ when $x = 2$, find
   a the formula for y in terms of x **b** y when $x = 7$ **c** x when $y = 66$

2 p varies as the square of q. If $p = 20$ when $q = 2$, find
   a the formula for p in terms of q **b** p when $q = 10$ **c** q when $p = 605$

3 The cost, $\$c$, of laying floor tiles is directly proportional to the square of the area, $a\,m^2$, to be covered. If a $40\,m^2$ kitchen floor costs $\$1200$ to tile, find
   a the formula for c in terms of a
   b the cost of tiling a floor of area $30\,m^2$
   c the area of floor covered by these tiles costing $\$600$

4 The time taken, t hours, to make a set of 20 curtains is inversely proportional to the number of people, n, who work on them. One person would take 80 hours to finish the task. Copy and complete this table.

n	1	2	4	
t	80			10

5 a is directly proportional to b. Copy and complete the following table.

b	10	15		
a	200		600	

6 The vertical distance a stone falls from a cliff, d m, is directly proportional to the square of the time, t seconds. If the stone falls 45 metres in 3 seconds, find
   a the formula for d in terms of t **b** d when $t = 2$ seconds
   c t when $d = 90$ m.

7 The cost in C of a circular wedding cake is directly proportional to its diameter d cm. If a 40 cm diameter cake costs \$60, find

   a the formula for C in terms of d
   b the cost of a cake with a diameter of 650 mm
   c the diameter of an \$80 cake.

8 The price of an art book, \$$p$ varies directly as the square of the number of coloured pictures, n, that it contains. If a \$150 art book contains ten coloured pictures, find
   a the formula for p in terms of n
   b the price of an art book containing 12 coloured pictures
   c the number of coloured pictures contained in a \$600 art book.

Exercise 125* (Revision)

1 y squared varies as z cubed. If $y = 20$ when $z = 2$, find
   a the formula relating y to z **b** y when $z = 4$
   c z when $y = 100$

2 m is inversely proportional to the square root of n.
   If $m = 2.5 \times 10^7$ when $n = 1.25 \times 10^{-7}$, find
   a the formula for m in terms of n **b** m when $n = 7.5 \times 10^{-4}$
   c n when m is one million

3 The frequency of radio waves, f MHz, varies inversely as their wavelengths, μ metres. If Radio Cayman has $f = 99$ and $\mu = 3$, what is the wavelength of the BBC World Service on 198 kHz?

4 If y is inversely proportional to the nth power of x, copy and complete this table.

x	0.25	1	4	25
y		10	5	

Find the formula for y in terms of x.

5 a is inversely proportional to the cube root of b. Copy and complete the following table.

b	125	8	
a	2		10

6 The energy of a meteorite, e kilo-joules (kJ) is directly proportional to the square of its speed, v m/s. If a meteorite travelling at 10 m/s has energy of 50 kJ, find
  a the formula for e in terms of v **b** e when $v = 50$ km/h
  c v when the energy is 10^6 J

7 The population of a termite hill, n thousand, is inversely proportional to the square of its age, t years. If a termite hill has a population of one million after six months, find
  a the formula for n in terms of t **b** n after two years
  c t when the population is one thousand.

8 Tidal waves (tsunamis) are the result of earthquakes in the sea bed. Their speed, v m/s, is directly proportional to the square root of the ocean depth, d m. If a tsunami travels at 9.8 m/s at an ocean depth of 10 m, find
  a the formula for v in terms of d
  b v when   **(i)** $d = 50$ m   **(ii)** $d = 1$ km
  c d when $v = 1$ m/s.
  d the ocean depth at the point where the fastest ever tsunami was recorded at 790 km/h.

Relations

Sets are often related to each other. The relationship can be shown in many ways:

Example 1

a A headed table

Child	Father
Nathan	Kavi
Puteri	Calvin
Maisie	Elton

b A mapping diagram using sets

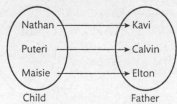

c A table

-2	0
-1	1
0	2
1	3

d Ordered pairs

$(-2, 0), (-1, 1), (0, 2), (1, 3)$

e A mapping diagram using number lines

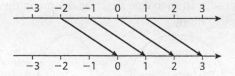

f A graph

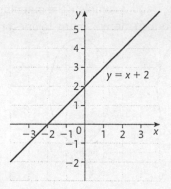

g A box diagram

h An algebraic relationship

$y = x + 2$

a and **b** represent the relationship between two sets of people.

c, d, e, f, g and **h** represent the same relationship between sets of numbers. In each case one set is two more than the other set. However, in cases **c, d** and **e** the sets are just a few integers, while in cases **f, g** and **h** the sets are any number you can think of. The relationships in Example 1 are called 'one to one' because there is only one connection between the members of each set.

Sometimes the relationship can be more complex.

Example 2

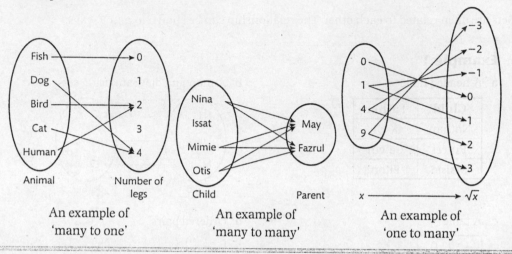

| An example of 'many to one' | An example of 'many to many' | An example of 'one to many' |

Complex relationships are usually shown more clearly on mapping diagrams.

Exercise 126

For Questions 1 and 2 complete the mapping diagrams. In each case describe the mapping as 'one to one', 'one to many', 'many to one' or 'many to many'.

1

2

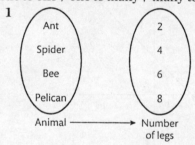

For Questions 3–6 copy this mapping diagram

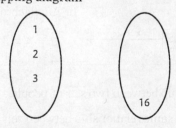

and use the given relationship to complete the mapping diagram.

3 $x \to x + 3$ **4** $x \to x - 1$ **5** $x \to 3x + 1$ **6** $x \to 2x - 2$

In Questions 7–10 a relationship is given by ordered pairs. Find a and b and describe the relationship in words.

7 $(0, 3), (1, 4), (2, a), (3, 6), (b, 7)$ **8** $(2, a), (3, 8), (4, 9), (b, 10), (6, 11)$

9 $(4, 0), (5, a), (7, 3), (b, 5), (11, 7)$ **10** $(6, 4), (8, 6), (a, 9), (12, 10), (15, b)$

In Questions 11–14 use the graphs to find a, b, c, and d in the ordered pairs $(-1, a)$, $(0, b,)$, $(1, c)$ and $(2, d)$.

11 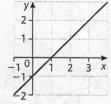 **12** 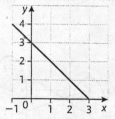 **13** 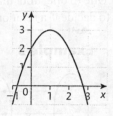 **14**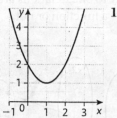

Exercise 126*

For Questions 1 and 2 complete the mapping diagrams. In each case describe the mapping as 'one to one', 'one to many', 'many to one' or 'many to many'.

1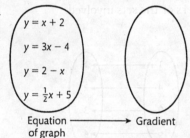

$y = x + 2$

$y = 3x - 4$

$y = 2 - x$

$y = \frac{1}{2}x + 5$

Equation of graph ⟶ Gradient

2

$\frac{1}{2}(x - 1) = 4$

$x^2 - 9 = 0$

$x^2 + 9 = 0$

$3x + 5 = 11$

Equation ⟶ Number of solutions

For Questions 3–6 copy this mapping diagram

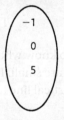

-1

0

5

7

and use the given relationship to complete the mapping diagram.

3 $x \rightarrow \frac{1}{2}(x - 1)$ **4** $x \rightarrow \frac{1}{3}(x + 1)$ **5** $x \rightarrow x^2 - 2$ **6** $x \rightarrow x^3 - 1$

In Questions 7–10 a relationship is given by ordered pairs. Find a and b and describe the relationship in the form $x \rightarrow$ ......

7 $(1, 2)$, $(2, a)$, $(3, 6)$, $(b, 10)$, $(8, 16)$ **8** $(-1, -3)$, $(1, 3)$, $(a, 9)$, $(5, 15)$, $(7, b)$

9 $(-6, -2)$, $(3, a)$, $(12, 4)$, $(b, 6)$, $(24, 8)$ **10** $(-8, -4)$, $(-4, a)$, $(b, 0)$, $(8, 4)$, $(14, 7)$

In Questions 11–14 use the graphs to find a, b, c, and d in the ordered pairs $(-1, a)$, $(0, b,)$, $(1, c)$ and $(2, d)$.

11 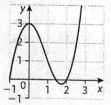 **12** 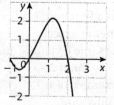 **13** 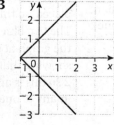 **14**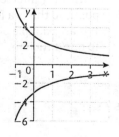

Functions

'One to one' and 'many to one' relations are called functions. A mapping diagram makes it **easy** to decide if a relation is a function or not.

Example 3

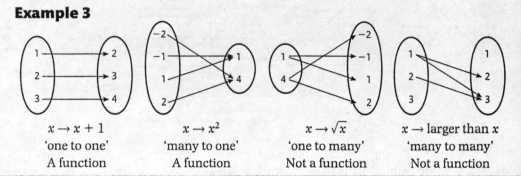

$x \rightarrow x + 1$
'one to one'
A function

$x \rightarrow x^2$
'many to one'
A function

$x \rightarrow \sqrt{x}$
'one to many'
Not a function

$x \rightarrow$ larger than x
'many to many'
Not a function

Whether a relation is a function can depend on the members of the sets involved.

Example 4

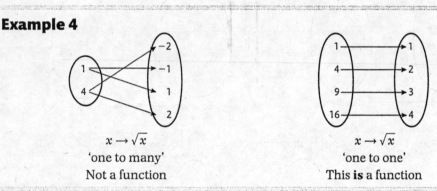

$x \rightarrow \sqrt{x}$
'one to many'
Not a function

$x \rightarrow \sqrt{x}$
'one to one'
This **is** a function

When deciding if a relation is a function you need to know what sets are involved.
A mapping diagram can be used if the sets involved are small. If the sets involved are infinite (for example the set of all real numbers) then the vertical line test on a graph is used.

Example 5

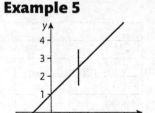

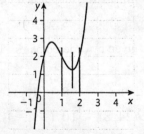

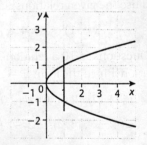

Wherever the red vertical line is placed on the graph, it will only intersect at one point, so this is a function. It is an example of a 'one to one' function.

The vertical lines only intersect the graph at one point wherever they are placed, so this is a function. The two blue lines intersect the graph at the same y value, showing that $1 \rightarrow 2$ and $2 \rightarrow 2$ so this is an example of a 'many to one' function.

The red vertical line intersects the graph at **two** points, showing, for example, that $1 \rightarrow 1$ and $1 \rightarrow -1$, so this is **not** a function.

Exercise 127

In Questions 1 and 2 state, giving a reason, whether the mapping shows a function or not.

1 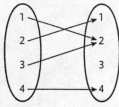 **2**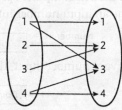

In questions 3–6 draw the graph for $-3 \leqslant x \leqslant 3$ and use the vertical line test to decide if it is a function or not.

3 $y = 3 - x$ **4** $y = x - 1$ **5** $y = x^2 + 1$ **6** $y = x^2 + x$

In questions 7–10 decide, giving a reason, if the relationship shown is a function or not.

7 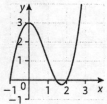 **8** 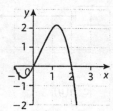 **9** 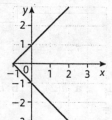 **10**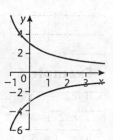

Exercise 127*

In Questions 1 and 2 complete the mapping for the sets shown (using your calculator if necessary). State, giving a reason, whether each mapping shows a function or not.

1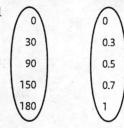

$$x \rightarrow \sin x°$$

2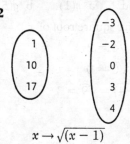

$$x \rightarrow \sqrt{(x - 1)}$$

In Questions 3–6 draw the graph for $-3 \leqslant x \leqslant 3$ and use the vertical line test to decide if it is a function or not.

3 $y = x^2 - x$ **4** $y = 4 - x^2$ **5** $y = +\sqrt{(x + 1)}$ **6** $y = \pm\sqrt{(x + 3)}$

In Questions 7 and 8 decide, giving a reason, if the relationship shown is a function or not.

7 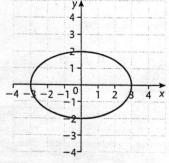 **8**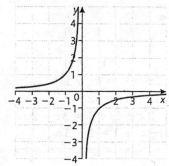

Function notation

A function is a set of rules for turning one number into another. Functions are very useful, for example, they are much used in computer spreadsheets. In effect a function is a computer, an imaginary box that turns an input number into an output number.

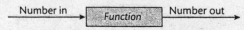

Number in → | Function | → Number out

If the function doubles every number input, then

2 → | Double input | → 4

A letter can be used to stand for the rule. If we call the doubling function f, then

5 → | f | → 10

f has operated on 5 to give 10, so we write $f(5) = 10$.

If x is input then $2x$ is output, so we write $f(x) = 2x$ or $f: x \rightarrow 2x$.

x → | f | → $2x$

Example 6

If $h: x \rightarrow 3x - 2$, find   **a** $h(2)$ **b** $h(\frac{1}{3})$ **c** $h(y)$

a $h(2) = 3 \times 2 - 2 = 4$

b $h(\frac{1}{3}) = 3 \times \frac{1}{3} - 2 = -1$

c $h(y) = 3y - 2$

Example 7

If $g(x) = \sqrt{5 - x}$, find   **a** $g(1)$ **b** $g(5)$ **c** $g(-4)$

($\sqrt{x}$ means the *positive* square root of x.)

a $g(1) = \sqrt{5 - 1} = \sqrt{4} = 2$

b $g(5) = \sqrt{5 - 5} = \sqrt{0} = 0$

c $g(-4) = \sqrt{5 - (-4)} = \sqrt{9} = 3$

Example 8

a If $f: x \rightarrow 2 + 3x$ and $f(x) = 8$, find x.

b If $g: x \rightarrow \dfrac{1}{x - 3}$ and $g(x) = \dfrac{1}{2}$, find x.

a $2 + 3x = 8 \Rightarrow 3x = 6 \Rightarrow x = 2$

b $\dfrac{1}{x - 3} = \dfrac{1}{2} \Rightarrow x - 3 = 2 \Rightarrow x = 5$

Example 9

If $f(x) = 3x + 1$, find:

a $f(2x)$ **b** $2f(x)$ **c** $f(x - 1)$ **d** $f(-x)$

a $f(2x) = 3(2x) + 1 = 6x + 1$

b $2f(x) = 2(3x + 1) = 6x + 2$

c $f(x - 1) = 3(x - 1) + 1 = 3x - 2$

d $f(-x) = 3(-x) + 1 = 1 - 3x$

Exercise 128

For Questions 1–4, calculate

a $f(2)$ **b** $f(-3)$ **c** $f(0.5)$ **d** $f(0)$

1 $f(x) = 2x + 1$ **2** $f: x \rightarrow 3x - 2$ **3** $f(x) = x^2 + 2x$ **4** $f: x \rightarrow x^3 + 1$

5 If $g(x) = \sqrt{x + 1}$, calculate **a** $g(3)$ **b** $g(-1)$ **c** $g(99)$

6 If $f: x \rightarrow \dfrac{1}{1 + 2x}$, calculate **a** $f(2)$ **b** $f(-1)$ **c** $f(a)$

7 If $h(x) = 2 - \dfrac{1}{x}$, calculate **a** $h(2)$ **b** $h(-2)$ **c** $h(y)$

8 If $f: x \rightarrow 2x + 2$ and $f(x) = 8$, find x.

9 If $p(x) = \dfrac{1}{x + 1}$ and $p(x) = \frac{1}{4}$, find x.

10 If $g(x) = \sqrt{5x + 1}$ and $g(x) = 4$, find x.

11 If $f: x \rightarrow 2x + 1$, find **a** $f(-x)$ **b** $f(x + 2)$ **c** $f(x) + 2$

12 If $f(x) = 4x - 3$, find **a** $f(x + 1)$ **b** $f(2x)$ **c** $2f(x)$

Exercise 128*

For Questions 1–4, calculate

a $f(-2)$ **b** $f(0.5)$ **c** $f(0)$ **d** $f(p)$

1 $f: x \rightarrow 8 - 2x$ **2** $f(x) = x^2 - 2x + 3$ **3** $f: x \rightarrow x(x + 2)$ **4** $f(x) = x^3 + 2$

5 If $g(x) = \dfrac{1}{3 - 2x}$, calculate **a** $g(3)$ **b** $g(-1)$ **c** $g(99)$

6 If $f: x \rightarrow \sqrt{x^2 + 2x}$, calculate **a** $f(2)$ **b** $f(-2)$ **c** $f(2a)$

7 If $h(x) = \dfrac{3x + 2}{x - 4}$, calculate **a** $h(2)$ **b** $h(-2)$ **c** $h(3y)$

8 If $f: x \rightarrow \dfrac{2}{3x + 1}$ and $f(x) = \frac{1}{2}$, find x.

9 If $p(x) = x^2 - x - 4$ and $p(x) = 2$, find x.

10 If $g(x) = \dfrac{2x + 17}{5 - x}$ and $g(x) = 4$, find x.

11 If $f: x \rightarrow 2 - x$, find **a** $f(-x)$ **b** $f(-2x)$ **c** $-2f(x)$

12 If $f(x) = x^2 + 1$, find **a** $f(x + 2)$ **b** $f(x) + 2$ **c** $f(-x)$

Domain and range

One way to picture a function is as a **mapping** from one set to another.

In this example the only numbers the function can use are from the set $\{1, 2, 4, 7\}$.
This set is called **domain** of the function.

The set $\{3, 4, 6, 9\}$ produced by the function is called the **range** of the function.

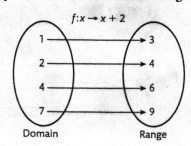

If the set on the right contains numbers with no arrows going to them, then this set is called the **co-domain**. The range is a subset of the co-domain.

The co-domain is {3, 4, 6, 9, 11, 13}.
The range is still {3, 4, 6, 9}.

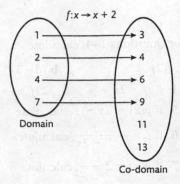

Example 10

Find the range of the function $f: x \rightarrow 2x + 1$ if the domain is $\{-1, 0, 1, 2\}$.

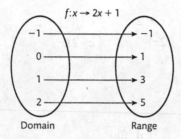

The diagram shows that the range is {−1, 1, 3, 5}.

Example 11

Find the range of the function $g(x) = x + 2$ if the domain is $\{x: x \geqslant -2, x \text{ is an integer}\}$.

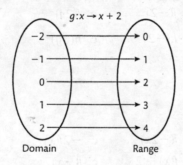

The diagram shows some of the domain. As the function changes an integer into another integer, the range is $\{y: y \geqslant 0, y \text{ is an integer}\}$.

Example 12

If $f(x) = 2x + 5$ has a domain of $-2 \leqslant x \leqslant 2$ find the range of $f(x)$.

The graph shows the range of $f(x)$ is $1 \leqslant f(x) \leqslant 9$

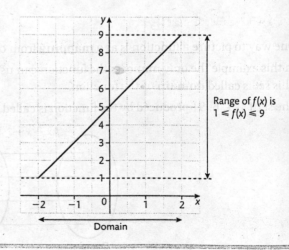

Range of $f(x)$ is $1 \leqslant f(x) \leqslant 9$

Example 13

Find the range of:

a $h: x \rightarrow x^2$ **b** $f: x \rightarrow x^2 + 2$

if the domain of both functions is all real numbers.

a When a positive number is squared, the answer is positive.
When zero is squared, the answer is zero.
When a negative number is squared, the answer is positive.
The function cannot produce negative numbers.
So the range is $\{y: y \geqslant 0, y$ a real number$\}$.

b Since $x^2 \geqslant 0$, then $x^2 + 2 \geqslant 2$, so the range is $\{y: y \geqslant 2, y$ a real number$\}$.

The graph of the function gives a useful picture of the domain and range. The domain corresponds to the x-axis, and the range to the y-axis. The graphs of the functions used in Example 13 are shown below.

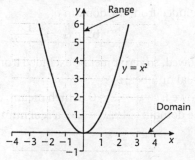

The first graph ($y = x^2$) shows that all the y values are greater than or equal to zero; that is, the range is $\{y: y \geqslant 0, y$ a real number$\}$.

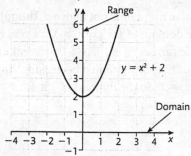

The second graph ($y = x^2 + 2$) shows that all the y values are greater than or equal to 2; that is, the range is $\{y: y \geqslant 2, y$ a real number$\}$.

Exercise 129

For each function, find the range of the given domains.

	Function	Domain for part a	Domain for part b
1	x	$0 \leqslant x \leqslant 2$	$-2 \leqslant x \leqslant 2$
2	$x + 1$	$0 \leqslant x \leqslant 3$	$-3 \leqslant x \leqslant 3$
3	$x - 1$	$0 \leqslant x \leqslant 1$	$-1 \leqslant x \leqslant 1$
4	$4 - x$	$0 \leqslant x \leqslant 4$	$-4 \leqslant x \leqslant 4$
5	$\frac{1}{2}x + \frac{1}{2}$	$-2 \leqslant x \leqslant 2$	$-10 \leqslant x \leqslant 10$

Exercise 129*

For each function, find the range of the given domains.

	Function	Domain for part a	Domain for part b
1	$2 - 3x$	$\{-2, -1, 0, 1\}$	All real numbers
2	$x^2 + 2x$	$\{-2, 0, 2, 4\}$	$\{x: x \geq 0, x \text{ a real number}\}$
3	$(x - 1)^2 + 2$	$\{-4, -2, 0, 2\}$	All real numbers
4	$x^3 + x$	$\{-2, 0, 2, 4\}$	$\{x: x \geq 1, x \text{ a real number}\}$
5	$\dfrac{1}{x + 1}$	$\{0, 1, 2, 3\}$	$\{x: x \geq 0, x \text{ a real number}\}$

Sometimes there are numbers which cannot be used for the domain as they lead to impossible operations, usually division by zero or taking the square root of a negative number.

Example 14

Which numbers must be excluded from the domain of

a $f(x) = \dfrac{1}{x}$ **b** $g(x) = \dfrac{1}{x - 2}$?

a Division by zero is not allowed, so zero must be excluded from the domain of f. The domain is $\{x: x \neq 0, x \text{ a real number}\}$.

b Division by zero is not allowed, so $x - 2 \neq 0$, which means $x = 2$ must be excluded from the domain of g. The domain is $\{x: x \neq 2, x \text{ a real number}\}$.

Example 15

State the domain and range of these functions:

a $f(x) = \sqrt{x}$ **b** $g(x) = 1 + \sqrt{x - 2}$

a The square root of a negative number is not allowed, though it is possible to take the square root of zero. So the domain of f is $\{x: x \geq 0, x \text{ a real number}\}$. Since $\sqrt{\ }$ means the *positive* square root, the range of f is $\{y: y \geq 0, y \text{ a real number}\}$.

b The number under the square root sign must be greater than or equal to zero, so $x - 2 \geq 0$, which means $x \geq 2$. So the domain of g is $\{x: x \geq 2, x \text{ a real number}\}$. The range is $\{y: y \geq 1, y \text{ a real number}\}$.

Exercise 130

State which values (if any) cannot be included in the domain of the following functions.

1 $f: x \to \dfrac{1}{x + 1}$ **2** $g: x \to \dfrac{1}{x - 1}$ **3** $h: x \to \sqrt{x - 2}$ **4** $f: x \to \sqrt{2 - x}$

5 $g(x) = x - \dfrac{1}{x}$ **6** $h(x) = x + \dfrac{1}{x^2}$ **7** $p(x) = x^2 + 3$ **8** $q(x) = 5x - 1$

9 $r: x \to \sqrt{x^2 - 4}$ **10** $s(x) = \sqrt{9 - x^2}$

Exercise 130*

State which values (if any) cannot be included in the domain of the following functions.

1 $h(x) = \dfrac{5}{2x - 1}$ **2** $g(x) = \dfrac{3}{4x - 3}$ **3** $f: x \to \sqrt{9 - x}$ **4** $h: x \to \sqrt{x + 4}$

5 $p(x) = \dfrac{1}{(x + 1)^2}$ **6** $q(x) = \dfrac{1}{(1 - x)^2}$ **7** $r: x \to \dfrac{1}{x^2 - 1}$ **8** $s: x \to \dfrac{1}{x^2 + 1}$

9 $f(x) = \dfrac{1}{\sqrt{x + 2}}$ **10** $g(x) = \dfrac{1}{\sqrt{2 - x}}$

Composite functions

When one function is followed by another, the result is a composite function.

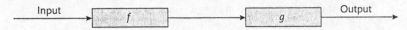

If $f: x \to 2x$ and $g: x \to x + 3$, then

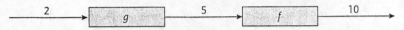

If the order of these functions is changed, then the output is different:

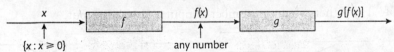

If x is input, then:

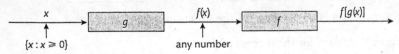

$g[f(x)]$ is usually written without the square brackets as $gf(x)$

Example 16

If $f(x) = x^2$ and $g(x) = x + 2$, find:

a $fg(3)$ **b** $gf(3)$ **c** $fg(x)$ **d** $gf(x)$

a $g(3) = 5$, so $fg(3) = f(5) = 25$
b $f(3) = 9$, so $gf(3) = g(9) = 11$
c $g(x) = x + 2$, so $fg(x) = f(x + 2) = (x + 2)^2$
d $f(x) = x^2$, so $gf(x) = g(x^2) = x^2 + 2$

Example 17

If $f(x) = x - 2$ and $g(x) = \sqrt{3x}$, what is the domain of:

a $gf(x)$ **b** $fg(x)$?

a The domain of $g(x)$ is $\{x: x \geqslant 0, x \text{ a real number}\}$, so the range of $f(x)$ is $\{x: x \geqslant 0, x \text{ a real number}\}$. This means the domain is of f is $\{x: x \geqslant 2, x \text{ a real number}\}$.

Alternative solution:

$gf(x) = g(x - 2) = \sqrt{3(x - 2)}$ which means the domain is $\{x: x \geqslant 2, x \text{ a real number}\}$.

b The domain of $f(x)$ is any real number, so the range of $g(x)$ is any real number. This means the domain is of g is $\{x: x \geqslant 0, x \text{ a real number}\}$.

Alternative solution:

$fg(x) = f(\sqrt{3x}) = \sqrt{3x} - 2$, which means the domain is $\{x: x \geqslant 0, x \text{ a real number}\}$.

Key Point

$gf(x)$ means do f **first**, followed by g.

The domain of g is the range of f.

In the same way, $fg(x)$ means do g **first**, followed by f.

$fg(x)$ and $gf(x)$ are not necessarily the same.

Exercise 131

1 Find $fg(3)$ and $gf(3)$ if $f(x) = x + 5$ and $g(x) = x - 2$.

2 Find $fg(1)$ and $gf(1)$ if $f(x) = x^2$ and $g(x) = x + 2$.

3 Find $fg(4)$ and $gf(4)$ if $f(x) = \dfrac{1}{x}$ and $g(x) = \dfrac{1}{x + 1}$.

For Questions 4–7, find   **a** $fg(x)$ **b** $gf(x)$ **c** $ff(x)$ **d** $gg(x)$

4 $f(x) = x - 4$, $g(x) = x + 3$ **5** $f(x) = 2x$, $g(x) = x + 2$

6 $f(x) = x^2$, $g(x) = x + 2$ **7** $f(x) = x - 6$, $g(x) = x + 6$

8 $f(x) = \dfrac{x}{2}$ and $g(x) = x + 1$. Find x if   **a** $fg(x) = 4$ **b** $gf(x) = 4$

Exercise 131*

1 Find $fg(-3)$ and $gf(-3)$ if $f(x) = 2x + 3$ and $g(x) = 5 - x$.

2 Find $fg(2)$ and $gf(2)$ if $f(x) = x^2 + 1$ and $g(x) = (x + 1)^2$.

3 Find $fg(-3)$ and $gf(-3)$ if $f(x) = x + \dfrac{2}{x}$ and $g(x) = \dfrac{2}{x - 1}$.

For Questions 4–7, find   **a** $fg(x)$ **b** $gf(x)$ **c** $ff(x)$ **d** $gg(x)$

4 $f(x) = \dfrac{x - 4}{2}$, $g(x) = 2x$ **5** $f(x) = 2x^2$, $g(x) = x - 2$

6 $f(x) = \dfrac{1}{x - 2}$, $g(x) = x + 2$ **7** $f(x) = 4x$, $g(x) = \sqrt{\dfrac{x}{4} + 4}$

8 $f(x) = 1 + \dfrac{x}{2}$ and $g(x) = 4x + 1$. Find x if   **a** $fg(x) = 4$ **b** $gf(x) = 4$

9 $f(x) = 1 + x^2$ and $g(x) = \dfrac{1}{x - 5}$. What is the domain of   **a** $fg(x)$ **b** $gf(x)$?

10 $f(x) = \sqrt{2x + 4}$ and $g(x) = 4x + 2$. What is the domain of   **a** $fg(x)$ **b** $gf(x)$?

Inverse functions

Consider the functions $f: x \rightarrow x + 1$ and $g: x - 1$.

If f is followed by g, then whatever number is input is also the output.

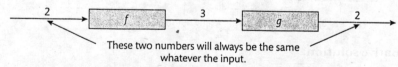

These two numbers will always be the same whatever the input.

If x is the input, then x is also the output.

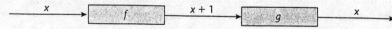

> **Key Point**
> The inverse of f is the function that undoes whatever f has done.

The function g is called the **inverse** of the function f.

The notation f^{-1} is used for the inverse of f.

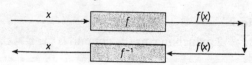

Graphically $f^{-1}(x)$ is a reflection of $f(x)$ in $y = x$

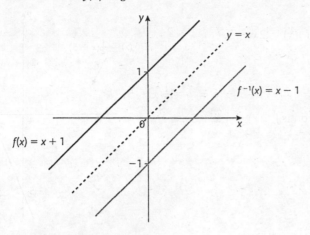

A function must be one to one to have an inverse.

A many to one function does not have an inverse as we cannot undo the function.

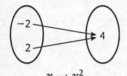

$x \rightarrow x^2$

A many to one function

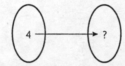

No inverse as 4 could
map to either -2 or $+2$

Activity 32

a Using the functions f and g defined above, show that f is the inverse of g.

b If $f(x) = 2x$, show that $g(x) = \dfrac{x}{2}$ is the inverse of f. Is f also the inverse of g?

c If $f(x) = 4 - x$, show that f is the inverse of f. (Functions like this are called 'self inverse'.)

Example 18

$f(x) = \sqrt{3x - 1}$

Find   **a** $f^{-1}f(4)$ **b** $ff^{-1}(4)$

a As f^{-1} undoes what f has done, $f^{-1}f(4) = 4$.

b Similarly, f undoes what f^{-1} has done, so $ff^{-1} = 4$.

Note: There was no need to find the inverse function.

Finding the inverse function

If the inverse function is not obvious, the following steps will find it.

Step 1   Write the function as $y = ...$

Step 2   Change any x to y and any y to x.

Step 3   Make y the subject, giving the inverse function.

Example 19

Find the inverse of $f(x) = 2x - 5$.

Step 1 $y = 2x - 5$
Step 2 $x = 2y - 5$
Step 3 $x = 2y - 5 \Rightarrow 2y = x + 5 \Rightarrow y = \frac{1}{2}(x + 5)$

The inverse function is $f^{-1}(x) = \frac{1}{2}(x + 5)$.

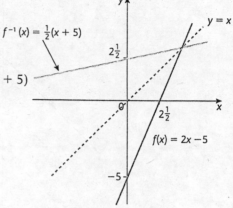

Example 20

Find the inverse of $g(x) = 2 + \frac{3}{x}$.

Step 1 $y = 2 + \frac{3}{x}$

Step 2 $x = 2 + \frac{3}{y}$

Step 3 $x - 2 = \frac{3}{y} \Rightarrow y = \frac{3}{x - 2}$ so $g^{-1}(x) = \frac{3}{x - 2}$

check: $g(1) = 2 + 3 = 5$

$g^{-1}(5) = \frac{3}{5 - 2} = 1$

Exercise 132

1 If $f(x) = 3x - 4$, find $f^{-1}f(7)$.   **2** If $f(x) = x^2 - 4x + 1$, find $ff^{-1}(4)$.

For Questions 3–8, find the inverse of the function given.

3 $f(x) = 6x + 4$ **4** $f(x) = 9 - \frac{1}{3}x$ **5** $f(x) = 3(x - 6)$

6 $g(x) = \frac{1}{3x + 4}$ **7** $p(x) = 4 - \frac{3}{x}$ **8** $f(x) = x^2 + 7$

9 If $f(x) = 2x - 5$, find   **a** $f^{-1}(3)$ **b** $f^{-1}(0)$ **c** $f^{-1}(-3)$

10 $f(x) = 2x + 5$. Solve the equation $f(x) = f^{-1}(x)$.

Exercise 132*

1 If $f(x) = 7(2x + 3)$, find $f^{-1}f(17)$.   **2** If $f(x) = (x - 7)^2 + 3x$, find $ff^{-1}(3.2)$.

For Questions 3–8, find the inverse of the function given.

3 $f(x) = 8(4 - 3x)$ **4** $f(x) = \frac{3}{4 - 2x}$ **5** $f(x) = 4 - \frac{7}{x}$

6 $g(x) = \sqrt{x^2 + 7}$ **7** $p(x) = 2x^2 + 16$ **8** $r(x) = \frac{2x + 3}{4 - x}$

9 If $f(x) = x^2 - 5$, find   **a** $f^{-1}(11)$ **b** $f^{-1}(44)$ **c** $f^{-1}(-5)$

10 $f(x) = 2(4x - 7)$. Solve the equation $f(x) = f^{-1}(x)$.

11 $f(x) = 3 - \frac{2}{x}$. Solve the equation $f(x) = f^{-1}(x)$.

Exercise 133 (Revision)

1 A relationship is given by ordered pairs. Find x and y and describe the relationship in words.

   a $(10, 9), (9, 8), (8, y), (7, 6), (x, 5)$ **b** $(0, y), (1, 2), (2, 4), (x, 8), (5, 10)$

2 Draw a mapping diagram for $x \to 2x - 1$ with domain $\{-1, 0, 1, 2\}$. What is the range?

3 Draw arrows on four copies of this diagram

   to show the following relationships.

   a One to one   **b** One to many   **c** Many to one   **d** Many to many

4 Complete the following mapping diagram.

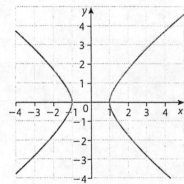

   Is this mapping a function? Give a reason. Would your conclusions change if the words changed?

5 Which graph shows a function? Give a reason.

a 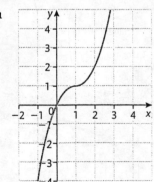   **b**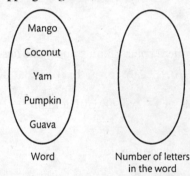

6 If $g: x \to 3x + 7$, calculate   **a** $g(2)$ **b** $g(-3)$ **c** $g(0)$

7 If $g(x) = 3 - 4x$, calculate x if   **a** $g(x) = 5$ **b** $g(x) = -2$

8 If f: $x \to 5x - 2$, find   **a** $f(x) + 1$ **b** $f(x + 1)$

9 State which values of x cannot be included in the domain of:

   a $f(x) = \dfrac{1}{x - 1}$ **b** $g(x) \to \dfrac{3}{2x - 1}$ **c** $h(x) = \sqrt{x + 1}$ **d** $p: x \to \sqrt{3x - 6}$

10 Find the range of the following functions if the domain is all real numbers.

   a $f(x) = 2x + 1$ **b** $g(x) = x^2 + 1$ **c** $h(x) = (x + 1)^2$ **d** $f(x) = x^3$

11 $f(x) = x^2 + 1$ and $g(x) = \dfrac{1}{x}$.

    a Find   **(i)** $fg(x)$ **(ii)** $gf(x)$

    b What values should be excluded from the domain of: **(i)** $fg(x)$ **(ii)** $gf(x)$?

    c Find and simplify $gg(x)$.

12 Find the inverse of

    a $p: x \rightarrow 4(2x + 3)$ **b** $q(x) = 7 - x$ **c** $r: x \rightarrow \dfrac{1}{x + 3}$ **d** $s(x) = x^2 + 4$

13 $f(x) = 4x - 3$ and $g(x) = \dfrac{x + 3}{4}$.

    a Find the function $fg(x)$.

    b Hence describe the relationship between the functions f and g.

    c Write down the exact value of $fg(\sqrt{3})$.

Exercise 133* (Revision)

1 A relationship is given by ordered pairs. Find x and y and describe the relationship in words.

    a $(1, 1), (2, y), (3, 9), (x, 16), (5, 25)$ **b** $(0, 1), (1, 3), (x, 5), (3, 7), (4, y)$

2 Draw a mapping diagram for $x \rightarrow 2x^2 - 4$ with domain $\{-2, -1, 0, 1, 2\}$. Is it a function? What is the range?

3 Using domain $\{0, 1, 2, 3\}$ and co-domain $\{4, 5, 6, 7\}$ draw mapping diagrams to show the following relationships:

    a One to one   **b** One to many   **c** Many to one   **d** Many to many

4 Complete the following mapping diagram.
Is this mapping a function? Give a reason.
Would your conclusions change if the
solids were different?

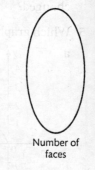

Tetrahedron

Cube

Octahedron

Triangular prism

Dodecahedron

Solid   Number of
faces

5 Which graph shows a function? Give a reason.

    a 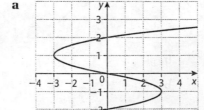   **b**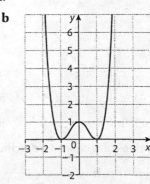

6 If $h: x \rightarrow \sqrt{x + 9}$, calculate   **a** $h(7)$ **b** $h(0)$ **c** $h(-9)$

7 If $g(x) = x^2 - 2x$, calculate x if   **a** $g(x) = 6$ **b** $g(x) = 56$

8 If $f: x \to 5 - 2x$, find   **a** $f(x) - 1$ **b** $f(x - 1)$

9 State which values of x cannot be included in the domain of:

    a $f(x) = \dfrac{5}{4 - 3x}$ **b** $g(x) \to \dfrac{7}{(x + 2)^2}$ **c** $h(x) = \sqrt{2 + 5x}$ **d** $p: x \to \sqrt{x^2 - 9}$

10 Find the range of the following functions if the domain is all real numbers.

    a $f(x) = 2x^2 + 3$ **b** $g(x) = (x - 2)^2$ **c** $h(x) = \sqrt{x + 2}$ **d** $f(x) = x^3 - 1$

11 $f(x) = x^3$ and $g(x) = \dfrac{1}{x - 8}$.

    a Find   **(i)** $fg(x)$ **(ii)** $gf(x)$

    b What values should be excluded from the domain of: **(i)** $fg(x)$ **(ii)** $gf(x)$?

    c Find and simplify $gg(x)$.

12 Find the inverse of

    a $p: x \to 4(1 - 2x)$ **b** $q(x) = 2 - \dfrac{3}{4 - x}$ **c** $r: x \to \sqrt{2x - 3}$ **d** $s(x) = (x - 2)^2$

13 $p(x) = \dfrac{1}{x - 2}$ and $q(x) = \dfrac{1}{x} + 2$.

    a Find the function $pq(x)$.

    b Hence describe the relationship between the functions p and q.

    c Write down the exact value of $pq(\sqrt{7})$.

Remember

Radii, tangent, chord

- AB is a chord
- The letter O will always indicate the centre of a circle
- △OAB is isosceles
- XY is a tangent to the circle at T
- ∠YTO = 90°

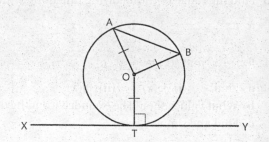

Angle at the centre is twice the angle at the circumference

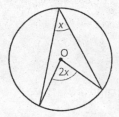
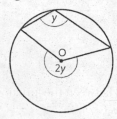
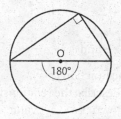

Opposite angles of a cyclic quadrilateral sum to 180°

$a° + b° = 180°$
$x° + y° = 180°$

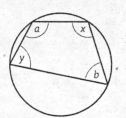

Angles in the same segment are equal

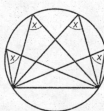

Activity 33

Use the diagram to prove that the angle at the centre is twice the angle at the circumference, and use this to prove that

- angles in the same segment are equal
- opposite angles of a cyclic quadrilateral sum to 180°.

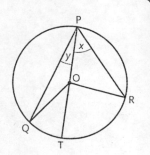

Example 1

Find ∠PNM.

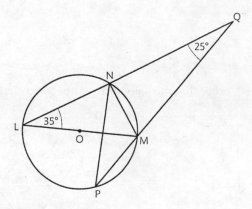

∠NPM = 35° (Angles in the same segment)

∠LNM = 90° (Angle at centre is 180°)

∠MNQ = 90° (△LMN is right-angled)

so ∠PNM = 30° (Angle sum of triangle PNQ)

Example 2

Prove that XY meets OZ at right angles.

∠OXY = 30° (Alternate to ∠ZYX)

∠OYX = 30° (△OXY is isosceles)

∠YZO = 60° (△YZO is isosceles)

∠ZNY = 90° (Angle sum of triangle ZNY)

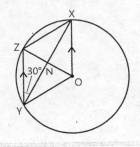

Exercise 134

For Questions 1–8, find the coloured angles, fully explaining your reasoning.

1

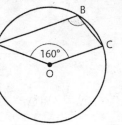

2

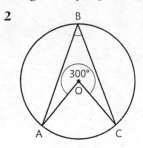

3

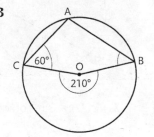

4

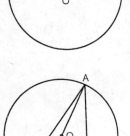

5

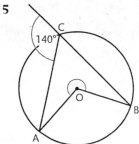

6

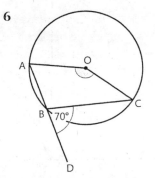

7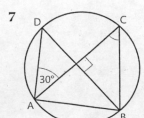

8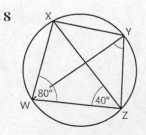

For Questions 9 and 10, prove that the points ABCD are concyclic.

9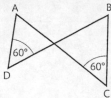

10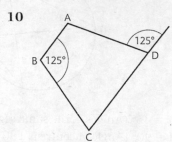

Exercise 134

For Questions 1–6, find the coloured angles, fully explaining your reasoning.

1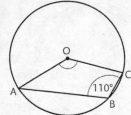

2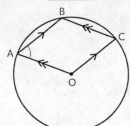

3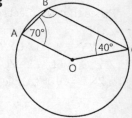

4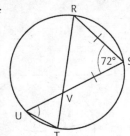

5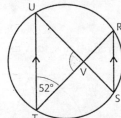

6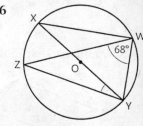

7 Find, in terms of x, $\angle$AOX.

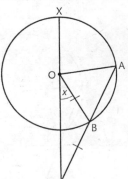

8 Prove that $\angle$CEA = $\angle$BDA.

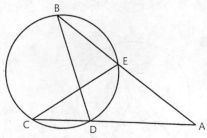

9 WXYZ is a cyclic quadrilateral. The sides XY and WZ produced meet at Q. The sides XW and YZ produced meet at P. ∠WPZ = 30° and ∠YQZ = 20°. Find the angles of the quadrilateral.

10 PQ and PR are any two chords of a circle, centre O. The diameter, perpendicular to PQ, cuts PR at X. Prove that the points Q, O, X and R are concyclic.

Angles in the alternate segment

Activity 34

Copy and complete the table one row at a time by calculating the sizes of the angles on each diagram.

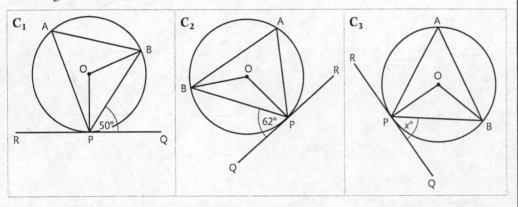

	QP̂B	OP̂B	OB̂P	BÔP	BÂP
C_1	50°				
C_2	62°				
C_3	$x°$				

Add reasons for your answers in circle C_3 to form a proof of the Alternate Segment Theorem.

Key Point

The angle between chord and tangent is equal to the angle in the alternate segment.

This is called the 'Alternate Segment Theorem'.

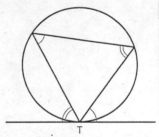

Exercise 135

In this exercise, O indicates the centre of a circle and T indicates a tangent to the circle.

For Questions 1–5, find the coloured angles, fully explaining your reasons.

1

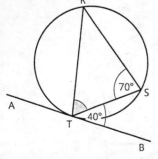

2

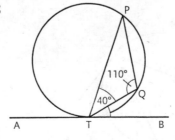

3

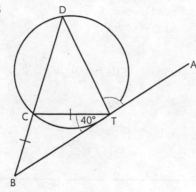

4

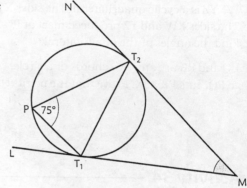

5

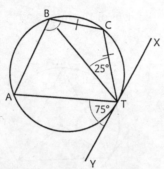

6 Find   **a** ∠OTX   **b** ∠TOB
            c ∠OBT   **d** ∠ATY

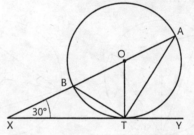

7 Copy and complete these two statements to prove
∠NPT = ∠PLT.

∠NTM = ... (Alternate segment)
∠PLT = ... (Corresponding angles)

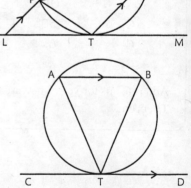

8 Copy and complete these two statements to
prove ∠ATC = ∠BTD.

∠ATC = ... (Alternate segment)

∠ABT = ∠BTD (.....)

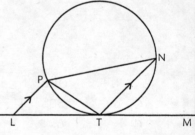

Exercise 135*

In this exercise, O indicates the centre of a circle and T indicates a tangent to the circle.
For Questions 1–2, find the coloured angles, fully explaining your reasons.

1

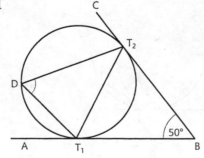

2

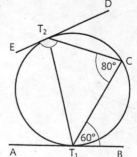

3 Prove that AB is the diameter.

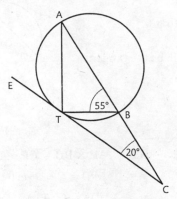

4 The inscribed circle of triangle XYZ touches XY at A, YZ at B and XZ at C. If ∠ZXY = 68° and ∠ZYX = 44°. find

   a ∠ABC   **b** ∠ACB

5 Find

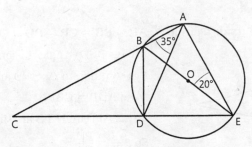

   a ∠DAE   **b** ∠BED

   c Prove that triangle ACD is isosceles.

6 CD and AB are tangents at T. Find ∠ETF.

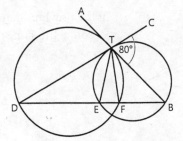

7

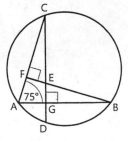

   a Explain why ∠ACG = ∠ABF = 15°.

   b Prove that the points CFGB are concyclic.

8 a Giving reasons, find, in terms of x, the angles EOC and CAE.

Use your answers to show that triangle ABE is isosceles.

b If BE = CE prove that BE will be the tangent to the larger circle at E.

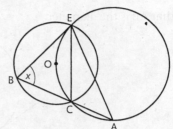

Intersecting chords theorem

Two chords intersecting inside a circle

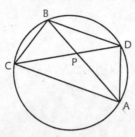

$\mathbf{AP \times PB = CP \times PD}$

$A\hat{P}D = B\hat{P}C$ (Vertically opposite angles)

$C\hat{D}A = C\hat{B}A$ (Angles in same segment; chord AC)

(And $B\hat{A}D = B\hat{C}D$)

So APD and CPB are similar.

So $\dfrac{AP}{CP} = \dfrac{PD}{PB}$

So $AP \times PB = CP \times PD$

Two chords intersecting outside a circle

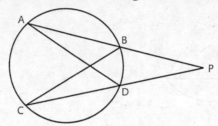

$\mathbf{AP \times PB = CP \times PD}$

$B\hat{A}D = B\hat{C}D$ (Angles in same segment; chord BD)

$\hat{P}$ is common to $\triangle$s APD and CPB

(And $A\hat{D}P = C\hat{B}P$)

So $\triangle$s APD and CPB are similar.

So $\dfrac{AP}{CP} = \dfrac{PD}{PB}$

So $AP \times PB = CP \times PD$

When one chord becomes a tangent

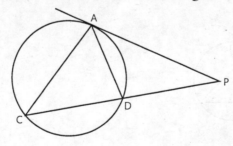

$\mathbf{AP^2 = CP \times PD}$

In the previous diagram, A and B 'become' the same point; so...

$AP \times PB$ becomes $AP \times PA = AP^2$

When both chords become tangents

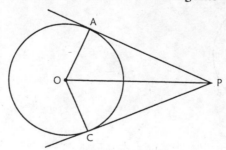

$\mathbf{AP = CP}$

Now, C and D 'become' the same point; so

$CP \times PD$ becomes $CP \times PC = CP^2$

$AP^2 = CP^2$

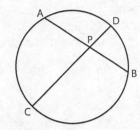

Example 3

AP = 12 cm, PD = 8 cm and CP = 9 cm. Find BP.

Let BP = x.
Now, AP × PB = CP × PD

so $\quad 12 \times x = 9 \times 8$

so $\qquad x = \dfrac{9 \times 8}{12}$

so $\qquad x = 6$ cm

Example 4

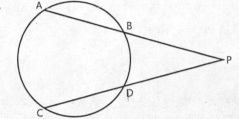

CD = 7 cm, DP = 5 cm and BP = 6 cm. Find AB.

Let AB = x.
Now, AP × PB = CP × PD

AP = AB + BP $= x + 6$
CP = CD + DP $= 12$
So $(x + 6) \times 6 = 12 \times 5$
So $\qquad x + 6 = 10$
So $\qquad\qquad x = 4$ cm

Exercise 136

1 AP = 15 cm, PB = 6 cm and DP = 4 cm. Find CP.

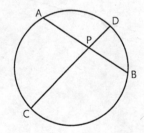

2 AB = 10 cm, PB = 3 cm and DP = 2 cm. Find CP.

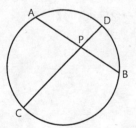

3 AP = 15 cm, BP = 6 cm and DP = 5 cm. Find CP.

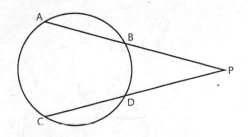

4 EX = 8 cm, HX = 6 cm and GH = 10 cm. Find EF.

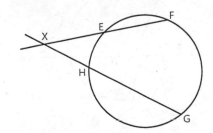

5 PT is a tangent to the circle, PS = 9 cm and SR = 7 cm. Calculate the length PT.

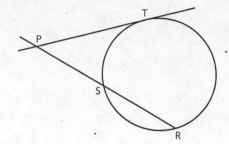

6 PQ is a tangent to the circle. PQ = 20 cm and QR = 16 cm. Calculate the length SR.

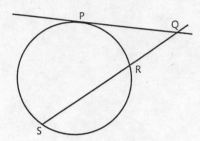

Example 5

AP = 8 cm, PB = 5 cm and CD = 14 cm. Find PD.

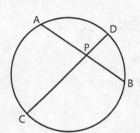

Let PD = x, then CP = $14 - x$.
Now, $AP \times PB = CP \times PD$
so $8 \times 5 = (14 - x)x$
$40 = 14x - x^2$
$x^2 - 14x + 40 = 0$
$(x - 10)(x - 4) = 0$
so $x = 4$ or 10 cm

Example 6

AB = 7 cm, PB = 5 cm and CD = 4 cm. Find PD.

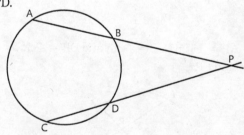

Let PD = x, then CP = $4 + x$.
Now, $AP \times PB = CP \times PD$
so $12 \times 5 = (4 + x)x$
$60 = 4x + x^2$
$x^2 + 4x - 60 = 0$
$(x - 6)(x + 10) = 0$
so $x = 6$ cm

Exercise 136*

1 PR = 10 cm, RT = 4 cm and QT = 8 cm. Find ST.

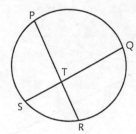

2 PR = 23 cm, PT = 8 cm, SQ = 22 cm and QT = x. Form a quadratic equation and find x.

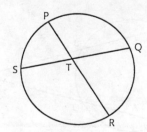

3 BC = 9 cm, AB = 7 cm and DE = 10 cm. Find CD.

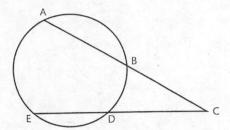

4 AC = 30 cm and AB = 14 cm. Calculate CE when DE = 4 cm.

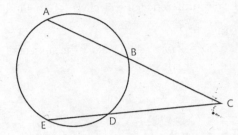

5 GT is a tangent to the circle. GT = 6 cm, IH = 3.5 cm and GH = x. Find x.

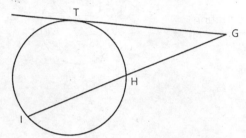

6 PQ is a tangent to the circle. PS passes through the centre of the circle at O. SR = 64 mm and RQ = 36 mm.

a Calculate the length PQ.

b Hence calculate the radius of the circle.

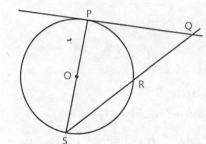

Circles and sectors

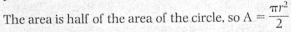

Remember

The perimeter of a shape is the distance all the way round the shape.

Circle

The perimeter of a circle is called the circumference.
If C is the circumference, A the area and r the radius, then

$C = 2\pi r$

$A = \pi r^2$

Semicircle

A semicircle is half a circle cut along a diameter.
The perimeter is half the circumference of the circle plus the diameter, so $P = \pi r + 2r$

The area is half of the area of the circle, so $A = \dfrac{\pi r^2}{2}$

Quadrant

A quadrant is quarter of a circle.
The perimeter is a quarter of the circumference of the circle plus twice the radius, so $P = \dfrac{\pi r}{2} + 2r$

The area is a quarter of the area of the circle, so $A = \dfrac{\pi r^2}{4}$

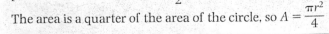

Example 7

The circumference of a circle is 10 cm. Find the radius.

Using $C = 2\pi r$

$\quad\quad 10 = 2\pi r$ $\quad\quad$ (Make r the subject of the equation)

$\quad\quad r = \dfrac{10}{2\pi}$

$\quad\quad\quad = 1.59$ cm to 3 s.f.

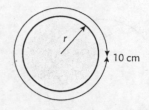

Example 8

The area of a circle is 24 cm². Find the radius.

Using $A = \pi r^2$

$\quad\quad 24 = \pi r^2$ $\quad\quad$ (Make r the subject of the equation)

$\quad\quad r^2 = \dfrac{24}{\pi}$

$\quad\quad r = \sqrt{\dfrac{24}{\pi}}$

$\quad\quad\quad = 2.76$ cm to 3 s.f.

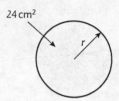

Example 9

Find the perimeter and area of the shape shown.

The radius of the quadrant BCD is 3 cm, so BC = 3 cm.

The perimeter = AB + BC + arc CD + DE + EA
$\quad\quad\quad = 4 + 3 + \dfrac{3\pi}{2} + 4 + 3$
$\quad\quad\quad = 18.7$ cm (to 3 s.f.)

The area = area of quadrant BCD + area of rectangle ABDE
$\quad\quad\quad = \dfrac{9\pi}{2} + 12$
$\quad\quad\quad = 19.1$ cm² (to 3 s.f.)

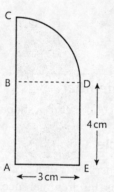

Exercise 137

Find the perimeter and area of each of the following shapes, giving answers to 3 s.f.

All dimensions are in cm. All arcs are parts of circles.

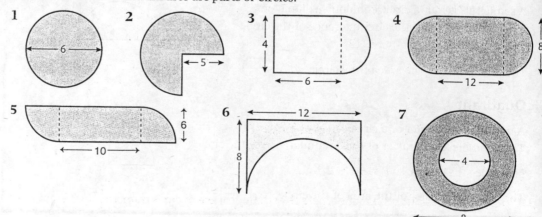

For Questions 8–11, fill in all the gaps in the following table:

Question	Radius in cm	Circumference in cm	Area in cm²
8		6	
9			14
10		52	
11			84

12 A bicycle wheel has a diameter of 66 cm. How many km does the bicycle travel if the wheel rotates 1000 times?

13 A car wheel has a diameter of 48 cm. On a journey the wheel rotates 5000 times. How long is the journey in km?

14 A CD is 120 mm in diameter. A speck of dust is on the edge of the CD. How many kilometres does the speck of dust travel when the CD rotates 10 000 times?

Exercise 137*

Find the perimeter and area of each of the following shapes, giving answers to 3 s.f.

All dimensions are in cm. All arcs are parts of circles.

1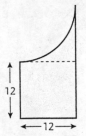

2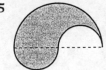

3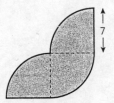

4

5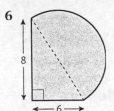

6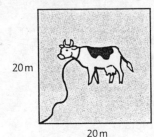

7 The area of a quadrant of a circle is 8 cm². Find the radius and perimeter.

8 A cow is tethered by a rope to one corner of a 20 m square field.
The cow can graze half the area of the field.
How long is the rope?

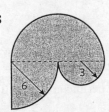

9 A goat is tethered to the outside corner of a 10 m square enclosure by a 15 m long rope. What area can the goat graze? (Assume the goat cannot jump into the enclosure.)

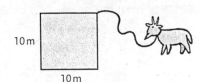

10 A new coin has just been made where the circumference in cm is numerically the same value as the area in cm². What is the radius of the coin?

11 The radius of the earth is 6380 km.
   a How far does a point on the equator travel in 24 hours?
   b Find the speed of a point on the equator in m/s.

12 A hot-air balloon travels round the Earth 1 km above the surface, following the equator. How much further does it travel than the distance around the equator?

13 The perimeter of this shape is 12 cm. Find r and the area.

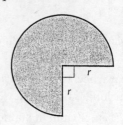

Arc of a circle

An **arc** is part of the circumference of a circle.

The arc shown is the fraction $\frac{x}{360}$ of the whole circumference.

So the arc length is

$$\frac{x}{360} \times 2\pi r$$

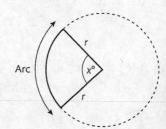

Example 10

Find the perimeter of the shape shown.

Using $\text{Arc length} = \frac{x}{360} \times 2\pi r$

$\text{Arc length} = \frac{80}{360} \times 2\pi \times 4 = 5.585\,\text{cm}$

The perimeter $= 5.585 + 4 + 4 = 13.6\,\text{cm}$ to 3 s.f.

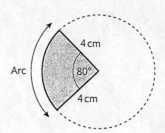

Example 11

Find the angle marked x.

Using $\text{Arc length} = \frac{x}{360} \times 2\pi r$

$12 = \frac{x}{360} \times 2\pi \times 9$ (Make x the subject of the equation)

$x = \frac{12 \times 360}{2\pi \times 9}$

$= 76.4°$ to 3 s.f.

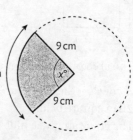

Example 12

Find the radius r.

Using $\text{Arc length} = \frac{x}{360} \times 2\pi r$

$20 = \frac{50}{360} \times 2\pi r$ (Make r the subject of the equation)

$r = \frac{20 \times 360}{50 \times 2\pi}$

$= 22.9\,\text{cm}$ to 3 s.f.

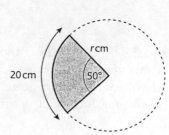

Exercise 138

In Questions 1–4, find the perimeter of the shape. Give answers to 3 s.f.

1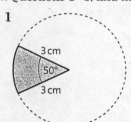
3 cm
50°
3 cm

2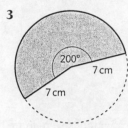
6 cm
130°
6 cm

3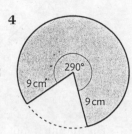
200°
7 cm
7 cm

4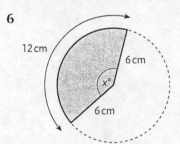
290°
9 cm
9 cm

In Questions 5 and 6, find the angle marked x.

5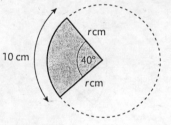
5 cm
3 cm
$x°$
5 cm

6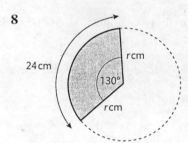
12 cm
6 cm
$x°$
6 cm

In Questions 7 and 8, find the radius r.

7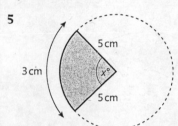
r cm
10 cm
40°
r cm

8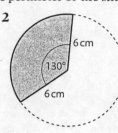
24 cm
r cm
130°
r cm

Exercise 138*

In Questions 1 and 2, find the perimeter of the shape. Give answers to 3 s.f.

1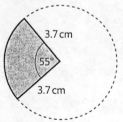
3.7 cm
55°
3.7 cm

2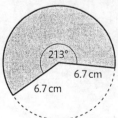
213°
6.7 cm
6.7 cm

In Questions 3 and 4, find the angle marked x.

3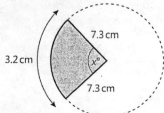
7.3 cm
3.2 cm
$x°$
7.3 cm

4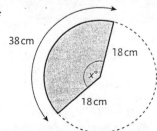
38 cm
18 cm
$x°$
18 cm

In Questions 5 and 6, find the radius r.

5

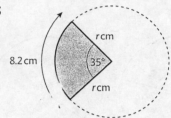

6

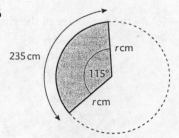

7 The minute hand of a watch is 9 cm long. How far does the tip travel in 35 minutes?

8 Find the perimeter of the shape to 3 s.f.

9 The perimeter of the shape is 28 cm. Find the value of r.

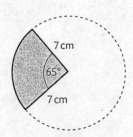

Sectors of circles

A **sector** of a circle is a region whose boundary is an arc and two radii.

The sector shown is the fraction $\dfrac{x}{360}$ of the whole circle.

So the sector area is

$$\dfrac{x}{360} \times \pi r^2$$

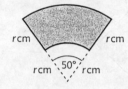

Example 13

Find the area of the sector shown.

Using Sector area $= \dfrac{x}{360} \times \pi r^2$

$$A = \dfrac{65}{360} \times \pi 7^2$$

$$= 27.8 \text{ cm}^2 \text{ to 3 s.f.}$$

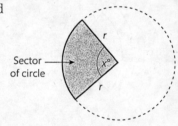

Example 14

Find the angle marked x.

Using Sector area $= \dfrac{x}{360} \times \pi r^2$

$$12 = \dfrac{x}{360} \times \pi \times 5^2 \qquad \text{(Make } x \text{ the subject of the equation)}$$

$$x = \dfrac{12 \times 360}{\pi \times 5^2}$$

$$= 55.0° \text{ to 3 s.f.}$$

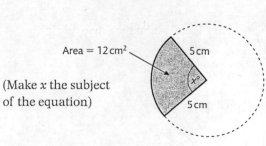

Example 15

Find the radius of the sector shown.

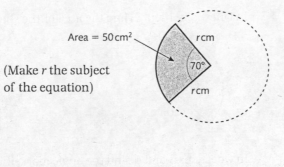

Area = 50 cm²
r cm
70°
r cm

Using   Sector area $= \dfrac{x}{360} \times \pi r^2$

$$50 = \dfrac{70}{360} \times \pi r^2$$ (Make r the subject of the equation)

$$r^2 = \dfrac{50 \times 360}{70 \times \pi}$$

$$r = \sqrt{\dfrac{50 \times 360}{70 \times \pi}}$$

$$= 9.05 \text{ cm (to 3 s.f.)}$$

Exercise 139

In Questions 1–4, find the area of the shape. Give answers to 3 s.f.

1

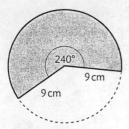

6 cm
40°
6 cm

2

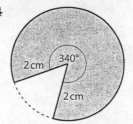

8 cm
110°
8 cm

3

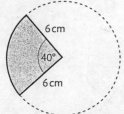

240°
9 cm
9 cm

4

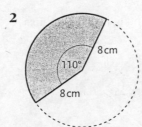

2 cm
340°
2 cm

In Questions 5 and 6, find the angle marked x.

5

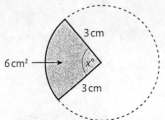

3 cm
6 cm²
$x°$
3 cm

6

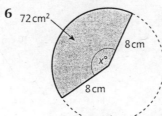

72 cm²
8 cm
$x°$
8 cm

In Questions 7 and 8, find the radius r.

7

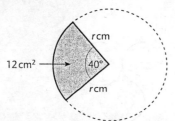

r cm
12 cm²
40°
r cm

8

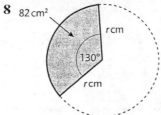

82 cm²
r cm
130°
r cm

Exercise 139*

In Questions 1 and 2, find the area of the shape. Give answers to 3 s.f.

1

7.2 cm
35°
7.2 cm

2

221°
18 cm 18 cm

In Questions 3 and 4, find the angle marked x.

3

9.5 cm
42 cm² → $x°$
9.5 cm

4

38 cm² →
6.5 cm
$x°$
6.5 cm

In Questions 5 and 6, find the radius r.

5

r cm
5.2 cm² → 25°
r cm

6

423 cm²
r cm
125°
r cm

7 Find the area of the shape to 3 s.f.

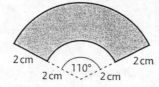

2 cm   2 cm
2 cm 110° 2 cm

8 The area of the shape is 54 cm².
Find the value of r.

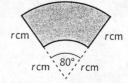

r cm r cm
r cm 80° r cm

9 Find the shaded area.

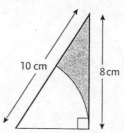

10 cm
8 cm

10 Three circular pencils, each with a diameter of 1 cm,
are held together by an elastic band.
What is the (stretched) length of the band?

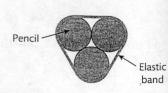

Pencil

Elastic
band

11 Three place mats, each with a diameter of 8 cm,
are placed on a table as shown.
Find the blue shaded area.

Exercise 140 (Revision)

In Questions 1–16 find the angle marked *a*.

1

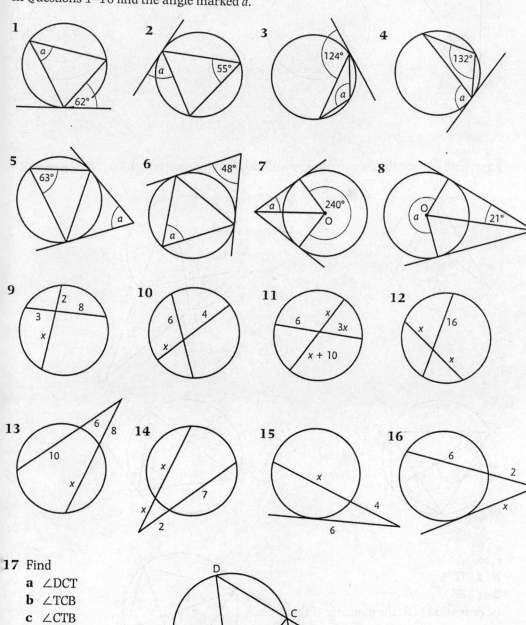

a

62°

2

a

55°

3

124°

a

4

132°

a

5

63°

a

6

48°

a

a

7

a

240°

O

8

a O

21°

9

2   8

3

x

10

6   4

x

11

x

6   3*x*

x + 10

12

x   16

x

13

6   8

10

x

14

x

7

x

2

15

x

4

6

16

6

2

x

17 Find
  a ∠DCT
  b ∠TCB
  c ∠CTB
  d ∠TDC

D

C

80°   30°

A   T   B

18 Calculate the area and perimeter of the following shapes.

a

6

b

3

3

c

4

4

d

6

4

1   1

19 Calculate the area and perimeter of the following shapes.

a

b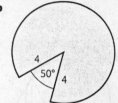

c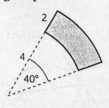

d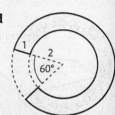

Exercise 140* (Revision)

For Questions 1–4, find the coloured angles, fully explaining your reasoning.

1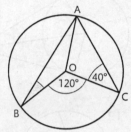

2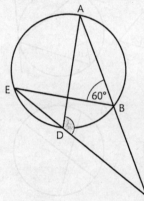

3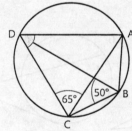

4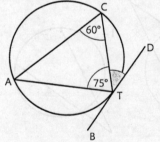

5 Find
  a ∠ETD
  b ∠TEB
  c Prove that EC is the diameter of the circle.

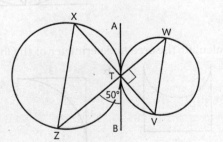

6 AB is the common tangent.
  a Find ∠XZT.
  b Find ∠WVT.
  c Prove that XZ is parallel to WV.

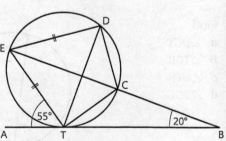

In Questions 7–14 find the length marked x.

7
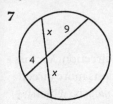

8
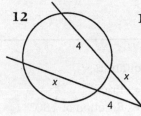

9
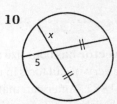

10
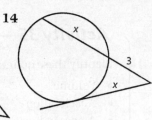

11
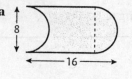
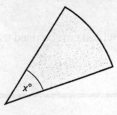

12

13
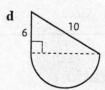

14
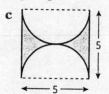

15 Calculate the area and perimeter of the following shapes.

a
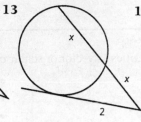

b

c

d

16 a If the perimeter = 14 cm, find x and the area.

b If the area = 35 cm², find x and the perimeter.

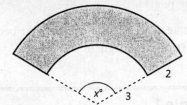

17 ABC is any triangle inscribed in a circle. The bisector of angle A meets CB at X and the circle at Y. Draw a neat diagram, then prove that

a triangle BCY is isosceles

b $\angle ABY = \angle BXY$

Unit 6 : Vectors and matrices

Vectors

A **vector** has both size and direction. In contrast, a **scalar** has size but no direction. Vectors are very useful tools in mathematics and physics, helping to make calculations more direct. In 1881, American mathematician J. W. Gibbs published the book *Vector Analysis*, which established vectors as they are known today.

Activity 35

Identify these quantities as vector or scalar quantities:

- Volume
- Acceleration
- A pass in hockey
- Area
- Temperature
- Velocity

- Price
- Rotation of 180°
- Force
- Length
- Density
- 10 km on a bearing of 075°

Notation

In this book, vectors are written in bold letters (**a**, **p**, **x**, ...) or capitals covered by an arrow ($\overrightarrow{AB}$, $\overrightarrow{PQ}$, $\overrightarrow{XY}$, ...). In other books, you might find vectors written as bold italic letters (*a*, *p*, *x*, ...). When hand-writing vectors, they are written with a wavy or straight underline (a͝, p͝, x͝, or a̲, p̲, x̲).

On coordinate axes, a vector can be described by a **column vector**, which can be used to find the **magnitude** and **angle** of the vector.

Example 1

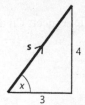

a Express vector **s** as a column vector.

$$\mathbf{s} = \begin{pmatrix} 3 \\ 4 \end{pmatrix}$$

b Find the magnitude of vector **s**.

Length of $\mathbf{s} = \sqrt{3^2 + 4^2} = \sqrt{25} = 5$

c Calculate the size of angle x.

$\tan x = \frac{4}{3} \Rightarrow x = 53.1°$ (to 3 s.f.)

Example 2

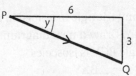

a Express vector $\overrightarrow{PQ}$ as a column vector:

$$\overrightarrow{PQ} = \begin{pmatrix} 6 \\ -3 \end{pmatrix}$$

b Find the magnitude of vector $\overrightarrow{PQ}$.

Length of $\overrightarrow{PQ} = \sqrt{6^2 + 3^2}$
$= \sqrt{45}$
$= 6.71$ (to 3 s.f.)

c Calculate the size of angle y.

$\tan y = \frac{3}{6} \Rightarrow y = 26.6°$ (to 3 s.f.)

Activity 36

Franz and Nina play golf. Their shots to the hole are shown as vectors.

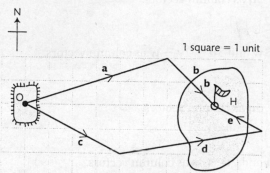

1 square = 1 unit

- Copy and complete this table by using the grid.

Vector	Column vector	Magnitude (to 3 s.f.)	Bearing
a	$\binom{6}{2}$	6.32	072°
b			
c			
d			
e			

- Write down the vector $\overrightarrow{OH}$ and state if there is a connection between $\overrightarrow{OH}$ and the vectors **a**, **b** and vectors **c**, **d**, **e**.

More addition, subtraction and multiplication

Vectors can be added, subtracted and multiplied using their components.

Example 3

$\mathbf{s} = \binom{1}{2}$, $\mathbf{t} = \binom{3}{0}$ and $\mathbf{u} = \binom{-2}{5}$

a Express in column vectors: $\mathbf{p} = \mathbf{s} + \mathbf{t} + \mathbf{u}$, $\mathbf{q} = \mathbf{s} - 2\mathbf{t} - \mathbf{u}$ and $\mathbf{r} = 3\mathbf{s} + \mathbf{t} - 2\mathbf{u}$

b Sketch the resultants **p**, **q** and **r** accurately.

c Find their magnitudes.

a Calculation	b Sketch	c Magnitude
$\mathbf{p} = \mathbf{s} + \mathbf{t} + \mathbf{u}$ $= \binom{1}{2} + \binom{3}{0} + \binom{-2}{5}$ $= \binom{2}{7}$	$\mathbf{p} = \binom{2}{7}$	Length of **p** $= \sqrt{2^2 + 7^2}$ $= \sqrt{53}$ $= 7.3$ to 1 d.p.
$\mathbf{q} = \mathbf{s} - 2\mathbf{t} - \mathbf{u}$ $= \binom{1}{2} - 2\binom{3}{0} - \binom{-2}{5}$ $= \binom{1}{2} + \binom{-6}{0} + \binom{2}{-5}$ $= \binom{-3}{-3}$	$\mathbf{q} = \binom{-3}{-3}$	Length of **q** $= \sqrt{3^2 + 3^2}$ $= \sqrt{18}$ $= 4.2$ to 1 d.p.
$\mathbf{r} = 3\mathbf{s} + \mathbf{t} - 2\mathbf{u}$ $= 3\binom{1}{2} + \binom{3}{0} - 2\binom{-2}{5}$ $= \binom{3}{6} + \binom{3}{0} + \binom{4}{-10}$ $= \binom{10}{-4}$	$\mathbf{r} = \binom{10}{-4}$	Length of **r** $= \sqrt{10^2 + (-4)^2}$ $= \sqrt{116}$ $= 10.8$ to 1 d.p.

Exercise 141

1 Given that $\mathbf{p} = \begin{pmatrix} 2 \\ 3 \end{pmatrix}$ and $\mathbf{q} = \begin{pmatrix} 4 \\ 5 \end{pmatrix}$,

simplify and express $\mathbf{p} + \mathbf{q}$, $\mathbf{p} - \mathbf{q}$ and $2\mathbf{p} + 3\mathbf{q}$ as column vectors.

2 Given that $\mathbf{u} = \begin{pmatrix} 1 \\ 2 \end{pmatrix}$, $\mathbf{v} = \begin{pmatrix} -4 \\ 3 \end{pmatrix}$ and $\mathbf{w} = \begin{pmatrix} 2 \\ -5 \end{pmatrix}$,

simplify and express $\mathbf{u} + \mathbf{v} + \mathbf{w}$, $\mathbf{u} + 2\mathbf{v} - 3\mathbf{w}$ and $3\mathbf{u} - 2\mathbf{v} - \mathbf{w}$ as column vectors.

3 Given that $\mathbf{p} = \begin{pmatrix} 1 \\ 2 \end{pmatrix}$ and $\mathbf{q} = \begin{pmatrix} 3 \\ 4 \end{pmatrix}$,

simplify and express $\mathbf{p} + \mathbf{q}$, $\mathbf{p} - \mathbf{q}$ and $2\mathbf{p} + 5\mathbf{q}$ as column vectors.

4 Given that $\mathbf{s} = \begin{pmatrix} 1 \\ -3 \end{pmatrix}$, $\mathbf{t} = \begin{pmatrix} 2 \\ 3 \end{pmatrix}$ and $\mathbf{u} = \begin{pmatrix} 4 \\ -5 \end{pmatrix}$,

simplify and express $\mathbf{s} + \mathbf{t} + \mathbf{u}$, $2\mathbf{s} - \mathbf{t} + 2\mathbf{u}$ and $2\mathbf{u} - 3\mathbf{s}$ as column vectors.

5 Two vectors are defined as $\mathbf{v} = \begin{pmatrix} 3 \\ 1 \end{pmatrix}$ and $\mathbf{w} = \begin{pmatrix} 1 \\ 4 \end{pmatrix}$.

Express $\mathbf{v} + \mathbf{w}$, $2\mathbf{v} - \mathbf{w}$ and $\mathbf{v} - 2\mathbf{w}$ as column vectors, find the magnitude and draw the resultant vector triangle for each vector.

6 Two vectors are defined as $\mathbf{p} = \begin{pmatrix} 2 \\ -1 \end{pmatrix}$ and $\mathbf{q} = \begin{pmatrix} 3 \\ 5 \end{pmatrix}$.

Express $\mathbf{p} + \mathbf{q}$, $3\mathbf{p} + \mathbf{q}$ and $\mathbf{p} - 3\mathbf{q}$ as column vectors, find the magnitude and draw the resultant vector triangle for each vector.

Exercise 141*

1 Given that $\mathbf{p} = \begin{pmatrix} 2 \\ 1 \end{pmatrix}$ and $\mathbf{q} = \begin{pmatrix} 3 \\ -1 \end{pmatrix}$,

find the magnitude and bearing of the vectors $\mathbf{p} + \mathbf{q}$, $\mathbf{p} - \mathbf{q}$ and $2\mathbf{p} - 3\mathbf{q}$.

2 Given that $\mathbf{r} = \begin{pmatrix} 1 \\ -3 \end{pmatrix}$ and $\mathbf{s} = \begin{pmatrix} 4 \\ 1 \end{pmatrix}$,

find the magnitude and bearing of the vectors $2(\mathbf{r} + \mathbf{s})$, $3(\mathbf{r} - 2\mathbf{s})$ and $(4\mathbf{r} - 6\mathbf{s}) \sin 30°$.

3 Given that $\mathbf{t} + \mathbf{u} = \begin{pmatrix} 1 \\ 1 \end{pmatrix}$, where $\mathbf{t} = \begin{pmatrix} m \\ 3 \end{pmatrix}$, $\mathbf{u} = \begin{pmatrix} 2 \\ n \end{pmatrix}$ and m and n are constants, find the values of m and n.

4 Given that $2\mathbf{v} - 3\mathbf{w} = \begin{pmatrix} 2 \\ -3 \end{pmatrix}$, where $\mathbf{v} = \begin{pmatrix} 5 \\ -m \end{pmatrix}$, $\mathbf{w} = \begin{pmatrix} n \\ 4 \end{pmatrix}$ and m and n are constants, find the values of m and n.

5 These vectors represent journeys undertaken by crows in km.
Express each one in column vector form.

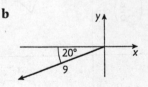

6 $\mathbf{s} = \begin{pmatrix} 2 \\ 3 \end{pmatrix}$ and $\mathbf{t} = \begin{pmatrix} -6 \\ 1 \end{pmatrix}$

a If $m\mathbf{s} + n\mathbf{t} = \begin{pmatrix} -16 \\ 6 \end{pmatrix}$, solve this vector equation to find the constants m and n.

b If $p\mathbf{s} + \mathbf{t} = \begin{pmatrix} 0 \\ q \end{pmatrix}$, solve this vector equation to find the constants p and q.

Matrices

Basic principles

Information is stored in many ways. People often make lists to make calculations easier (shopping, football league tables...). Computers like to work in ordered information stored in a **matrix**.

In an athletics match two countries' results are shown: 1st = 4pts, 2nd = 3pts

Country	100 m	400 m	Long jump
Jamaica	4	3	4
USA	3	4	3

This data can be displayed in a matrix. $\begin{pmatrix} 4 & 3 & 4 \\ 3 & 4 & 3 \end{pmatrix}$

This matrix has 6 **elements** (numbers).

The **order of the matrix** is 2×3 which means 2 rows and 3 columns.

There are different types of matrices:

Zero matrix: All elements are 0. $\quad \begin{pmatrix} 0 & 0 \\ 0 & 0 \end{pmatrix}, \begin{pmatrix} 0 & 0 & 0 \\ 0 & 0 & 0 \end{pmatrix}$

Identity matrix: All elements on leading diagonal are 1. $\begin{pmatrix} 1 & 0 \\ 0 & 1 \end{pmatrix}, \begin{pmatrix} 1 & 0 & 0 \\ 0 & 1 & 0 \\ 0 & 0 & 1 \end{pmatrix}$

Column matrix: All elements in a column. $\quad \begin{pmatrix} 1 \\ 2 \end{pmatrix}, \begin{pmatrix} 1 \\ 2 \\ 3 \end{pmatrix}$

Row matrix: All elements in a row. $\quad (1 \ 2), (1 \ 2 \ 3)$

Square matrix: Number of rows = Number of columns. $\begin{pmatrix} 1 & 3 \\ 2 & 4 \end{pmatrix}, \begin{pmatrix} 1 & 4 & 7 \\ 2 & 5 & 8 \\ 3 & 6 & 9 \end{pmatrix}$

> **Key Point**
> If a matrix has **m rows** and **n columns** it is of **order m × n**

Addition and subtraction

Only matrices of the same order (shape) can be added or subtracted term by term.

Example 4

$\begin{pmatrix} 1 & 3 \\ 2 & 4 \end{pmatrix} + \begin{pmatrix} 5 & 7 \\ 6 & 8 \end{pmatrix} = \begin{pmatrix} 6 & 10 \\ 8 & 12 \end{pmatrix}$

Example 5

$\begin{pmatrix} 1 & 3 \\ 2 & 4 \end{pmatrix} - \begin{pmatrix} 2 & 6 \\ 4 & 8 \end{pmatrix} = \begin{pmatrix} -1 & -3 \\ -2 & -4 \end{pmatrix}$

Example 6

$5 \times \begin{pmatrix} 1 & 3 & 5 \\ 2 & 4 & 6 \end{pmatrix} = \begin{pmatrix} 5 & 15 & 25 \\ 10 & 20 & 30 \end{pmatrix}$ The elements are each $\times 5$ as this is the same as adding the matrix five times.

Example 7

$2\begin{pmatrix} 1 & 4 \\ 2 & 5 \\ 3 & 6 \end{pmatrix} + 3\begin{pmatrix} 2 & 8 \\ 4 & 10 \\ 6 & 12 \end{pmatrix} - 4\begin{pmatrix} -1 & 2 \\ 3 & 2 \\ -2 & 1 \end{pmatrix} = \begin{pmatrix} 2 & 8 \\ 4 & 10 \\ 6 & 12 \end{pmatrix} + \begin{pmatrix} 6 & 24 \\ 12 & 30 \\ 18 & 36 \end{pmatrix} - \begin{pmatrix} -4 & 8 \\ 12 & 8 \\ -8 & 4 \end{pmatrix} = \begin{pmatrix} 12 & 24 \\ 4 & 32 \\ 32 & 44 \end{pmatrix}$

Be very careful of the double negative in these cases.

Questions in both Exercise 142 and Exercise 142* refer to the following matrices:

$$\mathbf{A} = \begin{pmatrix} 1 \\ 2 \end{pmatrix}, \mathbf{B} = \begin{pmatrix} 3 \\ 4 \end{pmatrix}, \mathbf{C} = (1 \ \ 2), \mathbf{D} = (3 \ \ 4), \mathbf{E} = \begin{pmatrix} 0 & -1 \\ 2 & 4 \\ -3 & 7 \end{pmatrix}, \mathbf{F} = \begin{pmatrix} 3 & 5 \\ -6 & 0 \\ 9 & 11 \end{pmatrix}$$

$$\mathbf{G} = \begin{pmatrix} 0 & 1 & 2 \\ 2 & 1 & 0 \end{pmatrix}, \mathbf{H} = \begin{pmatrix} -3 & 1 & -7 \\ 5 & 8 & 21 \end{pmatrix}, \mathbf{I} = \begin{pmatrix} 1 & 0 \\ 0 & 1 \end{pmatrix}, \mathbf{J} = \begin{pmatrix} -12 & 17 \\ 34 & 21 \end{pmatrix}$$

Exercise 142

1 State the order of matrices

   a A   **b** C

Simplify the following matrices where possible.

2 A + B   **3** E + F   **4** I + J   **5** B − A

6 F − E   **7** J − A   **8** (A + 3B)

Exercise 142*

1 If $\mathbf{A} + p\mathbf{B} = \begin{pmatrix} 7 \\ 10 \end{pmatrix}$, where p is a constant, find x.

2 If $\mathbf{E} + r\mathbf{F} = \begin{pmatrix} 6 & 9 \\ -10 & 4 \\ 15 & 29 \end{pmatrix}$, where r is a constant, find r.

3 If $2(\mathbf{I} + t\mathbf{J}) = \begin{pmatrix} 26 & -34 \\ -68 & -40 \end{pmatrix}$, where t is a constant, find t.

4 If $\mathbf{C} + \mathbf{D} + \mathbf{X} = \mathbf{0}$, find the matrix $\mathbf{X}$.

5 If $\begin{pmatrix} 2 & a \\ 1 & -2 \end{pmatrix} + \begin{pmatrix} 1 & -3 \\ 5 & b \end{pmatrix} = \begin{pmatrix} 3 & 9 \\ 6 & 8 \end{pmatrix}$, find constants a and b.

6 An inter-school sports day results are shown
for St Cuthbert's against St Mary's.
Both enter an A team and a B team and there
are 12 separate events for these two teams.

St Cuthbert's
$$\begin{array}{c} \\ \text{1st} \\ \text{2nd} \\ \text{3rd} \end{array} \begin{array}{cc} A & B \\ \begin{pmatrix} 5 & 6 \\ 3 & c \\ a & 4 \end{pmatrix} \end{array}$$

St Mary's
$$\begin{array}{c} \\ \text{1st} \\ \text{2nd} \\ \text{3rd} \end{array} \begin{array}{cc} A & B \\ \begin{pmatrix} 7 & 6 \\ 9 & d \\ b & 8 \end{pmatrix} \end{array}$$

Given that there are 3 points for 1st, 2 for 2nd and 1 for 3rd.

a If the A team match was won by St Mary's by 10 points find the value of a and b.

b If the B team match was won by St Cuthbert's by 12 points find the value of b and c.

c Which School won the combined A and B team match?

Matrix multiplication

Three American basketball teams enter a pre-season
tournament where there are 3 points for a win (W), 1 for a
draw (D) and 0 for a loss (L). The results are shown in the
matrix below. Each team plays the other twice.

$$\begin{array}{c} \\ \textit{Boston Knicks} \\ \textit{LA Lakers} \\ \textit{Washington Wizards} \end{array} \begin{array}{ccc} W & D & L \\ \begin{pmatrix} 1 & 2 & 1 \\ 1 & 1 & 2 \\ 2 & 1 & 1 \end{pmatrix} \end{array}$$

The points rewarded can be written as a 3 × 1 matrix $\begin{pmatrix} 3 \\ 1 \\ 0 \end{pmatrix}$

The total points for each team can be found by multiplying both matrices.

$$\begin{pmatrix} 1 & 2 & 1 \\ 1 & 1 & 2 \\ 2 & 1 & 1 \end{pmatrix}\begin{pmatrix} 3 \\ 1 \\ 0 \end{pmatrix} = \begin{pmatrix} 1 \times 3 + 2 \times 1 + 1 \times 0 \\ 1 \times 3 + 1 \times 1 + 2 \times 0 \\ 2 \times 3 + 1 \times 1 + 1 \times 0 \end{pmatrix} = \begin{pmatrix} 5 \\ 4 \\ 7 \end{pmatrix} \begin{array}{l} \textit{Boston Knicks} \\ \textit{LA Lakers} \\ \textit{Washington Wizards} \end{array}$$

The Washington Wizards win!

Matrices can only be multiplied if they are **compatible**. The number of columns of the left-side
matrix must be the same as the number of rows of the right-side matrix.

Key Points

If matrix **A** is of order $m \times n$ and matrix **B** is of order $n \times p$ the matrix **AB** can be found. It will be of order $m \times p$.

If the orders of the matrix products are written it can be determined whether they are compatible and the order of the resultant matrix can be found.

Let **A** be of order $m \times n$ and **B** be of order $n \times p$.

These must be equal

$$\mathbf{A} \times \mathbf{B} = (m \times n) \times (n \times p)$$

These give the order of the resultant

Note: Matrix multiplication is not **commutative**, namely **AB** does not always equal **BA**

Example 8

If $\mathbf{A} = \begin{pmatrix} 1 & 2 & 3 \\ 4 & 5 & 6 \end{pmatrix}$, $\mathbf{B} = \begin{pmatrix} 2 & -3 \\ 4 & 5 \\ 2 & 1 \end{pmatrix}$ evaluate matrix AB.

$A \times B = (2 \times 3) \times (3 \times 2)$, so the matrices are compatible and the resultant matrix **AB** will be of order 2×2.

$$\mathbf{A} \times \mathbf{B} = \begin{pmatrix} 1 & 2 & 3 \\ 4 & 5 & 6 \end{pmatrix} \times \begin{pmatrix} 2 & -3 \\ 4 & 5 \\ 2 & 1 \end{pmatrix}$$

$$= \begin{pmatrix} 1 \times 2 + 2 \times 4 + 3 \times 2, & 1 \times -3 + 2 \times 5 + 3 \times 1 \\ 4 \times 2 + 5 \times 4 + 6 \times 2, & 4 \times -3 + 5 \times 5 + 6 \times 1 \end{pmatrix} = \begin{pmatrix} 16 & 10 \\ 40 & 19 \end{pmatrix}$$

Exercise 143

$\mathbf{A} = \begin{pmatrix} 1 \\ 2 \end{pmatrix}$, $\mathbf{B} = \begin{pmatrix} 3 \\ -1 \end{pmatrix}$, $\mathbf{C} = (1 \ -2)$, $\mathbf{D} = (1 \ 4)$, $\mathbf{E} = \begin{pmatrix} 0 & -1 \\ 1 & 4 \\ 3 & 2 \end{pmatrix}$, $\mathbf{F} = \begin{pmatrix} 3 & -2 \\ 1 & 0 \\ 2 & 2 \end{pmatrix}$

$\mathbf{G} = \begin{pmatrix} 0 & 1 & 3 \\ 2 & 2 & 0 \end{pmatrix}$, $\mathbf{H} = \begin{pmatrix} 1 & 1 & 2 \\ 5 & -2 & 1 \end{pmatrix}$, $\mathbf{I} = \begin{pmatrix} 1 & 0 \\ 0 & 1 \end{pmatrix}$, $\mathbf{J} = \begin{pmatrix} -1 & 4 \\ 3 & 7 \end{pmatrix}$

Find resultant matrices where possible and state their order before evaluating them.

1 AI   **2** IA   **3** GI   **4** GE

5 HF   **6** I(E + F)   **7** $\mathbf{J}^2$ **8** E(G − H)

Exercise 143*

For questions 1–6 find the value of the letters in the matrix products:

1 $\begin{pmatrix} 1 & 0 \\ 3 & 0 \end{pmatrix} \begin{pmatrix} x \\ 3 \end{pmatrix} = \begin{pmatrix} 10 \\ y \end{pmatrix}$ **2** $\begin{pmatrix} 1 & 2 \\ 0 & 5 \end{pmatrix} \begin{pmatrix} 4 \\ x \end{pmatrix} = \begin{pmatrix} y \\ 15 \end{pmatrix}$

3 $\begin{pmatrix} x & 3 & 1 \\ y & -1 & y \end{pmatrix} \begin{pmatrix} 1 \\ 2 \\ 3 \end{pmatrix} = \begin{pmatrix} 11 \\ 2 \end{pmatrix}$ **4** $\begin{pmatrix} 1 & x & -4 \\ y & y & y \end{pmatrix} \begin{pmatrix} 1 \\ 2 \\ 3 \end{pmatrix} = \begin{pmatrix} 5 \\ 24 \end{pmatrix}$

5 $\begin{pmatrix} -1 & 3 \\ 4 & 5 \\ 2 & -1 \end{pmatrix} \begin{pmatrix} x \\ 4 \end{pmatrix} = \begin{pmatrix} 2 \\ y \\ 16 \end{pmatrix}$ **6** $\begin{pmatrix} 2 & 1 \\ -1 & 2 \\ 3 & 1 \end{pmatrix} \begin{pmatrix} 5 \\ x \end{pmatrix} = \begin{pmatrix} 9 \\ -7 \\ y \end{pmatrix}$

7 If $\mathbf{A} = \begin{pmatrix} 1 & 1 \\ -3 & -3 \end{pmatrix}$, find the value of the constant m if $\mathbf{A}^2 = m\mathbf{A}$.

8 An international sailing competition between Jamaica, Mexico and Cuba in various categories gave the following results for Gold (**G**), Silver (**S**) and Bronze (**B**).
The points awarded by these medals is shown in the second 3×1 matrix.

$$\begin{matrix} & \begin{matrix} G & S & B \end{matrix} \\ \begin{matrix} Jamaica \\ Mexico \\ Cuba \end{matrix} & \begin{pmatrix} 3 & 1 & 4 \\ 2 & 6 & 3 \\ 3 & x & 3 \end{pmatrix} \end{matrix} \times \begin{pmatrix} x \\ 2 \\ 1 \end{pmatrix} = \mathbf{P}$$

a If the matrix **P** represents the total points scored per country, state the order of **P**.

b If Jamaica scored a total of 18 points, find x.

c Find the total points scored by all three countries.

Inverse of a 2 × 2 matrix

The **Unit matrix** or **Identity matrix**, **I** represents 'no change'.
Multiplying by matrix **A** and then its inverse $\mathbf{A}^{-1}$ will result in no change.
The inverse matrix 'undoes what has been done'.

In general:

$$\mathbf{AA}^{-1} = \mathbf{A}^{-1}\mathbf{A} = \mathbf{I} = \begin{pmatrix} 1 & 0 \\ 0 & 1 \end{pmatrix}$$

Key Points

If $\mathbf{A} = \begin{pmatrix} a & c \\ b & d \end{pmatrix}$, then $\mathbf{A}^{-1} = \dfrac{1}{ad - bc}\begin{pmatrix} d & -c \\ -b & a \end{pmatrix}$, where $ad - bc$ is called the **determinant** and written as $|\mathbf{A}|$.
If $|\mathbf{A}| = 0$ the matrix is called **singular** and it has no inverse.

Example 9

Find the inverse of matrix $\mathbf{A} = \begin{pmatrix} 2 & -1 \\ 4 & 3 \end{pmatrix}$

$\mathbf{A}^{-1} = \dfrac{1}{2(3) - 4(-1)}\begin{pmatrix} 3 & 1 \\ -4 & 2 \end{pmatrix} = \dfrac{1}{10}\begin{pmatrix} 3 & 1 \\ -4 & 2 \end{pmatrix}$

Check that $\mathbf{A}^{-1}\mathbf{A} = \mathbf{I} = \dfrac{1}{10}\begin{pmatrix} 3 & 1 \\ -4 & 2 \end{pmatrix}\begin{pmatrix} 2 & -1 \\ 4 & 3 \end{pmatrix} = \dfrac{1}{10}\begin{pmatrix} 10 & 0 \\ 0 & 10 \end{pmatrix}$
$= \begin{pmatrix} 1 & 0 \\ 0 & 1 \end{pmatrix}$

Exercise 144

Find the inverse of the following matrices.

1 $\begin{pmatrix} 3 & -1 \\ 4 & 1 \end{pmatrix}$

2 $\begin{pmatrix} 1 & -1 \\ 1 & 5 \end{pmatrix}$

3 $\begin{pmatrix} 3 & 1 \\ -2 & -2 \end{pmatrix}$

4 $\begin{pmatrix} 6 & 7 \\ -1 & 1 \end{pmatrix}$

5 $\begin{pmatrix} 5 & -1 \\ 7 & 1 \end{pmatrix}$

6 $\begin{pmatrix} 2 & 0 \\ 1 & 1 \end{pmatrix}$

7 $\begin{pmatrix} 2 & -1 \\ -5 & 2 \end{pmatrix}$

8 $\begin{pmatrix} -2 & -1 \\ 0 & 3 \end{pmatrix}$

9 $\begin{pmatrix} 10 & -12 \\ 11 & 9 \end{pmatrix}$

10 $\begin{pmatrix} 20 & -10 \\ 40 & 30 \end{pmatrix}$

Exercise 144*

1 If matrix $\begin{pmatrix} 2 & x \\ 4 & 3 \end{pmatrix}$ has no inverse find x.

2 If matrix $\begin{pmatrix} 4 & 2 \\ y & 8 \end{pmatrix}$ has no inverse find y.

3 If matrix **A** represents a reflection in the x-axis, state what transformation matrix $\mathbf{A}^{-1}$ represents.

4 If matrix **B** represents an enlargement of scale factor 5 about 0, state what transformation matrix $\mathbf{B}^{-1}$ represents.

5 If $\mathbf{N} = \begin{pmatrix} 2 & -1 \\ -5 & 3 \end{pmatrix}$ and $\mathbf{MN} = \mathbf{I}$, find **M**.

6 If $\mathbf{Q} = \begin{pmatrix} 3 & 1 \\ 5 & 2 \end{pmatrix}$ and $\mathbf{PQ} = \mathbf{I}$, find **P**.

7 If $\mathbf{X} \begin{pmatrix} 1 & -2 \\ -4 & 7 \end{pmatrix} = \begin{pmatrix} 1 & 0 \\ 0 & 1 \end{pmatrix}$, find the matrix **X**.

8 If $\mathbf{Y} \begin{pmatrix} 1 & -2 \\ -4 & 9 \end{pmatrix} = \begin{pmatrix} 1 & 0 \\ 0 & 1 \end{pmatrix}$, find the matrix **Y**.

9 If $\mathbf{A} = \begin{pmatrix} 2 & -3 \\ -1 & 1 \end{pmatrix}$ and $\mathbf{B} = \begin{pmatrix} 1 & 5 \\ -2 & -2 \end{pmatrix}$ show that: $(\mathbf{AB})^{-1} = \mathbf{B}^{-1}\mathbf{A}^{-1}$.

10 If $\mathbf{P} = \begin{pmatrix} 1 & 4 \\ -2 & -2 \end{pmatrix}$ and $\mathbf{Q} = \begin{pmatrix} 2 & -1 \\ -3 & -1 \end{pmatrix}$ show that: $\mathbf{P}^{-1}\mathbf{Q}^{-1} = (\mathbf{PQ})^{-1}$.

Exercise 145 (Revision)

1 Given that $\mathbf{p} = \begin{pmatrix} 3 \\ 4 \end{pmatrix}$ and $\mathbf{q} = \begin{pmatrix} -2 \\ 1 \end{pmatrix}$

simplify $\mathbf{p} + \mathbf{q}$, $\mathbf{p} - \mathbf{q}$ and $3\mathbf{p} - 2\mathbf{q}$ as column vectors and find the magnitude of each vector.

2 Given that $\mathbf{r} = \begin{pmatrix} 2 \\ -5 \end{pmatrix}$ and $\mathbf{s} = \begin{pmatrix} 3 \\ 4 \end{pmatrix}$,

simplify $\mathbf{r} + \mathbf{s}$, $\mathbf{r} - \mathbf{s}$ and $3\mathbf{s} - 2\mathbf{r}$ as column vectors and find the magnitude of each vector.

3 $\mathbf{r} = \begin{pmatrix} 1 \\ 3 \end{pmatrix}$ and $\mathbf{s} = \begin{pmatrix} -2 \\ 5 \end{pmatrix}$

a Calculate $2\mathbf{r} - \mathbf{s}$.

b Calculate $2(\mathbf{r} - \mathbf{s})$.

c Calculate the length of vector **s** in root form.

d $v\mathbf{r} + w\mathbf{s} = \begin{pmatrix} -3 \\ 13 \end{pmatrix}$

What are the values of the constants v and w?

4 Given that $2\mathbf{p} - 3\mathbf{q} = \begin{pmatrix} 5 \\ 15 \end{pmatrix}$, where $\mathbf{p} = \begin{pmatrix} 4 \\ m \end{pmatrix}$, $\mathbf{q} = \begin{pmatrix} n \\ -3 \end{pmatrix}$ and m and n are constants, find the values of m and n.

5 If $\mathbf{r} = \begin{pmatrix} 4 \\ -1 \end{pmatrix}$, $\mathbf{s} = \begin{pmatrix} 3 \\ 7 \end{pmatrix}$ and $m\mathbf{r} + n\mathbf{s} = \begin{pmatrix} 7 \\ 37 \end{pmatrix}$, find constants m and n.

6 If $\mathbf{A} = \begin{pmatrix} 2 & 1 \\ 3 & 1 \end{pmatrix}$, $\mathbf{B} = \begin{pmatrix} 0 & -4 \\ -3 & 1 \end{pmatrix}$ find the value of

   a $\mathbf{A} + \mathbf{B}$ **b** $\mathbf{A} - \mathbf{B}$ **c** $4(\mathbf{A} + 2\mathbf{B})$ **d** $2(\mathbf{A} - 4\mathbf{B})$

7 If $\mathbf{M} = \begin{pmatrix} -1 & 1 \\ 2 & 1 \end{pmatrix}$, $\mathbf{N} = \begin{pmatrix} 1 & -4 \\ 0 & 1 \end{pmatrix}$ find the value of

   a $\mathbf{MN}$ **b** $\mathbf{NM}$ **c** $\mathbf{M}^2$ **d** $\mathbf{N}^{-1}$

Exercise 145* (Revision)

1 Given that $\mathbf{u} = \begin{pmatrix} 1 \\ 2 \end{pmatrix}$, $\mathbf{v} = \begin{pmatrix} -2 \\ 3 \end{pmatrix}$ express $\mathbf{p}$, $\mathbf{q}$, $\mathbf{r}$ and $\mathbf{s}$ as column vectors and find their magnitudes where

   a $\mathbf{p} = \mathbf{u} + \mathbf{v}$ b $\mathbf{q} = \mathbf{u} - \mathbf{v}$ c $\mathbf{r} = 2\mathbf{u} + 3\mathbf{v}$ d $\mathbf{s} = 3\mathbf{u} - 2\mathbf{v}$

2 Given that $\mathbf{u} = \begin{pmatrix} 3 \\ 1 \end{pmatrix}$, $\mathbf{v} = \begin{pmatrix} 2 \\ 5 \end{pmatrix}$ and $\mathbf{w} = \begin{pmatrix} 0 \\ -4 \end{pmatrix}$ express $\mathbf{p}$, $\mathbf{q}$, $\mathbf{r}$ and $\mathbf{s}$ as column vectors and find their magnitudes where

   a $\mathbf{p} = \mathbf{u} + \mathbf{v} + \mathbf{w}$ b $\mathbf{q} = \mathbf{u} - \mathbf{v} - \mathbf{w}$

   c $\mathbf{r} = 2\mathbf{u} + 3\mathbf{v} + 4\mathbf{w}$ d $\mathbf{s} = 2(\mathbf{u} - 2\mathbf{v} + 3\mathbf{w})$

3 Esther walks a route described using column vectors $\mathbf{p}$ and $\mathbf{q}$, where the units are in km. $\mathbf{p} = \begin{pmatrix} 1 \\ 2 \end{pmatrix}$ and $\mathbf{q} = \begin{pmatrix} -3 \\ 5 \end{pmatrix}$. Her journey from O is given by vector $\mathbf{w}$ where $\mathbf{w} = 2\mathbf{p} + 3\mathbf{q}$. Find

   a the vector $\mathbf{w}$

   b the magnitude and bearing of $\mathbf{w}$

   c the speed of Esther's journey from O if she takes 4 hours to complete it.

4 $\mathbf{p} = \begin{pmatrix} 3 \\ 2 \end{pmatrix}$ and $\mathbf{q} = \begin{pmatrix} 3 \\ -5 \end{pmatrix}$. Find

   a (i) $\mathbf{p} + \mathbf{q}$ (ii) $2\mathbf{p} - \mathbf{q}$

   b the values of constants m and n such that $m\mathbf{p} + n\mathbf{q} = \begin{pmatrix} 12 \\ -13 \end{pmatrix}$

   c the values of constants r and s such that $\mathbf{p} + r\mathbf{q} = \begin{pmatrix} s \\ -8 \end{pmatrix}$

   d the values of constants u and v such that $u(\mathbf{p} + \mathbf{q}) + v(2\mathbf{p} - \mathbf{q}) = \begin{pmatrix} 0 \\ 21 \end{pmatrix}$.

5 Given that $\mathbf{a} = \begin{pmatrix} 1 \\ 2 \end{pmatrix}$, $\mathbf{b} = \begin{pmatrix} 4 \\ 3 \end{pmatrix}$ and $m\mathbf{a} + n\mathbf{b} = \begin{pmatrix} -2 \\ 1 \end{pmatrix}$ where m and n are constants find the values of m and n.

6 If $\mathbf{X} = \begin{pmatrix} -1 & 1 \\ 5 & 2 \end{pmatrix}$, $\mathbf{Y} = \begin{pmatrix} 1 & 4 \\ -2 & -2 \end{pmatrix}$ find the value of

   a $\mathbf{XY}^2$ b $(\mathbf{XY})^{-1}$ c $(\mathbf{YX})^{-1}$ d $\mathbf{X}^4$

7 If $\mathbf{A} = \begin{pmatrix} 1 & -1 \\ -1 & 2 \\ 3 & 5 \end{pmatrix}$, $\mathbf{B} = \begin{pmatrix} 0 & 1 & -2 \\ x & y & 1 \end{pmatrix}$ find the value of

   a $\mathbf{AB}$ b $\mathbf{BA}$ c x and y if $\mathbf{AB} = \begin{pmatrix} -2 & -2 & -3 \\ 4 & 5 & 4 \\ 10 & 18 & -1 \end{pmatrix}$.

1 Yana earns \$18/h as a gardener for a 30 hour week. Her overtime rate is \$$x$/h. For a 36 hour week she earns \$660. The value of x is

   A 18.33   **B** 22   **C** 20   **D** 27

2 y is directly proportional to x squared. If $y = 20$ when $x = 2$, the value of y when $x = 4$ is:

   A 10   **B** 20   **C** 40   **D** 80

3 If $f(x) = x(x + 3)$, the value of x if $f(x) = -2$ is

   A 1 or 2   **B** -1 or -2 **C** -1 or 2   **D** 1 or -2

4 The perimeter of a quarter circle of area 100 cm² is approximately

   A 17.7 cm   **B** 28.5 cm   **C** 40.3 cm   **D** 80.6 cm

5 If $\mathbf{a} = \begin{pmatrix} 1 \\ 3 \end{pmatrix}$, $\mathbf{b} = \begin{pmatrix} -1 \\ 7 \end{pmatrix}$ the magnitude of the vector $\mathbf{c}$ if $\mathbf{c} = \mathbf{a} - \mathbf{b}$ is:

   A $\sqrt{20}$ **B** $\sqrt{10}$ **C** 2   **D** $\sqrt{2}$

6 If $f(x) = x + 4$, $g(x) = \dfrac{x}{2}$ and $fg(x) = 10$ then the value of x is:

   A 12   **B** 16   **C** 24   **D** 28

7 If the height h m, of a palm tree is proportional to the square root of its age, t years. Given that $h = 4$ when $t = 25$, the age of an 8 m tree is

   A 3.2   **B** 10   **C** 20   **D** 100

8 If $f(x) = \dfrac{3x}{2}$ and $g(x) = \dfrac{4}{x}$, the value of x if $f(x) = g(x) - 1$ is given by

   A -2 or $\frac{4}{3}$ **B** 2 or $-\frac{4}{3}$ **C** 2 or 4   **D** -2 or 4

9 If a quarter circle has a perimeter of 100 cm and its area is p cm², the value of p is approximately

   A 308   **B** 616   **C** 1012   **D** 3183

10 If $\mathbf{M} = \begin{pmatrix} 1 & 3 \\ -1 & 2 \end{pmatrix}$, $\mathbf{N} = \begin{pmatrix} 2 & 1 \\ 0 & 4 \end{pmatrix}$ the value of $(\mathbf{MN})^{-1}$ is

   A $\dfrac{1}{40}\begin{pmatrix} 8 & -8 \\ 4 & 1 \end{pmatrix}$ **B** $\dfrac{1}{-12}\begin{pmatrix} 7 & -13 \\ 2 & 2 \end{pmatrix}$ **C** $\dfrac{1}{-24}\begin{pmatrix} 1 & -8 \\ 4 & 8 \end{pmatrix}$ **D** $\dfrac{1}{40}\begin{pmatrix} 7 & -13 \\ 2 & 2 \end{pmatrix}$

1 Kalisa is paid $12 per hour for a 35 hour week, with overtime paid at $18 per hour. One week she works for 42 hours. How much does she get paid?

2 Amin has US $1200 and travels through China, Malaysia and India. He wishes to convert all his money to the three currencies in the ratio $3:2:1$ respectively. If $1 is worth 6.83 Chinese Yuan; 3.99 Malaysian Ringitts and 46.11 Indian Rupees, how much of each currency will he have?

3 The variable y is directly proportional to x, and $y = 24$ when $x = 3$. Find
   a the formula for y in terms of x
   b the value of y when $x = 10$
   c the value of x when $y = 40$.

4 x varies directly as the cube root of y. If $x = 12$ when $y = 27$, find
   a the formula for y in terms of x
   b the value of y when $x = 32$.

5 y is inversely proportional to x, and $y = 16$ when $x = 3$. Find
   a the formula for y in terms of x
   b the value of y when $x = 8$
   c the value of x when $y = 12$.

6 A machine produces coins, of a fixed thickness, from a given volume of metal. The number of coins, N, produced is inversely proportional to the square of the diameter, d. If 4000 coins of diameter 1.5 cm are made, find
   a the formula for N in terms of d
   b the number of coins that can be produced of diameter 2 cm
   c the diameter if 1000 coins are produced.

7 The surface area, $A\,\text{m}^2$, of an inflatable toy is proportional to the square of its height, $h\,\text{m}$. The surface area $= 60\,\text{m}^2$ when the height $= 2\,\text{m}$. Find
   a the formula for A in terms of h
   b the value of A when $h = 3\,\text{m}$
   c the value of h when $A = 540\,\text{m}^2$.

8 Solve the following to find the angles x, y and z.
   a
   b

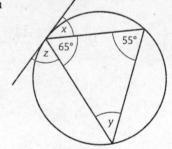
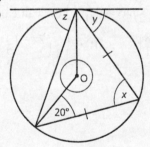

9
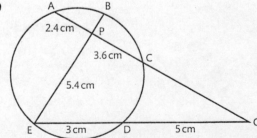

AP $= 2.4\,\text{cm}$, CP $= 3.6\,\text{cm}$, EP $= 5.4\,\text{cm}$, DE $= 3\,\text{cm}$ and DQ $= 5\,\text{cm}$.
   a Work out the length BP.
   b Calculate the length CQ.

10 A shape is made up of part of a circle of radius 3 cm and a square of side 3 cm. The circle's centre is at the corner of the square. Find the perimeter and area of the shape.

11 a The perimeter of the shape is 10 cm. Calculate the value of x.

b The area of the shape is 24 cm². Calculate the value of r.

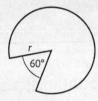

12 Draw a mapping diagram for $x \rightarrow 12 - 5x$ with domain $\{-2, -1, 0, 1, 2\}$. What is the range?

13 If $f(x) = 2x + 3$ and $g(x) = x(x + 3)$, find the value of
  a fg(10)   **b** gf(10)   **c** fg(x)   **d** gf(x)

14 If $f(x) = 3x - 1$ and $g(x) = \dfrac{1}{x}$, find the value of
  a $f^{-1}(x)$ **b** $g^{-1}(x)$
  c the value of x if $f^{-1}(x) = g^{-1}(x)$ **d** $(fg(x))^2$

15 If $f(x) = 10x - 1$ find
  a f(10)   **b** $f\left(\frac{1}{2}\right)$ **c** x if $f(x) = 199$

16 If $x = \begin{pmatrix} 1 \\ 2 \end{pmatrix}$, $y = \begin{pmatrix} -2 \\ 3 \end{pmatrix}$ find:
  a $x + y$ **b** $x - y$ **c** the magnitude of $x - 2y$

17 If $a\mathbf{p} + 3\mathbf{q} = \begin{pmatrix} 16 \\ -8 \end{pmatrix}$ where $\mathbf{p} = \begin{pmatrix} 1 \\ 1 \end{pmatrix}$ and $\mathbf{q} = \begin{pmatrix} 5 \\ b \end{pmatrix}$ find constants a and b.

18 If $\mathbf{A} = \begin{pmatrix} 1 & -4 \\ 2 & 1 \end{pmatrix}$ find
  a $2\mathbf{A}$ **b** $\mathbf{A}^2$ **c** $\mathbf{A}^{-1}$

Negative and fractional indices

> ◆ Rules of indices
>
> $3^2 \times 3^3 = 3^{2+3} = 3^5 = 243$ (Add the indices: $a^m \times a^n = a^{m+n}$)
>
> $\dfrac{8^6}{8^4} = 8^6 \div 8^4 = 8^{6-4} = 8^2 = 64$ (Subtract the indices: $a^m \div a^n = a^{m-n}$)
>
> $(2^3)^2 = 2^6 = 64$ (Multiply the indices: $(a^m)^n = a^{m \times n} = a^{mn}$)
>
> ◆ $10^{-2} = \dfrac{1}{10^2}$ $\left(a^{-n} = \dfrac{1}{a^n}\right)$
>
> ◆ $3^{\frac{1}{2}} \times 3^{\frac{1}{2}} = 3^1$ so $3^{\frac{1}{2}} = \sqrt{3}$ ($a^{\frac{1}{2}} = \sqrt{a}$ and similarly $a^{\frac{1}{3}} = \sqrt[3]{a}$)

Single negative and fractional indices

Example 1

Without using a calculator, evaluate these and, where appropriate, leave the answer as a fraction.

a $4^{-3} = \dfrac{1}{4^3} = \dfrac{1}{64}$

b $125^{\frac{1}{3}} = \sqrt[3]{125} = 5$

c $3^{-3} \times 3^2 = 3^{(-3+2)} = 3^{-1} = \dfrac{1}{3}$

d $6^{-4} \div 6^{-2} = 6^{(-4--2)} = 6^{-2} = \dfrac{1}{6^2} = \dfrac{1}{36}$

e $(3^{-1})^2 = 3^{(-1 \times 2)} = 3^{-2} = \dfrac{1}{3^2} = \dfrac{1}{9}$

Example 2

Use a calculator to work out these, correct to 3 significant figures.

a $4^{-3} = 0.0156$

b $60^{\frac{1}{3}} = 2.27$

c $(5^{-2})^{-2} = 5^{(-2 \times -2)} = 5^4 = 625$

Example 3

Simplify these.

a $a^{\frac{1}{2}} \times a^{\frac{1}{2}} \times a = a^{\left(\frac{1}{2} + \frac{1}{2} + 1\right)} = a^2$

b $3a^2 \times 2a^{-3} = 6 \times a^{2+(-3)} = 6a^{-1} = \dfrac{6}{a}$

c $(6a^{-2}) \div (2a^2) = 3a^{(-2-2)} = 3a^{-4} = \dfrac{3}{a^4}$

Alternatively $\dfrac{6}{a^2} \div 2a^2 = \dfrac{6}{a^2} \times \dfrac{1}{2a^2} = \dfrac{3}{a^4}$

Exercise 146

Without using a calculator, evaluate these. Where appropriate, leave the answer as a fraction.

1 3^{-2} **2** 10^{-1} **3** 2^{-3} **4** 8^{-1}

5 $81^{\frac{1}{2}}$ **6** 4^{-3} **7** $27^{\frac{1}{3}}$ **8** $2^2 \times 2^{-1}$

9 $2^{-1} \div 2^3$ **10** $(3^{-1})^2$ **11** $2^{-2} \times 2^{-2}$ **12** $2^{-2} \div 2^{-1}$

13 $(2^{-2})^{-2}$

Use your calculator to work out these, correct to 3 significant figures.

14 5^4 **15** $23^{\frac{1}{3}}$ **16** $6^2 \times 6^{-3}$ **17** $3^{-1} \times 3^{-3}$

18 $6^2 \div 6^{-4}$ **19** $3^{-1} \div 3^2$ **20** $(3^2)^{-1}$ **21** $(8^{-2})^{-1}$

Simplify these.

22 $a^2 \times a^{-1}$ **23** $c^{-1} \div c^2$ **24** $e^2 \times e^3 \times e^{-4}$ **25** $(a^2)^{-1}$

Exercise 146*

Without using a calculator, evaluate these. Where appropriate, leave the answer as a fraction.

1 2^{-6} **2** $125^{\frac{1}{3}}$ **3** 6^0 **4** $2^{10} \times 2^3 \times 2^{-4}$

5 $200 \times (4^2)^{-1}$ **6** $\dfrac{4^{-2}}{4^{-3}}$ **7** 0.1×0.1^{-2} **8** $\left(\frac{1}{2}\right)^{-3}$

9 $\left(\frac{1}{4}\right)^{\frac{1}{2}}$ **10** $(-6)^{-2}$

Use your calculator to work out these, correct to 3 significant figures.

11 1.4^{-3} **12** $362^{\frac{1}{9}}$ **13** $3^{-3} \times 3^{-2} \times 3^{-1}$

Simplify these.

14 $a^{-2} \times a^2 \div a^{-2}$ **15** $2(c^2)^{-2}$ **16** $a^{-2} + a^{-2} + a^{-2}$

17 $3a^{-2} \times 4a$ **18** $a^{\frac{1}{2}} \times a^{\frac{1}{2}}$ **19** $(c^{-2})^{\frac{1}{2}}$

Solve these for x.

20 $9^{\frac{1}{x}} = 3$ **21** $81^{\frac{1}{2}} = 3^x$ **22** $2^x = \frac{1}{8}$

23 Find k if $x^k = \sqrt[3]{x} \div \dfrac{1}{x^2}$

Simplify these.

24 $1 \div a^3$ **25** $b^{-2} \div \dfrac{1}{b^3}$ **26** $a^{-2} + \dfrac{1}{a^2}$ **27** $\dfrac{3}{a^2} + 2a^{-2}$

28 $(3a)^{-2} \div \dfrac{1}{3a^2}$ **29** $(-3a)^3 \div (3a^{-3})$ **30** $\left(\dfrac{a}{b}\right)^{-1}$

Combined negative and fractional indices

$16^{\frac{3}{4}}$ can be written as $\quad \left(16^{\frac{1}{4}}\right)^3 = \left(\sqrt[4]{16}\right)^3 = 2^3 = 8$

This shows that $\qquad 16^{-\frac{3}{4}} = \dfrac{1}{16^{\frac{3}{4}}} = \dfrac{1}{2^3} = \dfrac{1}{8}$

Exercise 147

Without using a calculator, evaluate these and, where appropriate, leave the answer as a fraction.

1 $9^{-\frac{1}{2}}$ **2** $100^{-\frac{1}{2}}$ **3** $81^{-\frac{1}{2}}$ **4** $64^{-\frac{1}{3}}$ **5** $9^{\frac{3}{2}}$ **6** $9^{-\frac{3}{2}}$

Use your calculator to work out these, correct to 3 significant figures.

7 $20^{-\frac{1}{2}}$ **8** $50^{-\frac{1}{4}}$ **9** $11^{\frac{3}{2}}$ **10** $7^{-\frac{3}{2}}$

Simplify these.

11 $b^{-\frac{1}{3}} \times b^{-\frac{1}{3}} \times b^{-\frac{1}{3}}$ **12** $d^{\frac{2}{3}} \div d^{-\frac{1}{3}}$ **13** $(f^2)^{-\frac{1}{2}}$

Exercise 147*

Without using a calculator, evaluate these and, where appropriate, leave the answer as a fraction.

1 $36^{-\frac{1}{2}}$ **2** $1 \div 216^{-\frac{1}{3}}$ **3** $8^{\frac{4}{3}}$ **4** $8^{-\frac{4}{3}}$

Use your calculator to work out these, correct to 3 significant figures.

5 $0.6^{-\frac{1}{2}}$ **6** $1.4^{-\frac{1}{3}}$ **7** $2.01^{\frac{1}{2}}$ **8** $2.01^{-\frac{3}{2}}$

Simplify these.

9 $a^{\frac{1}{2}} \times a^{-2\frac{1}{2}}$ **10** $c^{-2\frac{1}{3}} \div c^{-\frac{1}{3}}$ **11** $\left(e^{-\frac{1}{2}}\right)^{-2}$ **12** $(27^2)^{\frac{1}{3}}$ **13** $216^{-\frac{2}{3}}$

Exercise 148 (Revision)

Work out these and, where appropriate, leave the answer as a fraction.

1 2^{-4} **2** $100^{\frac{1}{2}}$ **3** $64^{\frac{1}{3}}$ **4** $64^{\frac{1}{2}}$ **5** $25^{\frac{1}{2}}$

6 $125^{\frac{1}{3}}$ **7** 8^{-2} **8** $16^{-\frac{1}{2}}$ **9** $121^{-\frac{1}{2}}$ **10** $3^3 \times 3^{-1}$

11 $3^{-1} \div 3^3$ **12** $(3^{-3})^{-1}$ **13** Show that $36^{-\frac{1}{2}} = \frac{1}{6}$ **14** Show that $9^{\frac{3}{2}} = 27$

Simplify these.

15 $a^3 \times a^{-1}$ **16** $a^3 \div a^{-1}$ **17** $(d^{-1})^2$ **18** $b^{\frac{1}{2}} \times b^{-2}$ **19** $b^{\frac{1}{2}} \div b^{-2}$ **20** $\left(c^{-\frac{1}{2}}\right)^{-2}$

Exercise 148* (Revision)

Work out these and, where appropriate, leave the answer as a fraction.

1 2^{-5} **2** $216^{\frac{1}{3}}$ **3** 9.7^0 **4** $27 \times (3^{-1})^2$ **5** $2^8 \times 2^{-2} \times 2^{-4}$

6 $49^{-\frac{1}{2}}$ **7** $\left(\frac{1}{2}\right)^{-3} \div 2^{-3}$ **8** $16^{\frac{3}{4}}$ **9** $\left(\frac{1}{27}\right)^{\frac{2}{3}}$ **10** $(125)^{-\frac{4}{3}}$

11 Show that $(0.125)^{-\frac{2}{3}} = 4$ **12** Show that $\left(4^{\frac{1}{3}}\right)^{-1\frac{1}{2}} = \frac{1}{2}$

Simplify these.

13 $3c^3 \times c^{-2}$ **14** $b^{-2} + 4b^{-2}$ **15** $a^3 \times a^{-2} \div a^{-1}$ **16** $3 \times (c^{-1})^2$

17 $2\left(a^{\frac{1}{3}}\right)^{-3}$ **18** $d^{\frac{1}{2}} \times d^{\frac{2}{3}}$ **19** $(-3a)^3 \div (3a^{-3})$ **20** $(2b^2)^{-1} \div (-2b)^{-2}$

Solving quadratic equations

Quadratic equations can be written as $ax^2 + bx + c = 0$ where a, b and c are constants ($a \neq 0$).

Solving quadratic equations by factorising

Quadratic equations can often be solved by factorising.

> ### Remember
> There are three types of quadratic equations with $a = 1$.
> - If $b = 0$
> $$x^2 - c = 0$$
> $$x^2 = c$$
> $$x = \pm\sqrt{c}$$
> - If $c = 0$
> $$x^2 + bx = 0$$
> $$x(x + b) = 0$$
> $$x = 0 \text{ or } x = -b$$
> - If $b \neq 0$ and $c \neq 0$
> $$x^2 + bx + c = 0$$
> $$(x + p)(x + q) = 0$$
> $$x = -p \text{ or } x = -q$$
>
> where $p \times q = c$ and $p + q = b$.
> If c is positive then p and q have the same sign as b.
> If c is negative then p and q have opposite signs to each other.

Example 1

Solve these quadratic equations.

a $\quad x^2 - 81 = 0$
$$x^2 = 81$$
$$x = -9 \text{ or } x = 9$$

b $\quad x^2 - 7x = 0$
$$x(x - 7) = 0$$
$$x = 0 \text{ or } x = 7$$

c $\quad x^2 - 10x + 21 = 0$
$$(x - 7)(x - 3) = 0$$
$$x = 7 \text{ or } x = 3$$
(*Note*: there are *two* solutions)

Exercise 149

Solve the equations by factorising.

1 $\ x^2 + 3x + 2 = 0$ 2 $\ x^2 + x - 6 = 0$ 3 $\ x^2 + 7x + 10 = 0$

4 $\ x^2 - 2x - 15 = 0$ 5 $\ x^2 - 6x + 9 = 0$ 6 $\ x^2 + 4x - 12 = 0$

7 $\ x^2 + x = 0$ 8 $\ x^2 - 4x = 0$ 9 $\ x^2 - 4 = 0$

10 $\ x^2 - 36 = 0$

Solve the equations by factorising.

1 $x^2 + 6x - 5 = 0$ **2** $x^2 - 3x - 4 = 0$ **3** $x^2 + 15x + 56 = 0$

4 $x^2 - 4x - 45 = 0$ **5** $x^2 - 14x + 49 = 0$ **6** $x^2 - 3x - 40 = 0$

7 $x^2 - 13x = 0$ **8** $x^2 + 17x = 0$ **9** $x^2 - 81 = 0$

10 $x^2 - 121 = 0$

More difficult quadratic equations

When $a \neq 1$, factorisation may be harder. *Always* take out any common factors first.

Example 2

a $9x^2 - 25 = 0$ **b** $3x^2 - 12x = 0$

$\quad\quad 9x^2 = 25$ $3x(x - 4) = 0$

$\quad\quad x^2 = \dfrac{25}{9}$ $x = 0$ or $x = 4$

$\quad\quad x = \pm\dfrac{5}{3}$

c $12x^2 - 24x - 96 = 0$

$\quad\quad 12(x^2 - 2x - 8) = 0$

$\quad\quad 12(x + 2)(x - 4) = 0$

$\quad\quad\quad\quad\quad x = -2$ or $x = 4$

If there is no simple number factor, then the factorisation is harder.

Example 3

Solve $3x^2 - 13x - 10 = 0$.

$3x^2 - 13x - 10 = 0$

$(3x + 2)(x - 5) = 0$

$\quad\quad\quad x = -\dfrac{2}{3}$ or $x = 5$

Solve the equations by factorising.

1 $4x^2 - 49 = 0$ **2** $16x^2 - 81 = 0$ **3** $3x^2 + 6x = 0$

4 $5x^2 - 5x = 0$ **5** $2x^2 - 10x + 12 = 0$ **6** $2x^2 - 5x + 2 = 0$

7 $2x^2 + 5x + 3 = 0$ **8** $3x^2 + 9x + 6 = 0$ **9** $2x^2 - 18 = 0$

10 $3x^2 - 6x = 0$ **11** $3x^2 + 7x + 2 = 0$ **12** $3x^2 - 5x - 2 = 0$

13 $4x^2 - 4x = 24$ **14** $3x^2 + 8x + 4 = 0$ **15** $3x^2 + 10x = 8$

Solve the equations by factorising.

1 $49x^2 - 25 = 0$ **2** $128 - 18x^2 = 0$ **3** $10x + 5x^2 = 0$

4 $6x^2 - 9x = 0$ **5** $2x^2 - 6x + 4 = 0$ **6** $2x^2 - 7x + 6 = 0$

7 $3x^2 + 31x + 36 = 0$ **8** $6x^2 - 7x - 3 = 0$ **9** $8x^2 + 6x + 1 = 0$

10 $5x^2 - 27x + 10 = 0$ **11** $10x^2 - 23x + 12 = 0$ **12** $3x^2 = 17x + 28$

13 $3x^2 - 48 = 0$ **14** $7x^2 - 21x = 0$ **15** $4x^2 + 40x + 100 = 0$

16 $4x^2 = 29x - 7$ **17** $x(6x - 13) = -6$ **18** $9x^2 + 25 = 30x$

Completing the square

If a quadratic equation cannot be factorised, then the method of completing the square can be used. Completing the square can also provide further information about the quadratic function.

Example 4

Solve $x^2 + 4x + 4 = 9$.

$x^2 + 4x + 4 = 9 \Rightarrow (x + 2)^2 = 9$ The LHS is a perfect square so square root both sides.

$\Rightarrow x + 2 = \pm 3$

$\Rightarrow x = 3 - 2 \text{ or } x = -3 - 2$

$\Rightarrow x = 1 \text{ or } x = -5$

Completing the square involves writing quadratic expressions as perfect squares so the method of Example 4 can be used. Diagrams are a good way to complete the square.

Example 5

Complete the square for $x^2 + 2x$.

The diagram shows part of the square is missing.
The area of the missing part is 1.

So $x^2 + 2x = x^2 + 2x + 1 - 1$

           $= (x + 1)^2 - 1$

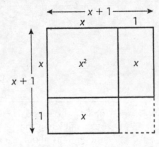

Example 6

Complete the square for $x^2 + 6x$.

The diagram shows part of the square is missing.
The area of the missing part is 9.

So $x^2 + 6x = x^2 + 6x + 9 - 9$

           $= (x + 3)^2 - 9$

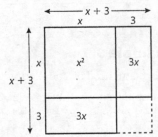

If the coefficient of x is not even, the working is a little more complex.

Example 7

Complete the square for $x^2 + 5x$.

The diagram shows part of the square is missing.
The area of the missing part is $\frac{25}{4}$.

So $x^2 + 5x = x^2 + 5x + \frac{25}{4} - \frac{25}{4}$

           $= (x + \frac{5}{2})^2 - \frac{25}{4}$

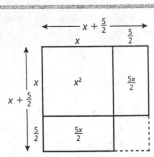

With imagination, the diagram can be used with negative signs.

Example 8

Complete the square for $x^2 - 6x$.

The diagram shows part of the square is missing.
The area of the missing part is $+9$.
So $\qquad x^2 - 6x = x^2 - 6x + 9 - 9$
$\qquad\qquad = (x - 3)^2 - 9$

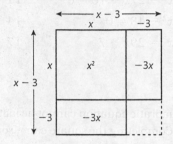

Key Point

Completing the square is done as follows:

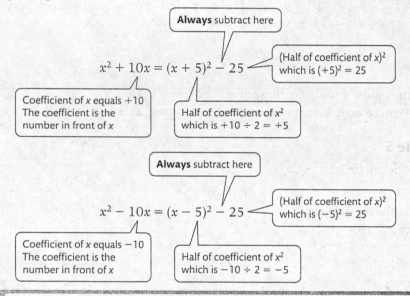

Example 9

Use the method of completing the square to solve $x^2 + 2x - 2 = 0$

Example 5 shows that $x^2 + 2x = (x + 1)^2 - 1$
So $x^2 + 2x - 2 = 0 \Rightarrow (x + 1)^2 - 1 - 2 = 0$
$\qquad\qquad\qquad \Rightarrow (x + 1)^2 = 3$
$\qquad\qquad\qquad \Rightarrow x + 1 = \pm\sqrt{3}$
$\qquad\qquad\qquad \Rightarrow x = \sqrt{3} - 1 \text{ or } x = -\sqrt{3} - 1$
$\qquad\qquad\qquad \Rightarrow x = 0.732 \text{ or } x = -2.73 \text{ to 3 s.f.}$

Exercise 151

Solve the following equations by completing the square.

1 $x^2 + 2x - 5 = 0$ **2** $x^2 - 2x - 6 = 0$ **3** $x^2 + 4x = 8$

4 $x^2 + 10x + 15 = 0$ **5** $x^2 + 14x - 3 = 0$ **6** $x^2 - 20x - 33 = 0$

7 $x^2 - 4x - 20 = 0$ **8** $x^2 - 10x = 120$ **9** $x^2 + 3x - 2 = 0$

10 $x^2 - 5x - 3 = 0$

Exercise 151*

Solve the following equations by completing the square.

1 $x^2 - 6x + 1 = 0$ **2** $x^2 - 16x + 3 = 0$ **3** $x^2 + 6x - 12 = 0$

4 $x^2 - 13 = 6x$ **5** $x^2 - 2x = 1$ **6** $2x^2 - 16x + 4 = 0$

7 $2x^2 - 5x = 7$ **8** $x(5x + 12) = -5$ **9** $3 - 10x - 4x^2 = 0$

10 $7x^2 = 4 + 4x$

The quadratic formula

The quadratic formula is used to solve quadratic equations that may be awkward to solve by other means.

Key Point

If $ax^2 + bx + c = 0$ then

$$x = \frac{-b \pm \sqrt{b^2 - 4ac}}{2a}$$

Example 10

Solve $3x^2 - 8x + 2 = 0$ giving your solution correct to 3 significant figures.

Here $a = 3$, $b = -8$ and $c = 2$.   *Note*: b is a negative number.

Substituting into the formula gives

$$x = \frac{-(-8) \pm \sqrt{(-8)^2 - 4 \times 3 \times 2}}{2 \times 3} = \frac{8 \pm \sqrt{64 - 24}}{6}$$

So $x = \dfrac{8 + \sqrt{40}}{6} = 2.39$ or $x = \dfrac{8 - \sqrt{40}}{6} = 0.279$

Example 11

Solve $2.3x^2 + 3.5x - 4.8 = 0$ giving your solution correct to 3 significant figures.

Here $a = 2.3$, $b = 3.5$ and $c = -4.8$. Substituting into the formula gives

$$x = \frac{-3.5 \pm \sqrt{(12.25 - 4 \times 2.3 \times (-4.8))}}{2 \times 2.3} = \frac{-3.5 \pm \sqrt{56.41}}{4.6}$$

So $x = \dfrac{-3.5 + \sqrt{56.41}}{4.6} = 0.872$ or $x = \dfrac{-3.5 - \sqrt{56.41}}{4.6} = -2.39$

The solutions are $x = 0.872$ or $x = -2.39$.

Exercise 152

Solve these equations using the quadratic formula.
Give your solutions correct to 3 significant figures.

1 $x^2 + 2x - 5 = 0$ **2** $x^2 - 2x - 6 = 0$ **3** $x^2 + 4x = 8$

4 $x^2 - 10x + 15 = 0$ **5** $x^2 + 14x - 3 = 0$ **6** $x^2 - 20x - 33 = 0$

7 $x^2 - 4x - 20 = 0$ **8** $x^2 - 10x = 120$ **9** $x^2 + 3x - 2 = 0$

10 $x^2 - 5x - 3 = 0$ **11** $x^2 + x - 8 = 0$ **12** $x^2 - 2x - 7 = 0$

Exercise 152*

Solve these equations using the quadratic formula.
Give your solutions correct to 3 significant figures.

1 $x^2 - 6x + 1 = 0$ **2** $x^2 - 16x + 3 = 0$ **3** $x^2 + 6x - 12 = 0$

4 $x^2 + 13x + 4 = 0$ **5** $x^2 - 6x + 7 = 0$ **6** $3x^2 - 5x = 2$

7 $x^2 - 13 = 6x$

8 $x^2 - 2x = 1$

9 $2x^2 - 16x + 4 = 0$

10 $10 + 3x - 2x^2 = 0$

11 $2.3x^2 - 12.6x + 1.3 = 0$

12 $x(x + 1) + (x - 1)(x + 2) = 3$

Proof of the quadratic formula

The following proof is for the case when $a = 1$, i.e. for the equation $x^2 + bx + c = 0$.

$$x^2 + bx + c = 0 \Rightarrow \left(x + \frac{b}{2}\right)^2 - \left(\frac{b}{2}\right)^2 + c = 0 \text{ (by completing the square)}$$

$$\Rightarrow \left(x + \frac{b}{2}\right)^2 = \left(\frac{b}{2}\right)^2 - c$$

$$\Rightarrow \left(x + \frac{b}{2}\right)^2 = \frac{b^2}{4} - c$$

$$\Rightarrow \left(x + \frac{b}{2}\right)^2 = \frac{b^2 - 4c}{4}$$

$$\Rightarrow x + \frac{b}{2} = \frac{\pm\sqrt{b^2 - 4c}}{2}$$

$$\Rightarrow x = \frac{-b \pm \sqrt{b^2 - 4c}}{2}$$

Investigate

Modify the proof above to prove the general quadratic formula for $ax^2 + bx + c = 0$. Start by dividing both sides of the equation by a.

The quadratic formula gives an easy way of finding how many solutions there are to a quadratic equation.

Activity 37

- Use the graphs to find how many solutions there are to each of these equations.
  $$x^2 + 8x + 15 = 0$$
  $$x^2 + 8x + 16 = 0$$
  $$x^2 + 8x + 17 = 0$$
- For each of the quadratic equations, work out the values of $b^2 - 4ac$.
- Try to find a rule involving $b^2 - 4ac$ that tells you how many solutions a quadratic equation has.

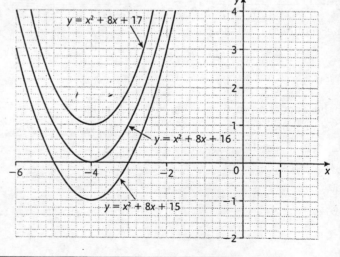

Investigate

For what values of k does the equation $x^2 + 8x + k = 0$ have real solutions?

Exercise 153

State how many solutions there are to these equations. Do *not* solve them.

1 $x^2 - 2x + 1 = 0$ **2** $x^2 - 9 = 0$ **3** $x^2 + 4 = 0$

4 $x^2 + 2x + 1 = 0$ **5** $x^2 - 2x + 5 = 0$ **6** $x^2 - 4x + 4 = 0$

7 $x^2 + 6x + 1 = 0$ **8** $x^2 - 2x - 3 = 0$ **9** $x^2 - x + 1 = 0$

10 $x^2 + 8x + 12 = 0$

Exercise 153*

State how many solutions there are to these equations. Do *not* solve them.

1 $x^2 - 3 = 0$ **2** $x^2 + 3x + 3 = 0$ **3** $x^2 - x - 1 = 0$

4 $4x^2 - 4x + 5 = 0$ **5** $4x^2 - 4x + 1 = 0$ **6** $2x^2 + 3x + 2 = 0$

7 $4x^2 - 7x + 2 = 0$ **8** $2x^2 - 4x + 9 = 0$ **9** $3x^2 + 8x + 3 = 0$

10 $9x^2 + 6x + 1 = 0$

Exercise 154 (Revision)

1 Solve these quadratic equations.

  a $x^2 = 16$ **b** $x^2 - 25 = 0$ **c** $x^2 + 4x = 0$ **d** $x^2 = 3x$

2 Solve these quadratic equations by factorisation.

  a $x^2 - 4x + 3 = 0$ **b** $x^2 + 7x + 12 = 0$ **c** $x^2 + x - 12 = 0$ **d** $x^2 - 2x - 8 = 0$

3 Solve these quadratic equations by factorisation.

  a $5x^2 - 5x - 30 = 0$ **b** $2x^2 + 6x + 4 = 0$

  c $3x^2 + x - 2 = 0$ **d** $7x^2 + 7x = 0$

4 Solve these equations by using the quadratic formula, giving your answers to 3 s.f.

  a $x^2 + 4x + 1 = 0$ **b** $x^2 - 5x + 3 = 0$ **c** $x^2 - 2x - 4 = 0$ **d** $4x^2 + 10x - 3 = 0$

5 Solve these quadratic equations.

  a $x^2 - 3x - 10 = 0$ **b** $3x^2 + 3x - 18 = 0$ **c** $2x^2 - 8x = 0$ **d** $3x^2 - 5x + 1 = 0$

Exercise 154* (Revision)

1 Solve these quadratic equations.

  a $x^2 - 169 = 0$ **b** $5x^2 - 20 = 0$ **c** $x^2 = 9x$ **d** $3x^2 + 108x = 0$

2 Solve these quadratic equations by factorisation.

  a $x^2 + x - 72 = 0$ **b** $4 + 3x - x^2 = 0$ **c** $x^2 - 2x - 48 = 0$ **d** $18 - x^2 - 3x = 0$

3 Solve these quadratic equations by factorisation.

  a $5x^2 + 35x + 60 = 0$ **b** $7x^2 - 28x + 28 = 0$

  c $7x^2 = 14x + 168$ **d** $2x(4x + 7) = 15$

4 Solve these equations by using the quadratic formula, giving your answers to 3 s.f.

  a $2x^2 - 6x - 3 = 0$ **b** $x(x + 1) + (x - 1)(x + 2) = 1$

  c $3x^2 = 7x + 5$ **d** $2.1x^2 + 8.4x = 4.3$

5 Solve these quadratic equations.

  a $x^2 + 5x - 14 = 0$ **b** $4x^2 - 8x - 32 = 0$

  c $x(3x - 7) + 2 = 0$ **d** $1.2x^2 - 3.5x - 1.5 = 0$

Cubic graphs $y = ax^3 + bx^2 + cx + d$

In cubic curves the highest power of x is x^3.

Those curves have distinctive shapes and can be used to model real-life situations.

Example 1

Draw the graph of $y = 2x^3 + 2x^2 - 4x$ for $-2 \leqslant x \leqslant 2$ by compiling a suitable table.

x	-2	-1	0	$\frac{1}{2}$	1	2
$2x^3$	-16	-2	0	$\frac{1}{4}$	2	16
$2x^2$	8	2	0	$\frac{1}{2}$	2	8
$-4x$	8	4	0	-2	-4	-8
y	0	4	0	$-1\frac{1}{4}$	0	16

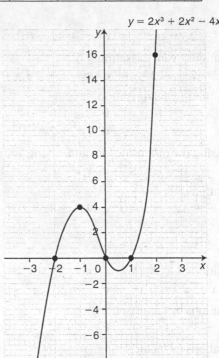

$y = 2x^3 + 2x^2 - 4x$

Remember

Cubic graphs have distinctive shapes that depend on the value of a.

a is positive a is negative

Exercise 155

Draw the graphs of these equations between the stated x values after compiling a suitable table.

1 $y = x^3 + 2$ $-3 \leqslant x \leqslant 3$

2 $y = x^3 - 2$ $-3 \leqslant x \leqslant 3$

3 $y = x^3 + 3x$ $-3 \leqslant x \leqslant 3$

4 $y = x^3 - 3x$ $-3 \leqslant x \leqslant 3$

5 $y = x^3 + x^2 - 2x$ $-3 \leqslant x \leqslant 3$

6 $y = x^3 - 3x^2 + x$ $-2 \leqslant x \leqslant 3$

7 This water tank has dimensions as shown, in metres.

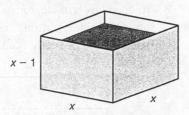

  a Show that the volume of the tank, $V\,\text{m}^3$, is given by the formula $V = x^3 - x^2$.

  b Draw the graph of V against x for $2 \leqslant x \leqslant 5$ by first constructing a suitable table of values.

  c Use your graph to estimate the volume of a tank for which the base area is $16\,\text{m}^2$.

  d What are the dimensions of a tank of volume $75\,\text{m}^3$?

8 The cross-section of a hilly region can be drawn as the graph of $y = x^3 - 8x^2 + 16x + 8$ for $0 \leqslant x \leqslant 5$, where x is measured in kilometres and y is the height above sea level in metres.

  a Draw the cross-section by first constructing a suitable table for $0 \leqslant x \leqslant 5$.

  b The peak is called Triblik and at the base of the valley is Vim Tarn. Mark these two features on your cross-section, and estimate the height of Triblik above Vim Tarn.

Exercise 155*

Draw the graphs of these equations between the stated x values after compiling a suitable table of values.

1 $y = 2x^3 - x^2 + x - 3$ $-3 \leqslant x \leqslant 3$

2 $y = 2x^3 - 2x^2 - 24x$ $-4 \leqslant x \leqslant 4$

3 $y = -2x^3 + 3x^2 + 4x$ $-3 \leqslant x \leqslant 3$

4 $y = -x^3 - 2x^2 + 11x + 12$ $-4 \leqslant x \leqslant 4$

5 A firework is fired and its velocity, v metres per second, t seconds later is $v = 27t - t^3$, where $0 \leqslant t \leqslant 5$.

  a Draw the graph of v after first compiling a table for the given values of t.

  b Use your graph to estimate the greatest velocity of the firework, and the time at which this occurs.

  c For how long does the firework travel faster than $30\,\text{m/s}$?

6 A toy is made that comprises a cylinder of diameter $2x\,\text{cm}$ and height $x\,\text{cm}$ upon which is fixed a right circular cone of base radius x centimetres and height $6\,\text{cm}$.

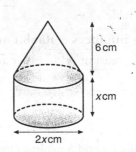

  a Volume of a right circular cone $= \frac{1}{3} \times$ base area $\times$ height. Show that the total volume V, in cubic centimetres, of the toy is given by $V = \pi x^2(x + 2)$.

  b Draw the graph of V against x for $0 \leqslant x \leqslant 5$ after compiling a suitable table.

  c Use your graph to find the volume of the toy of diameter $7\,\text{cm}$.

  d What is the curved surface area of the cylinder if the total volume of the toy is $300\,\text{cm}^3$?

7 A closed cylindrical can of height h cm and radius r cm is made from a thin sheet of metal. The *total* surface area is 100π cm².

a Show that $h = \dfrac{50}{r} - r$.

Hence show that the volume of the can, V cm³, is given by $V = 50\pi r - \pi r^3$.

b Draw the graph of V against r for $0 \le r \le 7$ by first compiling a suitable table of values.

c Use the graph to estimate the greatest possible volume of the can.

d What is the diameter and height of the can of maximum volume?

8 An open box is made from a thin square metal sheet measuring 10 cm by 10 cm. Four squares of side x centimetres are cut away, and the remaining sides are folded upwards to make the box of depth x centimetres.

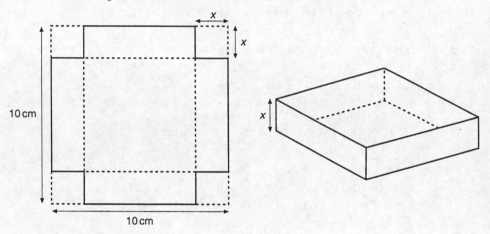

a Show that the side length of the box is $(10 - 2x)$ cm.

b Show that the volume V in cm³ of the box is given by the formula $V = 100x - 40x^2 + 4x^3$, for $0 \le x \le 5$.

c Draw the graph V against x by first constructing a table of suitable values.

d Use your graph to estimate the maximum volume of the box, and state its dimensions.

Activity 38

A forest contains F foxes and R rabbits. Their numbers change throughout the course of a given year as shown in the graph of F against R. t is the number of months after 1 January.

- Copy and complete this table. ·

Year interval	Fox numbers	Rabbit numbers	Reason
Jan–Mar (A–B)	Decreasing	Increasing	Fewer foxes to eat rabbits
Apr–June (B–C)			
Jul–Sep (C–D)			
Oct–Dec (D–A)			

Sketch two graphs of F against t and R against t for the interval $0 \le t \le 12$, placing the horizontal axes as shown for comparison.

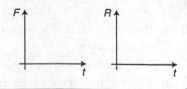

Reciprocal graphs $y = \dfrac{a}{x}$

Reciprocal graphs have x as the denominator, and they produce another type of curve called a **hyperbola**.

Investigate

- Copy and complete these tables of values for the equations.

$y = \dfrac{3}{x}$

x	-3	-2	-1	0	1	2	3
y	-1				3		

$y = -\dfrac{3}{x}$

x	-3	-2	-1	0	1	2	3
y	1				-3		

Why are there no values for y when $x = 0$?

- Draw one set of x and y axes, and use them to plot the graphs of both equations, labelling each curve.

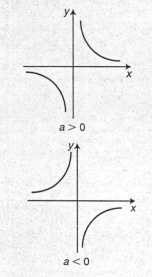
Other reciprocal graphs which involve division by an x term include, for example, $y = 1 + \dfrac{4}{x}$, $y = 2x - \dfrac{1}{x}$ and $y = \dfrac{12}{x-1}$.

Exercise 156

Draw the graphs between the stated x values after compiling a suitable table.

1 $y = \dfrac{4}{x}$ $-4 \leqslant x \leqslant 4$ **2** $y = -\dfrac{4}{x}$ $-4 \leqslant x \leqslant 4$

3 $y = \dfrac{10}{x}$ $-5 \leqslant x \leqslant 5$ **4** $y = -\dfrac{8}{x}$ $-4 \leqslant x \leqslant 4$

5 An insect colony decreases after the spread of a virus.
Its population y after t months is given by the equation

$$y = \dfrac{2000}{t}$$

valid for $1 \leqslant t \leqslant 6$.

a Copy and complete this table.

t (months)	1	2	3	4	5	6
y	2000					

b Draw a graph of y against t.
c Use your graph to estimate when the population decreases by 70% from its size after 1 month.
d How long does it take for the population to decrease from 1500 to 500?

6 A water tank springs a leak. The volume v, in m³, at time t hours after the leak occurs is given by the equation.

$$v = \dfrac{1000}{t}$$

valid for $1 \leqslant t \leqslant 20$.

a Copy and complete this table.

t (hours)	1	5		15	
v (m³)			100		50

b Draw the graph of v against t.

c Use the graph to estimate when the volume of water is reduced by 750 m³ from its value after 1 hour.

d How much water has been lost between 8 hours and 16 hours?

7 Lily calculates that the temperature of a cup of tea is t in degrees Celsius, m minutes after it has been made, given by the equation

$$t = \frac{k}{t}$$

where k is a constant and the equation is valid for $5 \leqslant m \leqslant 10$.

a Use the figures in this table to find the value of k, and hence copy and complete the table.

m (minutes)	5	6	7	8	9	10
t (°C)						40

b Draw the graph of t against m.

c Use your graph to estimate the temperature of a cup of tea after 450 s.

d After how long is the temperature of the cup of tea 60 °C?

e Lily is fussy, and she only drinks tea at a temperature of between 50 °C and 75 °C. Use your graph to find between what times Lily will drink her cup of tea.

Example 2

Draw the graph of $y = \frac{10}{x} + x - 5$ for $1 \leqslant x \leqslant 6$.

Construct a table and plot a graph from it.

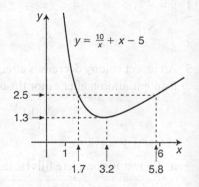

x	1	2	3	4	5	6
$\frac{10}{x}$	10	5	$3\frac{1}{3}$	$2\frac{1}{2}$	2	$1\frac{2}{3}$
x	1	2	3	4	5	6
−5	−5	−5	−5	−5	−5	−5
y	6	2	$1\frac{1}{3}$	$1\frac{1}{2}$	2	$2\frac{2}{3}$

Use the graph to estimate the smallest value of y, and the value of x where this occurs.

The graph shows that the minimum value of $y \simeq 1.3$ when $x \simeq 3.2$

For what values of x is y = 2.5? When y = 2.5, $x \simeq 1.7$ or 5.8.

Exercise 156*

Draw these graphs between the stated x values after compiling a suitable table.

1 $y = 1 + \frac{4}{x}$ $-4 \leqslant x \leqslant 4$ 2 $y = \frac{6}{x-2}$ $-3 \leqslant x \leqslant 6$

3 $y = x^2 + \frac{2}{x}$ $-4 \leqslant x \leqslant 4$ 4 $y = \frac{8}{x} + x - 4$ $1 \leqslant x \leqslant 6$

5 Jacqui is training for an athletics match at school. She experiments with various angles of projection x for putting the shot, and finds that for $30° \leqslant x \leqslant 60°$ the horizontal distance d, in metres, is given by

$$d = 100 - x - \frac{2000}{x}$$

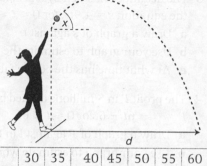

a Copy and complete this table. Then draw a graph of d against x for $30° \leqslant x \leqslant 60°$.

b Use your graph to estimate the greatest distance obtained by Jacqui, and the angle necessary to achieve it.

c What values can x take if Jacqui is to put the shot at least 9 m?

x (°)	30	35	40	45	50	55	60
100			100				
$-x$			-40				
$-\dfrac{2000}{x}$			-50				
d (m)			10				

6 A snowboard company called Zoom hires out x hundred boards per week. The amount received R and the costs C, both measured in \$1000s, are given by

$$R = \frac{6x}{x+1} \quad \text{and} \quad C = x + 1$$

both valid for $0 \leqslant x \leqslant 5$.

a Compile a table for $0 \leqslant x \leqslant 5$ and then draw graphs of R against x and C against x on a single set of axes.

b Zoom's profit P in \$1000s per week is given by $P = R - C$. Use your graphs to estimate how many boards must be loaned out per week for Zoom to make a profit.

c What is the greatest weekly profit that Zoom can make, and the number of boards that must be hired out per week for this to be achieved?

7 The Dimox paint factory wants to store $50\,\text{m}^3$ of paint in a closed cylindrical tank. To reduce costs, it wants to use the minimum possible surface area (including the top and bottom).

a If the total surface area of the tank is $A\,\text{m}^2$, and the radius is $r\,\text{m}$, show that the height $h\,\text{m}$ of the tank is given by

$$h = \frac{50}{\pi r^2}$$

and use this to show that

$$A = 2\pi r^2 + \frac{100}{r}$$

b Construct a suitable table of values for $1 \leqslant r \leqslant 5$ and draw a graph of A against r.

c Use your graph to estimate the value of r that produces the smallest surface area, and this value of A.

Exercise 157 (Revision)

Draw the graphs of these equations between the stated x values by first compiling suitable tables of values.

1 $y = x^3 + x - 3$ $-3 \leqslant x \leqslant 3$

2 $y = x^3 + x^2 + 3$ $-3 \leqslant x \leqslant 3$

3 The distance s m fallen by a pebble from a clifftop t seconds after the pebble falls is given by the equation $s = 4.9t^2$, for $0 \leqslant t \leqslant 4$.

  a Draw a graph of s against t.

  b Use your graph to estimate the distance fallen by the pebble after 2.5 s.

  c At what time has the pebble fallen 50 m?

4 The profit P in \$ millions earned by Pixel Internet after t years is given by the equation $P = \quad -6t - 6$, for $0 \leqslant t \leqslant 6$.

  a Draw a graph of P against t in the given range by first compiling a suitable table.

  b Use your graph to estimate when Pixel Internet first made a profit.

  c How long will it take for a profit of \$100 000 000 to be made?

5 Nick goes on a strict diet, which claims that his weight w kg after t weeks between weeks 30 and 40 will be given by

$$w = \frac{k}{t}$$

where k is a constant whole number.

  a Use the data in this table to find the value of k, and hence copy and complete the table, giving your answers correct to 2 significant figures.

t (weeks)	30	32	34	36	38	40
w (kg)						70

  b Draw a graph of w against t.

  c Use your graph to estimate when Nick's weight should reach 80 kg.

  d Why is there only a limited range of t values for which the given equation works?

6 The temperature of Kim's cup of coffee, t °C, m minutes after it has been poured into her mug, is given by the equation

$$t = \frac{425}{m} \qquad \text{valid for } 5 \leqslant m \leqslant 10$$

  a Copy and complete this table and use it to draw the graph of t against m for $5 \leqslant m \leqslant 10$.

m	5	6	7	8	9	10
t	85			53.1		

  b Kim likes her coffee between the temperatures of 50 °C and 70 °C. Use your graph to find at what times Kim should drink her coffee.

Exercise 157* (Revision)

Draw the graphs of these equations between the stated x values by first compiling tables of values.

  1 $y = 2x^3 - x^2 - 3x$ $-3 \leqslant x \leqslant 3$ **2** $y = 3x(x + 2)^2 - 5$ $-3 \leqslant x \leqslant 2$

3 The equation for the flight path of a golfer's shot is $y = 0.2x - 0.001x^2$, for $0 \leqslant x \leqslant 200$, where y m is the ball's height, and x m is the horizontal distance moved by the ball.

  a Draw a graph of y against x by first compiling a suitable table of values between the stated x values.

  b Use your graph to estimate the maximum height of the ball.

  c Between what distances is the ball at least 5 m above the ground?

4 The flow $Q\,\text{m}^3/\text{s}$ of a small river t hours after midnight is monitored after a storm, and is given by the equation $Q = t^3 - 8t^2 + 14t + 10$, for $0 \leqslant t \leqslant 5$.

  a Draw a graph of Q against t by first constructing a suitable table between the stated t values.

  b Use your graph to estimate the maximum flow and the time when this occurs.

  c Between what times does the river flood, if this occurs when the flow exceeds $10\,\text{m}^3/\text{s}$?

5 A goat farmer wants to enclose $600\,\text{m}^2$ for a rectangular goat pen. Three of the sides will consist of fencing, while the remaining side will consist of an existing stone wall. Two sides of the fence will be of length x metres.

  a Write the length of the third side in terms of x.

  b Show that the total length L in metres of the fence is given by

$$L = 2x + \frac{600}{x}$$

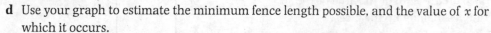

  c Construct a suitable table of values for $5 \leqslant x \leqslant 40$. Use this table to draw a graph of L against x.

  d Use your graph to estimate the minimum fence length possible, and the value of x for which it occurs.

  e What values can x take, if the fence length is not to exceed $75\,\text{m}$?

6 The graph of $y = x^3 - 5x^2 + ax + b$, where a and b are constants, passes through the points $P(0, 5)$ and $Q(1, 6)$.

  a Show that the values of a and b are both 5 by making sensible substitutions using the co-ordinates of P and Q.

  b Copy and complete this table for $-1 \leqslant x \leqslant 4$.

x	-1	0	1	2	3	4
y		5	6			9

  c Draw the graph of $y = x^3 - 5x^2 + 5x + 5$

  d The curve represents a cross-section of a hillside. The top of the curve (R) represents the top of a hill and the bottom of the curve (S) represents the bottom of a valley. State the co-ordinates of R and S.

Surface areas and volumes of solids

Prisms

> **Remember**
>
> Any solid with parallel sides that has a constant cross-section is called a prism.
> Volume of a prism = cross-sectional area × height
>
> A **cuboid** is a prism with a rectangular cross-section.
> Volume of a cuboid = width × depth × height
> Surface area = sum of the area of the six rectangles making up the faces.
>
> A **cylinder** is a prism with a circular cross-section.
> If the height is h and the radius r, then
> Volume of a cylinder = $\pi r^2 h$
> Curved surface area of a cylinder = $2\pi rh$

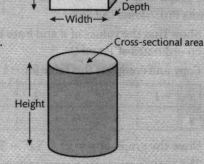

Example 1

Calculate the volume and surface area of the prism shown.

The cross-section is a right-handed triangle.

> Cross-sectional area
> $= \frac{1}{2} \times 3 \times 4 = 6\,\text{cm}^2$

So   Volume = $6 \times 8 = 48\,\text{cm}^3$

> Surface area
> = two end triangles plus three rectangles
> $= 2 \times 6 + 5 \times 8 + 3 \times 8 + 4 \times 8$
> $= 108\,\text{cm}^2$

Example 2

Calculate the volume and surface area of a cola can that is a cylinder with diameter 6 cm and height 11 cm.

Using $V = \pi r^2 h$ with $r = 3$ and $h = 11$
> $V = \pi \times 3^2 \times 11$
> $= 311\,\text{cm}^3$ to 3 s.f.

Surface area = two ends plus curved surface area
> $A = 2 \times \pi r^2 + 2\pi rh$
> $= 2 \times \pi \times 3^2 + 2 \times \pi \times 3 \times 11$
> $= 264\,\text{cm}^2$ to 3 s.f.

Exercise 158

1 Find the volume of the prism shown.

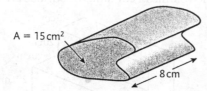

A = 15 cm²

8 cm

2 Find the volume of the prism shown.

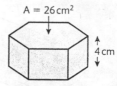

A = 26 cm²

4 cm

3 Find the volume and surface area of this wedge of cheese.

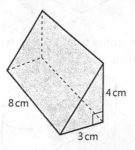

4 cm

8 cm

3 cm

4 Find the volume and surface area of this pack of butter.

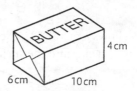

BUTTER

4 cm

6 cm   10 cm

5 Find the volume and surface area of this fuel tank.

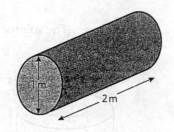

1 m

2 m

6 Find the volume and surface area of this can of drink.

16 cm

ORANGE

6 cm

7 A swimming pool has the dimensions shown. Find the volume in m³.

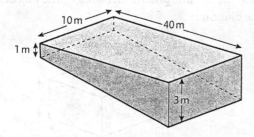

10 m

40 m

1 m

3 m

8 A penthouse shed has the dimensions shown. Find the volume in m³.

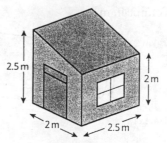

2.5 m

2 m

2 m   2.5 m

9 A barn has the dimensions shown.
The volume is 686 m³.
Find the length in m to 3 s.f.

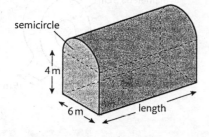

semicircle

4 m

6 m   length

10 A carton contains 1 litre of orange juice.
The carton is 10 cm wide and 6 cm deep.
How tall is it?

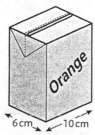

Orange

6 cm   10 cm

Exercise 158*

1 The diagram shows an extrusion for a conservatory roof. Find the volume of the extrusion in cm³.

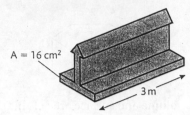

A = 16 cm²

3 m

2 The diagram shows some stage steps. Find the volume in cm³ and the surface area in cm².

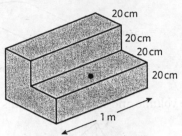

20 cm
20 cm
20 cm
20 cm
1 m

3 The diagram shows a metal 'T' girder. Find the volume in cm³ and the surface area in cm².

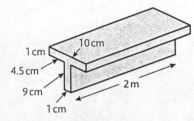

1 cm
10 cm
4.5 cm
9 cm
2 m
1 cm

4 The diagram shows a can of food with semicircular ends. Find the volume and surface area.

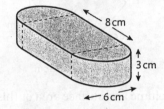

8 cm
3 cm
6 cm

5 The diagram shows a CD player with semicircular ends. Find the volume and surface area.

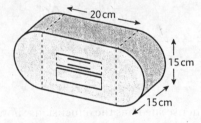

20 cm
15 cm
15 cm

6 The diagram shows a sweet. The volume is 1.18 cm³. Find the height in mm to 2 s.f.

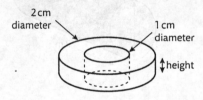

2 cm diameter
1 cm diameter
height

7 The diagram shows a hollow concrete pipe. Find the volume in m³.

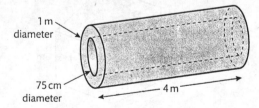

1 m diameter
75 cm diameter
4 m

8 Find the volume and surface area of the object shown in the diagram.

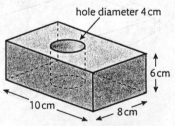

hole diameter 4 cm
6 cm
10 cm
8 cm

9 A toilet roll has the dimensions shown. If the thickness of the paper is $\frac{1}{5}$ mm, find the length of paper on the roll.

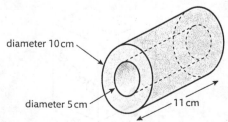

diameter 10 cm
diameter 5 cm
11 cm

10 A reel of sticky tape has the dimensions shown. If the tape is 25 m long, how thick is the tape?

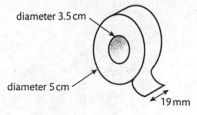

diameter 3.5 cm
diameter 5 cm
19 mm

Pyramids, cones and spheres

Remember

Volume of a pyramid $= \frac{1}{3} \times$ base area $\times$ vertical height

Surface area = area of the base plus the triangular faces.

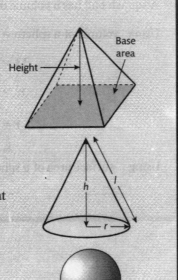

A cone is a pyramid with a circular base.

Volume of a cone $= \frac{1}{3} \times \pi r^2 \times h$

Curved surface area of a cone $= \pi r l$, where l is the slant height

Volume of a sphere $= \frac{4}{3}\pi r^3$

Surface area of a sphere $= 4\pi r^2$

Example 3

Find the volume of the rectangular-based pyramid shown.

Using $V = \frac{1}{3} \times$ base area $\times$ vertical height

$V = \frac{1}{3} \times 8 \times 10 \times 12$

$= 320 \text{ cm}^3$

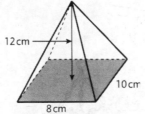

Example 4

Find the total surface area of the cone shown.

Use Pythagoras' Theorem to work out l.

$l^2 = 5^2 + 12^2$

$l^2 = 169$

$l = 13$

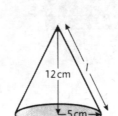

Curved surface area of a cone $= \pi r l$

$= \pi \times 5 \times 13$

$= 204 \text{ cm}^2$

The base is a circle with area $\pi r^2 = \pi \times 5 \times 5$

$= 78.5 \text{ cm}^2$

Total surface area = curved surface area plus base area

$= 204 + 78.5$

$= 283 \text{ cm}^2$ to 3 s.f.

Example 5

A squash ball has a volume of 33 cm^3. Find the radius and surface area.

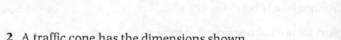

Using   volume of a sphere $= \frac{4}{3}\pi r^3$

$$33 = \frac{4}{3}\pi r^3 \qquad \text{(Make } r \text{ the subject of the equation)}$$

$$r^3 = \frac{33 \times 3}{4 \times \pi}$$

$$r = \sqrt[3]{\frac{33 \times 3}{4 \times \pi}}$$

$$= 1.99 \text{ cm}$$

Using   surface area of a sphere $= 4\pi r^2$

$$A = 4 \times \pi \times 1.99^2$$

$$= 49.8 \text{ cm}^2 \text{ to 3 s.f.}$$

Exercise 159

1 The Great Pyramid of Giza is a square-based pyramid with dimensions as shown. Find the volume in m^3.

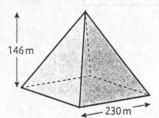

146 m
230 m

2 A traffic cone has the dimensions shown. Find the volume in cm^3 and the curved surface area in cm^2.

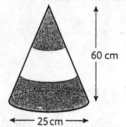

60 cm
25 cm

3 A hanging flower basket is a hemisphere with diameter 30 cm. Find the volume and the external curved surface area.

4 A scoop for ground coffee is a hollow hemisphere with diameter of 4 cm.
When full, the coffee forms a cone on top of the scoop. Find the volume of coffee.

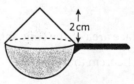

2 cm

5 A grain silo is a cylinder with a hemisphere on top, with the dimensions shown.
Find the volume and surface area including the base.

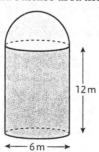

12 m
6 m

6 The volume of a football is 7240 cm^3. Find the radius and surface area.

7 A flat roof is a rectangle 6 m by 8 m.
The rain drains into a cylindrical water butt with radius 50 cm.
By how much does the water level in the butt rise if 1 cm of rain falls?
(Assume the butt does not overflow.)

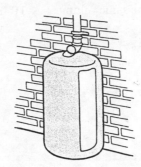

Exercise 159*

1 A crystal consists of two square-based pyramids as shown.
Calculate the volume of the crystal.

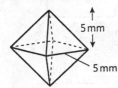

5 mm

5 mm

2 An ice-cream cone is full of ice-cream as shown.
What is the volume of ice cream?

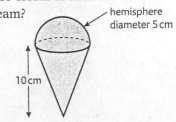

hemisphere
diameter 5 cm

10 cm

3 A water bottle is a cylinder with a cone at one end and a hemisphere at the other.
Find the volume and surface area.

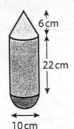

6 cm

22 cm

10 cm

4 A monument in South America is in the shape of a truncated pyramid.
Find the volume of the monument.

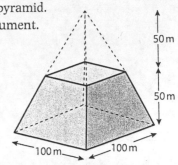

50 m

50 m

100 m   100 m

5 A vase is a truncated cone.
Find the volume of the vase.

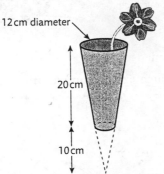

12 cm diameter

20 cm

10 cm

6 The volume of the Earth is 1.09×10^{12} km^3.
Find the surface area of the Earth assuming it is a sphere.

7 A stone ball is dropped into a barrel of water and sinks to the bottom. The ball is completely covered by water.
By how much does the water rise in the barrel?

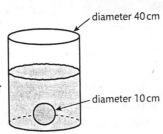

diameter 40 cm

diameter 10 cm

8 A cuboid of chocolate measures 12 cm by 8 cm by 6 cm.
It is melted down and cast into chocolates that are spheres with diameter 2 cm. How many spheres can be made?

9 The sphere and the cone shown have the same volume. Calculate the height of the cone.

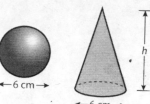

10 A spherical drop of oil with diameter 3 mm falls onto a water surface and produces a circular oil film of radius 10 cm. Calculate the thickness of the oil film.

Areas of similar shapes

When a shape doubles in size, then the area does **NOT** double, but increases by a factor of four.

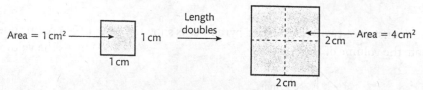

The Length Scale Factor is 2, and the Area Scale Factor is 4.

If the shape triples in size, then the area increases by a factor of nine.

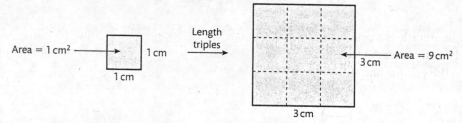

If a shape increases by a Length Scale Factor of k, then the Area Scale Factor is k^2.

This applies even if the shape is irregular.

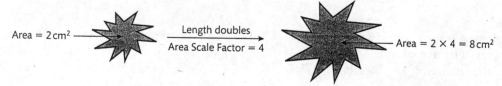

Note: The two shapes must be similar.

Two similar shapes that are the same size are **congruent**.

Example 6

The two shapes shown are similar.
The area of the smaller shape is $10\,cm^2$.
Find the area of the larger shape.

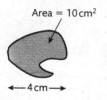

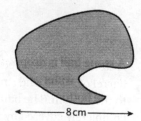

The Length Scale Factor $k = \dfrac{8}{4} = 2$

(Note: Divide the length of the second shape by the length of the first shape.)

The Area Scale Factor $k^2 = 2^2 = 4$

So the area of the larger shape is $10 \times 4 = 40\,cm^2$.

Example 7

The two shapes shown are similar.
The area of the larger shape is
18 cm². Find the area of the
smaller shape.

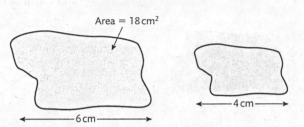

Area = 18 cm²

The Length Scale Factor $k = \frac{4}{6} = \frac{2}{3}$

(Note: Divide the length of the second
shape by the length of the first shape.)

The Area Scale Factor $k^2 = \left(\frac{2}{3}\right)^2 = \frac{4}{9}$

So the area of the smaller shape is $18 \times \frac{4}{9} = 8$ cm².

Example 8

The two triangles are similar, with
dimensions and areas as shown.
What is the value of x?

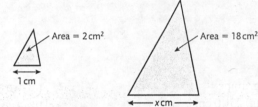

The Area Scale Factor $k^2 = \frac{18}{2} = 9$

(Note: Divide the area of the second
shape by the area of the first shape.)

The Length Scale Factor $k = \sqrt{9} = 3$

So $x = 1 \times 3 = 3$ cm.

Exercise 160

1 A and B are similar shapes. The area of A is 4 cm².
Find the area of B.

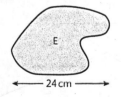

2 a Why are the two triangles shown similar?
b If the area of T_1 is 3.8 cm², find the area of T_2.

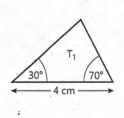

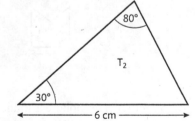

3 E and F are similar shapes. The area of E is 480 cm².
Find the area of F.

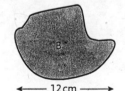

4 I and J are similar shapes. The area of I is 150 cm².
Find the area of J.

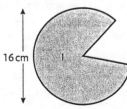

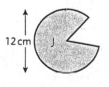

5 The shapes M and N are similar. The area of M is 8 cm²
and the area of N is 32 cm².
Find x.

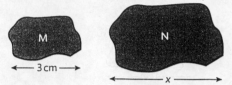

6 Q and R are similar shapes. The area of Q is 5 cm²
and the area of R is 11.25 cm².
Find x.

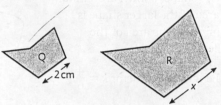

7 U and V are similar shapes.
The area of U is 48 cm² and the area of V is 12 cm².
Find x.

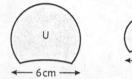

8 Y and Z are similar ellipses.
The area of Y is 225 cm² and the area of Z is 100 cm².
Find x.

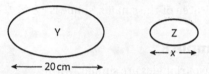

Exercise 160*

1 The two stars are similar in shape.
The area of the smaller star is 300 cm².
Find the area of the larger star.

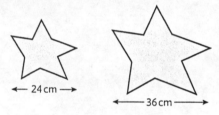

2 The two shapes shown are similar.
The area of the larger shape is 125 cm².
Find the area of the smaller shape.

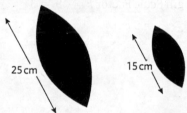

3 The two shapes are similar.
Find x.

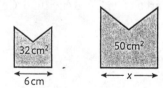

4 The two leaves shown are similar in shape.
Find x

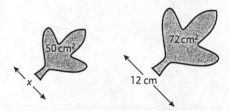

5 A model aeroplane is made to a scale of $\frac{1}{20}$ of the size of
the real plane. The area of the wings of the real plane are
40 m². Find the area of the wings of the model in cm².

6 An oil slick increases in length by 20%. Assuming the
new shape is similar to the original shape, what is the
percentage increase in area?

7 Meera washes some napkins in hot water and they shrink
by 10%. What is the percentage reduction in area?

8 Calculate the shaded area A.

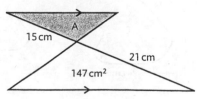

Volumes of similar shapes

When a solid doubles in size, the volume does **NOT** double, but increases by a factor of eight.

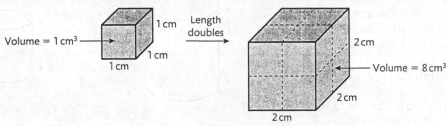

The Length Scale Factor is 2, and the Volume Scale Factor is 8.

If the solid triples in size, then the volume increases by a factor of 27.

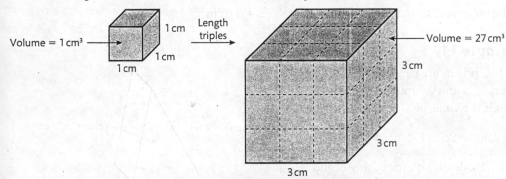

If a solid increases by a Length Scale Factor of k, then the Volume Scale Factor is k^3.

This applies even if the solid is irregular.

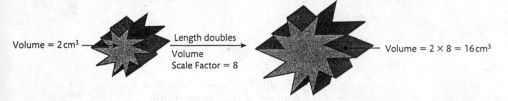

Note: The two solids must be similar.

Example 9

The two solids shown are similar.
The volume of the smaller solid is $20\,cm^3$.
Find the volume of the larger solid.

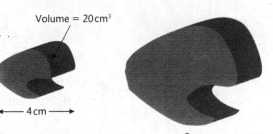

The Length Scale Factor $k = \dfrac{8}{4} = 2$ (Note: Divide the length of the second solid by the length of the first solid.)

The Volume Scale Factor $k^3 = 2^3 = 8$

So the volume of the larger solid is $20 \times 8 = 160\,cm^3$.

Key Point
If the Length Scale Factor
is k, then the Volume
Scale Factor is k^3.

Example 10

The two cylinders shown are similar. The volume of the larger cylinder is 54 cm³. Find the volume of the smaller cylinder.

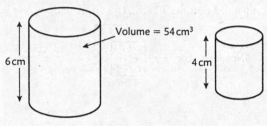

The Length Scale Factor $k = \frac{4}{6} = \frac{2}{3}$

(Note: Divide the height of the second cylinder by the height of the first cylinder.)

The Volume Scale Factor $k^3 = \left(\frac{2}{3}\right)^2 = \frac{8}{27}$

So the volume of the smaller cylinder is $54 \times \frac{8}{27} = 16$ cm³.

Example 11

The two prisms are similar, with dimensions and volumes as shown. What is the value of x?

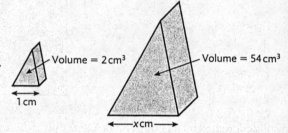

The Volume Scale Factor $k^3 = \frac{54}{2} = 27$

(Note: Divide the volume of the second solid by the volume of the first solid.)

The Length Scale Factor $k = \sqrt[3]{27} = 3$

So $x = 1 \times 3 = 3$ cm.

Exercise 161

1 The cylinders shown are similar. Find the volume of the larger cylinder.

2 The statues shown are similar. Find the volume of the larger statue.

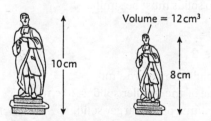

3 Find the volume of the smaller sphere. (Note: All spheres are similar.)

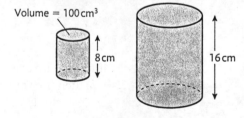

4 The two mobile phones are similar. Find the volume of the smaller phone.

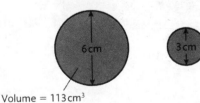

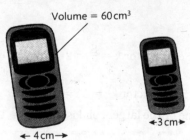

5 The two glasses are similar.
Find the height of the larger glass.

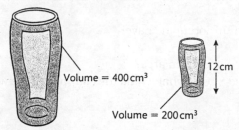

Volume = 400 cm³
Volume = 200 cm³
12 cm

6 The two radios are similar.
Find the width of the larger radio.

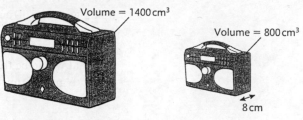

Volume = 1400 cm³
Volume = 800 cm³
8 cm

7 The two eggs are similar.
Find the height of the smaller egg.

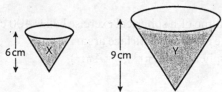

6 cm
Volume = 100 cm³   Volume = 60 cm³

8 The two buckets are similar.
Find the height of the smaller bucket.

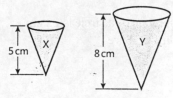

30 cm
Volume = 10 000 cm³   Volume = 8000 cm³

Exercise 161*

1 X and Y are similar shapes. The volume of X is 40 cm³.
Find the volume of Y.

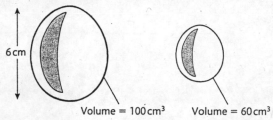

6 cm   X   9 cm   Y

2 X and Y are similar shapes. The volume of Y is 128 cm³.
Find the volume of X.

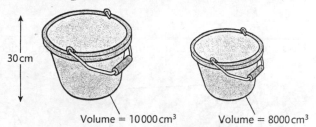

5 cm   X   8 cm   Y

3 The two candlesticks are similar.
Find the height of the larger candlestick.

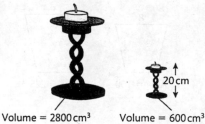

20 cm
Volume = 2800 cm³   Volume = 600 cm³

4 The two bottles of shampoo are similar.
Find the height of the smaller bottle.

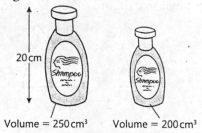

20 cm
Shampoo   Shampoo
Volume = 250 cm³   Volume = 200 cm³

5 The manufacturers of a chocolate bar decide to produce a similar bar by increasing all dimensions by 20%. What would be a fair percentage increase in price?

6 'ET Pizza' produces two pizzas that are similar in shape. The smaller pizza is 20 cm in diameter and costs $10. The larger pizza is 30 cm in diameter. What is a fair cost for the larger pizza?

7 A supermarket stocks similar small and large cans of beans. The areas of their labels are 63 cm² and 112 cm² respectively.
a The weight of the large can is 640 g. What is the weight of the small can?
b The height of the small can is 12 cm. What is the height of the large can?

8 A solid sphere weighs 10 g.
a What will be the weight of another sphere made from the same material but having three times the diameter?
b The surface area of the 10 g sphere is 20 cm². What is the surface area of the larger sphere?

9 Suppose that a grown-up porcupine is an exact enlargement of a baby porcupine on a scale of $3:2$, and that the baby porcupine has 2000 quills with a total length of 15 m and a skin area of 360 cm².

   a How many quills would the grown-up porcupine have?

   b What would be their total length?

   c What would be the grown-up porcupine's skin area?

   d If the grown-up porcupine weighed 810 g, what would the baby porcupine weigh?

Exercise 162 (Revision)

1 Find the surface area and volume of the following objects.

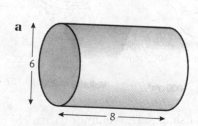
a

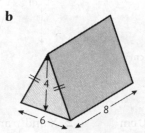
b

2 The circumference of a circle is 12 cm. Find the radius and area.

3 The volume of a spherical exercise ball is 400 cm³. Find the radius and surface area. (For a sphere $V = \frac{4}{3}\pi r^3$, $A = 4\pi r^2$)

4 The two triangles are similar. Find the perimeter of the smaller triangle and the area of the larger triangle.

4   5

Area = 12   Perimeter = 25

5 The two shapes are similar. Find the length marked x.

8   x

Area = 100   Area = 225

6 The two cylinders are similar. Find the volume of the larger cylinder and the surface area of the smaller cylinder.

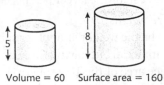

Volume = 60   Surface area = 160

7 The two yoghurt pots are similar. Find the height of the larger pot and the surface area of the smaller pot.

 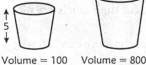

Volume = 100   Volume = 800

Surface area = 480

Exercise 162* (Revision)

1 Find the surface area and volume of the following objects. All dimensions are in centimetres.

a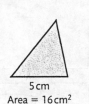
hole $r = 3$
depth $= 2$
5
$r = 6$

b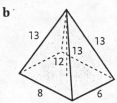
13   13
13
12
8   6

2 The perimeter of a semicircle is 24 cm.
Find the radius and area.

3 The surface area of a solid hemisphere is 76 cm². Find the radius and volume.
(For a sphere $V = \frac{4}{3}\pi r^3$, $A = 4\pi r^2$)

4 The two triangles are similar. Find the perimeter of the smaller triangle and the area of the larger triangle.

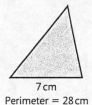

5 cm
Area = 16 cm²

7 cm
Perimeter = 28 cm

5 The two shapes are similar.
Find the length marked x.

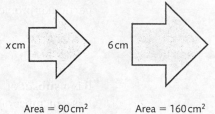

x cm   6 cm

Area = 90 cm²   Area = 160 cm²

6 The two packets of cereal are similar. Find the volume of the smaller packet and the surface area of the larger packet.

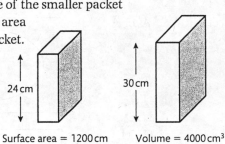

24 cm   30 cm

Surface area = 1200 cm   Volume = 4000 cm³

7 The volume of the Earth is 1.08×10^{12} km³ and its diameter is 12 740 km. The volume of the Moon is 2.2×10^{10} km³ and its surface area is 3.8×10^7 km². Assuming the Earth and Moon are similar, calculate the diameter of the Moon and the surface area of the Earth.

Problems involving sets

Remember

A **set** is a collection of objects, described by a list or a rule.

$A = \{1, 3, 5\}$

Each object is an **element** or **member** of the set.

$1 \in A, 2 \notin A$

Sets are **equal** if they have exactly the same elements.

$B = \{5, 3, 1\}, B = A$

The **number of elements** of set A is given by $n(A)$.

$n(A) = 3$

The **empty set** is the set with no members.

$\{ \}$ or $\varnothing$

The **universal set** contains all the elements being discussed in a particular problem.

$\mathcal{E}$

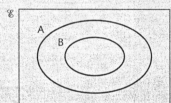

B is a **subset** of A if every member of B is a member of A.

$B \subset A$

The **complement** of set A is the set of all elements not in A.

A' or $\overline{A}$

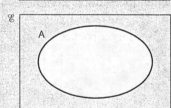

The **intersection** of A and B is the set of elements which are in both A and B.

$A \cap B$

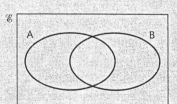

The **union** of A and B is the set of elements which are in A or B or both.

$A \cup B$

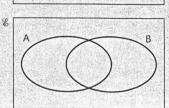

Entering the information from a problem into a Venn diagram often means the numbers in the sets can be worked out. Sometimes it is easier to use some algebra as well.

Example 1

In a class of 23 students, 15 like coffee, 13 like tea and 4 students don't like either drink. How many like

a tea only   **b** coffee only   **c** both drinks?

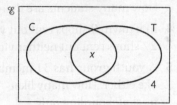

Enter the information into a Venn diagram in stages. Let C be the set of coffee drinkers and T the set of tea drinkers. Let x be the number of students who like both. The 4 students who don't like either drink can be put in, along with x for the students who like both.

Now the other information can be added.

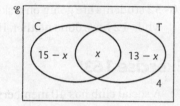

The number who like coffee, tea or both is $23 - 4 = 19$.
This means $n(C \cup T) = 19$.
So $(15 - x) + x + (13 - x) = 19$
$28 - x = 19$
$x = 9$

So 9 students like both.
The number liking tea only is $13 - x = 4$.
The number liking coffee only is $15 - x = 6$.

Example 2

For two sets A and B, $n(A) = 12$, $n(B) = 8$, $n(A \cap B) = 5$.
Show that $n(A \cup B) = n(A) + n(B) - n(A \cap B)$

The information is shown on a Venn diagram.

From the diagram, $n(A \cup B) = 7 + 5 + 3 = 15$

Substituting into $n(A \cup B) = n(A) + n(B) - n(A \cap B)$ gives $15 = 12 + 8 - 5$ which is true.

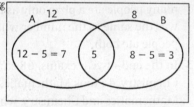

Activity 39

Use the method of Example 2 to show that $n(A \cup B) = n(A) + n(B) - n(A \cap B)$ when
a $n(A) = 12, n(B) = 8, n(A \cap B) = x$ **b** $n(A) = 12, n(B) = y, n(A \cap B) = x$
c $n(A) = z, n(B) = y, n(A \cap B) = x$

Key Point
For two sets A and B,
$n(A \cup B) =$
$n(A) + n(I \qquad n(A \cap B)$

Exercise 163

1 In a class of 40 students, 18 had watched 'Next Door' last night, 23 had watched 'Westenders' and 7 had watched both programmes. How many students did not watch either programme?

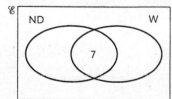

2 There are 182 spectators at a football match. 79 are wearing a hat, 62 are wearing a scarf and 27 are wearing a hat but not a scarf. How many are wearing neither a hat nor a scarf?

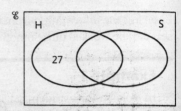

3 In one form in a school, 13 students are studying media, 12 students are studying sociology, 8 students are studying both and 5 students are doing neither subject. How many students are there in the form?

4 In a town, 9 shops rent out videos, 12 shops rent out DVDs, 7 shops rent out both and 27 shops rent out neither videos nor DVDs. How many shops are there in the town?

5 A youth group has 31 members. 15 like skateboarding, 13 like roller-skating and 8 don't like either. How many like

    a skateboarding only   **b** roller-skating only   **c** both?

6 52 students are going on a skiing trip. 28 have skied before, 30 have snowboarded before, while 12 have done neither. How many have done both sports before?

Exercise 163*

1 A social club has 40 members. 18 like singing, 7 like both singing and dancing, while 6 like neither. How many like dancing?

2 At Siti's party there were both pizzas and burgers to eat. Some people had one of each. The number who had a burger only was seven more than the number who had both. Twice as many people ate a pizza only as the number who had both. The number who ate neither was the same as the number who had both. If there were 57 people at the party, how many people ate both?

3 In the end-of-year exams, 68 students took Mathematics, 72 took Physics and 77 took Chemistry. 44 took Mathematics and Physics, 55 took Physics and Chemistry, 50 took Mathematics and Chemistry, while 32 took all three subjects. Draw a Venn diagram to represent this information and hence calculate how many students took these three exams.

4 A group of 40 teenagers have all seen the film 'Parry Hotter'. 22 have seen it on DVD, 23 have seen it on video and 17 have seen it at the cinema. 12 have seen it on DVD and video, 6 have seen it on video and at the cinema, and 7 have seen it on DVD and at the cinema. How many have seen it on DVD, on video and at the cinema? (Hint – in the Venn diagram let x be the number who have seen it on DVD, on video and at the cinema.)

5 In a form of 25 students, 19 have scientific calculators and 14 have graphic calculators. If x students have both and y students have neither, what are the largest and smallest possible values of x and y?

6 It is claimed that 75% of teenagers can ride a bike and 65% can swim. What can be said about the percentage who do both?

Identifying sets by shading

Sometimes it can be difficult to find the intersection or union of sets in a Venn diagram. If one set is shaded in one direction and the other set in another direction, then the intersection is given wherever there is cross shading; the union is given by any shading at all.

Example 3

Show on a Venn diagram

a $A' \cap B$ **b** $A' \cup B$

The diagrams show first the sets A and B, then the set A' shaded one way, then the set B shaded the other way.

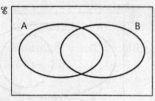
Sets A and B

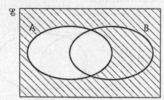
Set A' shaded one way

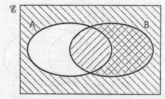
Set B shaded the other way

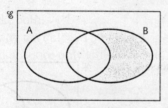
Shading shows $A' \cap B$

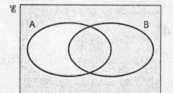
Shading shows $A' \cap B$

Exercise 164

1 On copies of Diagram 1 shade the following sets:

  a $A \cap B'$

  b $A \cup B'$

  c $A' \cap B'$

  d $A' \cup B'$

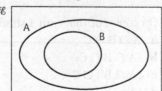
Diagram 1

2 On copies of Diagram 2 shade the following sets:

  a $A \cap B'$

  b $A \cup B'$

  c $A' \cap B'$

  d $A' \cup B'$

Diagram 2

3 On copies of Diagram 3 shade the following sets:

  a $A \cap B \cap C$

  b $A' \cup (B \cap C)$

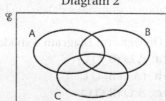
Diagram 3

4 On copies of Diagram 3 shade the following sets:

   a $(A \cup B') \cap C$

   b $A \cup B \cup C'$

5 Describe the shaded sets using set notation.

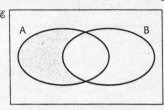 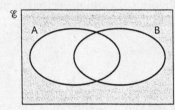 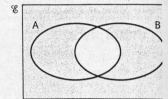

6 Describe the shaded sets using set notation.

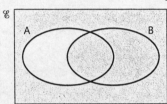 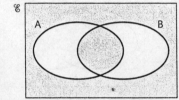 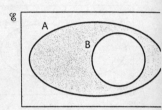

Exercise 164*

1 On copies of Diagram 1 shade the following sets:

   a $(A' \cap B)'$

   b $(A \cup B')'$

   c $(A \cap B')'$

   d $(A \cup B)'$

Diagram 1

2 On copies of Diagram 2 shade the following sets:

   a $(A \cap B')'$

   b $(A \cup B')'$

   c $(A \cap B)'$

   d $(A' \cup B)'$

Diagram 2

3 On copies of Diagram 3 shade the following sets:

   a $A \cap B \cap C$

   b $(A \cap B \cap C)'$

   c $(A \cap B') \cup C$

   d $(A \cup B)' \cap C$

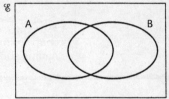

Diagram 3

4 On copies of Diagram 3 shade the following sets:

   a $A \cup (B' \cap C)$

   b $(A \cap B) \cup C'$

   c $A \cup (B \cap C)$

   d $(A \cup B) \cap C$

5 Describe the shaded sets using set notation.

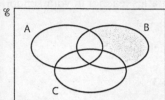 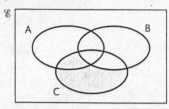 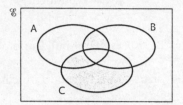

6 Describe the shaded sets using set notation.

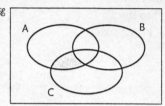 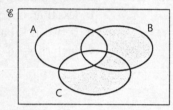 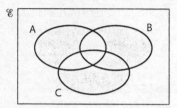

7 This question is about **De Morgan's Laws**.

These state that $(A \cup B)' = A' \cap B'$ and $(A \cap B)' = A' \cup B'$.

Shade copies of Diagram 4 to show the following sets:
$A \cup B, (A \cup B)', A', B'$ and $A' \cap B'$ and thus prove the
first law. Use a similar method to prove the second law.

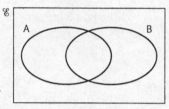

Diagram 4

Set-builder notation

Sets can be described using **set-builder notation**:
$A = \{x \text{ such that } x > 2\}$ means 'A is the set of all x such that x is greater than 2'.
Rather than write '**such that**', the notation $A = \{x: x > 2\}$ is used.

$B = \{x: x > 2\}$, x is a positive integer} means the set of positive integers x such that x is greater
than 2. This means $B = \{3, 4, 5, 6, ...\}$.

Certain sets of numbers are used so frequently that they are given special symbols.

$\mathbb{N}$ is the set of natural numbers or positive integers $\{1, 2, 3, 4, ...\}$.
$\mathbb{Z}$ is the set of integers $\{..., -2, -1, 0, 1, 2, ...\}$.
$\mathbb{Q}$ is the set of rational numbers.
$\mathbb{R}$ is the set of real numbers.

Example 4

The set $B = \{3, 4, 5, 6, ...\}$ is written as $B = \{x: x > 2, x \in \mathbb{N}\}$.
The set $A = \{x: x > 2, x \in \mathbb{R}\}$ is the set of all real numbers greater than two.
Note that $3.2 \in A$ but $3.2 \notin B$, so $A \neq B$.
The set $\{x: x \text{ is even}, x \in \mathbb{N}\}$ is the set $\{2, 4, 6, ...\}$.
The set $\{x: x = 3y, y \in \mathbb{N}\}$ is the set $\{3, 6, 9, 12, ...\}$.

Exercise 165

1 List the following sets:
  a $\{x: x \text{ is a weekday beginning with T}\}$ **b** $\{z: z \text{ is a colour in traffic lights}\}$
  c $\{x: x < 7, x \in \mathbb{N}\}$ **d** $\{x: -2 < x < 7, x \in \mathbb{Z}\}$

2 List the following sets:

  a {x: x is a continent}
  b {y: y is a Mathematics teacher in your school}
  c {x: $x \leqslant 5, x \in \mathbb{N}$}
  d {x: $-4 < x \leqslant 2, x \in \mathbb{Z}$}

3 Express in set-builder notation the set of natural numbers which are

  a less than 7
  b greater than 4
  c between 2 and 11 inclusive
  d between -3 and 3
  e odd
  f prime

4 Express in set-builder notation the set of natural numbers which are

  a greater than -3
  b less than or equal to 9
  c between 5 and 19
  d between -4 and 31 inclusive
  e multiples of 5
  f factors of 48

Exercise 165*

1 A = {x: $x \leqslant 6, x \in \mathbb{N}$}, B = {$x$: $x = 2y, y \in$ A}, C = {1, 3, 5, 7, 9, 11}
List the following sets:

  a B
  b {x: $x = 2y + 1, y \in$ C}
  c A $\cap$ B
  d Give a rule to describe B $\cup$ C

2 A = {x: $-2 \leqslant x \leqslant 2, x \in \mathbb{Z}$}, B = {$x$: $x = y^2, y \in$ A}, C = {x: $x = 2^y, y \in$ A}
List the following sets:

  a B
  b C
  c A $\cap$ B $\cap$ C
  d {(x, y): $x = y, x \in$ A, $y \in$ C}

3 List the sets

  a {x: $2^x = -1, x \in \mathbb{R}$}
  b {2^{-x}: $0 \leqslant x < 5, x \in \mathbb{Z}$}
  c {x: $x^2 + x - 6 = 0, x \in \mathbb{N}$}
  d {x: $x^2 + x - 6 = 0, x \in \mathbb{Z}$}

4 List the sets

  a {x: $x^2 + 1 = 0, x \in \mathbb{R}$}
  b {2^x: $0 \leqslant x \leqslant 5, x \in \mathbb{Z}$}
  c {x: $x^2 + 2x - 6 = 0, x \in \mathbb{Q}$}
  d {x: $x^2 + 2x - 6 = 0, x \in \mathbb{R}$}

5 Show the sets $\mathbb{N}, \mathbb{Z}, \mathbb{Q}, \mathbb{R}$ in a Venn diagram.

Exercise 166 (Revision)

1 $n(A) = 20$, $n(A \cap B) = 7$ and $n(A' \cap B) = 10$.

  a Draw a Venn diagram to show this information.
  b Find $n(B)$.
  c Find $n(A \cup B)$.

2 In a class of 20 students, 16 drink tea, 12 drink coffee and 2 students drink neither drink.

  a How many drink tea only?
  b How many drink coffee only?
  c How many drink both drinks?

3 In an island in the Caribbean, all the inhabitants speak either French or English.
69% speak French and 48% speak English.

  a What percentage speak both languages?
  b What percentage speak French only?
  c What percentage speak English only?

4 On copies of the diagram, shade the following sets:

  a $A' \cap B$

  b $(A \cup B)'$

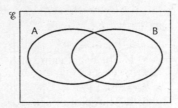

5 Describe the shaded set using set notation.

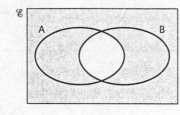

6 List the following sets:

  a $\{x: -3 < x < 4, x \in \mathbb{Z}\}$ **b** $\{x: x < 5, x \in \mathbb{N}\}$ **c** $\{x: -6 < x < -1, x \in \mathbb{N}\}$

7 Express in set-builder notation the set of natural numbers which are

  a even   **b** factors of 24   **c** between -1 and 4 inclusive

Exercise 166* (Revision)

1 $n(A \cap B \cap C) = 2$, $n(A \cup B \cup C)' = 5$, $n(A) = n(B) = n(C) = 15$
and $n(A \cap B) = n(A \cap C) = n(B \cap C) = 6$. How many are in the universal set?

2 A youth club has 140 members. 80 listen to pop music, 40 to rock music, 75 to heavy metal and 2 members don't listen to any music. 15 members listen to pop and rock only. 12 to rock and heavy metal only, while 10 listen to pop and heavy metal only. How many listen to all three types of music?

3 In a group of 50 students at a summer school, 15 play tennis, 20 play cricket, 20 swim and 7 students do nothing. 3 students play tennis and cricket, 6 students play cricket and swim, while 5 students play tennis and swim. How many do all three sports?

4 On copies of the diagram, shade the following sets:

  a $(A' \cap B) \cup C$

  b $A \cap (B \cup C)$

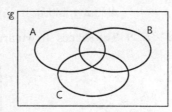

5 Describe the shaded sets using set notation.

  a

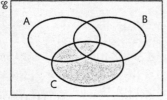

  b

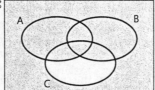

6 List the sets

  a $\{x: x^2 - 1 = 0, x \in \mathbb{R}\}$

  b $\{x: x^2 + 4x = 0, x \in \mathbb{Q}\}$

  c $\{x: x^2 + 2x + 2 = 0, x \in \mathbb{R}\}$

7 Express in set-builder notation the set of natural numbers which are

  a greater than -5 **b** between 4 and 12   **c** multiples of 3

1 The exact value of 8^{-2} is

    A -16 **B** -64 **C** $\frac{1}{64}$ **D** $-\frac{1}{64}$

2 The solutions to $(x - 1)(2x - 3) = 0$ are

    A $-1, 3$ **B** $1, 3$ **C** $1, \frac{3}{2}$ **D** $1, \frac{2}{3}$

3 The graph below is

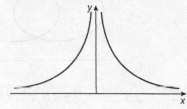

    A $y = \dfrac{1}{x}$ **B** $y = \dfrac{1}{x^2}$

    C $y = -\dfrac{1}{x}$ **D** $y = x^3$

4 A cylinder of volume 100π cm^3 and height 4 cm has a radius of

    A 5 cm   **B** 12.5 cm   **C** 20 cm   **D** 25 cm

5 The statement which describes the shaded region in the Venn diagram below is

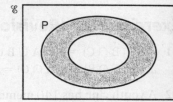

    A $P \cap Q$ **B** $P \cup Q$

    C $P' \cap Q$ **D** $P \cap Q'$

6 If $P = \{\text{Prime numbers}\}$, which statement is false?

    A $11 \in P$ **B** $4 \cap P = \varnothing$ **C** $\{23, 31\} \subset P$ **D** $13^0 \in P$

7 The point $(-1, p)$ lies on the curve $y = 2x^3 + x^2 - \dfrac{1}{x}$. The value of p is

    A 0   **B** 2   **C** -2 **D** -6

8 The quadratic equation $2x^2 + 3x - 4 = 0$ has solutions to 2 s.f. of

    A $-2.4, 0.85$ **B** $-0.85, 0.24$ **C** $-6.2, 0.20$ **D** $-6.2, 1.7$

9 The shapes shown are geometrically similar.

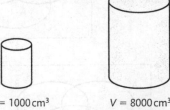

$v = 1000 \text{ cm}^3$ $V = 8000 \text{ cm}^3$
$a = ? \text{ cm}^2$ $A = 400 \text{ cm}^2$

The value of the smaller surface area (a cm^2) is

    A 20 cm^2 **B** 40 cm^2 **C** 50 cm^2 **D** 100 cm^2

10 If $x^{-\frac{3}{5}} = \frac{1}{8}$, the exact value of x is

    A $\frac{1}{32}$ **B** 6   **C** 10   **D** 32

1 Express in their simplest form:

a $x^2 \times x^5$ **b** $x^5 \div x^2$ **c** $(x^2)^5$ **d** $(x^{-1})^3$

2 Work these out and, where appropriate, leave your answer as a fraction.

a 3^{-2} **b** $8^{\frac{1}{3}}$ **c** $2^{-3} \times 2^2$ **d** $5^{-4} \div 5^{-2}$ **e** $(4^{-1})^2$

3 Simplify these.

a $b^2 \times b^{-1}$ **b** $c^{-1} \div c^2$ **c** $(b^2)^{-1}$ **d** $c^{\frac{1}{2}} \times c^{\frac{1}{2}}$ **e** $a^{\frac{1}{4}} \div a^{\frac{1}{4}}$

4 Solve for x:

a $2^x = 128$ **b** $4^x = \frac{1}{4}$ **c** $81^x = 9$ **d** $125^x = 5$

5 Solve the following equations by factorisation.

a $x^2 - 7x + 12 = 0$ **b** $x^2 + 7x + 10 = 0$
c $x^2 - 7x = 0$ **d** $5x^2 - 25x = 0$

6 Solve the following equations by factorisation.

a $2x^2 - 5x + 2 = 0$ **b** $3x^2 + 14x + 8 = 0$
c $2 - x - 3x^2 = 0$ **d** $x(2x + 1) = 10$

7 Solve the following equations by completing the square.

a $x^2 - 6x - 16 = 0$ **b** $x^2 + 4x - 8 = 0$
c $2x^2 - 16x + 12 = 0$ **d** $4x^2 + 8x - 12 = 0$

8 Solve the following equations by use of the formula to 3 s.f.

a $3x^2 - 4x - 5 = 0$ **b** $5x^2 - 8x + 2 = 0$
c $(2x - 1)(x + 3) = 8$ **d** $x + \frac{1}{x} = 5$

9 Shape I is geometrically similar to shape II.
Find the value of a.

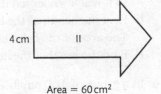

2 cm   I

4 cm   II

Area = a cm²

Area = 60 cm²

10 Shape I is geometrically similar to shape II.
Find the value of a and b.

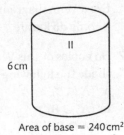

b cm   I

6 cm   II

Area of base = a cm²
Volume = 180 cm³

Area of base = 240 cm²
Volume = 1440 cm³

11 A solid sphere of radius 10 cm is melted down and cast into a cube.
Find the surface area of the cube.

12 A hemisphere of total surface area 1000 cm² is melted down and cast into a cube.
Find the total length of all its edges.

13 Find the surface area and volume of the following objects. All dimensions are in centimetres.

a

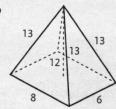

hole $r = 3$
depth $= 2$

5

$r = 6$

b

13   13

13

12

8   6

14 Carina calculates that the temperature of a cup of tea is t in degrees Celsius, m minutes after it has been made, given by the equation

$$t = \frac{k}{m}$$

where k is a constant and the equation is valid for $5 \leqslant m \leqslant 10$.

a Use the figures in this table to find the value of k, and hence copy and complete the table.

m (minutes)	5	6	7	8	9	10
t (°C)						40

b Draw the graph of t against m.

c Use your graph to estimate the temperature of a cup of tea after 450 s.

d After how long is the temperature of the cup of tea 60 °C?

e Carina is fussy, and she only drinks tea at a temperature of between 50 °C and 75 °C. Use your graph to find between what times Carina will drink her cup of tea.

15 Nina is making three different sizes of candle for Valentine's Day. The candles are heart shaped prisms and are all similar.

The smallest candle is 5 cm tall and has a volume of 40 cm³. The medium sized candle is 7.5 cm tall.

a Find the volume of the medium sized candle.

Nina wants the largest candle to have a volume of 400 cm³ so it will burn for ten times longer than the smallest candle.

b Find the height of the largest candle.

The area of the heart shaped face of the medium sized candle is 27 cm².

c Find the areas of the heart shaped faces of the other two candles.

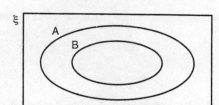

16 In a group of 100 pupils, 40 play bowls, 55 play golf, 30 do karate and 8 do nothing. 12 pupils play bowls and golf, 18 play golf and do karate, whilst 10 do karate and bowls. Use a Venn diagram to find out how many pupils

a only do karate   **b** do only bowls and golf   **c** do all three sports.

17 On copies of this Venn diagram, shade the following sets.

a $(A \cap B)'$

b $(A' \cup B)'$

c $(A \cap B')'$

d $(A \cup B)'$

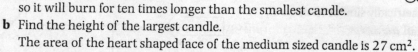

ξ
A
B

Continuing sequences

A set of numbers that follows a definite pattern is called a **sequence**. Many problems in mathematics can be solved by using sequences.

> **Activity 40**
>
> Seema is decorating the walls of a hall with balloons in preparation for a disco.
> She wants to place the balloons in a triangular pattern.
>
> To make a 'triangle' with one row, she needs **one** balloon.
>
> To make a triangle with two rows, she needs **three** balloons.
>
> To make a triangle with three rows, she needs **six** balloons.
>
>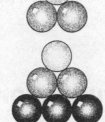
>
> The numbers of balloons needed form the sequence 1, 3, 6, ... Describe in words how to continue the sequence, and find the next three terms.
>
> Seema thinks she can work out how many balloons are needed for n rows by using the formula $\frac{1}{2}n(n + 1)$. Her friend Julia thinks the formula is $\frac{1}{2}n(n - 1)$. Which formula is correct? If Seema has 100 balloons, find, using 'trial and improvement', how many rows she can make and how many balloons will be left over.

> **Remember**
>
> These are some important sequences.
> - **Natural numbers**   1, 2, 3, 4, ...
> - **Even numbers**   2, 4, 6, 8, ...
> - **Odd numbers**   1, 3, 5, 7, ...
> - **Triangle numbers**   1, 3, 6, 10, ...   (Seema's first sequence)
> - **Square numbers**   1, 4, 9, 16, ...   (The squares of the natural numbers)
> - **Powers of 2**   1, 2, 4, 8, ...   (Numbers of the form 2^n)
> - **Powers of 10**   1, 10, 100, ...   (Numbers of the form 10^n)
> - **Prime numbers**   2, 3, 5, 7, ...   (Notice that 1 is not a prime number)
>
> Other sequences may be based on these important sequences.

Exercise 167

For Questions 1–5, write down the first four terms of the sequence.

1 Starting with 2, keep adding 2.

2 Starting with –9, keep adding 3.

3 Starting with 15, keep subtracting 5.

4 Starting with 2, keep multiplying by 2.

5 Starting with 12, keep dividing by 2.

For Questions 6–10, describe the rule for going from one term to the next, and write down the next three numbers in the sequence.

6 3, 7, 11, 15, ..., ..., ...   **7** 13, 8, 3, −2, ..., ..., ...   **8** 3, 6, 12, 24, ..., ..., ...

9 64, 32, 16, 8, ..., ..., ...   **10** 0.2, 0.5, 0.8, 1.1, ..., ..., ...

Exercise 167*

For Questions 1–4, write down the first four terms of the sequence.

1 Starting with −1, keep adding 1.5.   **2** Starting with 3, keep subtracting 1.25.

3 Starting with 1, keep multiplying by 2.5.   **4** Starting with 3, keep dividing by −3.

5 The first two terms of a sequence are 1, 1. The next term is found by adding together the last two terms. Find the first six terms of the sequence.

For Questions 6–10, describe the rule for going from one term to the next, and write down the next three numbers in the sequence.

6 $3, 5\frac{1}{2}, 8, 10\frac{1}{2}, ..., ..., ...$ **7** 243, 81, 27, 9, ..., ..., ...   **8** 2, 4, 16, 256, ..., ..., ...

9 $1, -\frac{1}{2}, \frac{1}{4}, -\frac{1}{8}, ..., ..., ...$ **10** 1, 3, 7, 15, 31, ..., ..., ...

Formulae for sequences

Sometimes the sequence is given by a formula. This means that any term can be found without working out all the previous terms.

Example 1

Find the sequence given by nth term = $2n - 1$. Also find the 100th term.

Substituting $n = 1$ into the formula gives the first term as	$2 \times 1 - 1 = 1$
Substituting $n = 2$ into the formula gives the second term as	$2 \times 2 - 1 = 3$
Substituting $n = 3$ into the formula gives the third term as	$2 \times 3 - 1 = 5$
Substituting $n = 4$ into the formula gives the fourth term as	$2 \times 4 - 1 = 7$

So the sequence is 1, 3, 5, 7... or the odd numbers.

Substituting $n = 100$ into the formula gives the 100th term as $2 \times 100 - 1 = 199$

Example 2

A sequence is given by nth term = $4n + 2$. Find the value of n for which the nth term equals 50.

$4n + 2 = 50 \Rightarrow 4n = 48 \Rightarrow n = 12$

So the 12th term equals 50.

Exercise 168

In Questions 1–6, find the first four terms of the sequence.

1 nth term = $2n + 1$ **2** nth term = $5n - 1$ **3** nth term = $33 - 3n$

4 nth term = $4 - 2n$ **5** nth term = $3n$ **6** nth term = $-4n$

In Questions 7–9, find the value of n for which the nth term has the value given in brackets.

7 nth term = $4n + 4$ (36)   **8** nth term = $6n - 12$ (30)   **9** nth term = $22 - 2n$ (8)

10 If the nth term = $12 - 2n$, which is the first term less than −3?

Exercise 168*

In Questions 1–6, find the first four terms of the sequence.

1 nth term $= 5n - 6$ **2** nth term $= 100 - 3n$ **3** nth term $= \frac{1}{2}(n + 1)$

4 nth term $= \frac{1}{3}(n + 2)$ **5** nth term $= n^2 + 1$ **6** nth term $= \frac{1}{n}$

In Questions 7–18, find the value of n for which the nth term has the value given in brackets.

7 nth term $= 7n + 9$ (65) **8** nth term $= 3n - 119$ (−83)

9 nth term $= 12 - 5n$ (−38)

10 If the nth term $= \frac{1}{2n - 1}$, which is the first term less than 0.01?

> **Investigate**
>
> A sequence is given by the formula nth term $= an + b$.
> Investigate the connection between a and b and the numbers in the sequence.

The difference method

When it is difficult to spot a pattern in a sequence, the difference method can often help.
Underneath the sequence, write down the differences between each pair of terms.
If the differences show a pattern, then the sequence can be extended.

Example 3

Find the next three terms in the sequence 2, 5, 10, 17, 26, ..., ..., ...

Sequence		2		5		10		17		26
Differences			3		5		7		9	

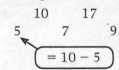

$= 5 - 2$ $= 10 - 5$

The differences increase by 2 each time, so the table can now be extended.

$= 26 + 11$ $= 37 + 13$

Sequence	2		5		10		17		26		**37**		**50**		**65**
Differences		3		5		7		9		**11**		**13**		**15**	

If the pattern in the differences is still not clear, add a third row giving the differences between the terms in the second row. More rows can then be added until a pattern is found, though not all sequences will result in a pattern.

Example 4

Find the next three terms in the sequence 3, 13, 29, 51, 79, ...

Sequence	3		13		29		51		79
Differences		10		16		22		28	
			6		6		6		

Now the table can be extended.

Sequence	3		13		29		51		79		**113**		**153**		**199**
Differences		10		16		22		28		**34**		**40**		**46**	
			6		6		6		**6**		**6**		**6**		

Exercise 169

Find the next three terms of the following sequences, using the difference method.

1 2, 5, 8, 11, 14   **2** 4, 9, 14, 19, 24   **3** 8, 5, 2, −1, −4   **4** 11, 7, 3, −1, −5

5 1, 6, 14, 25, 39   **6** 2, 8, 16, 26, 38   **7** 5, −2, −6, −7, −5   **8** 4, 0, −2, −2, 0

9 1, 4, 5, 4, 1   **10** 3, 5, 6, 6, 5

Exercise 169*

Find the next three terms of the following sequences, using the difference method.

1 1, 3, 8, 16, 27, 41   **2** 2, 4, 10, 20, 34, 52   **3** 1, 6, 9, 10, 9, 6

4 5, 8, 10, 11, 11, 10   **5** 1, −4, −7, −8, −7, −4   **6** 3, −1, −4, −6, −7, −7

7 1, 3, 6, 11, 19, 31, 48   **8** 3, 5, 10, 20, 37, 63, 100   **9** 1, 2, 4, 6, 7, 6, 2

10 2, 1, 1, 0, −4, −13, −29

Finding a formula for a sequence

Activity 41

Seema decides to decorate the walls of the hall with different patterns of sausage-shaped balloons.
She starts with a triangular pattern.

Seema wants to work out how many balloons she will need to make 100 triangles.

- If t is the number of triangles and b is the number of balloons, copy and complete this table.

t	1	2	3	4	5	6
b						

  Notice that the sequence in the row labelled b goes up by 2 each time. Add another row to the table labelled $2t$.

$t = 1$ $t = 2$

- Write down the formula that connects b and $2t$. How many balloons does Seema need to make 100 triangles?

$2t$	2	4	6	8	10	12

- Draw some other patterns that Seema might use, and find a formula for the number of balloons needed. Here are some possible patterns.

 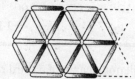 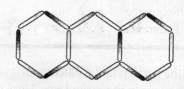

Activity 42

12 is a square perimeter number, because 12 pebbles can be arranged as the perimeter of a square.

The first square perimeter number is 4.

- Copy and complete this table, where s is the square perimeter number and p is the number of pebbles needed.

s	1	2	3	4	5	6
p	4					

- Use the method of Activity 41 to find a formula for the number of pebbles in the nth square perimeter number.

Investigate

Investigate triangular, pentagonal and hexagonal perimeter numbers, finding formulae for the nth perimeter number in each case.

Sometimes it is easy to see the formula.

Example 5

Find a formula for the nth term of the sequence 40, 38, 36, 34, ...

Table of differences   40   38   36   34

                         -2 -2 -2

The first row of differences is constant and equal to -2, so the formula is $-2n + b$.
When $n = 1$, the formula must give the first term as 40. So $-2 \times 1 + b = 40$, and $b = 42$.
The formula for the nth term is $42 - 2n$.

> **Key Point**
>
> If the first row of differences is constant and equal to a, then the formula for the nth term will be $an + b$, where b is another constant.

Exercise 170

For Questions 1–8, find a formula for the nth term of the sequence.

1   4, 7, 10, 13, ...   **2**   5, 7, 9, 11, ...   **3**   30, 26, 22, 18, ...   **4**   26, 23, 20, 17, ...

5   Nural has designed a range of Christmas candle decorations using a triangle of wood and some candles. She makes them in various sizes. The one shown here is the three-layer size, because it has three layers of candles.

    a   Copy and complete this table, where l is the number of layers and c is the number of candles.

l	1	2	3	4	5	6
c	1					

    b   Find a formula connecting l and c.
    c   Mr Rich wants a Christmas candle decoration with exactly 100 candles.
       Explain why this is impossible.
       What is the largest number of layers than can be made if 100 candles are available?

6   Jasmine is investigating rectangle perimeter numbers with one pair of constant sides of three pebbles.

    a   Copy and complete this table, where n is the number in the sequence and p is the number of pebbles

l	1	2	3	4	5	6
c	8					

    b   Find a formula connecting n and p.
    c   Given only 100 pebbles, what is the largest rectangle in her sequence that Jasmine can construct?

Exercise 170*

For Questions 1–8, find a formula for the *n*th term of the sequence.

1 3, 7, 11, 15, ... **2** 1, 4, 7, 10, ... **3** 6, 3, 0, −3, ... **4** 9, 5, 1, −3, ...

5 Poppy has designed a tessellation based on this shape.

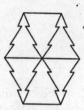
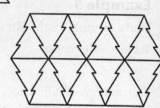

Here are the first three members of the tessellation sequence.

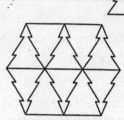

a Copy and complete this table, where *n* is the number in the sequence and *s* is the number of shapes used.

n	1	2	3	4	5	6
s	6					

b Find a formula giving *s* in terms of *n*.

c How many shapes will be needed to make the 50th member of the sequence?

6 Nashrin is using patterns of pebbles to investigate rectangle perimeter numbers where the inner rectangle is twice as long as it is wide.

a Copy and complete this table, where *n* is the number in the sequence and *p* is the number of pebbles.

n	1	2	3	4	5	6
s	10					

b Find a formula connecting *n* and *p*.

c What is the largest member of the sequence that Nashrin can build with only 200 pebbles?

Exercise 171 (Revision)

1 Find the first, tenth and hundredth terms of the sequence given by *n*th term = $5n - 7$.

2 Find the next three members of the sequence −7, −4, −1, 2, ...

3 a Find a formula for the *n*th term of the sequence 12, 9, 6, 3, ...
  b Find a formula for the *n*th term of the sequence 10, 16, 22, 28, ...

4 Khairal is training for a long-distance run.
On the first day of his training he runs 1000 m. Each day after that he runs an extra 200 m.

a How far does he run on the fifth day?
b How far does he run on the *n*th day?
c One day he runs 8 km. How many days has he been training?

5 a What name is given to the sequence 1, 3, 5, 7, 9, ...?

b Copy and complete:

$1 + 3 = ...$

$1 + 3 + 5 = ...$

$1 + 3 + 5 + 7 = ...$

$1 + 3 + 5 + 7 + 9 = ...$

c Find the sum $1 + 3 + 5 + ... + 19 + 21$

d Find a formula for the sum of the first n members of the sequence.

e The sum of m members of this sequence is 841. What is m?

6 a Find the nth term of the sequence $(1 + 2), (2 + 3), (3 + 4), ...$

b Find the nth term of the sequence $(2 + 1), (4 + 1), (6 + 1), ...$

c Explain why the two sequences are the same.

d Explain why every term is odd.

Exercise 171* (Revision)

1 Find the first, tenth and hundredth terms of the sequence given by nth term $= 12 - 7n$.

2 Find the next three members of the sequence 10, 7, 4, 1, ...

3 Find a formula for the nth term of the sequence $-6, -2, 2, 6, ...$

4 Find a formula for the nth term of the sequence 13, 8, 3, −2, ...

5 On Jamila's fifth birthday, she was given pocket money of 50c/month, to increase by 20c each month.

a How much pocket money did Jamila receive on her sixth birthday?

b Find a formula that gives Jamila's pocket money on the nth month after her fifth birthday.

c How old was Jamila when her pocket money became $17.30 per month?

6 a Find the next four terms in the sequence 1, 5, 9, 13, ...

b The first and third terms of this sequence are square numbers. Find the positions of the next two members of the sequence that are square numbers.

c Form a new sequence from the numbers giving the positions of the square numbers (that is, starting 1, 3, ...). Use this sequence to find the position of the fifth square number in the original sequence.

7 a Find the nth term of the sequence $(1 \times 2), (2 \times 3), (3 \times 4), ...$

b Find the nth term of the sequence $(1 + 1), (4 + 2), (9 + 3), ...$

c Explain why the two sequences are the same.

d Explain why every term is even.

8 Sequence A: 2, 3, 4, 5, 6, ...

Sequence B: 0, 1, 2, 3, 4, ...

Sequence C: 0, 3, 8, 15, 24, ...

a Find the nth term of sequence A.

b Find the nth term of sequence B.

c Sequence C is obtained from sequences A and B. Find the nth term of sequence C.

d Show how sequence C can be obtained from the sequence whose nth term $= n^2$.

Factor theorem

Let $f(x) = x^2 + x - 2$

Substituting $x = 1$ gives $f(1) = 1^2 + 1 - 2 = 0$

Substituting $x = -2$ gives $f(-2) = (-2)^2 + (-2) - 2 = 4 - 2 - 2 = 0$

When $f(x)$ is factorised we get $f(x) = (x - 1)(x + 2)$

It is no coincidence that $f(1) = 0$ and that $(x - 1)$ is a factor, or that $f(-2) = 0$ and that $(x - -2) = (x + 2)$ is a factor.

The **Factor Theorem** states that if $f(a) = 0$ then $(x - a)$ is a factor of $f(x)$.
$f(x)$ must be a polynomial; it is composed of terms like px^n where n is a positive integer.

Example 1

Show that $(x + 5)$ and $(x - 2)$ are factors of $f(x) = x^2 + 3x - 10$

$f(-5) = (-5)^2 + 3 \times -5 - 10 = 25 - 15 - 10 = 0 \Rightarrow (x + 5)$ is a factor

$f(2) = 2^2 + 3 \times 2 - 10 = 4 + 6 - 10 = 0 \Rightarrow (x - 2)$ is a factor

Example 1 shows that care must be taken over the signs. To show $(x + 5)$ is a factor -5 must be substituted for x in $f(x)$. Similarly to show $(x - 2)$ is a factor, $-(-2) = 2$ must be substituted in $f(x)$.

Choosing the numbers to substitute

Example 1 shows that $(x + 5)(x - 2) = x^2 + 3x - 10$
Notice that the numbers in the brackets multiply to give -10, i.e. they are factors of -10.
The numbers in the brackets will always be factors of the number in the polynomial.

Example 2

What numbers can be chosen to factorise $x^2 - x - 6$?
Use these numbers to find the two factors of $x^2 - x - 6$

The numbers chosen must multiply together to give -6.
The factors of -6 are ± 1 with ± 6 and ± 2 with ± 3.

$f(1) = -6 \Rightarrow (x - 1)$ is not a factor so $(x + 6)$ is not a factor either.

$f(-1) = -4 \Rightarrow (x + 1)$ is not a factor so $(x - 6)$ is not a factor either.

$f(2) = -4 \Rightarrow (x - 2)$ is not a factor.

$f(-2) = 0 \Rightarrow (x + 2)$ is a factor.

As the numbers chosen must multiply together to give -6, the other number must be 3
$f(3) = 0 \Rightarrow (x - 3)$ is a factor.

So the factors of $x^2 - x - 6$ are $(x + 2)$ and $(x - 3)$.

Example 2 probably seemed a rather long way to factorise a quadratic, and quadratics are usually factorised by other methods. However, when trying to factorise a cubic or higher powered polynomial the factor theorem is invaluable.

Exercise 172

1 Show that $(x + 2)$ and $(x - 4)$ are factors of $x^2 - 2x - 8$

2 Show that $(x + 4)$ and $(x + 6)$ are factors of $x^2 + 10x + 24$

3 Is $(x - 1)$ a factor of $x^2 - 7x - 8$?

4 Is $(x + 2)$ a factor of $x^2 - 3x - 10$?

5 Show that $(x - 6)$ is a factor of $x^2 - 9x + 18$ and use the Factor Theorem to find the other factor.

6 Show that $(x + 3)$ is a factor of $x^2 - 4x - 21$ and use the Factor Theorem to find the other factor.

7 Use the Factor Theorem to factorise $x^2 + 8x - 9$, showing your working.

8 Use the Factor Theorem to factorise $x^2 - 6x - 16$, showing your working.

9 $(x + 3)$ is a factor of $x^2 + px - 12$. Use the Factor Theorem to find p.

10 $(x - 2)$ is a factor of $x^2 + qx - 14$. Use the Factor Theorem to find q.

Exercise 172*

1 Is $(x + 9)$ a factor of $x^2 + 4x - 285$?

2 Is $(x - 8)$ a factor of $x^2 + 17x - 192$?

3 Is $(x + 7)$ a factor of $x^2 - 29x - 259$?

4 Is $(x - 14)$ a factor of $x^2 - 31x + 238$?

5 Use the Factor Theorem to factorise $x^2 - 21x + 38$, showing your working.

6 Use the Factor Theorem to factorise $x^2 - 22x - 23$, showing your working.

7 $(x + 12)$ is a factor of $x^2 + px + 288$. Use the Factor Theorem to find p.

8 $(x - 15)$ is a factor of $x^2 + qx - 135$. Use the Factor Theorem to find q.

9 $(x + 5)$ is a factor of $x^2 - 2x + r$. Use the Factor Theorem to find r.

10 $(x - 8)$ is a factor of $x^2 - 21x + s$. Use the Factor Theorem to find s.

Proving the factor theorem

It is not difficult to prove the factor theorem. The proof will first be given for a quadratic.

Assume that $f(x)$ has a factor $(x - a) \Rightarrow f(x) = (x - a)(bx + c)$
Substituting a for x gives $f(a) = (a - a)$ (some number) $= 0 \times$ (some number) $= 0$
It doesn't matter what number is in the second bracket as $0 \times$ number always equals 0
So if $f(a) = 0$ then $(x - a)$ is a factor of $f(x)$.

For a cubic the proof is:
Assume that $f(x)$ has a factor $(x - a) \Rightarrow f(x) = (x - a)(bx^2 + cx + d)$
Substituting a for x gives $f(a) = (a - a)$ (some number) $= 0 \times$ (some number) $= 0$
Again it doesn't matter what number is in the second bracket
So if $f(a) = 0$ then $(x - a)$ is a factor $f(x)$.

The proof can be extended to cope with polynomials of any power.

Example 3

$(x - 2)$ is a factor of $f(x) = x^3 + px - 18x + 40$. Find the value of p.

$f(2) = 0 \Rightarrow (2)^3 + p(2)^2 - 18 \times 2 + 40 = 0 \Rightarrow 4p = -12 \Rightarrow p = -3$
So $f(x) = x^3 - 3x^2 - 18x + 40$

Example 4

Show that $(x + 5)$ and $(x - 2)$ are factors of $f(x) = x^3 + 6x^2 - x - 30$.
Hence factorise $x^3 + 6x^2 - x - 30$ completely and solve the equation
$x^3 + 6x^2 - x - 30 = 0$

$f(-5) = -125 + 6 \times 25 + 5 - 30 = 0 \Rightarrow (x + 5)$ is a factor.
$f(2) = 8 + 6 \times 4 - 2 - 30 = 0 \Rightarrow (x - 2)$ is a factor.

The numbers in the brackets multiply to give -30, so the other factor is $(x + 3)$ as
$5 \times -2 \times 3 = -30$
Check: $f(-3) = -27 + 6 \times 9 + 3 - 30 = 0$ so $(x + 3)$ is a factor.
So $x^3 + 6x^2 - x - 30 = (x + 5)(x - 2)(x + 3)$

$x^3 + 6x^2 - x - 30 = 0 \Rightarrow (x + 5)(x - 2)(x + 3) = 0$
Using the same method for solving factorised quadratics $\Rightarrow x = -5, x = 2$ or $x = -3$

> **Key Point**
> The **Factor Theorem**
> states that if $f(a) = 0$ then
> $(x - a)$ is a factor $f(x)$.
> The numbers to
> substitute as a must be
> factors of the number
> term of $f(x)$.

Exercise 173

1 $(x + 1)$ is a factor of $x^3 + px^2 - 13x - 15$. Find the value of p.

2 $(x - 1)$ is a factor of $x^3 + 6x^2 + qx - 12$. Find the value of q.

3 $(x + 2)$ is a factor of $x^3 + 10x^2 + 28x + r$. Find the value of r.

4 Show that $(x + 2)$, $(x - 3)$ and $(x + 4)$ are factors of $f(x) = x^3 + 3x^2 - 10x - 24$.
   Hence solve $f(x) = 0$.

5 Show that $(x + 1)$, $(x + 5)$ and $(x - 4)$ are factors of $f(x) = x^3 + 2x^2 - 19x - 20$.
   Hence solve $f(x) = 0$.

6 Show that $(x + 2)$ and $(x + 3)$ are factors of $f(x) = x^3 + 4x^2 + x - 6$. Hence solve $f(x) = 0$.

7 Show that $(x - 2)$ and $(x + 4)$ are factors of $f(x) = x^3 + 3x^2 - 6x - 8$. Hence solve $f(x) = 0$.

8 Show that $(x + 5)$ is a factor of $f(x) = x^3 + 5x^2 - x - 5$. Hence solve $f(x) = 0$.

9 Show that $(x - 4)$ is a factor of $f(x) = x^3 - 5x^2 + 2x + 8$. Hence solve $f(x) = 0$.

10 Factorise $f(x) = x^3 + 3x^2 - x - 3$ completely. Hence solve $f(x) = 0$.

Exercise 173*

1 $(x - 3)$ is a factor of $x^3 + 3x^2 - 10x + p$. Find the value of p.

2 $(x + 6)$ is a factor of $x^3 + 7x^2 + qx - 36$. Find the value of q.

3 $(x + 5)$ and $(x - 1)$ are factors of $x^3 + px^2 + 7x + q$. Find the values of p and q.

4 $(x - 4)$ and $(x + 2)$ are factors of $x^3 + rx^2 + sx + 64$. Find the values of r and s.

5 Show that $(x + 3)$ and $(x - 4)$ are factors of $f(x) = x^3 - 3x^2 - 10x + 24$.
   Hence solve $f(x) = 0$.

6 Show that $(x - 5)$ and $(x + 4)$ are factors of $f(x) = x^3 - 2x^2 - 19x + 20$.
   Hence solve $f(x) = 0$.

7 Show that $(x + 6)$ is a factor of $f(x) = x^3 + x^2 - 24x + 36$. Hence solve $f(x) = 0$.

8 Show that $(x - 7)$ is a factor of $f(x) = x^3 - 4x^2 - 19x - 14$. Hence solve $f(x) = 0$.

9 Factorise $f(x) = x^3 + x^2 - 4x - 4$ completely. Hence solve $f(x) = 0$.

10 Factorise $f(x) = x^3 + 2x^2 - 9x - 18$ completely. Hence solve $f(x) = 0$.

No number term

If there is no number term, then x must be a factor. If this is taken out, then the Factor Theorem can be applied to the polynomial that is left.

Example 5

Factorise $f(x) = x^4 + 6x^3 - x^2 - 30x$.

Taking x out as a common factor gives $f(x) = x(x^3 + 6x^2 - x - 30)$

The Factor Theorem can be used to factorise $x^3 + 6x^2 - x - 30$ (See Example 4)

So $f(x) = x(x + 5)(x - 2)(x + 3)$

Repeated Factors

Let $f(x) = x^2 - 2x + 1$.

To use the Factor Theorem to find factors try the factors of 1 which are ± 1.

$f(-1) \neq 0 \Rightarrow (x + 1)$ is not a factor.

$f(1) = 0 \Rightarrow (x - 1)$ is a factor.

There are no other numbers to try, but there must be two factors.
The reason is that $f(x) = (x - 1)(x - 1) = (x - 1)^2$.
$(x - 1)$ is called a **repeated factor**.

Example 6

Factorise $x^3 - 3x^2 + 3x - 1$

The factors of -1 are ± 1

$f(-1) \neq 0 \Rightarrow (x + 1)$ is not a factor.　　　$f(1) = 0 \Rightarrow (x - 1)$ is a factor.

There are no other numbers to try, but there are probably three factors.

So probably $f(x) = (x - 1)^3$, but this should be checked by multiplying out.

Check: $(x - 1)^3 = (x - 1)(x - 1)^2 = (x - 1)(x^2 - 2x + 1) = x^3 - 3x^2 + 3x - 1$

Example 7

Factorise $f(x) = x^3 - 3x - 2$

The factors of -2 are ± 1 and ± 2

$f(-1) = 0 \Rightarrow (x + 1)$ is a factor.　　　$f(1) \neq 0 \Rightarrow (x - 1)$ is not a factor.

$f(-2) \neq 0 \Rightarrow (x + 2)$ is not a factor.　　　$f(2) = 0 \Rightarrow (x - 2)$ is a factor.

$(x - 2)$ can't be a repeated factor. If it was then the number term in $f(x)$ would be 4.

So $f(x) = (x - 2)(x + 1)^2$.

Check: $(x - 2)(x + 1)^2 = (x - 2)(x^2 + 2x + 1) = x^3 - 3x - 2$

In Example 6 is was stated that there were probably three factors.
It is possible for a cubic to have only two factors, a linear factor and a quadratic factor that doesn't factorise, for example $x^3 - x^2 + x - 1 = (x - 1)(x^2 + 1)$. This is why it is important to check when there are repeated factors.

Coefficient of x^3 greater than 1

If the coefficient of x^3 is greater than 1 then there are two ways to proceed as shown in the next two examples.

Example 8

Factorise $f(x) = 2x^3 + 6x^2 - 2x - 6$

As 2 is a common factor write $f(x)$ as $f(x) = 2(x^3 + 3x^2 - x - 3)$

Factorise $x^3 + 3x^2 - x - 3$ using the Factor Theorem to give $(x + 1)(x - 1)(x + 3)$

Hence $f(x) = 2(x + 1)(x - 1)(x + 3)$

Example 9

Factorise $f(x) = 2x^3 + 3x^2 - 2x - 2$

The Factor Theorem establishes that $(x - 1)$ and $(x + 2)$ are factors.
The third factor must be of the form $(2x + p)$ giving $f(x) = (2x + p)(x - 1)(x + 2)$
Comparing number terms gives $p \times -1 \times 2 = -2 \Rightarrow p = 1$
So $f(x)$ is probably $(2x + 1)(x - 1)(x + 2)$, but this should be checked by multiplying out.
Check: $(2x + 1)(x - 1)(x + 2) = (2x + 1)(x^2 + x - 2) = 2x^3 + 3x^2 - 3x - 2$

Exercise 174

Factorise the following cubics.

1 $x^3 - x^2 - 2x$ **2** $x^3 - 4x^2 + 3x$

3 $x^3 - 3x^2 + 2x$ **4** $x^3 + 3x^2 + 3x + 1$

5 $2x^3 + 12x^2 + 22x + 12$ **6** $3x^3 - 21x + 18$

7 $(x + 3)$ is a factor of $f(x) = 2x^3 + 7x^2 + px - 9$.
Find the value of p, factorise $f(x)$ and hence solve $f(x) = 0$.

8 $(x + 2)$ is a factor of $f(x) = 3x^3 + x^2 + qx - 4$.
Find the value of q, factorise $f(x)$ and hence solve $f(x) = 0$.

Exercise 174*

Factorise the following cubics.

1 $2x^3 - 5x^2 - 3x$ **2** $3x^3 - 8x^2 + 4x$

3 $x^3 - 3x^2 + 4$ **4** $x^3 + 8x^2 + 21x + 18$

5 $3x^3 - 3x^2 - 15x - 9$ **6** $2x^3 + 9x + 7x - 6$

7 $(x + 4)$ is a factor of $f(x) = 3x^3 + px^2 - 10x + 16$.
Find the value of p, factorise $f(x)$ and hence solve $f(x) = 0$.

8 $(x + 3)$ is a factor of $f(x) = 2x^3 + 3x^2 + qx - 15$.
Find the value of q, factorise $f(x)$ and hence solve $f(x) = 0$.

Problems leading to quadratic equations

Key Points

- Where relevant, draw a clear diagram and put all the information on it.
- Let x stand for what you are trying to find.
- Form a quadratic equation in x and simplify it.
- Solve the equation by either factorising or using the formula.
- Check that the answers make sense.

Example 10

The width of a rectangular photograph is 4 cm more than the height.
The area is 77 cm².
Find the height of the photograph.

Let x be the height in cm.
Then the width is $x + 4$ cm.
The diagram is shown on the right.

As the area is 77 cm²,

$$x(x + 4) = 77$$
$$x^2 + 4x = 77$$
$$x^2 + 4x - 77 = 0$$
$$(x - 7)(x + 11) = 0$$

So $\qquad x = 7 \text{ or } -11 \text{ cm}$

The height cannot be negative, so the height is 7 cm.

Example 11

A rectangular fish pond is 6 m by 9 m. The pond is surrounded by a concrete path of constant width. The area of the pond is the same as the area of the path. Find the width of the path.

Let x be the width of the path.

The diagram is shown on the right.

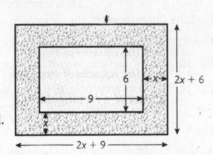

The area of the path is $(2x + 9)(2x + 6) - 9 \times 6$
$$= 4x^2 + 30x + 54 - 54$$
$$= 4x^2 + 30x$$

The area of the pond is $9 \times 6 = 54 \text{ m}^2$.
As the area of the path equals the area of the pond,

$$4x^2 + 30x = 54$$
$$4x^2 + 30x - 54 = 0$$
$$2x^2 + 15x - 27 = 0$$
$$(2x - 3)(x + 9) = 0$$

So $\qquad x = 1.5 \text{ or } -9 \text{ m}$

As x cannot be negative, the width of the path is 1.5 m.

Exercise 175

1. One number is four more than another number. The product of the numbers is 96. Find the numbers.

2. One number is two less than another number. The product of the numbers is 63. Find the numbers.

3 The width of a rectangle is 2 cm more than the height.
The area is 12 cm².
Find the height of the rectangle.

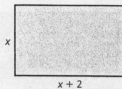

x

$x + 2$

4 The height of a rectangle is 3 cm more than the width.
The area is 30 cm².
Find the width of the rectangle.

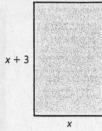

$x + 3$

x

5 The height of a right-angled triangle is
3 cm more than the width.
The area is 10 cm².
Find the width of the triangle.

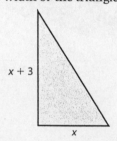

$x + 3$

x

6 The height of a right-angled triangle is
1 cm more than the width.
The area is 12 cm².
Find the width of
the triangle.

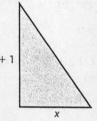

$x + 1$

x

7 The sum of the squares of two consecutive numbers is 113. Find the numbers.

8 The sum of the squares of two consecutive numbers is 181. Find the numbers.

Exercise 175*

1 The height of a triangle is 3 cm more than the width. The area is 14 cm².
Find the width of the triangle.

2 A rectangular classroom has a perimeter of 28 m and an area of 48 m².
Find the dimensions of the classroom.

3 The sum of the squares of two consecutive integers is 145. Find the integers.

4 The sum of the squares of two consecutive odd integers is 130. Find the integers.

5 The perimeter of a rectangular room is 32 m. The length of a diagonal is 8 m more than the width. Find the dimensions of the room.

6 The sum of the first n integers $1 + 2 + 3 + \ldots + n = \dfrac{n(n + 1)}{2}$

  a How many numbers must be taken to have a sum greater than one million?

  b Why cannot the sum ever equal 100 000?

7 An n-sided polygon has $\dfrac{n(n - 3)}{2}$ diagonals.

  a How many sides has a polygon with 665 diagonals?

  b Why cannot a polygon have 406 diagonals?

8 Lee spent \$1200 on holiday. If he had spent \$50 less per day, he would have been able to stay an extra two days. How long was his holiday?

9 When a sheet of A4 paper is cut in half as shown,
the result is a sheet of A5 paper. The two
rectangles formed by A4 and A5 paper are similar.
Find x.

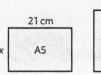

21 cm

x A5

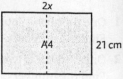

$2x$

A4

21 cm

Exercise 176 (Revision)

1. $(x - 2)$ is a factor of $x^3 - 3x^2 + px + 4$. Find the value of p.

2. Show that $(x + 3)$ is a factor of $f(x) = x^3 + x^2 - 9x - 9$. Find the other two factors and hence the equation $f(x) = 0$.

3. Factorise $f(x) = x^3 - x^2 - 4x + 4$ completely. Hence solve the equation $f(x) = 0$.

4. Factorise $f(x) = x^3 - 3x^2 - 28x$ completely. Hence solve the equation $f(x) = 0$.

5. $(x + 4)$ is a factor of $f(x) = 2x^3 + 7x^2 - px - 4$.
   Find the value of p, factorise $f(x)$ and hence solve $f(x) = 0$.

6. The height of a rectangle is 1.5 cm more than the width. The area is 10 cm². Find the width of the rectangle.

7. The base of a triangle is 3 cm more than the height, and the area is 25 cm².

   Find the height of the triangle.

8. A right-angled triangle has sides of length x, $x + 1$ and $x + 2$.
   Find the value of x.

9. The sum of the squares of two consecutive numbers is 145. Find the numbers.

10. A piece of wire 60 cm long is bent to form the perimeter of a rectangle of area 210 cm². Find the dimensions of the rectangle.

Exercise 176* (Revision)

1. $(x + 1)$ and $(x - 2)$ are factors of $x^3 + px^2 + qx - 8$. Find the values of p and q.

2. Factorise $f(x) = x^3 + 7x^2 + 7x - 15$ completely. Hence solve the equation $f(x) = 0$.

3. Solve the equation $2x^3 - 16x^2 + 10x + 28 = 0$.

4. Solve the equation $x^3 - 7x^2 + 15x - 9 = 0$.

5. $(x + 5)$ is a factor of $f(x) = 2x^3 + 11x^2 + px - 30$.
   Find the value of p, factorise $f(x)$ and hence solve $f(x) = 0$.

6. A cereal packet is a cuboid with height 12 cm. The depth of the box is 4 cm more than the width, and the volume is 480 cm³. Find the width of the box.

7. The area of a rectangular lawn is 30 m². During landscaping the length was decreased by 1 m and the width increased by 1 m, but the area did not change. Find the original dimensions of the lawn.

8. A right-angled triangle has sides of length x, $x + 7$ and $x + 8$.

   Find the lengths of the sides.

9. The sum of the squares of two consecutive odd numbers is 202. Find the numbers.

10. The hypotenuse of a right-angled triangle is 4 cm long, and the perimeter is 9 cm. Find the lengths of the other two sides.

11. A circular fish pond of radius 2 m is surrounded by a path of constant width. The area of the path is the same as the area of the pond. Find the width of the path.

12. Aran spent $1200 on holiday. If he had spent $20 less each day he would have been able to stay an extra three days. How long was his holiday?

Tangents to a curve

Remember

For a straight line the gradient = $\frac{\text{'rise'}}{\text{'run'}}$

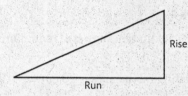

Rise

Run

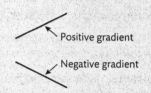

Positive gradient

Negative gradient

The gradient of a graph often provides useful information.

Graph	Gradient
Distance (m) – time (s)	Speed (m/s)
Speed (m/s) – time (s)	Acceleration (m/s²)
Temperature (°C) – time (min)	Rate of change of temperature (°C/min)
Population (ants) – time (weeks)	Rate of change of population (ants/week)
Financial profit ($) – time (years)	Rate of change of profit ($/year)

Most graphs of real-life situations are curves rather than straight lines, but information on rates of change can still be found by drawing a tangent to the curve and using this to estimate the gradient of the curve at that point.

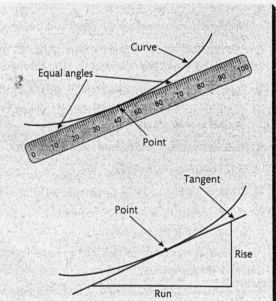

Remember

To find the gradient of a curve at a point: Draw the tangent to the curve at the point. Do this by pivoting a ruler about the point until the angles between the curve are as equal as possible.

Curve

Equal angles

Point

The gradient of the tangent is the gradient of the curve at the point.

The gradient of the tangent is found by working out $\frac{\text{'rise'}}{\text{'run'}}$

Tangent

Point

Rise

Run

Example 1

Find the gradient of the curve $y = x^2 - x$
at the point where $x = 2$.
First draw the graph for x values around 2.

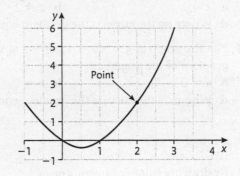

Next draw in the tangent.
(Be careful finding the rise and the run
when the scales on the axis are different,
as in this example.)

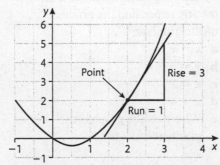

The gradient of the curve at $x = 2$ is 3.

Because the tangent is judged by eye, different people will get different answers for the gradient.
The answers given are calculated using a different technique which you will learn in the next
unit, so do not expect your answers to be exactly the same as those given in the book.

Exercise 177

1 By drawing suitable tangents on
  tracing paper, find the gradient of
  the graph at:

  a $x = 2$ **b** $x = 2\frac{1}{2}$
  c $x = \frac{1}{2}$ **d** $x = 0$
  e Where on the graph is the gradient
  equal to 0?

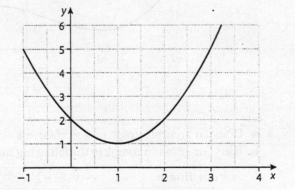

2 By drawing suitable tangents on tracing
  paper, find the gradient of the graph at:

  a $x = 0$ **b** $x = 2$
  c $x = 0.4$ **d** $x = 2.5$
  e Where on the graph is the gradient
  equal to 1?

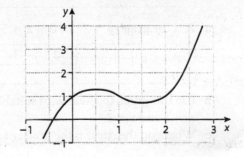

3 Plot the graph of $y = x(6 - x)$ for $0 \leqslant x \leqslant 6$. By drawing suitable tangents, find the gradient of the graph at:

   a $x = 1$ **b** $x = 2$ **c** $x = 5$

   d Where on the curve is the gradient equal to 0?

4 Plot the graph of $y = x(x - 4)$ for $-1 \leqslant x \leqslant 5$. By drawing suitable tangents, find the gradient of the graph at:

   a $x = 0$ **b** $x = 2$ **c** $x = 3$

   d Where on the curve is the gradient equal to 4?

Exercise 177*

1 By drawing suitable tangents on tracing paper, find the gradient of the graph at:

   a $x = \frac{1}{2}$

   b $x = -\frac{1}{2}$

   c $x = 2$

   d $x = 3$

   e Where on the graph is the gradient equal to 0?

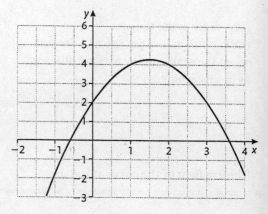

2 By drawing suitable tangents on tracing paper, find the gradient of the graph at:

   a $x = -1\frac{1}{2}$ **b** $x = -\frac{1}{2}$ **c** $x = \frac{1}{2}$ **d** $x = 1$

   e Where on the graph is the gradient equal to 0?

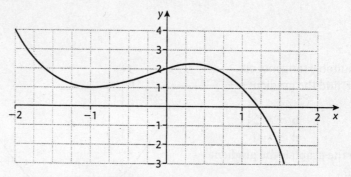

3 **a** Draw an accurate graph of $y = x^2$ for $-4 \leqslant x \leqslant 4$.

   b Use your graph to complete the following table:

x-coordinate	−4	−3	−2	−1	0	1	2	3	4
Gradient	−8								8

4 **a** Plot the graph $y = 2^x$ for $0 \leqslant x \leqslant 5$ by first copying and completing the table.

x	0	1	2	3	4	5
y	1					32

   b Use this graph together with suitable tangents to estimate the gradient of the curve at $x = 1$ and $x = 3$.

   c Where on the curve is the gradient equal to 12?

Example 2

A dog is running in a straight line away from its owner. Part of the distance–time graph describing the motion is shown.

a Describe how the dog's speed varies.

b Estimate the dog's speed after 30 seconds.

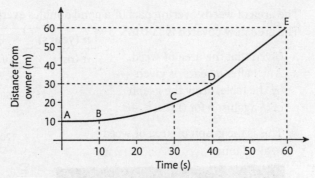

Remember that the gradient of a distance–time graph gives the speed.

a A to B: The gradient is zero, so the speed is zero. The dog is stationary for the first 10 seconds, 10 metres away from its owner.

B to D: The gradient is gradually increasing, so the speed is gradually increasing. For the next 30 seconds the dog runs with increasing speed.

D to E: The gradient is constant and equal to $\frac{30}{20}$ or 1.5, so the dog is running at a constant speed of 1.5 m/s.

b Draw a tangent at C and calculate the gradient of the tangent.

The gradient is $\dfrac{15\,\text{m}}{20\,\text{s}} = 0.75$ m/s, so the

speed of the dog is approximately 0.75 m/s.

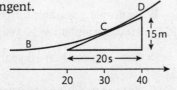

Exercise 178

1 The graph shows part of the distance–time graph for a car caught in a traffic jam.

a By drawing suitable tangents on tracing paper, estimate the speed of the car when

   (i) $t = 15\,\text{s}$
   (ii) $t = 25\,\text{s}$
   (iii) $t = 45\,\text{s}$

b Describe the car's journey.

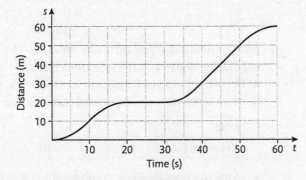

2 The velocity, v m/s of a vintage aircraft t seconds after it starts to take off is given by $v = 0.025t^2$.

a Draw a graph of v against t for $0 \leqslant t \leqslant 60$.

b Use the graph to draw suitable tangents to estimate the acceleration of the aircraft when:

   (i) $t = 10$ (ii) $t = 30$ (iii) $t = 50$

3 The distance, s m, fallen by a stone t seconds after being dropped down a well is given by $s = 5t^2$.

a Draw a graph of s against t for $0 \leqslant t \leqslant 4$.

b Use your graph to draw suitable tangents to estimate the velocity of the stone when:

   (i) $t = 1$ (ii) $t = 2$ (iii) $t = 3$

4 a Draw the curve $y = x(x - 5)$ for $0 \leqslant x \leqslant 6$.

b By drawing suitable tangents, find the gradient of the curve at the points $x = 1$, $x = 2.5$ and $x = 5$.

Example 3

The area of weed covering part of a pond doubles every 10 years.
The area now covered is $100\,\text{m}^2$.

n (years)	0	10	20	30	40
n (m²)	100	200	400	800	1600

a Given that the area of weed, $A\,\text{m}^2$, after n years, is given by the table, draw the graph of A against n for $0 \leqslant n \leqslant 40$.

This is the graph of area of weed against time.

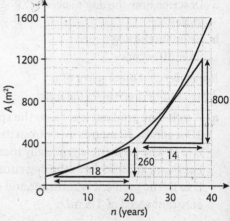

b By drawing suitable tangents, find the rate of growth of the weed in m^2 per year after 10 years and after 30 years.

Rate of growth at 10 years $\simeq \dfrac{260\,\text{m}^2}{18\,\text{years}} \simeq 14\,\text{m}^2/\text{year}$.

Rate of growth at 30 years $\simeq \dfrac{800\,\text{m}^2}{14\,\text{years}} \simeq 57\,\text{m}^2/\text{year}$.

The rate of growth is clearly increasing with time.

Exercise 178*

1 A catapult fires a stone vertically upwards. The height, h metres, of the stone, t seconds after firing, is given by the formula $h = 40t - 5t^2$.

   a Draw the graph of h against t for $0 \leqslant t \leqslant 8$.

   b Draw tangents to this curve and measure their gradients. Then copy and complete this table

t (s)	0	1	2	3	4	5	6	7	8
Velocity (m/s)	40		20			−10			−40

   c Draw the velocity–time graph for the stone for $0 \leqslant t \leqslant 8$.

   d What can you say about the stone's acceleration?

2 The point $A(1, 3)$ is on the graph of $y = x(4 - x)$ as shown.

   a By drawing a suitable tangent on tracing paper, find the gradient of the curve at point A.

   b What is the equation of the tangent to the curve at A?

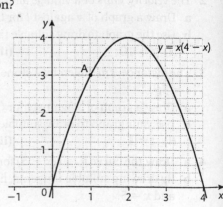

3 The height of a palm tree, y cm, after t months, is given in the table.

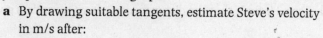

t **(months)**	0	6	12	18	24	30	36
y **(cm)**	1	2.5	6.3	15.8	39.8	100	251

   a Draw the graph of y against t.

   b By drawing suitable tangents to the curve, estimate the rate of growth of the palm tree in cm *per year* when $t = 15$ and when $t = 30$.

4 Steve performs a 'bungee-jump' from a platform above a river.

His height, h metres, above the river t seconds after he jumps is shown on the graph.

   a By drawing suitable tangents, estimate Steve's velocity in m/s after:

      (i) 1 s

      (ii) 8 s

      (iii) 14 s

   b Estimate Steve's maximum velocity and the time it occurs.

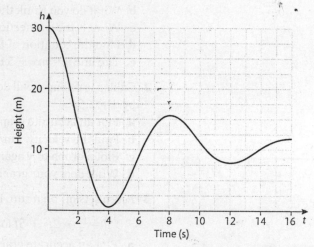

Exercise 179 (Revision)

1 Plot an accurate graph of $y = \frac{1}{2}x^2$ for $-4 \le x \le 4$. By drawing suitable tangents, find the gradient of the graph at

   a (i) $x = 1$ **(ii)** $x = -2$ **(iii)** $x = 3$

   b Where on the curve is the gradient equal to -2?

   c Find the equation of the tangent to the curve at the point where $x = 2$.

2 The velocity v m/s of a skydiver for the first 5 seconds of her fall is given by $v = 10t - t^2$.

   a Draw an accurate graph of v against t for $0 \le t \le 5$.

   b Find the acceleration of the skydiver after   **(i)** 1 s   **(ii)** 3 s

   c When was the skydiver's acceleration 5 m/s²?

3 The distance–time graph shows a bee's journey from its hive.

   a Find the velocity of the bee at

      (i) 5 s

      (ii) 15 s

      (iii) 22 s

   b When was the bee flying fastest?

   c Describe the bee's journey.

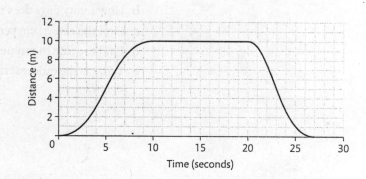

4 A population of eagles is increasing by 50% every 10 years.

   a Copy and complete the table and use it to plot the graph of N against t.

t **(years)**	0	10	20	30	40	50	60	70
N **(eagles)**	100		225					

   b Estimate the rate of growth of the population in eagles **per year** after

      i 30 years   **ii** 50 years

   c When was the rate of growth 20 eagles per year?

Exercise 179* (Revision)

1 a Plot an accurate graph of $y = 4 - \frac{1}{2}x^2$ for $-4 \leqslant x \leqslant 4$. By drawing suitable tangents, find the gradient of the graph at

    (i) $x = -1$ **(ii)** $x = 2$ **(iii)** $x = -3$

  b What do you think the gradient at $x = 10$ will be?

  c What is the connection between the gradient at $x = a$ and $x = -a$?

  d Find the equation of the tangent to the curve at the point where $x = -2$.

  e The line $y = mx + 5$ is a tangent to the curve. Find the values of m.

2 Galileo is rolling a ball down a slope. He finds that the distance, d m, after t seconds is given by $d = 2.5t^2$.

  a Plot an accurate graph of d against t for $0 \leqslant t \leqslant 4$.

  b Find the velocity, v m/s, after   **(i)** 1 s   **(ii)** 2 s   **(iii)** 3 s   **(iv)** 4 s   **(v)** 0 s

  c Plot a graph of v against t.

  d What does your graph tell you about the acceleration?

3 Pedro sets off on a run. His distance, d m, from his starting point after t seconds is given by

$$d = \frac{t^3}{5} - 2t^2 + 5t \text{ for } 0 \leqslant t \leqslant 8.$$

  a Plot an accurate graph of distance against time for $0 \leqslant t \leqslant 8$.

  b Describe Pedro's run for $0 \leqslant t \leqslant 8$.

  c What was his velocity after   **i** 1 second   **ii** 3 seconds?

  d When was his velocity 6 m/s?

4 Mala fills a flower vase with water to a depth of 20 cm.
The level of water drops every day by 15% due to evaporation.

  a Copy and complete the table, and use it to plot the graph of depth against time for $0 \leqslant t \leqslant 8$.

t (days)	0	1	2	3	4	5	6	7	8
d (cm)	20		14.45						5.45

  b How many days does it take for the water level to drop by half?

  c Find the rate in cm per day that the water is dropping when $t = 2$.

  d Find the rate in **cm per hour** that the water is dropping when $t = 6$.

  e What is the quickest rate in cm per day at which the water drops, and when does this occur?

Trigonometric ratios for angles up to 180°

This section extends the sin x, cos x and tan x ratios beyond acute angles (where $0° \leqslant x \leqslant 90°$) to consider angles where $0° \leqslant x \leqslant 180°$.

Remember

- To work out the sine of 100°, press

  [sin] [1] [0] [0] [=]

  sin 100° = 0.985 (to 3 s.f.)

- To work out x in sin x = 0.766, press

  [SHIFT] [sin] [0] [.] [7] [6] [6] [=]

  x = 50.0° (to 3 s.f.)

Activity 43

Sine ratio

This is the graph of $y = \sin x$, drawn for $0° \leqslant x \leqslant 180°$.

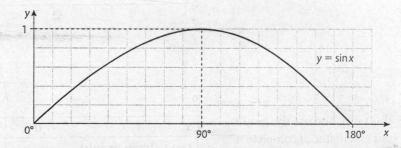

- Copy this table, and use the graph and your calculator to complete it.

x	36°	72°	108°	144°	180°
sin x (from graph)					
sin x (calc. to 3 s.f.)					

- Copy and complete this table, for $0° \leqslant x \leqslant 90°$.

sin x	0	0.2	0.5	0.8	1.0
x (from graph)					
x (calc. to 3 s.f.)					

Cosine ratio

This is the graph of $y = \cos x$, drawn for $0° \leqslant x \leqslant 180°$.

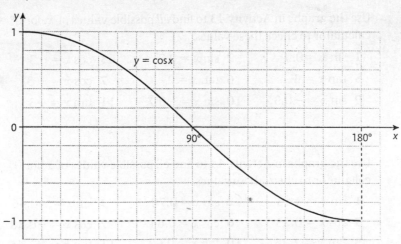

- Copy this table, and use the graph and your calculator to complete it.

x	36°	72°	108°	144°	180°
cos x (from graph)					
cos x (calc. to 3 s.f.)					

- Copy and complete this table.

cos x	0	0.2	0.5	0.8	−1.0
x (from graph)					180°
x (calc. to 3 s.f.)					

Tangent ratio

This is the graph of $y = \tan x$, drawn for $0° \leqslant x \leqslant 180°$.

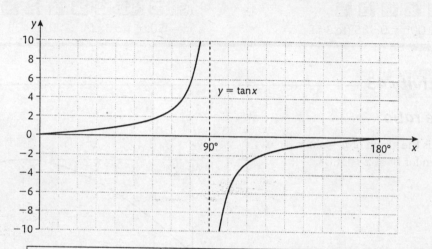

- Copy this table, and use the graph and your calculator to complete it.

x	0°	60°	85°	95°	180°
tan x (from graph)					
tan x (calc. to 3 s.f.)					

- Why does tan x become so large as x approaches 90°?
- Copy and complete this table.

tan x	0	0.5	5	10	−10
x (from graph)					
x (calc. to 3 s.f.)					

Exercise 180

Use the graphs in Activity 43 to find *all* possible values of x, for $0° \leqslant x \leqslant 180°$. Use your calculator to check these values.

1 $\sin x = 0.3$ **2** $\sin x = 0.7$ **3** $\cos x = 0.3$ **4** $\cos x = 0.6$

5 $\tan x = 3$ **6** $\tan x = 2$ **7** $\cos x = -3$ **8** $\cos x = -0.6$

9 $\sin x = -0.3$ **10** $\sin x = -0.7$ **11** $\tan x = -3$ **12** $\tan x = -2$

Exercise 180*

Use the graphs in Activity 43 to find all possible values of x, for $0° \leqslant x \leqslant 360°$. Use your calculator to check these values.

1 $\sin x = 0.46$ **2** $\sin x = 0.70$ **3** $\sin x = -0.46$ **4** $\sin x = -0.65$

5 $\cos x = 0.25$ **6** $\cos x = 0.64$ **7** $\cos x = -0.38$ **8** $\cos x = -0.84$

9 $\tan x = 2.46$ **10** $\tan x = 0.68$ **11** $\tan x = -1.45$ **12** $\tan x = -0.73$

Sine rule

The **sine rule** is a method to calculate sides and angles for any triangle. It is useful for finding the length of one side when all angles and one other side are known, or finding **an angle when two sides and the included angle are known.**

Key Points

The sine rule

In triangle ABC, notice that side a is opposite angle A, side b is opposite angle B, etc.

- To find a side, use the sine rule written as

$$\frac{a}{\sin A} = \frac{b}{\sin B} = \frac{c}{\sin C}$$

- To find an angle, use the sine rule written as

$$\frac{\sin A}{a} = \frac{\sin B}{b} = \frac{\sin C}{c}$$

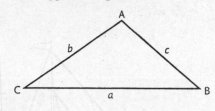

Example 1

In triangle ABC, find the length of side a, correct to 3 significant figures.

$$\frac{a}{\sin 48°} = \frac{7.4}{\sin 50°}$$

$$a = \frac{7.4}{\sin 50°} \times \sin 48°$$

$$a = 7.1787 \dots \text{ cm}$$

$$a = 7.18 \text{ cm} \quad \text{(to 3 s.f.)}$$

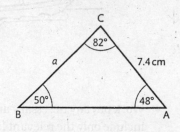

Example 2

In triangle ABC, find angle B correct to 3 significant figures.

$$\frac{\sin B}{6.8} = \frac{\sin 70°}{8.7} \quad \text{(Multiply both sides by 6.8)}$$

$$\sin B = \frac{\sin 70°}{8.7} \times 6.8$$

$$B = 47.262 \dots$$

$$B = 47.3° \quad \text{(to 3 s.f.)}$$

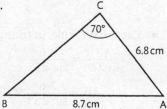

Remember

- Bearings are measured clockwise from North.
- When drawing bearings, start by drawing an arrow to indicate North.
- When calculating angles on a bearings diagram, look for 'alternate angles'.

N

-  e is angle of elevation.
- d is angle of depression.

Example 3

A yacht crosses the start line of a race at C, on a bearing of 026°. After 2.6 km, it rounds a buoy B and sails on a bearing of 335°. When it is due North of its start, at A, how far has sailed altogether?

- Draw a diagram and include all the facts.

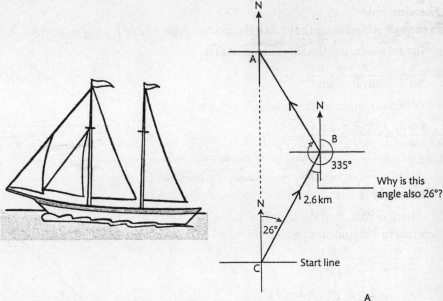

- Work out any necessary angles and redraw the triangle and include only the relevant facts.

$$\angle CBA = 335° - 26° - 180° = 129°$$

so $\angle BAC = 25°$

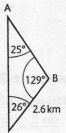

- Use the sine rule: in triangle ABC, the length AB has to be calculated.

$$\frac{AB}{\sin 26°} = \frac{2.6}{\sin 25°} \qquad \text{(Multiply both sides by } \sin 26°\text{)}$$

$$AB = \frac{2.6}{\sin 25°} \times \sin 26°$$

$$AB = 2.6969 \ldots \text{ km}$$

total distance travelled = CB + BA

$$= 2.6 \text{ km} + 2.697 \text{ km} = 5.30 \text{ km (to 3 s.f.)}$$

Exercise 181

Write your answers correct to 3 significant figures.

1 Find x.

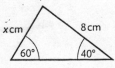

2 Find y.

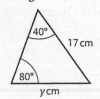

3 Find MN.

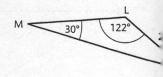

4 Find RT.

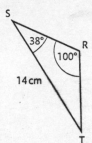

5 Find AC.

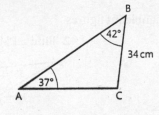

6 Find YZ.

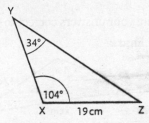

7 Find x.

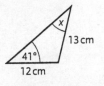

8 Find y.

9 Find $\angle$ABC.

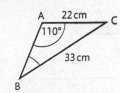

10 Find $\angle$XYZ.

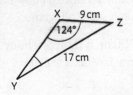

11 Find $\angle$ACB.

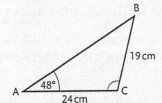

12 Find $\angle$DCE.

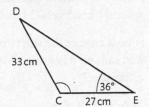

Activity 44

The ambiguous case

- Use a pair of compasses to construct triangle ABC, with AB = 7 cm, BC = 6 cm and $\angle$BAC = 50°.

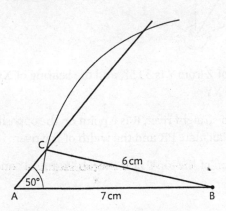

- Note that two triangles can be constructed: ABC and ABC′. The arc centred at B cuts the line from A at C *and* C′. This is an example of the 'ambiguous' case where two triangles can be constructed from the same facts.
- Show by calculation that AC ≈ 1.8 cm and AC′ ≈ 7.2 cm.

Exercise 181*

Write your answers correct to 3 significant figures.

1 Find x.

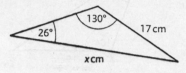

2 Find $\angle$LMN.

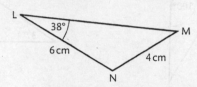

3 Find EF, $\angle$DEF and $\angle$FDE.

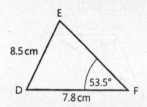

4 Find MN, $\angle$MLN and $\angle$LNM.

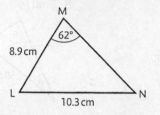

5 A yacht crosses the start line of a race on a bearing of 031°. After 4.3 km, it rounds a buoy and sails on a bearing of 346°. When it is due North of its start, how far has it sailed altogether?

6 A boat crosses the start line of a race on a bearing of 340°. After 700 m, it rounds a buoy and sails on a bearing of 038°. When it is due North of its start, how far has it sailed altogether?

7 The bearing of B from A is 065°. The bearing of C from B is 150°, and the bearing of A from C is 305°. If AC = 300 m, find BC.

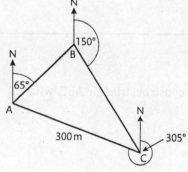

8 The bearing of Y from X is 205°. The bearing of Z from Y is 315°, and the bearing of X from Z is 085°. If XZ = 4 km, find the distance XY.

9 P and Q are points 80 m apart on the bank of a straight river. R is a point on the opposite bank where $\angle$QPR = 76° and $\angle$PQR = 65°. Calculate PR and the width of the river.

10 From two points X and Y, the angles of elevation of the top T of a church TZ are 14° and 19°, respectively.

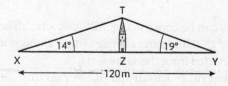

IF XYZ is a horizontal straight line and XY = 120 m, find YT and the height of the tower.

Cosine rule

The **cosine rule** is another method to calculate sides and angles of any triangle. It is used either to find the third side when two sides and the included angle are given, or to find an angle when all three sides are given.

Key Points

The cosine rule

In triangle ABC:

$$a^2 = b^2 + c^2 - 2bc \cos A$$

- To find side a, use the cosine rule written as

$$a^2 = b^2 + c^2 - 2bc \cos A$$

- To find angle A, use the cosine rule written as $\cos A = \dfrac{b^2 + c^2 - a^2}{2bc}$

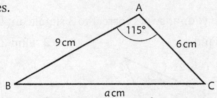

Example 4

In triangle ABC find a correct to 3 significant figures.

$$
\begin{aligned}
a^2 &= b^2 + c^2 - 2bc \cos A \\
&= 6^2 + 9^2 - 2 \times 6 \times 9 \times \cos 115° \\
&= 36 + 81 - 108 \times (-0.4226) \\
&= 36 + 81 + 45.64 \\
a &= \sqrt{162.6} \\
&= 12.8 \text{ (to 3 s.f.)}
\end{aligned}
$$

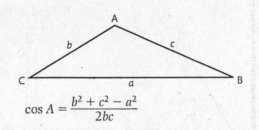

Example 5

In triangle ABC, find angle A correct to 3 significant figures.

$$
\begin{aligned}
\cos A &= \frac{b^2 + c^2 - a^2}{2bc} \\
&= \frac{8^2 + 11^2 - 9^2}{2 \times 8 \times 11} \\
&= \frac{104}{176} \\
A &= 53.8° \text{ (to 3 s.f.)}
\end{aligned}
$$

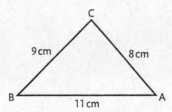

Exercise 182

Write your answers correct to 3 significant figures.

1 Find a.

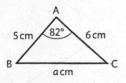

2 Find b.

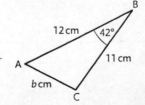

3 Find AB.

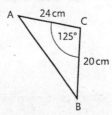

4 Find AB.

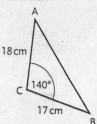

5 Find RT.

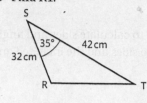

6 Find MN.

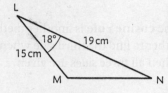

7 Find Y.

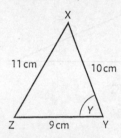

8 Find ∠ABC

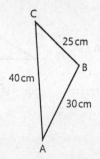

9 Find ∠XYZ.

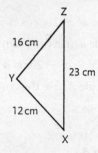

Exercise 182*

Write your answers correct to 3 significant figures.

1 Find x.

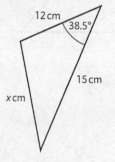

2 Find ∠XYZ.

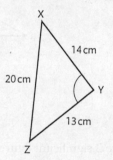

3 Find ∠BAC.

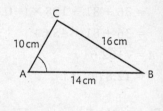

4 Find ∠CAB.

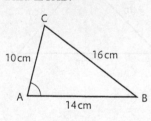

5 Find QR, ∠PQR and ∠QRP.

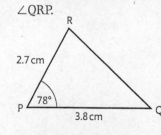

6 Find LM, ∠NLM and ∠LMN.

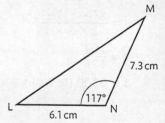

7 From S, a yacht sails on a bearing of 040°. After 3 km, at buoy A, it sails on a bearing of 140°. After another 4 km, at buoy B, it heads back to the start S. Calculate the total length of the course.

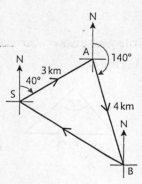

8 Copy and label this diagram using the facts.
VW = 50 km, WU = 45 km, VU = 30 km
  a Find ∠VWU.
  b If the bearing of V from W is 300°,
     find the bearing of U from W.

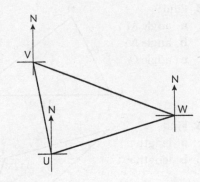

Mixed questions

Sometimes both the sine and cosine rules have to be used or a careful choice has to be made of which method will produce the most efficient solution.

Example 6

A motorboat, M, is 8 km from a harbour, H, on a bearing of 125°, whilst a rowing boat, R, is 16 km from the harbour on a bearing of 074°.

Find the distance and bearing of the rowing boat from the motorboat.

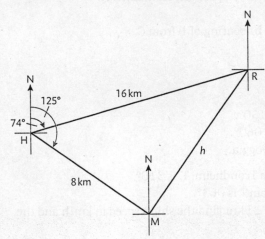

Cosine rule:

∠RHM = 125° − 74° = 51°

$$h^2 = 8^2 + 16^2 - 2 \times 8 \times 16 \times \cos 51°$$

$$= 158.89 \dots$$

$$\Rightarrow h = 12.6 \text{ km (3 s.f.)}$$

Sine rule:

$$\frac{\sin 51°}{12.6} = \frac{\sin M}{16}$$

$$\sin M = 0.986 \dots$$

$$M = 80.7° \text{ (3 s.f.)}$$

Bearing of R from M = 80.7° − 55°

$$= 025.7° \text{ (3 s.f.)}$$

Key Points

If the triangle is right-angled, it is quicker to solve using ordinary trigonometry rather than using sine rule or cosine rule.

Exercise 183

Write your answers correct to 3 significant figures.

1 Find
  a length a
  b angle C.

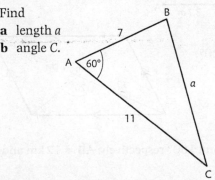

2 Find
  a length r
  b angle Q.

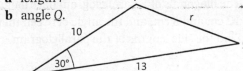

3 Find
   a angle M
   b angle N
   c angle O.

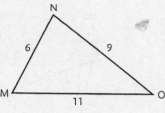

4 Find
   a angle X
   b angle Y
   c angle Z.

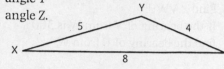

5 Find
   a length p
   b length r.

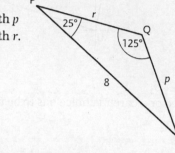

6 Find
   a angle Z
   b length x.

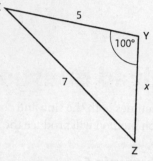

7 Point P is on a bearing of 060° from port O. Point Q is on a bearing of 130° from port O. OQ = 17 km, OP = 11 km. Find the
   a distance PQ
                                    b bearing of Q from P.

8 Towns B and C are on bearings of 140° and 200° respectively from town A. AB = 7 km and AC = 10 km. Find the
   a distance BC
                                      b bearing of B from C.

Exercise 183*

Write answers to 3 significant figures.

1 Napoli is 170 km from Rome on a bearing of 130°.
Foggia is 130 km from Napoli on a bearing of 060°.
Find the distance and bearing of Rome from Foggia.

2 At 12:00, a ship is at X where its bearing from Trondheim, T, is 310°.
At 14:00, the ship is at Y where its bearing from T is 063°.
If XY is a straight line, TX = 14 km and TY = 21 km, find the ship's speed in km/h and the bearing of the ship's journey from X to Y.

3 Calculate
   a angle BAE
   b length CD
   c angle ACD.

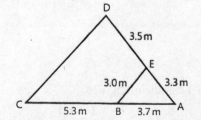

4 The diagonals of a parallelogram have lengths 12 cm and 8 cm, and the angle between them is 120°. Find the side lengths of the parallelogram.

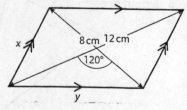

5 From a point A, the bearings of B and C are 040° and 160° respectively. AB = 12 km and AC = 15 km. Find BC and the bearing of C from B.

6 SBC is a triangular orienteering course, with details as in this table.

Find the distance CS, and the bearing of S from C.

	Bearing	Distance
Start S to B	195°	2.8 km
B to C	305°	1.2 km

7 In the figure, find
a ∠XZY **b** WX

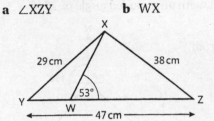

8 In the figure, find
a BD **b** ∠BCD

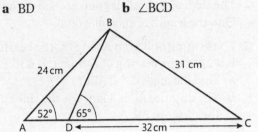

Exercise 184 (Revision)

For Questions 1–6, find the side marked with a letter. Write your answers correct to 3 significant figures.

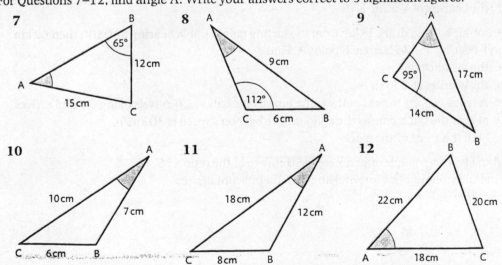

For Questions 7–12, find angle A. Write your answers correct to 3 significant figures.

Exercise 184* (Revision)

Where appropriate, give answers correct to 3 significant figures.

1 A yacht at point A is due West of a headland H. It sails on a bearing of 125°, for 800 m, to a point B. If the bearing of H from B is 335°, find the distance BH.

2 The sides of a parallelogram are 4.6 cm and 6.8 cm, with an included angle of 116°. Find the length of each diagonal.

3 The bearing of B from A is 150°, the bearing of C from B is 280°, the bearing of A from C is 030°.

    a Find the angles of △ABC.

    b The distance AC is 4 km. Use the sine rule to find the distance AB.

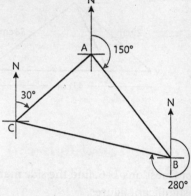

4 In △ABC, AB = 6 cm, BC = 5 cm and AC = 10 cm. Use the cosine rule to find ∠A.

5 Find side a and angle C for

    a

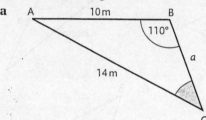

    b

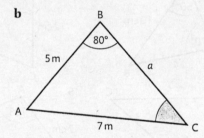

6 A yacht, Y, is 10 km from a port, P, on a bearing of 156°, whilst fishing boat F, is 20 km from the port on a bearing of 044°. Find

    a the distance of the yacht from the fishing boat

    b the bearing of F from Y

    c the bearing of Y from F.

7 A hot-air balloon drifts 28 km from its starting point O, on a bearing of 080°, then 62 km on a bearing of 225° to reach point A. Find

    a the distance OA

    b the bearing of O from A.

    c A truck departs from O at the same time as the balloon. It travels along OA and arrives at A at the same time as the balloon. The balloon's speed is 30 km/h. Find the speed of the truck.

8 The sides of a triangle of perimeter 150 cm are in the ratio 4 : 5 : 6. Find the value of each internal angle to 3 significant figures.

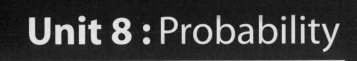

Compound probability

Laws of probability

Example 1

Calculate the probability that a prime number is not obtained when a fair 10-sided spinner that is numbered from 1 to 10 is spun.

Let A be the event that a prime number is obtained.

$p(\overline{A}) = 1 - p(A)$

$= 1 - \frac{4}{10}$ (There are 4 prime numbers: 2, 3, 5, 7)

$= \frac{6}{10} = \frac{3}{5}$

> **Remember**
> - $p(A)$ means the probability of event A occuring
> - $p(\overline{A})$ means the probability of event A *not* occuring
> - $0 \leqslant p(A) \leqslant 1$
> - $p(A) + p(\overline{A}) = 1$, so $p(\overline{A}) = 1 - p(A)$

Independent events

If two events have no effect on each other, they are independent.

If it snows in Moscow, it would be unlikely that this event would have any effect on your teacher winning the lottery on the same day. These events are said to be *independent*.

Mutually exclusive events

Some events exclude the outcome of another. A number rolled on a dice cannot be both odd and even. These events are said to be *mutually exclusive*.

Combined events

Multiplication ('and') rule

Two dice are thrown together.
One is a fair die numbered 1 to 6.
On the other, each face is of a different colour: red, yellow, blue, green, white and purple.

All possible outcomes are shown in this sample space diagram.

	R	Y	B	G	W	P
1	•	•	•	•	•	⊙
2	•	•	•	•	•	•
3	•	•	•	•	•	⊙
4	•	•	•	•	•	•
5	•	•	•	•	•	⊙
6	•	•	•	•	•	•

What is the probability that the die will show an odd number and a purple face?

Let this event be A. By inspection of the possibility space, $p(A) = \frac{3}{36} = \frac{1}{12}$.

Event O is that an odd number is thrown. $p(O) = \frac{1}{2}$.

Event P is that a purple colour is thrown. $p(P) = \frac{1}{6}$.

$p(A) = p(O \text{ and } P) = p(O) \times p(P) = \frac{1}{2} \times \frac{1}{6} = \frac{1}{12}$

> **Remember**
> For A and B, two independent events, the probability of both events occurring is:
> $p(A \text{ and } B) = p(A) \times p(B)$

Addition ('or') rule

A card is randomly selected from a pack of 52 playing cards.
What is the probability that it is an ace or a king?

Let this event be E. There are 8 cards which are either aces or kings, so $p(E) = \frac{8}{52}$.

Event A is that an ace is selected. $p(A) = \frac{4}{52}$. Event K is that a king is selected. $p(K) = \frac{4}{52}$.

$p(E) = p(\ \text{ or } K) = p(A) + p(K) = \frac{4}{52} + \frac{4}{52} = \frac{8}{52}$

Tree diagrams

Tree diagrams show all the possible outcomes. Together with the 'and' and 'or' rules, they can make problems easier to solve.

Example 2

A litter of Border Collie puppies contains four females and two males. A vet randomly removes one from its basket, and it is *not* replaced before another is chosen.
What is the probability that the vet removes two males?

Let event F be that a female is picked.
Let event M be that a male is picked.

Notice that when the second puppy is taken, there are only five left in the basket.

First puppy   Second puppy

M $\frac{2}{6}$ — M $\frac{1}{5}$
M — F $\frac{4}{5}$
F $\frac{4}{6}$ — M $\frac{2}{5}$
F — F $\frac{3}{5}$

Let event A be that two males (dogs) are chosen.

$p(A) = p(M_1 \text{ and } M_2)$ (M_1 means the first puppy is a male.
M_2 means the second puppy is a male)

$= p(M_1) \times p(M_2)$ (given by tree diagram route MM)

$= \frac{2}{6} \times \frac{1}{5} = \frac{2}{30} = \frac{1}{15}$

What is the probability that the vet removes one male and one female?

Let event B be that a male and a female are chosen.

$p(B) = p(M_1 \text{ and } F_2) \text{ or } p(F_1 \text{ and } M_2)$
    (given by tree diagram routes MF and FM)

$= \frac{2}{6} \times \frac{4}{5} + \frac{4}{6} \times \frac{2}{5}$

$= \frac{8}{30} + \frac{8}{30} = \frac{16}{30} = \frac{8}{15}$

Exercise 185

Use tree diagrams to solve these problems.

1 A fair six-sided die is thrown twice. Calculate the probability of obtaining these scores.
   a Two sixes
   b No sixes
   c A six and not a six, in that order
   d A six and not a six, in any order

2 A box contains two red and three green beads.
   One is randomly chosen, and replaced before another is chosen.
   Calculate the probability of obtaining these combinations of beads.
   a Two red beads
   b Two green beads
   c A red bead and a green bead, in that order
   d A red bead and a green bead, in any order

3 A chest of drawers contains four yellow ties and six blue ties.
One is randomly selected and replaced before another is chosen.
Calculate the probability of obtaining these ties.

  a Two yellow ties   **b** Two blue ties

  c A yellow tie and a blue tie, in that order

  d A yellow tie and a blue tie, in any order

4 A spice rack contains three jars of chilli and four jars of mint.
One is randomly selected, and replaced before another is chosen.

  a Calculate the probability of selecting two jars of chilli.

  b What is the probability of selecting a jar of chilli and a jar of mint?

5 In a game of basketball the probability of scoring from a free shot is $\frac{2}{3}$.
A player has two consecutive free shots.

  a Calculate the probability that he scores two baskets.

  b What is the probability that he scores one basket?

  c What is the probability that he scores no baskets?

6 Each evening Dina either reads a book or watches television.
The probability that she watches television is $\frac{3}{4}$, and if she does this, the probability that
she will fall asleep is $\frac{4}{7}$. If she reads a book, the probability that she will fall asleep is $\frac{3}{7}$.

  a Calculate the probability that she does not fall asleep.

  b What is the probability that she *does* fall asleep?

7 Two cards are picked at random from a pack of 52 playing cards with replacement.
Calculate the probability that these cards are picked.

  a Two kings   **b** A red card and a black card

  c A picture card and an odd number card   **d** A heart and a diamond

8 Amy oversleeps one day in 5, and when this happens she breaks her shoelace 2 out of 3
times. When she does *not* oversleep, she breaks her shoelace only 1 out of 6 times.
If she breaks her shoelace, she is late for school.

  a Calculate the probability that Amy is late for school.

  b What is the probability that she is not late for school?

  c In the space of 30 school days, how many times would you expect Amy to be late?

The 'at least' situation

Example 3

A fruit basket contains two oranges (O) and three apples (A).
A fruit is selected at random and *not* replaced before another is randomly selected.
Calculate the probability of choosing *at least* one apple.

Let the number of apples be a. 'At least one apple' means that $a \geqslant 1$.

One method of calculating $p(a \geqslant 1)$ is to find all the possibilities.

$$p(a \geqslant 1) = p(AO) + p(OA) + p(AA)$$
$$= \frac{3}{5} \times \frac{2}{4} + \frac{2}{5} \times \frac{3}{4} + \frac{3}{5} \times \frac{2}{4}$$
$$= \frac{3}{10} + \frac{3}{10} + \frac{3}{10} = \frac{9}{10}$$

A quicker method uses the rule $p(E) + p(\overline{E}) = 1$.

$$p(a \geqslant 1) = 1 - p(a = 0)$$
$$= 1 - p(OO)$$
$$= 1 - \frac{2}{5} \times \frac{1}{4} = 1 - \frac{1}{10} = \frac{9}{10}$$

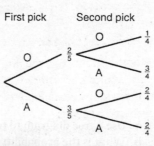

Exercise 185*

1 A box contains two black stones and four white stones.
   One is randomly selected and *not* replaced before another is randomly taken out.
   Use a tree diagram to help you calculate the probability of selecting these stones:
   a A black stone and a white stone, in that order.
   b A black stone and a white stone, in any order.
   c At least one black stone.

2 A bag contains three orange counters and four purple counters. One is randomly selected and
   replaced *together with a counter of the colour not picked out* (orange or purple). Another counter
   is then randomly selected. Use a tree diagram to help you to calculate these probabilities:
   a p(Two orange counters)
   b p(One counter of each colour)
   c p(At least one purple counter)

3 An archer fires his arrows at a target.
   The probability that he scores a bullseye in any one
   attempt is $\frac{1}{3}$. If he fires twice, calculate these probabilities:
   a p(Two bullseyes)
   b p(No bullseyes)
   c p(At least one bullseye)

4 A netball shooter has two free shots.
   The probability that she scores (or misses) with each shot can be found from this table.

	First shot	Second shot
Scores	$\frac{2}{3}$	
Misses		$\frac{3}{7}$

   Copy and complete the table, and use it to construct a tree diagram to help you to calculate
   these probabilities:
   a p(Both shots missed)   b p(Scores once)   c p(Scores at least once)

5 The spinner is spun twice.
   Use tree diagrams to help you to calculate these probabilities:
   a p(Two even numbers)
   b p(An even number and an odd number)
   c p(A black number and a white number)
   d p(A white even number and a black odd number)

6 A marble is randomly taken from bag A and is then placed in bag B. A marble is then
   randomly selected from bag B.

Bag A   Bag B

   a Use a tree diagram to help you to find the probability that this marble is black.
   b What is the probability that the marble is white?

7 A box contains two red sweets and three green sweets. A sweet is selected at random and *not* replaced. If a red sweet is picked on the first attempt, then two extra reds are placed in the box. If a green sweet is picked on the first attempt, then three extra greens are placed in the box. Calculate these probabilities of selecting from two picks:

a p(Two red sweets)

b p(A red sweet and a green sweet)

c p(At least one green sweet)

8 Suzi has just taken up golf, and she buys a golf bag containing five different clubs.

Unfortunately, she does not know when to use each club, and so chooses them randomly for each shot. The probabilities for each shot that Suzi makes are shown in this table.

	Good shot	Bad shot
Right club	$\frac{2}{3}$	
Wrong club		$\frac{3}{4}$

a Copy and complete the table, and use it to construct a tree diagram.

b At one short hole, she can reach the green in one shot if it is 'good'. If her first shot is 'bad', it takes one more 'good' shot to reach the green. Find the probability that she reaches the green in at most two shots.

Activity 45

Network XAB: XAB is a simple one-way road system. When there is a choice at a junction, a driver is *equally likely* to turn down any accessible road.

p(*A*) is the probability of reaching A starting at X.
p(*B*) is the probability of reaching B starting at X.

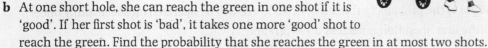

- Copy the network XAB.
  Write down the probability of going from X to B and A to B.
  Calculate p(*A*) and p(*B*).

- 60 vehicles pass through X.
  How many vehicles would you expect to pass through junction A?
  How many would you expect to pass through junction B?
  How many would you expect to go through B not via A?

Network XABCD: The road system is now extended as shown.

- Copy the network XABCD, and write down the probability of turning into each of the roads.
  Calculate p(*A*), p(*B*), p(*C*) and p(*D*).

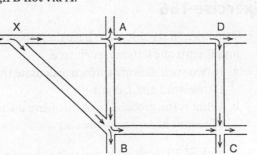

- 60 vehicles pass through X.
  How many vehicles would you expect to pass through each junction: A, B, C and D?

- Compare these theoretical answers by devising a method to simulate 60 vehicles passing through the road network XABCD with each vehicle starting at X (hint: dice, counters or computers).

Further problems

Example 4

A vet has a 90% probability of detecting a particular virus in a horse.
If he detects the virus, the operation has an 80% success rate the first time it is attempted.
If this operation is unsuccessful, it can be repeated, but with a success rate of only 40%.
Any subsequent operations have such a low chance of success that a vet will not attempt further operations.

Let event D be that the virus is detected.
Let event S be that the operation succeeds.
Let event F be that the operation fails.

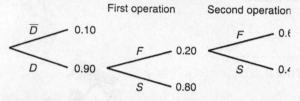

What is the probability that an infected horse will be operated on successfully once?

p(the first operation is successful) = p(D and S)
$$= p(D) \times p(S)$$
$$= 0.90 \times 0.80 = 0.72$$

What is the probability that an infected horse will be cured at the second attempt?

p(the second operation is successful) = p(D and F and S)
$$= p(D) \times p(F) \times p(S)$$
$$= 0.90 \times 0.20 \times 0.40 = 0.072$$

What is the probability that an infected horse will be cured?

p(the horse is cured) = p(the first operation is successful or the second operation is successful)
$$= \text{p(the first operation is success)} + \text{p(the second operation is successful)}$$
$$= 0.72 + 0.072 = 0.792$$

Exercise 186

1 Two normal six-sided dice have their spots covered and replaced by the letters A, B, C, D, E and F, with one letter on each face.
   a If two such dice are thrown, calculate the probability that they show two vowels.
      (Vowels are a, e, i, o, u.)
   b What is the probability of throwing a vowel and a consonant?
      (Consonants are non-vowels.)

2 A pack of 20 cards is formed using the ace, ten, jack, queen and king of each of the four suits from an ordinary full pack of playing cards. This reduced pack is shuffled and then dealt one at a time *without replacement*. Calculate these probabilities:
   a p(The first card dealt is a king)
   b p(The second card dealt is a king)
   c p(At least one king is dealt in the first three cards)

3 A box contains two white beads and five red beads.
A bead is randomly selected and its colour noted.
It is then returned to the box *together with two more beads of the same colour*.

  a If a second bead is now randomly selected from the box, calculate the probability that it is the same colour as the first bead.
  b What is the probability that the second bead is a different colour from the first bead?
  c Find the probability that the second bead drawn is white.
  d Nick says, 'However many beads of the same colour as the first bead withdrawn are added, the probability that the second bead selected is white stays the same!'
     True or false? Justify your answer.

4 The probability that a washing machine will break down in the first 5 years of use is 0.27.
The probability that a television will break down in the first 5 years of use is 0.17.
Mr Khan buys a washing machine and a television on the same day.
By using a tree diagram or otherwise, calculate the probability that, in the five years after that day:

  a both the washing machine and the television will break down;
  b at least one of them will break down.   © Edexcel Limited

Exercise 186*

1 A virus is present in 1 in 250 of a flock of sheep. To make testing for the virus possible, a quick test is used on each sheep. However, the test is not completely reliable. An infected sheep tests positive in 85% of cases and a healthy sheep tests positive in 5% of cases.

  a Use a tree diagram to help you to calculate the probability that a sheep will be infected and test positive.
  b What is the probability that a sheep will be infected but test negative?
  c What is the probability that a sheep will test positive?

2 An office block has five floors (ground, 1, 2, 3 and 4), all connected by a lift. When it goes up to any floor (except 4), the probability that after it has stopped it will continue to rise is $\frac{3}{4}$. When it goes down to any floor (except the ground floor), the probability that after it has stopped it will continue to go down is $\frac{1}{4}$. The lift stops at any floor it passes.
The lift is currently at the first floor having just descended. Calculate the probability of these events.

  a Its second stop is the third floor.
  b Its third stop is the fourth floor.
  c Its fourth stop is the first floor.

3 Two opera singers, Mario and Clarissa, both perform on the same night, in separate recitals. The independent probabilities that two newspapers X and Y publish reviews of their recitals are given in this table.

	Mario's recital	Clarissa's recital
Probability of review in newspaper X	$\frac{1}{2}$	$\frac{2}{3}$
Probability of review in newspaper Y	$\frac{1}{4}$	$\frac{2}{5}$

  a If Mario buys both newspapers, find the probability that both papers review his recital.
  b If Clarissa buys both newspapers, find the probability that only one paper reviews her recital.
  c Mario buys one of the newspapers at random. What is the probability that it has reviewed *both* recitals?

4 A school has an unreliable clock in its tower. The probabilities of gain or loss in the clock in any 24-hour period are given in this table.

	Gain of 1 minute	No change	Loss of 1 minute
Probability	$\frac{1}{2}$	$\frac{1}{3}$	$\frac{1}{6}$

If the clock is set to the correct time at noon on Sunday, find the probability of these events:
a The clock is correct at noon on Tuesday.
b The clock is *not* slow at noon on Wednesday.

5 A card is randomly taken from an ordinary pack of cards and not replaced. This process is repeated again and again. Explain, with calculations, why these probabilities are found:
a p(First card is a heart) = $\frac{1}{4}$ **b** p(Second card is a heart) = $\frac{1}{4}$
c p(Third card is a heart) = $\frac{1}{4}$ **d** p(Fourth card is a heart) = $\frac{1}{4}$
                                     and so on.

6 Show that in a room of only 23 people, the probability of two of them sharing the same birth date is just over $\frac{1}{2}$.

Activity 46

The data in the table below shows the risks of death from various causes in the USA in 2000.

Cause	Risk of death per person per year ($\times 10^{-7}$)
Rock climbing	400
Football	400
Drinking (1 bottle of wine per day)	750
Smoking (20 cigarettes per day)	50 000
Motor racing	200 000
Being run over by a vehicle	500
Floods	22
Earthquakes	17
Tornadoes	22
Lightning	3
Falling aircraft	1
Bites of venomous creatures	2
Influenza	2000
Being struck by a meteorite	0.0006

- Rank these activities in order of safety.
  Comment.
- How do you think these figures were obtained?
  Comment on their accuracy.

Investigate

A television audience of 1024 people is mesmerised by a psychic who convinces them that there is someone with special telepathic powers among them. She sets about demonstrating this claim by simply tossing a fair coin.

The audience is split into two halves of 512. One half is asked to focus on heads, while the other half concentrates on tails. The coin is tossed, and the group who are wrong sit down. The audience is now split into two halves of 256, and this process is repeated time and again until just one person is left standing. This person is declared to be gifted owner of the telepathic powers!

1 Find the probability of this person being picked out as the 'gifted' one.

2 Is this process misleading? Explain.

Exercise 187 (Revision)

Use tree diagrams to answer these questions.

1 A box contains two maths books and three French books. A book is removed and replaced before another is taken. Find the probability of these events:
  a Two French books are selected.
  b A maths and a French book are selected, in any order.

2 A hockey penalty taker has a $\frac{3}{4}$ probability of scoring a goal.
  a If she takes two penalties, find the probability that she scores no goals.
  b What is the probability that she scores one goal from two penalties?

3 The probability that a biased coin shows tails is $\frac{2}{5}$.
  a Find the probability that, when the coin is thrown twice, it shows two heads.
  b What is the probability of exactly one tail in two throws?

4 Assume that it is either sunny or rainy in Italy. If the weather is sunny one day, the probability that it is sunny the day after is $\frac{1}{5}$. If it rains one day, the probability that it rains the next is $\frac{3}{4}$.
  a If it is sunny on Sunday, calculate the probability that it rains on Monday.
  b What is the probability that it is sunny on Sunday and Tuesday?

5 The letters of the word HYPOTHETICAL appear on plastic squares that are placed in a bag and jumbled up. A square is randomly selected and replaced before another is taken. Calculate these probabilities.
  a p(Two H) b p(Exactly one T) c p(A vowel)

6 All female chaffinches have the same patterns of laying eggs.
The probability that any female chaffinch will lay a certain number of eggs is given in the table.

Number of eggs	0	1	2	3	4 or more
Probability	0.1	0.3	0.3	0.2	x

  a Calculate the value of x.
  b Calculate the probability that a female chaffinch will lay less than 3 eggs.
  c Calculate the probability that two female chaffinches will lay a total of 2 eggs.

© Edexcel Limited

Exercise 187* (Revision)

Use tree diagrams where appropriate.

1 A chocolate box contains four milk chocolates and five Turkish delights. Gina loves milk chocolates, and hates Turkish delight. She takes one chocolate randomly, and *does not replace it* before picking out another one at random.
   a Calculate the probability that Gina is very happy.
   b What is the probability that Gina is very unhappy.
   c What is the probability that she has at least one milk chocolate?

2 The probability that Mr Glum remembers his wife's birthday and buys her a present is $\frac{1}{3}$.
   The probability that he does not lose the present on the way home is $\frac{2}{3}$.
   The probability that Mrs Glum likes the present is $\frac{1}{5}$.
   a Calculate the probability that Mrs Glum receives a birthday present.
   b What is the probability that Mrs Glum receives a birthday present but dislikes it?
   c What is the probability that she is happy on her birthday?
   d What is the probability that she is not happy on her birthday?

3 A sleepwalker gets out of bed and is five steps away from his bedroom door. He is equally likely to take a step forward as he is to take one backwards.
   a Calculate the probability that after five steps he is at his bedroom door.
   b What is the probability that by then he is only one step closer to his bedroom door?
   c What is the probability that, having taken five steps, he is closer to his bedroom door than to his bed?

4 There are 25 beads in a bag.
   Some of the beads are red.
   All the other beads are blue.
   Kate picks up two beads at random without replacement.
   The probability that she will pick two red beads is 0.07.
   Calculate the probability that the two beads she picks will be of different colours.

© Edexcel Limited

5 The diagram shows a shooting target that is divided into three regions by concentric circles with radii that are in the ratio $1:2:3$.

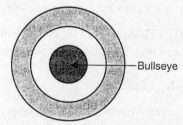

Bullseye

   a Find the ratio of the areas of the three regions in the form $a:b:c$, where a, b and c are integers.
   The probability that a shot will hit the target is $\frac{4}{5}$. If it does not the target, the probability of it hitting any region is proportional to the area of that region.
   b Calculate the probability that a shot will hit the bullseye.
   c Two shots are fired. Calculate the probability of these events:
      (i)   They both hit the bullseye.
      (ii) The first hits the bullseye and the second does not.
      (iii) At least one hits the bullseye.

6 Melissa is growing apple and pear trees for her orchard. She plants seeds in her greenhouse, but forgets to label the seedling pots. She knows that the types of apple and pear trees are in the numbers shown.

	Eating	Cooking
Apple	24	76
Pear	62	38

She picks out an apple seedling at random.

a Estimate the probability that it is an eater.

b Estimate the probability that it is an cooker.

She picks out an eater seedling at random.

c Estimate the probability that it is an apple.

d Estimate the probability that it is an pear.

e She picks out three seedlings at random without replacement. Find the probability that she ends up with at least one eating-apple seedling.

1 The 100th term of the sequence 3, 7, 11, 15 ... is

    A 398   **B** 399   **C** 400   **D** 401

2 $(x + a)$ is a factor of $x^3 - x^2 - 6$.

    The value of a is

    A 2   **B** -2 **C** 3   **D** -1

3 The gradient of the tangent of a speed–time graph is a measure of the

    A Distance   **B** Speed   **C** Mean speed   **D** Acceleration

4 The value of the angle A to 3 s.f. is

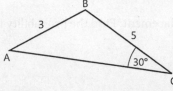

    A 56.4°   **B** 33.6°   **C** 17.5°   **D** 72.5°

5 The formula for the nth term of the sequence 14, 10, 6... is

    A $14 - 4n$ **B** $18 - 4n$ **C** $10 + 4n$ **D** $14 + 4n$

6 A fair coin is tossed three times in succession. The probability of obtaining three heads is

    A $\frac{1}{2}$ **B** $\frac{3}{2}$ **C** $\frac{1}{6}$ **D** $\frac{1}{8}$

7 $x - 2$ is a factor of $x^3 + ax^2 + 5x + 6$. The value of a is

    A -3 **B** 3   **C** -6 **D** 6

8 The maximum point of $y = \sin(x°)$ for $0° \leqslant x \leqslant 180°$ has coordinates of

    A $(1, 90°)$ **B** $(90°, -1)$ **C** $(-1, 90°)$ **D** $(90°, 1)$

9 The value of the angle A to 3 s.f. is

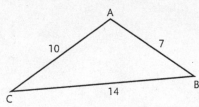

    A 109.6°   **B** 42.3°   **C** 28.1°   **D** 44.4°

10 A fair die is rolled twice in succession. The probability of obtaining at least one prime number is

    A $\frac{3}{4}$ **B** $\frac{8}{9}$ **C** $\frac{1}{2}$ **D** $\frac{1}{4}$

1 **a** Find the nth term for the following sequence.
$$5, \ 9, \ 13, \ 17, \ 21...$$
   b Use this formula to find the 100th term.
   c Find n if the nth term is 201.

2 **a** Find the nth term for the following sequence.
$$12, \ 6, \ 0, \ -6, \ -12...$$
   b Use this formula to find the 100th term.
   c Find n if the nth term is -132.

3 Show that $(x - 2)$ is a factor of $x^3 + 7x^2 - 36$

4 $(x + 3)$ is a factor of $x^3 + px^2 - x - 30$. Find the value of p.

5 Factorise $f(x) = x^3 - 5x^2 - x + 5$ completely. Hence solve $f(x) = 0$.

6 $(x + a)$ is a factor of $x^3 + ax^2 + ax + 4$. Find the two possible values of a.

7 $(x + 3)$ is a factor of $f(x) = 3x^3 + 7x^2 + px - 24$.
   Find the value of p, factorise $f(x)$ and hence solve $f(x) = 0$.

8 **a** Use the intersecting chords theorem to form an equation in x from the diagram.
   b Solve your equation to find x, giving your answer correct to 3 s.f.

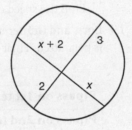

For Questions 9–12, form a quadratic equation and then solve.

9 The length of a rectangle is 1 m greater than its width and the area is $7\,\text{m}^2$. Find the dimensions of the rectangle.

10 The base of a triangle is 5 cm more than the height, and the area is $18\,\text{cm}^2$. Find the height of the triangle.

11 A right-angled triangle has sides of length x, x and $x + 2$.
   Find the value of x.

12 The sum of the squares of two consecutive numbers is 265. Find the numbers.

For Questions 13–15 refer to the graph shown.

13 Find the gradient when
   a $x = 1$ **b** $x = 3$ **c** $x = -2$

14 Where is the gradient equal to
   a 1   **b** -1 **c** 0?

15 The tangents at the point P and Q are parallel to the line $y = x$. Find the coordinates of P and Q.

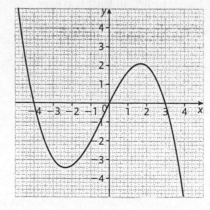

16 Find side *a* and angle *C* for

a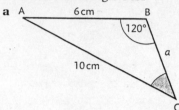

b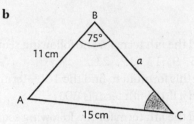

17 A ship, A, is 50 km from a port, P, on a bearing of 130°, whilst speed boat B, is 100 km from the port on a bearing of 075°. Find

a the distance of the ship from the speed boat

b the bearing of B from A

c the bearing of A from B.

18 A small aeroplane flies 56 km from its starting point O, on a bearing of 070°, then 40 km on a bearing of 210° to reach point A. Find

a the distance OA **b** the bearing of O from A.

c A truck departs from O at the same time as the aeroplane. It travels along OA and arrives at A at the same time as the aeroplane. The aeroplane's speed is 100 km/h. Find the speed of the truck.

19 Boris and Gunter are taking a scuba dive leader test. If they fail the first test only one re-test is permitted. Their probabilities of passing are given in the table.

	Boris	**Gunter**
P(pass on 1st test)	0.65	0.75
P(pass on 2nd test)	0.85	0.80

These probabilities are independent of each other. Find the probabilities that

a Gunter becomes a dive leader at the second test

b Boris passes at his first test whilst Gunter fails his first test

c only one of the two men becomes a scuba dive leader, assuming that a re-test is taken if the first test has been failed

20 The probability of a red kite laying a certain number of eggs is given in the table.
Find

a the value of *k*

b the probability that a red kite

   i lays at least two eggs

   ii lays at most two eggs.

c The probability that two red kites lay a total of at least three eggs.

Number of eggs	Probability
0	0.1
1	*k*
2	0.3
3	0.4
4	0.06
⩾5	0.04

Irrational numbers

Primitive people only needed the numbers 1, 2, 3, ... (called the set of natural numbers) to count objects. Later it was found that negative numbers and zero (called the set of integers) were needed. Also, fractions were required, for example to divide 4 loaves equally between 6 people. Early civilisations thought that all numbers could be expressed as fractions, i.e. the **ratio** of two integers, hence the word **rational** to describe these numbers.

In the sixth century BC, the ancient Greeks proved that $\sqrt{2}$, the length of the diagonal of a unit square, could not be written as a fraction. They soon found that other numbers such as $\sqrt{3}$ and $\sqrt{5}$ could not be written as fractions. These numbers are called **irrational** numbers. Later it was shown that π is also an irrational number.

π = 3.141 592 653 589 793 238 462 643 383 279 502 884 197 169 399 375 105, 820 974 944 592 307 816 406 286 208 998 628 034 825 342 117 067 982 148 086 513 282 ...

Investigate

The pursuit of the true value of π has fascinated many people for many years. Investigate the approximations shown here and, where possible, rank them in order of accuracy.

Approximation of π	Source
3	Bible 1 Kings 7:23
$3\frac{1}{8}$	Babylon 2000 BC; found on a clay tablet in 1936
$256 \div 81$	Egypt 2000 BC; found on the 'Rhind Papyrus'
$22 \div 7$	Syracuse 250 BC; Archimedes
$377 \div 120$	Greece 140 BC; Hipparchus
$355 \div 113$	China AD 450; Tsu Chung-Chih
$\sqrt{10}$	India AD 625; Brahmagupta
$864 \div 275$	Italy AD 1225; Fibonacci
$\dfrac{2 \times 2 \times 2 \times 4 \times 4 \times 6 \times 6 \times 8 \times 8 ...}{1 \times 3 \times 3 \times 5 \times 5 \times 7 \times 7 \times 9 ...}$	England AD 1655; Wallis
$\sqrt[4]{9^2 + \frac{19^2}{22}}$	India AD 1910; Ramanujan

The decimal expansion of irrational numbers is infinite. A decimal expansion that is infinite but where the digits recur is rational because it can be written as a fraction (see page 184).

Together the rational numbers and the irrational numbers form the set of real numbers. All these sets can be shown on a Venn diagram where:

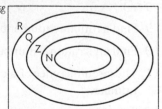

$\mathbb{N}$ is the set of natural numbers or positive integers {1, 2, 3, 4, ...}.

$\mathbb{Z}$ is the set of integers {..., −2, −1, 0, 1, 2, ...}.

$\mathbb{Q}$ is the set of rational numbers.

$\mathbb{R}$ is the set of real numbers.

Activity 47

If $\mathbb{W}$ is the set of whole numbers $\{0, 1, 2, 3, ...\}$
      $\mathbb{Z}^+$ is the set of positive integers $\{1, 2, 3, ...\}$
      $\mathbb{Z}^-$ is the set of negative integers $\{... -3, -2, -1\}$

draw Venn diagrams to show the relation between

a $\mathbb{W}$, $\mathbb{Z}^+$ and $\mathbb{Q}$

b $\mathbb{Z}^+$, $\mathbb{Z}^-$ and $\mathbb{R}$

Example 1

Which of these numbers are rational and which are irrational?

2   0.789 $0.\dot{6}\dot{7}$ $\sqrt{2} \times \sqrt{2}$ $\sqrt{2} + \sqrt{2}$

$2 = \frac{2}{1}$ so it is rational.

$0.789 = \frac{789}{1000}$ so it is rational.

$0.\dot{6}\dot{7} = \frac{67}{99}$ so it is rational.

$\sqrt{2} \times \sqrt{2} = 2$ so it is rational.

$\sqrt{2} + \sqrt{2} = 2\sqrt{2}$ so it is irrational.

Example 2

a Find an irrational number between 4 and 5.

   $4^2 = 16$ and $5^2 = 25$, so the square root of any integer between 17 and 24 inclusive will be irrational, for example $\sqrt{19}$.

b Find a rational number between $\sqrt{2}$ and $\sqrt{3}$.

   $\sqrt{2} = 1.414213...$ and $\sqrt{3} = 1.732050...$ so any terminating or recurring decimal between these numbers will be rational, for example 1.5

Remember

Good examples of irrational numbers are the square roots of prime numbers and π.

Exercise 188

For Questions 1–8, state which of the numbers are rational and which are irrational. Express the rational numbers in the form $\frac{a}{b}$ where a and b are integers.

1 5.7   2 $0.\dot{4}\dot{7}$ 3 $\sqrt{49}$ 4 2π

5 $\sqrt{3} \times \sqrt{3}$ 6 $\sqrt{3} \div \sqrt{3}$ 7 $\sqrt{5} + \sqrt{5}$ 8 $\sqrt{3} - \sqrt{3}$

9 Find a rational number between $\sqrt{5}$ and $\sqrt{7}$.

10 Find an irrational number between 7 and 8.

11 The circumference of a circle is 4 cm. Find the radius, giving your answer as an irrational number.

12 The area of a circle is 9 cm². Find the radius, giving your answer as an irrational number.

Exercise 188*

For Questions 1–7, state which of the numbers are rational and which are irrational. Express the rational numbers in the form $\frac{a}{b}$ where a and b are integers.

1 $\pi + 2$

2 $\sqrt{\dfrac{4}{25}}$

3 $\sqrt{0.36}$

4 $\sqrt{2\tfrac{1}{4}}$

5 $\sqrt{5} - \sqrt{3}$

6 $\sqrt{13} \div \sqrt{13}$

7 $\sqrt{3} \times \sqrt{3} \times \sqrt{3}$

8 Find a rational number between $\sqrt{11}$ and $\sqrt{13}$.

9 Find an irrational number between 2 and 3.

10 Find two *different* irrational numbers whose product is rational.

11 The circumference of a circle is 6 cm. Find the area, giving your answer as an irrational number.

12 Right-angled triangles can have sides with lengths that are rational or irrational numbers of units. Give an example of a right-angled triangle to fit each description below. Draw a separate triangle for each part.
 a All sides are rational.
 b The hypotenuse is rational and the other two sides are irrational.
 c The hypotenuse is irrational and the other two sides are rational.
 d The hypotenuse and one of the other sides is rational and the remaining side is irrational.

Surds

A **surd** is a root that is irrational. The roots of all prime numbers are surds.

$\sqrt{2}, \sqrt{5}$ and $\sqrt[3]{7}$ are all surds.

$\sqrt{4} = 2$, so $\sqrt{4}$ is *not* a surd.

Note that $\sqrt{2}$ means the positive square root of 2.

When an answer is given using a surd, it is an *exact* answer.

For example, the diagonal of a square of side 1 cm is *exactly* $\sqrt{2}$ cm long, or sin 60° is *exactly* $\dfrac{\sqrt{3}}{2}$.

Simplifying surds

Example 3

Simplify the following, giving the exact answer:

 a $2\sqrt{3} + 3\sqrt{3}$ **b** $(2\sqrt{3})^2$ **c** $2\sqrt{3} \times 3\sqrt{3}$ **d** $(\sqrt{3})^3$

 a $2\sqrt{3} + 3\sqrt{3} = 5\sqrt{3}$

 b $(2\sqrt{3})^2 = 2\sqrt{3} \times 2\sqrt{3} = 2 \times 2 \times \sqrt{3} \times \sqrt{3} = 4 \times 3 = 12$

 c $2\sqrt{3} \times 3\sqrt{3} = 2 \times 3 \times \sqrt{3} \times \sqrt{3} = 6 \times 3 = 18$

 d $(\sqrt{3})^3 = \sqrt{3} \times \sqrt{3} \times \sqrt{3} = 3\sqrt{3}$

Exercise 189

Simplify the following, giving the exact answer:

1 $2\sqrt{5} + 4\sqrt{5}$ **2** $5\sqrt{3} - \sqrt{3}$ **3** $(4\sqrt{2})^2$

4 $(2\sqrt{5})^2$ **5** $4\sqrt{2} \times \sqrt{2}$ **6** $5\sqrt{7} \times 3\sqrt{7}$

7 $(\sqrt{2})^3$ **8** $4\sqrt{2} \div \sqrt{2}$ **9** $8\sqrt{5} \div \sqrt{5}$

Exercise 189*

Simplify the following, giving the exact answer:

1 $4\sqrt{11} + \sqrt{11}$ **2** $6\sqrt{7} - 2\sqrt{7}$ **3** $(3\sqrt{11})^2$

4 $8\sqrt{3} \times 4\sqrt{3}$ **5** $(2\sqrt{7})^3$ **6** $3\sqrt{2} \times 4\sqrt{2} \times 5\sqrt{2}$

7 $(\sqrt{3})^4$ **8** $4\sqrt{7} \div \sqrt{7}$ **9** $6\sqrt{11} \div \sqrt{11}$

Example 3 part **b** showed that $(2\sqrt{3})^2 = 4 \times 3 = 12$.

This means that $\sqrt{12} = \sqrt{4 \times 3} = \sqrt{4} \times \sqrt{3} = 2\sqrt{3}$

This is because $\sqrt{a \times b} = \sqrt{a} \times \sqrt{b}$.

Example 4

Simplify   **a** $\sqrt{18}$ **b** $\sqrt{35}$

a Find two factors of 18, one of which is a perfect square, for example $18 = 9 \times 2$.
Then $\sqrt{18} = \sqrt{9} \times \sqrt{2} = 3\sqrt{2}$

b In this case a perfect square factor cannot be found.
As $35 = 5 \times 7$, then $\sqrt{35} = \sqrt{5} \times \sqrt{7} = \sqrt{5}\sqrt{7}$

Example 5

Simplify $\sqrt{72}$.

$72 = 36 \times 2$, so $\sqrt{72} = \sqrt{36} \times \sqrt{2} = 6\sqrt{2}$

In Example 5 you might have written $72 = 9 \times 8$ so $\sqrt{72} = 3\sqrt{8}$. This is correct, but it is not as simple as possible because $\sqrt{8}$ can be simplified as $\sqrt{8} = \sqrt{4} \times \sqrt{2} = 2\sqrt{2}$.

This gives $\sqrt{72} = 3 \times 2\sqrt{2} = 6\sqrt{2}$ as before.

So when finding factors, try to find the largest possible factor that is a perfect square.

Example 6

Simplify $\sqrt{18} + 2\sqrt{2}$

         $\sqrt{18} = 3\sqrt{2}$ (see Example 4)

So $\sqrt{18} + 2\sqrt{2} = 3\sqrt{2} + 2\sqrt{2} = 5\sqrt{2}$

Example 7

Express $5\sqrt{6}$ as the square root of a single number.

$5\sqrt{6} = \sqrt{25} \times \sqrt{6} = \sqrt{25 \times 6} = \sqrt{150}$

To simplify $\sqrt{\dfrac{16}{25}}$ note that $\left(\dfrac{4}{5}\right)^2 = \dfrac{4^2}{5^2} = \dfrac{16}{25}$ so this means that $\sqrt{\dfrac{16}{25}} = \dfrac{4}{5}$.

This is because $\sqrt{\dfrac{a}{b}} = \dfrac{\sqrt{a}}{\sqrt{b}}$.

Example 8

Simplify $\sqrt{\dfrac{81}{49}}$.

$$\sqrt{\dfrac{81}{49}} = \dfrac{\sqrt{81}}{\sqrt{49}} = \dfrac{9}{7}$$

Exercise 190

For Questions 1–6, simplify.

1 $\sqrt{12}$ **2** $\sqrt{18}$ **3** $\sqrt{48}$

4 $\sqrt{45}$ **5** $\sqrt{12} + \sqrt{3}$ **6** $\sqrt{32} + 2\sqrt{2}$

For Questions 7–9, express as the square root of a single number.

7 $5\sqrt{2}$ **8** $3\sqrt{3}$ **9** $3\sqrt{6}$

For Questions 10–12, simplify.

10 $\sqrt{\dfrac{1}{4}}$ **11** $\sqrt{\dfrac{4}{25}}$ **12** $\sqrt{\dfrac{36}{81}}$

13 A rectangle has sides of length $\sqrt{18}$ and $\sqrt{8}$. Find exact values for the area, the perimeter and the length of a diagonal.

Exercise 190*

For Questions 1–6, simplify.

1 $\sqrt{28}$ **2** $\sqrt{99}$ **3** $\sqrt{80}$

4 $\sqrt{117}$ **5** $\sqrt{48} + \sqrt{3}$ **6** $\sqrt{27} + \sqrt{12}$

For Questions 7–9, express as the square root of a single number.

7 $5\sqrt{3}$ **8** $4\sqrt{5}$ **9** $3\sqrt{7}$

For Questions 10–12, simplify.

10 $\sqrt{\dfrac{1}{36}}$ **11** $\sqrt{\dfrac{81}{100}}$ **12** $\sqrt{\dfrac{49}{169}}$

13 The length of the diagonal of a square is $4\,\text{cm}$. Find exact values for the area and the perimeter.

Activity 48

a The diagram shows an isosceles right-angled triangle.

The two shorter sides are $1\,\text{cm}$ in length.

(i) What is the exact value of x (the hypotenuse)?

(ii) What is the value of the angle marked a?

(iii) Use your answers to parts (i) and (ii) to fill in the table below, giving exact answers.

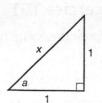

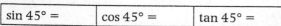

sin 45° =	cos 45° =	tan 45° =

b The diagram shows half an equilateral triangle ABC with sides of length 2 cm. D is the mid-point of AC.

(i) Find the exact value of the length BD.

(ii) What are the values of the angles BAD and ABD?

(iii) Use your answers to parts (i) and (ii) to fill in the table below, giving exact answers.

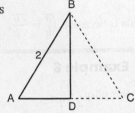

| sin 30° = | cos 30° = | tan 30° = |
| sin 60° = | cos 60° = | tan 60° = |

Rationalising the denominator

When writing fractions it is not usual to write surds in the denominator (see Activity 49 for a reason). The surds can be cleared by multiplying the top and bottom of the fraction by the same amount. This is equivalent to multiplying the fraction by 1 and so does not change its value. The process is called 'rationalising the denominator'.

Example 9

Express $\frac{4}{\sqrt{8}}$ in the form $a\sqrt{2}$.

$$\frac{4}{\sqrt{8}} = \frac{4}{\sqrt{8}} \times \frac{\sqrt{8}}{\sqrt{8}} \quad \text{(Multiply top and bottom by } \sqrt{8}\text{)}$$

$$= \frac{4\sqrt{8}}{8}$$

$$= \frac{4 \times 2\sqrt{2}}{8} = \sqrt{2}$$

Activity 49

a If you had no calculator but knew $\sqrt{2} = 1.414214$ to 6 d.p., how would you work out $\frac{1}{\sqrt{2}}$?

b Express $\frac{1}{\sqrt{2}}$ in the form $a\sqrt{2}$.

c Using your answer to part **b**, work out $\frac{1}{\sqrt{2}}$ to 6 d.p. without using your calculator.

Exercise 191

Express these with rational denominators.

1 $\frac{1}{\sqrt{3}}$

2 $\frac{1}{\sqrt{5}}$

3 $\frac{3}{\sqrt{3}}$

4 $\frac{4}{\sqrt{2}}$

5 $\frac{4}{\sqrt{12}}$

6 $\frac{6}{\sqrt{8}}$

7 $\frac{3}{2\sqrt{2}}$

8 $\frac{5}{2\sqrt{3}}$

9 $\frac{1+\sqrt{2}}{\sqrt{2}}$

10 $\frac{1+\sqrt{3}}{\sqrt{3}}$

Exercise 191*

Express these with rational denominators.

1 $\dfrac{1}{\sqrt{13}}$ **2** $\dfrac{1}{\sqrt{11}}$ **3** $\dfrac{a}{\sqrt{a}}$ **4** $\dfrac{b^2}{\sqrt{b}}$ **5** $\dfrac{6-\sqrt{3}}{\sqrt{3}}$

6 $\dfrac{4-\sqrt{2}}{\sqrt{2}}$ **7** $\dfrac{\sqrt{2}}{\sqrt{6}}$ **8** $\dfrac{\sqrt{8}}{\sqrt{12}}$ **9** $\dfrac{5+2\sqrt{5}}{\sqrt{5}}$ **10** $\dfrac{14+3\sqrt{7}}{\sqrt{7}}$

Exercise 192 (Revision)

1 Which of $0.\dot{3}$, π, $\sqrt{25}$ and $\sqrt{5}$ are rational?

2 Find a rational number between $\sqrt{3}$ and $\sqrt{5}$.

3 Find an irrational number between 3 and 4.

4 Write $3\sqrt{5}$ as the square root of a single number.

For Questions 5–12, simplify.

5 $3\sqrt{3}+2\sqrt{3}$ **6** $3\sqrt{3}-2\sqrt{3}$ **7** $3\sqrt{3}\times2\sqrt{3}$ **8** $3\sqrt{3}\div2\sqrt{3}$

9 $\sqrt{8}$ **10** $\sqrt{63}$ **11** $\sqrt{3}+\sqrt{12}$ **12** $\sqrt{\dfrac{4}{9}}$

For Questions 13–16, rationalise the denominator.

13 $\dfrac{5}{\sqrt{5}}$ **14** $\dfrac{6}{\sqrt{3}}$ **15** $\dfrac{\sqrt{27}}{\sqrt{12}}$ **16** $\dfrac{3+\sqrt{3}}{\sqrt{3}}$

17 A rectangle has sides of length $3\sqrt{2}$ and $5\sqrt{2}$. Find the exact values of the perimeter, the area and the length of a diagonal.

Exercise 192* (Revision)

1 Which of $(\sqrt{3})^2$, $\sqrt{13}$, $\sqrt{5}+\sqrt{5}$ and $0.\dot{2}\dot{3}$ are rational?

2 Find a rational number between $\sqrt{7}$ and $\sqrt{11}$.

3 Find an irrational number between 6 and 7.

4 Write $4\sqrt{11}$ as the square root of a single number.

For Questions 5–15, simplify.

5 $5\sqrt{5}+3\sqrt{5}$ **6** $5\sqrt{5}-3\sqrt{5}$ **7** $5\sqrt{5}\times3\sqrt{5}$ **8** $5\sqrt{5}\div3\sqrt{5}$

9 $\sqrt{48}$ **10** $\sqrt{242}$ **11** $7\sqrt{54}-\sqrt{24}$ **12** $\sqrt{\dfrac{18}{50}}$

For Questions 13–16, rationalise the denominator.

13 $\dfrac{1}{2\sqrt{5}}$ **14** $\dfrac{12}{\sqrt{6}}$ **15** $\dfrac{6+2\sqrt{6}}{\sqrt{6}}$ **16** $\dfrac{\sqrt{112}}{\sqrt{28}}$ **17** $\sqrt{\dfrac{1}{2}}+\sqrt{\dfrac{1}{4}}+\sqrt{\dfrac{1}{8}}$

18 The two shorter sides of a right-angled triangle are $3\sqrt{3}$ and $4\sqrt{3}$.
Find the exact values of the length of the hypotenuse and the area.

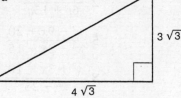

19 Find the exact height of an equilateral triangle with side length 2 units.
Hence find the exact values of cos 60° and sin 60°.

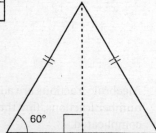

Unit 9 : Algebra

Algebraic fractions

Simplifying algebraic fractions

To simplify algebraic fractions, factorise as much as possible and then cancel common terms.

Example 1

Simplify $\dfrac{x^2 + x}{x + 1}$.

$$\dfrac{x^2 + x}{x + 1} = \dfrac{x(x + 1)}{x + 1}$$
$$= x$$

Example 2

Simplify $\dfrac{x^2 + x - 2}{x^2 - 4}$.

$$\dfrac{x^2 + x - 2}{x^2 - 4} = \dfrac{(x + 2)(x - 1)}{(x + 2)(x - 2)}$$
$$= \dfrac{x - 1}{x - 2}$$

Exercise 193

Simplify the following:

1 $\dfrac{3x + 12}{2x + 8}$

2 $\dfrac{8x + 4y}{6x + 3y}$

3 $\dfrac{x + x^2}{y + xy}$

4 $\dfrac{xy + y^2}{x^2 + xy}$

5 $\dfrac{x^2 - x - 6}{x - 3}$

6 $\dfrac{x^2 - 6x - 7}{x + 1}$

7 $\dfrac{x + 1}{x^2 + 3x + 2}$

8 $\dfrac{x + 2}{x^2 + 4x + 4}$

9 $\dfrac{x^2 - y^2}{(x - y)^2}$

Exercise 193*

Simplify the following:

1 $\dfrac{6a + 9b}{10a - 15b}$

2 $\dfrac{14p + 35q}{6p + 15q}$

3 $\dfrac{x^2y + xy^2}{x^2y + x^3}$

4 $\dfrac{x^3y + xy}{xy^2 + x^3y^2}$

5 $\dfrac{x^2 - 9x + 20}{x^2 - 2x - 15}$

6 $\dfrac{x^2 + 3x - 18}{x^2 + 11x + 30}$

7 $\dfrac{x^2 - x - 12}{x^2 - 16}$

8 $\dfrac{x^2 + 3x - 10}{x^2 - 25}$

9 $\dfrac{3r^2 + 6r - 45}{3r^2 + 18r + 15}$

Addition and subtraction

Algebraic fractions are added or subtracted in the same way as number fractions. As with number fractions, find the lowest common denominator, otherwise the working can become complicated.

Fractions with number denominators

Example 3

Express $\dfrac{x+1}{3} - \dfrac{x-3}{4}$ as a single fraction.

$$\frac{x+1}{3} - \frac{x-3}{4} = \frac{4(x+1)\ 2(x-3)}{12}$$

$$= \frac{4x + 4 - 3x + 9}{12}$$

$$= \frac{x + 13}{12}$$

Example 4

Express $\dfrac{3(4x-1)}{2} - \dfrac{2(5x+3)}{3}$ as a single fraction.

$$\frac{3(4x-1)}{2} - \frac{2(5x+3)}{3} = \frac{9(4x-1) - 4(5x+3)}{6}$$

$$= \frac{36x - 9 - 20x - 12}{6}$$

$$= \frac{16x - 21}{6}$$

Note how in both examples the brackets in the numerator are not multiplied out until the second step. This will help avoid mistakes with signs, especially when the fractions are subtracted.

Key Point

- Find the lowest common denominator when adding or subtracting fractions.

Exercise 194

Express as a single fraction:

1 $\dfrac{x}{3} + \dfrac{x+1}{2}$ **2** $\dfrac{x}{2} + \dfrac{x+2}{3}$ **3** $\dfrac{x}{3} - \dfrac{x+1}{4}$

4 $\dfrac{x+4}{5} + \dfrac{x}{3}$ **5** $x + \dfrac{x+3}{4}$ **6** $x + \dfrac{x-2}{3}$

7 $\dfrac{x+3}{2} - x$ **8** $\dfrac{x-1}{3} + \dfrac{x+2}{4}$ **9** $\dfrac{x-2}{3} + \dfrac{x+1}{2}$

Exercise 194*

Express as a single fraction:

1 $\dfrac{2x-1}{5} + \dfrac{x+3}{2}$ **2** $\dfrac{x-1}{4} + \dfrac{2x+2}{3}$ **3** $\dfrac{x-3}{5} - \dfrac{3x-2}{3}$

4 $\dfrac{2x+1}{7} - \dfrac{x-2}{2}$ **5** $\dfrac{2x+3}{4} + \dfrac{3-2x}{8}$ **6** $\dfrac{2(x-5)}{3} - \dfrac{3(x+20)}{5}$

7 $\dfrac{3(x+4)}{8} - \dfrac{2(x-3)}{3}$ **8** $\dfrac{x-3}{18} - \dfrac{x-2}{24}$ **9** $\dfrac{3x+1}{20} - \dfrac{2x+1}{25}$

Fractions with x in the denominator

The same method is used as for fractions with numbers in the denominator. It is more important to find the lowest common denominator, otherwise the working can become very complicated.

Example 5

Express $\dfrac{3}{2x} - \dfrac{4}{3x}$ as a single fraction.

Notice that the lowest common denominator is 6, not $6x^2$.

$$\dfrac{3}{2x} - \dfrac{4}{3x} = \dfrac{3 \times 3 - 2 \times 4}{6x}$$

$$= \dfrac{1}{6x}$$

Example 6

Express $\dfrac{1}{x} + \dfrac{1}{x+1}$ as a single fraction.

$$\dfrac{1}{x} + \dfrac{1}{x+1} = \dfrac{(x+1) + x}{x(x+1)}$$

$$= \dfrac{2x+1}{x(x+1)} \leftarrow \text{The denominator is best left factorised}$$

Example 7

Express $\dfrac{x+1}{x+2} - \dfrac{x-2}{x-1}$ as a single fraction.

$$\dfrac{x+1}{x+2} - \dfrac{x-2}{x-1} = \dfrac{(x-1)(x+1) - (x+2)(x-2)}{(x+2)(x-1)}$$

$$= \dfrac{(x^2 - 1) - (x^2 - 4)}{(x+2)(x-1)}$$

$$= \dfrac{x^2 - 1 - x^2 + 4}{(x+2)(x-1)}$$

$$= \dfrac{3}{(x+2)(x-1)} \leftarrow \text{The denominator is best left factorised}$$

Example 8

Express $\dfrac{2}{x-2} - \dfrac{6}{x^2 - x - 2}$ as a single fraction.

To find the lowest common denominator, factorise $x^2 - x - 2$ as $(x+1)(x-2)$, giving the lowest common denominator as $(x+1)(x-2)$. Because the denominator is left factorised it is easier to simplify the resulting expression.

$$\dfrac{2}{x-2} - \dfrac{6}{x^2 - x - 2} = \dfrac{2}{x-2} - \dfrac{6}{(x+1)(x-2)}$$

$$= \dfrac{2(x+1) - 6}{(x+1)(x-2)}$$

$$= \dfrac{2x - 4}{(x+1)(x-2)}$$

$$= \dfrac{2(x-2)}{(x+1)(x-2)}$$

$$= \dfrac{2}{x+1}$$

Exercise 195

Express as single fractions:

1 $\dfrac{1}{2x} + \dfrac{1}{3x}$

2 $\dfrac{3}{4x} - \dfrac{1}{2x}$

3 $\dfrac{1}{2} - \dfrac{1}{x-2}$

4 $\dfrac{1}{x+1} + \dfrac{1}{x-1}$

5 $\dfrac{3}{x-1} - \dfrac{2}{x+2}$

6 $\dfrac{4}{x+1} - \dfrac{3}{x-3}$

7 $\dfrac{x}{x+2} + \dfrac{1}{x}$

8 $\dfrac{1}{x} + \dfrac{x}{x-3}$

9 $\dfrac{x+2}{x+1} - \dfrac{x+1}{x+2}$

10 $\dfrac{x+1}{x+3} - \dfrac{x-3}{x+1}$

Exercise 195*

Express as single fractions:

1 $\dfrac{1}{x} + \dfrac{1}{3x} - \dfrac{1}{5x}$

2 $\dfrac{4}{x+3} - \dfrac{3}{x+2}$

3 $\dfrac{1}{1+x} + x$

4 $1 - \dfrac{1}{x+1}$

5 $\dfrac{1}{x} - \dfrac{1}{x(x+1)}$

6 $\dfrac{1}{x(x+1)} + \dfrac{1}{x+1}$

7 $\dfrac{x+2}{x+1} - \dfrac{x+1}{x+2}$

8 $\dfrac{4}{x^2+2x-3} + \dfrac{1}{x+3}$

9 $\dfrac{1}{x^2+2x+2} + \dfrac{3}{x+1}$

10 $\dfrac{x+1}{x^2-4x+3} - \dfrac{x-3}{x^2-1}$

Multiplication and division of algebraic fractions

Factorise as much as possible and then simplify.

Example 9

Simplify $\dfrac{x^2-2x}{x+3} \times \dfrac{3x+9}{x-2}$

$$\dfrac{x^2-2x}{x+3} \times \dfrac{3x+9}{x-2} = \dfrac{x(x-2)}{x+3} \times \dfrac{3(x+3)}{x-2} = 3x$$

If dividing, 'turn the second fraction upside down and multiply' in exactly the same way as number fractions are manipulated.

Example 10

Simplify $\dfrac{x^2-3x}{2x^2+7x+3} \div \dfrac{x^2-5x+6}{2x^2-3x-2}$

$$\dfrac{x^2-3x}{2x^2+7x+3} \div \dfrac{x^2-5x+6}{2x^2-3x-2} = \dfrac{x(x-3)}{(2x+1)(x+3)} \times \dfrac{(2x+1)(x-2)}{(x-3)(x-2)} = \dfrac{x}{x+3}$$

Exercise 196

Simplify

1 $\dfrac{4x+4}{x-3} \times \dfrac{x^2-3x}{2x+2}$

2 $\dfrac{x^2+2x}{x^2-x} \times \dfrac{6x-6}{x+2}$

3 $\dfrac{x-y}{x^2-xy} \times \dfrac{xy+y^2}{x+y}$

4 $\dfrac{2a+4b}{ab+2b^2} \times \dfrac{2a^2-ab}{2a-b}$

5 $\dfrac{p^2+p-2}{p^2-p-6} \times \dfrac{p-3}{p-2}$

6 $\dfrac{r^2-r-2}{r^2-3r+2} \times \dfrac{r+2}{r+1}$

7 $\dfrac{x+2}{x+4} \div \dfrac{x^2-2x-8}{x^2+2x-8}$

8 $\dfrac{x+3}{x+5} \div \dfrac{x^2-2x-15}{x^2+2x-15}$

9 $\dfrac{x^2-9}{x^2-6x+9} \div \dfrac{x+4}{x-3}$

10 $\dfrac{x^2+5x+6}{x^2-25} \div \dfrac{x+3}{x-5}$

Exercise 196*

Simplify

1 $\dfrac{4x+28}{x^2+2x} \times \dfrac{x^2-3x}{2x+14}$

2 $\dfrac{3x^2-12x}{3x+18} \times \dfrac{2x+12}{x^2-3x}$

3 $\dfrac{x^2+9x+20}{x^2-16} \times \dfrac{x-4}{x^2+6x+5}$

4 $\dfrac{x^2-x-6}{x^2-9} \times \dfrac{x^2+4x+3}{x+1}$

5 $\dfrac{x^2+x-2}{x^2+3x-4} \times \dfrac{x^2+x-12}{x^2-5x+6}$

6 $\dfrac{x^2-x-2}{x^2+x-6} \times \dfrac{x^2-x-12}{x^2+3x+2}$

7 $\dfrac{p^2+7p+12}{p^2-7p+10} \div \dfrac{p+3}{p-2}$

8 $\dfrac{q^2-5q-14}{q^2+3q-18} \div \dfrac{q-7}{q-3}$

9 $\dfrac{x^2+xy-2y^2}{x^2-4y^2} \div \dfrac{x^2+2xy-3y^2}{xy-2y^2}$

10 $\dfrac{x^2-2xy+y^2}{x^2-y^2} \div \dfrac{x^2+xy}{x^2+2xy+y^2}$

Equations with fractions

Equations with number denominators

If an equation involves fractions, clear the fractions by multiplying *both sides* of the equation by the lowest common denominator. Then simplify and solve in the usual way.

Example 11

Solve $\dfrac{3x}{7} = 2$.

$$\dfrac{3x}{7} = 2 \qquad \text{(Multiply both sides by 7)}$$

$$7 \times \dfrac{3x}{7} = 7 \times 2 \qquad \text{(Simplify)}$$

$$3x = 14 \qquad \text{(Divide both sides by 3)}$$

$$x = \dfrac{14}{3}$$

Example 12

Solve $\dfrac{x-1}{2} = \dfrac{x+2}{3}$.

$$\dfrac{x-1}{2} = \dfrac{x+2}{3} \qquad \text{(Multiply both sides by 6)}$$

$$6 \times \dfrac{x-1}{2} = 6 \times \dfrac{x+2}{3} \qquad \text{(Simplify)}$$

$$3(x-1) = 2(x+2) \qquad \text{(Multiply out brackets)}$$

$$3x - 3 = 2x + 4 \qquad \text{(Collect terms)}$$

$$3x - 2x = 4 + 3 \qquad \text{(Simplify)}$$

$$x = 7$$

Example 13

Solve $\dfrac{x+1}{3} - \dfrac{x-3}{4} = 1$.

$$\dfrac{x+1}{3} - \dfrac{x-3}{4} = 1 \qquad \text{(Multiply both sides by 12)}$$

$$12 \times \dfrac{x+1}{3} - 12 \times \dfrac{x-3}{4} = 12 \times 1 \qquad \text{(Notice is multiplied by 12)}$$

$$4(x+1) - 3(x-3) = 12 \qquad \text{(Multiply out, note sign change in 2nd bracket)}$$

$$4x + 4 - 3x + 9 = 12 \qquad \text{(Collect terms)}$$

$$4x - 3x = 12 - 4 - 9 \qquad \text{(Simplify)}$$

$$x = -1$$

Key Point

Clear the fractions by multiplying *both* sides of the equation by the lowest common denominator.

Exercise 197

Solve the following equations:

1 $\dfrac{x}{3} = 7$

2 $\dfrac{x}{2} = \dfrac{x+2}{4}$

3 $\dfrac{2-x}{3} = x$

4 $x = \dfrac{x+3}{2}$

5 $\dfrac{x-4}{6} = \dfrac{x+2}{3}$

6 $\dfrac{2-3x}{6} = \dfrac{2}{3}$

7 $\dfrac{x+1}{3} = \dfrac{2x+1}{4}$

8 $\dfrac{2(x-5)}{3} = \dfrac{3(x+20)}{5}$

9 $\dfrac{x+1}{7} - \dfrac{3(x-2)}{14} = 1$

10 $\dfrac{6-3x}{3} - \dfrac{5x+12}{4} = -1$

Exercise 197*

For Questions 1–10, solve the equation.

1 $\dfrac{3x}{7} = \dfrac{6}{35}$

2 $\dfrac{x-3}{3} = 4$

3 $\dfrac{x+1}{2} = 2x$

4 $\dfrac{3-x}{3} = \dfrac{2+x}{2}$

5 $\dfrac{x-3}{12} + \dfrac{x}{5} = 4$

6 $\dfrac{2x+1}{3} = x - 2$

7 $\dfrac{7x-1}{6} + 5x = 6$

8 $\dfrac{1+x}{2} = \dfrac{2-x}{3} + 1$

9 $4 - \dfrac{x-2}{2} = 3 + \dfrac{2-3x}{3}$

10 $\dfrac{1-x}{2} - \dfrac{2+x}{3} + \dfrac{3-x}{4} = 1$

11 Pedro does one-sixth of his journey to school by car and two-thirds by bus. He then walks the final kilometre. How long is his journey to school?

12 Meera is competing in a triathlon. She cycles half the course then swims one-thirtieth of the course. She then runs 14 km to the finish. How long is the course?

Equations with x in the denominator

These are solved in the same way as equations with numbers in the denominator. Clear the fractions by multiplying *both sides* of the equation by the lowest common denominator. Then simplify and solve in the usual way.

Key Point

Clear the fractions by multiplying *both* sides of the equation by the lowest common denominator.

Example 14

Solve the equation $x + 2 = \dfrac{8}{x}$.

The lowest common denominator is x.

$$x + 2 = \frac{8}{x} \Rightarrow x \times x + x \times 2 = x \times \frac{8}{x}$$
$$\Rightarrow \quad x^2 + 2x = 8$$
$$\Rightarrow \quad x^2 + 2x - 8 = 0$$
$$\Rightarrow (x - 2)(x + 4) = 0$$
$$\Rightarrow \quad x = 2 \text{ or } x = -4$$

Example 15

Solve the equation $\dfrac{1}{3} + \dfrac{1}{x + 1} = \dfrac{x}{3}$.

The lowest common denominator is $3(x + 1)$.

$$\frac{1}{3} + \frac{1}{x + 1} = \frac{x}{3} \Rightarrow 3(x + 1) \times \frac{1}{3} + 3(x + 1) \times \frac{1}{x + 1} = 3(x + 1) \times \frac{x}{3}$$
$$\Rightarrow \quad (x + 1) + 3 = x(x + 1)$$
$$\Rightarrow \quad x + 4 = x^2 + x$$
$$\Rightarrow \quad x^2 = 4$$
$$\Rightarrow \quad x = \pm 2$$

Example 16

Solve the equation $\dfrac{x}{x - 2} - \dfrac{2}{x + 1} = 3$.

The lowest common denominator is $(x - 2)(x + 1)$

$$\frac{x}{x - 2} - \frac{2}{x + 1} = 3 \Rightarrow (x - 2)(x + 1) \times \frac{x}{x - 2} - (x - 2)(x + 1) \times \frac{2}{x + 1} = (x - 2)(x + 1) \times 3$$
$$\Rightarrow \quad x(x + 1) - 2(x - 2) = 3(x - 2)(x + 1)$$
$$\Rightarrow \quad x^2 + x - 2x + 4 = 3x^2 - 3x - 6$$
$$\Rightarrow \quad 2x^2 - 2x - 10 = 0$$
$$\Rightarrow \quad x^2 - x - 5 = 0$$
$$\Rightarrow \quad x = \frac{1 \pm \sqrt{1 + 20}}{2}$$
$$\Rightarrow \quad x = 2.79 \text{ or}$$
$$\quad x = -1.79 \text{ to 3 s.f.}$$

Exercise 198

Solve the following equations:

1 $x + 5 = \dfrac{14}{x}$

2 $x + \dfrac{2}{x} = 3$

3 $\dfrac{1}{2} - \dfrac{1}{x - 2} = \dfrac{1}{4}$

4 $\dfrac{x}{x + 2} - \dfrac{1}{x} = 1$

5 $\dfrac{6}{x - 2} - \dfrac{6}{x + 1} = 1$

6 $\dfrac{x}{x - 1} + \dfrac{8}{x + 4} = 2$

7 $\dfrac{x - 2}{x - 1} = x + \dfrac{4}{2x + 4}$

8 $\dfrac{2x - 1}{x + 2} = \dfrac{4x + 1}{5x + 2}$

Exercise 198*

For Questions 1–7, solve the equation.

1 $2x + 3 = \dfrac{7}{x}$

2 $1 + \dfrac{3}{x} = \dfrac{4}{x^2}$

3 $\dfrac{7}{9} - \dfrac{x}{x + 5} = \dfrac{1}{3}$

4 $\dfrac{7}{x - 1} - \dfrac{4}{x + 4} = 1$

5 $\dfrac{6x}{x + 1} - \dfrac{5}{x + 3} = 3$

6 $\dfrac{2x + 1}{x + 1} = \dfrac{x + 2}{2x + 1}$

7 $\dfrac{3x + 2}{x + 1} + \dfrac{x + 2}{2x - 5} = 4$

8 Mala drives the first 30 km of her journey at x km/h. She then increases her speed by 20 km/h for the final 40 km of her journey. Her journey takes 1 hour. Find x.

9 Lucas is running a race. He runs the first 800 m at x m/s.
He then slows down by 3 m/s for the final 400 m.
His total time is 130 seconds. Find x.

For Questions 10–12, solve the equation.

10 $\dfrac{3}{x + 2} + \dfrac{4}{x + 3} = \dfrac{7}{x + 6}$

11 $\dfrac{4}{x - 3} + \dfrac{3x - 3}{x^2 - x - 6} = \dfrac{2 - 20x}{2x + 4}$

12 $y - 1 = \dfrac{y^2 + 3}{y - 1} + \dfrac{y - 2}{y - 6}$

Exercise 199 (Revision)

For Questions 1–4, simplify.

1 $\dfrac{3x + 6}{x + 2}$

2 $\dfrac{x^2 + 7x + 10}{x + 5}$

3 $\dfrac{x^2 + 6x + 9}{x^2 - 9}$

4 $\dfrac{x^2 - 2x + 1}{x^2 + 2x - 3}$

For Questions 5–10, express as a single fraction.

5 $\dfrac{x - 1}{2} - \dfrac{x}{3}$

6 $\dfrac{3x - 1}{4} - \dfrac{3x + 4}{5}$

7 $\dfrac{1}{x + 1} - \dfrac{1}{x - 2}$

8 $\dfrac{3}{x - 1} - \dfrac{2}{x + 1}$

9 $\dfrac{x}{x - 2} + \dfrac{2}{x}$

10 $\dfrac{x + 3}{x + 2} - \dfrac{x + 1}{x + 4}$

11 Simplify $\dfrac{x^2 - 1}{x} \times \dfrac{x^2}{x^2 + 2x + 1}$

12 Simplify $\dfrac{x + 3}{x - 3} \div \dfrac{x^2 + 2x - 3}{x^2 - 3x + 2}$

For Questions 13–20, solve the equation.

13 $\dfrac{3x}{4} = 2$

14 $\dfrac{x}{3} = \dfrac{x-2}{5}$

15 $1 + \dfrac{x}{2} = \dfrac{x-3}{4}$

16 $\dfrac{x+2}{3} - \dfrac{x-2}{4} = 1$

17 $x + 1 = \dfrac{2}{x}$

18 $\dfrac{x}{x-1} + \dfrac{1}{x} = 1$

19 $\dfrac{3}{x-1} - \dfrac{8}{x+2} = 1$

20 $\dfrac{x+3}{x+5} = \dfrac{x+1}{x+2}$

Exercise 199* (Revision)

For Questions 1–4, simplify.

1 $\dfrac{2x+14}{3x+21}$

2 $\dfrac{x^2 - 12x + 11}{x^2 + 4x - 5}$

3 $\dfrac{x^2 + 11x + 28}{x^2 - 49}$

4 $\dfrac{x^2 - 11x + 10}{x^2 - 9x - 10}$

For Questions 5–10, express as a single fraction.

5 $\dfrac{2x+5}{6} - \dfrac{x+7}{4}$

6 $\dfrac{3(x-1)}{6} - \dfrac{2(x+1)}{9}$

7 $\dfrac{x+1}{2} - \dfrac{x+2}{3} + \dfrac{x+3}{4}$

8 $\dfrac{2(3x-1)}{5} + 1 - \dfrac{x}{2}$

9 $\dfrac{x+2}{x+1} - \dfrac{x+1}{x+2}$

10 $\dfrac{x}{x^2 - 3x - 4} - \dfrac{1}{x-4}$

11 Simplify $\dfrac{x^2 + 6x + 5}{x^2 + 4x - 5} \times \dfrac{x^2 - 1}{x+1}$

12 Simplify $\dfrac{x^2 - 3x - 10}{x^2 - x - 6} \div \dfrac{x^2 - 4x - 5}{x^2 - 2x - 3}$

For Questions 13–20, solve the equation.

13 $\dfrac{x+3}{2} + \dfrac{x+2}{3} = \dfrac{4}{3}$

14 $\dfrac{4x+1}{3} - \dfrac{3x-1}{2} = 0$

15 $\dfrac{2x+1}{7} - x = \dfrac{3x-1}{8}$

16 $\dfrac{3x-1}{7} = \dfrac{2x+1}{11} + 1$

17 $\dfrac{x}{x-1} + \dfrac{1}{x} = \dfrac{5}{2}$

18 $\dfrac{2x-1}{x+3} = \dfrac{x}{2x+2}$

19 $\dfrac{x+1}{x-3} - \dfrac{1}{x} = 2$

20 $\dfrac{2x+3}{x-5} = \dfrac{x-4}{x-3}$

Differentiation

The gradient of a curve

The gradient (or the slope, or direction) of a curve is constantly changing.

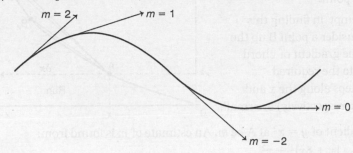

The gradient of the curve at any point is equal to the gradient of the tangent to the curve at that point.

Constructing the tangent

A tangent can be drawn 'by eye' by sliding a ruler up against the curve. However, this will only give an approximate result. The following process produces the exact tangent and thus the correct gradient.

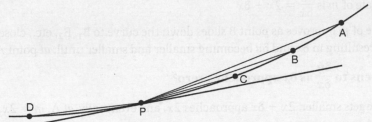

Place your ruler along line PA. Rotate the ruler around the point P, moving through positions PB, PC and PD. Your ruler will pass through the *tangent* at P, between the *chords* PC and PD.

gradient PA > gradient PB > gradient PC > tangent at P > gradient PD

This process can be developed into one that enables the gradient to be *precisely calculated*, without the need to draw the tangent.

The mathematicians Newton and Leibnitz first investigated this process in the seventeenth century. Newton called this branch of mathematics fluxions; nowadays, it is called calculus.

Calculus involves considering very small values (or increments) in x and y: δx and δy. δx is pronounced 'delta x' and means a very small distance (increment) along the x-axis. It does not mean δ multiplied by x. δx must be considered as one symbol.

The process of calculating a gradient (or a rate of change) is shown in Example 1.

Example 1

Consider the graph $y = x^2$.

The gradient of the curve at point A is equal to the tangent to the curve at this point.

A good 'first attempt' in finding this gradient is to consider a point B up the curve from A. The gradient of chord AB will be close to the required gradient if the steps along the x and y axes (δx and δy respectively) are small.

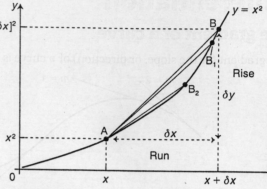

Let the *exact* gradient of $y = x^2$ at A be m. An estimate of m is found from:

$$\frac{\delta y}{\delta x} = \frac{\text{Rise}}{\text{Run}} = \frac{[x + \delta x]^2 - x^2}{\delta x} \qquad [\text{Expand } (x + \delta x)^2]$$

$$= \frac{[x^2 + 2x\delta x + (\delta x)^2] - x^2}{\delta x}$$

$$= \frac{2x\delta x + (\delta x)^2}{\delta x}$$

$$= \frac{\delta x(2x + \delta x)}{\delta x} = 2x + \delta x$$

So the estimate of m is $\dfrac{\delta y}{\delta x} = 2x + \delta x$

This estimate of m improves as point B slides down the curve to B_1, B_2, etc., closer and closer to A, resulting in δx and δy becoming smaller and smaller until, *at point A, $\delta x = 0$*.

What happens to $\dfrac{\delta y}{\delta x}$ as δx approaches zero?

Clearly, as δx gets smaller, $2x + \delta x$ approaches $2x$, and eventually, *at A, $m = 2x$*.

This is a beautifully simple result implying that the gradient of the curve $y = x^2$ at *any* point x is given by $2x$. So, at $x = 10$, the gradient of $y = x^2$ is 20, and so on.

Differentiating x^n with respect to x

This process of finding the gradient is called **differentiation** and can be applied to any power of x. In all cases the gradient of x^n is nx^{n-1}.

The result is called the **derivative** or the gradient function.

The derivative is written $\dfrac{dy}{dx}$

curve	$\dfrac{dy}{dx}$	Gradient of curve at $x = 10$
$y = x^2$	$2x$	20
$y = x^3$	$3x^2$	300
$y = x^4$	$4x^3$	4000
$y = x^5$	$5x^4$	50000

Example 2

Find $\dfrac{dy}{dx}$ when $y = 5x$.

In index form:
$$y = 5x^1$$

Differentiating:
$$\frac{dy}{dx} = 5 \times 1x^0 = 5$$

Note that $y = 5x$ is a straight line with gradient $= 5$.

Example 3

Find $\dfrac{dy}{dx}$ when $y = 5x + 2$.

In index form:
$$y = 5x^1 + 2x^0$$

Differentiating:
$$\frac{dy}{dx} = 5 \times 1x^0 + 2 \times 0x^{-1}$$

$$\frac{dy}{dx} = 5 + 0 = 5$$

The '+2' lifts the line up 2 units. It does not change the gradient.

Exercise 200

In Questions 1–12, differentiate using the correct notation.

1 $y = x^3$

2 $y = x^4$

3 $y = x^5$

4 $y = x^6$

5 $y = x^8$

6 $y = x^7$

7 $y = x^{10}$

8 $y = x^{11}$

9 $y = x^{-1}$

10 $y = x^{-2}$

11 $y = \dfrac{1}{x}$

12 $y = \dfrac{1}{x^2}$

In Questions 13–18, find the gradient of the curve at the point where $x = 2$.

13 $y = x^2$

14 $y = x^3$

15 $y = x^4$

16 $y = x^5$

17 $y = x^7$

18 $y = x^6$

Exercise 200*

In Questions 1–12, differentiate using the correct notation.

1 $y = x^{12}$

2 $y = x^9$

3 $y = x^1$

4 $y = x$

5 $y = x^{-3}$

6 $y = x^{-4}$

7 $y = x^{-6}$

8 $y = x^{-5}$

9 $y = x^{\frac{1}{2}}$

10 $y = x^{\frac{1}{3}}$

11 $y = \sqrt{x}$

12 $y = 1$

In Questions 13–20, find the gradient of the curve at the given value of x.

13 $y = x^{-3}$ at $x = 2$

14 $y = x^{-4}$ at $x = 2$

15 $y = \dfrac{1}{x}$ at $x = 2$

16 $y = \dfrac{1}{x^2}$ at $x = 2$

17 $y = x^{\frac{1}{2}}$ at $x = 4$

18 $y = \sqrt[3]{x}$ at $x = 8$

19 $y = x^0$ at $x = 3$

20 $y = x$ at $x = 2.5$

The rule for differentiation can also be extended when x^n terms are multiplied by a number and added together.

Key Point

$y = ax \equiv ax^1$

Differentiating:
$$\frac{dy}{dx} = ax^0 \equiv a$$

$$y = c \equiv cx^0$$

Differentiating:
$$\frac{dy}{dx} = 0$$

Example 4

Find $\dfrac{dy}{dx}$ when $y = 4x^6$.

Differentiating:
$$\frac{dy}{dx} = 4 \times 6x^5 = 24x^5$$

Example 5

Differentiate $y = 2x^2 + 3x^{-1}$.

Differentiating:
$$\frac{dy}{dx} = 2 \times 2x^1 + 3 \times -1x^{-2}$$

$$= 4x - 3x^{-2}$$

Key Point

Before differentiating:

- multiply or divide composite terms to give individual terms
- express all algebraic fractions in index form.

Example 6

Find $\dfrac{dy}{dx}$ when $y = (2x + 3)^2$.

Multiply through:

$y = 4x^2 + 12x + 9$

Differentiating:

$\dfrac{dy}{dx} = 8x + 12$

Example 7

Find $\dfrac{dy}{dx}$ when $y = \dfrac{2}{x^3}$.

Express in index form:

$y = 2x^{-3}$

Differentiating:

$\dfrac{dy}{dx} = -6x^{-4} \equiv -\dfrac{6}{x^4}$

Exercise 201

Using the correct notation, differentiate the following:

1 $y = x^5 + x^2$
2 $y = x^4 + x^3$
3 $y = 2x^3 + 4x$

4 $y = 5x^4 - 3x$
5 $y = 3x^3 + 2x^4$
6 $y = 5x^2 - x^3$

7 $y = 2x^2 + 3$
8 $y = 6x^3 - 5$
9 $y = 4 + x^{-1}$

10 $y = 2 + x^{-2}$
11 $y = 6 + 2x^{-3}$
12 $y = 1 + 5x^{-4}$

13 $y = x^2 + x^{-2}$
14 $y = x^3 + x^{-1}$
15 $y = x^3 - x^{-3}$

16 $y = x - x^{-1}$

Exercise 201*

For Questions 1–10, differentiate using the correct notation.

1 $y = x^2(x + 2)$
2 $y = x(x^2 + 2)$
3 $y = (x + 1)(x + 3)$

4 $y = (x + 2)(x + 4)$
5 $y = (x + 3)^2$
6 $y = (x + 5)^2$

7 $y = (2x - 1)^2$
8 $y = (5 - 3x)^2$
9 $y = \left(2x - \dfrac{1}{x}\right)^2$

10 $y = \left(\dfrac{1}{x} + x\right)^2$

11 Find the gradient of the curve $y = 2x^2 - 3x$ at the point where $x = 2$.

12 Find the gradient of the curve $y = x^4 + 3x$ at the point where $x = 1$.

13 Find the co-ordinates of the point on the curve $y = x^2 - 6x$ where the gradient is zero.

14 Find the co-ordinates of the point on the curve $y = 4x + x^{-1}$ where the gradient is zero.

15 Find the co-ordinates of the point on the curve $y = 10 - 6x + x^2$ where $\dfrac{dy}{dx} = 0$. Sketch the curve and mark this point on your diagram.

16 The diagram shows a sketch of the curve $y = 4x + \dfrac{1}{x}$.
    Find the co-ordinates of points A and B.

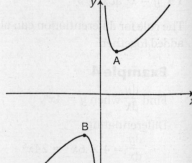

Tangents

For a given value of x:

- The equation gives the y value.
- $\dfrac{dy}{dx}$ gives the gradient of the tangent.
- Apply the technique for calculating the equation of a straight line.

Example 8

Calculate the equation of the tangent to the curve $y = x^3 - 2x^2 + 5$ at the point where $x = 2$.

The y co-ordinate:

When $x = 2$, $y = 8 - 8 + 5 = 5$.

The gradient:

$$\frac{dy}{dx} = 3x^2 - 4x$$

And, when $x = 2$, $\dfrac{dy}{dx} = 12 - 8 = 4$.

so gradient of tangent = 4

The equation:

$$\frac{y - 5}{x - 2} = 4 \Rightarrow y = 4x - 3$$

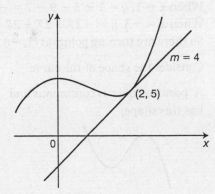

Turning points

Turning points are points on the curve where the gradient is zero, i.e. where $\dfrac{dy}{dx} = 0$.

Maximum point

A turning point.

Gradient $= \dfrac{dy}{dx} = 0$

Gradient is **decreasing** from +ve, through zero, to −ve.

Gradient is *positive* just before and *negative* just after.

Minimum point

A turning point.

Gradient $= \dfrac{dy}{dx} = 0$

Gradient is **increasing** from −ve, through zero, to +ve.

Gradient is *negative* just before and *positive* just after.

Key Point

At a turning point, $\dfrac{dy}{dx} = 0$.

A turning point can be classified as a **maximum** or **minimum** by:

Knowing the shape of a familiar curve.   Or, by looking at the gradient immediately before and immediately after the stationary point.

Example 9

Find and classify the turning points on the curve $y = x^3 + 3x^2 - 9x - 7$.

$$\frac{dy}{dx} = 3x^2 + 6x - 9$$

$$\frac{dy}{dx} = (x^2 + 2x - 3)$$

$$\frac{dy}{dx} = 3(x - 1)(x + 3)$$

$$\frac{dy}{dx} = 0 \text{ when } x = 1 \text{ and } x = -3.$$

When $x = 1$, $y = 1 + 3 - 9 - 7 = -12$.
When $x = -3$, $y = -27 + 27 + 27 - 7 = 20$.
So there are turning points at $(1, -12)$ and $(-3, 20)$.

Consider the shape of the curve.

A 'positive cubic' is continuous and has this shape:

The maximum comes first and has the higher y co-ordinate.

Or, if you do not know the shape of the curve, look at the gradient 'before and after'.

At $(1, -12)$

x	0.9	1	1.1
$\frac{dy}{dx}$	−ve	0	+ve
	\	−	/

minimum

At $(-3, 20)$

−3.1	−3	−2.9
+ve	0	−ve
/	−	\

maximum

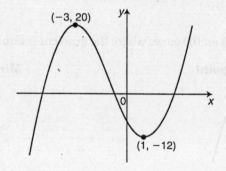

Exercise 202

1 For the curve $y = x^2 + 4x + 10$:

  a Work out $\frac{dy}{dx}$.

  b Solve the equation $\frac{dy}{dx} = 0$, and thus find the co-ordinates of the turning point.

2 For the curve $y = x^2 + 2x - 8$:

  a Work out $\frac{dy}{dx}$.

  b Solve the equation $\frac{dy}{dx} = 0$, and thus find the co-ordinates of the turning point.

3 For the curve $y = 15 + 6x - x^2$:

  a Solve the equation $\dfrac{dy}{dx} = 0$.

  b Thus find the maximum value of y.

4 For the curve $y = 6 - 10x - x^2$:

  a Solve the equation $\dfrac{dy}{dx} = 0$.

  b Thus find the maximum value of y.

5 For the curve $y = x^2 + 4x + 10$:

  a Work out the value of y when $x = 1$.

  b Work out the value of $\dfrac{dy}{dx}$ when $x = 1$.

  c Thus find the equation of the tangent to the curve at the point where $x = 1$.

6 For the curve $y = x^2 + 2x - 8$:

  a Work out the value of y when $x = 3$.

  b Work out the value of $\dfrac{dy}{dx}$ when $x = 3$.

  c Thus find the equation of the tangent to the curve at the point where $x = 3$.

7 A iron is being heated in a forge. The temperature, T, after t minutes is given by the formula
$T = 4t^2 + 8t + 20$ for $0 \leqslant t \leqslant 10$.

  a Work out $\dfrac{dT}{dt}$.

  b Calculate the rate, in degrees per minute, at which the temperature is increasing:

    (i) after one minute

    (ii) after five minutes.

8 The number of registered voters, E, on the electoral role of a regional district, t months after democracy was declared, is given by the formula
$E = 100t^2 + 200t + 200$ for $0 \leqslant t \leqslant 6$.

  a Work out $\dfrac{dE}{dt}$.

  b Calculate the rate at which voters were registering:

    (i) after one month

    (ii) after four months.

9 For the curve $y = x^3 - 12x^2 + 5$:

  a Work out $\dfrac{dy}{dx}$.

  b Solve the equation $\dfrac{dy}{dx} = 0$.

  c Work out the co-ordinates of the two turning points.

  d Determine whether each turning point is a maximum or minimum, and show your reasoning.

10 For the curve $y = x^3 - 6x^2 + 10$:

    a Work out $\dfrac{dy}{dx}$.   **b** Solve the equation $\dfrac{dy}{dx} = 0$.

    c Work out the co-ordinates of the two turning points.

    d Determine whether each turning point is a maximum or minimum, and show your reasoning.

11 For the curve $y = 11 + 6x - x^2$:

    a Find $\dfrac{dy}{dx}$.   **b** Thus work out the maximum value of y.

12 For the curve $y = x^2 - 10x + 10$:

    a Find $\dfrac{dy}{dx}$.   **b** Thus work out the minimum value of y.

13 For the curve $y = (4 + x)(2 - x)$:

    a Find $\dfrac{dy}{dx}$.   **b** Thus work out the maximum value of y.

14 For the curve $y = (3 + x)(5 + x)$:

    a Find $\dfrac{dy}{dx}$.   **b** Thus work out the minimum value of y.

Exercise 202*

1 For the curve $y = 3x^2 - 7x + 5$:

    a Find $\dfrac{dy}{dx}$.   **b** Thus work out the equation of the tangent at the point $(2, 3)$.

2 For the curve $y = x^2 - \dfrac{4}{x} + x$:

    a Find $\dfrac{dy}{dx}$.   **b** Thus work out the equation of the tangent at the point $(-1, 4)$.

3 Find the turning points of the function $y = 27x - x^3$ and determine their nature.

4 Find the turning points of the function $y = x^3 - 6x^2$ and determine their nature.

5 The population, P, of a new town is modelled by the formula $P = t^3 - t^2 + 25t + 10000$, where t is measured in years. t is set at 0 in the year 2000.

    Use the formula to predict

    a the population in the year 2010

    b the rate of population growth in 2010.

6 The output of a company, $\$P$ in month t, is modelled by the formula $P = 800t^3 - 4000t^2 + 2000t + 480\,000$. January is set as $t = 1$.

    Use the formula to predict

    a the output in May   **b** the rate at which output is growing in May.

7 The temperature of a piece of charcoal, T degrees, after t hours in the ashes of a fire is given by the formula $T = 270 + 80t - 20t^2$ for $0 \le t \le 6$.

a Work out $\dfrac{dT}{dt}$.

b Calculate the time at which the charcoal starts to cool down.

c Calculate the maximum temperature of the charcoal during the six hours.

d Calculate the rate, in degrees per hour, at which the charcoal is cooling after five hours.

8 The lead, L metres, of a runner in the last 75 minutes of a marathon is given by the formula $L = 1000 + 6t - \dfrac{t^2}{4}$, where t is the time in minutes.

a Work out $\dfrac{dL}{dt}$.

b Calculate the time at which the runner has the greatest lead.

c At what rate is the runner's lead being cut when $t = 60$ minutes?

9 For the curve $y = 2x^3 - x^2 - 4x + 10$:

a Work out $\dfrac{dy}{dx}$.

b Given that $\dfrac{dy}{dx} = 0$ when $x = 1$, find the other value of x for which $\dfrac{dy}{dx} = 0$.

c Work out the co-ordinates of the two turning points.

d Determine whether each turning point is a maximum or minimum, and show your reasoning.

10 For the curve $y = 2x^3 + 2x^2 - 16x - 12$:

a Work out $\dfrac{dy}{dx}$.

b Given that $\dfrac{dy}{dx} = 0$ when $x = -2$, find the other value of x for which $\dfrac{dy}{dx} = 0$.

c Work out the co-ordinates of the two turning points.

d Determine whether each turning point is a maximum or minimum, and show your reasoning.

11 The diagram shows a sketch of the graph of $y = x^2 + \dfrac{16}{x}$.

a Find $\dfrac{dy}{dx}$.

b Solve the equation $\dfrac{dy}{dx} = 0$.

c Calculate the co-ordinates of the turning point.

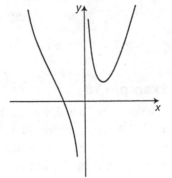

12 The diagram shows a sketch of the graph of $y = \left(x + \dfrac{1}{x}\right)^2$.

a Expand y, and then find $\dfrac{dy}{dx}$.

b Solve the equation $\dfrac{dy}{dx} = 0$.

c Calculate the co-ordinates of the turning points.

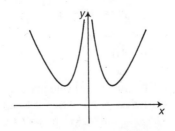

13 The temperature, $C°$, of the water in a lake, t months after it was constructed, is given by the formula $C = 3t + \dfrac{27}{t}$ for $1 \leqslant t \leqslant 6$.

a Work out $\dfrac{dC}{dt}$. b Find the coldest temperature of the lake.

c Calculate the rate at which the temperature is warming after five months.

14 The population, P, of an endangered species was monitored over t years. P is modelled by the equation $P = 3t^2 + \dfrac{48}{t}$ for $1 \leqslant t \leqslant 5$.

a Work out $\dfrac{dP}{dt}$. b Find the smallest population of the species.

c Calculate the rate at which the population is growing after four years.

Kinematics

Isaac Newton investigated the motion of particles, and his second law states that the acceleration of a particle is proportional to the force applied to it. The behaviour of a particle will depend on the direction and magnitude of the forces and the time for which they are applied. It is not sufficient just to consider distance and speed, because these quantities do not take into account the direction of motion.

Key Point

s	**Displacement**	Distance; $+$ or $-$ indicates direction
v	**Velocity**	Speed; $+$ or $-$ indicates direction
a	**Acceleration**	$+$ or $-$ indicates direction
t	**Time**	s, v and a are all expressed in terms of time, t

Velocity is the *rate* at which **displacement** changes over *time*. So $v = \dfrac{ds}{dt}$.

Acceleration is the *rate* at which **velocity** changes over *time*. So $a = \dfrac{dv}{dt}$.

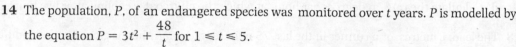

displacement   velocity   acceleration
s → *differentiate* → $v = \dfrac{ds}{dt}$ → *differentiate* → $a = \dfrac{dv}{dt}$

Example 10

The velocity of a ball, v m/s, after t seconds is given by $v = 8 + 10t - t^2$.

a Find the acceleration after t seconds.

b Work out when the acceleration is zero.

c Hence find the maximum velocity.

a $a = \dfrac{dv}{dt} = 10 - 2t$

b $0 = 10 - 2t$

$2t = 10$

$t = 5$ seconds

c The maximum velocity occurs when $a = 0$, when $t = 5$.

$v\,(\text{max}) = 8 + 50 - 25 = 33\,\text{m/s}$

Example 11

A particle moves according to the formula $s = 3t^2 - t^3$, where s is the distance from the starting point, O, at time t. Calculate:

a the time at which the particle momentarily comes to rest.

b the time at which the particle returns to O.

c the acceleration at time $t = 1$.

d Draw a velocity–time graph for $0 \leqslant t \leqslant 3$ and comment on the motion.

a 'comes to rest' $\Rightarrow v = 0$

$$v = \frac{ds}{dt} = 6t - 3t^2 = 3t(2 - t)$$

$v = 0$ when $t = 0$ or 2

so $t = 2$ or 0 seconds

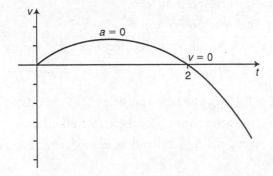

b $s = 3t^2 - t^3 = t^2(3 - t)$

$s = 0$ when $t = 0$ or 3

so $t = 3$ seconds

c $\frac{dv}{dt} = 6 - 6t$

so $a = 0$ when $t = 1$

d The particle moves away for two seconds ($v \geqslant 0$) before changing direction and then arriving back at $O(s = 0)$ after three seconds.

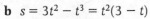

Exercise 203

1 The displacement, s metres, of a particle after t seconds is given by $s = 100 + 5t^2$.

Find an expression for the velocity, $v = \frac{ds}{dt}$.

2 The displacement, s metres, of a particle after t seconds is given by $s = 30 + 48t - 16t^2$.

Find an expression for the velocity, $v = \frac{ds}{dt}$.

3 The displacement, s metres, of a particle after t seconds is given by $s = 20 + 40t + 5t^2$.

   a Find an expression for the velocity, v.

   b Work out the velocity, in m/s, after 3 seconds.

4 The displacement, s metres, of a particle after t seconds is given by $s = 20 + 30t - 5t^2$.

   a Find an expression for the velocity, v.

   b Work out the velocity, in m/s, after 3 seconds.

5 The velocity, v m/s, of a particle after t seconds is given by $v = 32t + 100$.

Find an expression for the acceleration, $a = \frac{dv}{dt}$.

6 The velocity, v m/s, of a particle after t seconds is given by $v = 160 - 32t$.

Find an expression for the acceleration, $a = \frac{dv}{dt}$.

7 The displacement, s metres, of a particle after t seconds is given by $s = t^3 + 4t^2 - 5t + 2$.

   a Find an expression for v.

   b Find an expression for a.

   c Work out, giving the correct units, the velocity and acceleration of the particle after one second.

8 The displacement, s metres, of a particle after t seconds is given by $s = t^3 - 2t^2 + 3t + 1$.

   a Find an expression for v.

   b Find an expression for a.

   c Work out, giving the correct units, the velocity and acceleration of the particle after two seconds.

9 The velocity, v m/s, of a particle after t seconds is given by $v = t^2 + 10t + 5$.

   a Find an expression for the acceleration, a.

   b Work out the acceleration, in m/s^2, after 2 seconds.

10 The velocity, v m/s, of a particle after t seconds is given by $v = 24 + 6t - t^2$.

   a Find an expression for the acceleration, a.

   b Work out the acceleration, in m/s^2, after 2 seconds.

Exercise 203*

1 The displacement, s metres, of a particle after t seconds is given by $s = 4t^2 - \dfrac{2}{t} + 2$.

   a Find an expression for $v = \dfrac{\mathrm{d}s}{\mathrm{d}t}$.

   b Find an expression for $a = \dfrac{\mathrm{d}v}{\mathrm{d}t}$.

2 The displacement, s metres, of a particle after t seconds is given by $s = 6t + \dfrac{4}{t} + 10$.

   a Find an expression for $v = \dfrac{\mathrm{d}s}{\mathrm{d}t}$.

   b Find an expression for $a = \dfrac{\mathrm{d}v}{\mathrm{d}t}$.

3 The displacement, s metres, of a particle after t seconds is given by $s = 20t + 5t^2$.

   a Find $\dfrac{\mathrm{d}s}{\mathrm{d}t}$.

   b Calculate the velocity for $t = 1$, 2 and 3 seconds.

   c Find the acceleration of the particle.

4 The displacement, s metres, of a particle after t seconds is given by $s = 4t^2 + 5t$.

   a Find $\dfrac{\mathrm{d}s}{\mathrm{d}t}$.

   b Calculate the velocity for $t = 1$, 2 and 3 seconds.

   c Find the acceleration of the particle.

5 A ball is thrown vertically upwards with initial velocity $40\,\mathrm{m/s}$. The height or displacement, s metres, after t seconds, is given by $s = 40t - 5t^2$.

  a Find $\dfrac{ds}{dt}$.

  b Find the velocity, v, when $t = 1, 2, 3$ and 4 seconds.

  c Find the maximum height of the ball during the flight.

6 The displacement of a particle, s metres, from a point O at time t seconds is given by $s = 25t - t^2$, for $0 \leqslant t \leqslant 25$. Find

  a $\dfrac{ds}{dt}$

  b the value of t when $v = 0$

  c the displacement of the particle when it is furthest from O.

7 A particle is moving in a straight line in a force field. Its distance, s metres, from point O after t seconds is given by $s = 40 - 2t - \dfrac{18}{t}$ for $1 \leqslant t \leqslant 20$.

  a Find $\dfrac{ds}{dt}$ and $\dfrac{dv}{dt}$.

  b Find the time at which the particle stops momentarily.

  c Calculate the maximum distance from O during this period.

8 In a baseball game an outfielder moves in a straight line.

His distance from the striker, s metres, t seconds after the strike is given by $s = 50 - \dfrac{t^2}{2} - \dfrac{8}{t}$ for $1 \leqslant t \leqslant 8$.

  a Find $\dfrac{ds}{dt}$ and $\dfrac{dv}{dt}$.

  b How fast is the outfielder running one second after the strike?

  c Find the time at which the outfielder stops running.

  d If the outfielder stops to field the ball, how far did the striker hit it?

9 A ball is hit vertically upwards from a cliff, which is $80\,\mathrm{m}$ above the sea. The displacement from the edge of the cliff, s, measured upwards in metres, after t seconds is given by $s = 30t - 5t^2$. Calculate:

  a the time it takes for the ball to land in the sea

  b the time it takes to reach the highest point

  c the total distance travelled during the motion.

10 The displacement of a particle, s metres, from a point O at time t seconds is given by $s = 4t^2 - 20t$ for $t \geqslant 0$. It starts at O.

  a Calculate T, the time it takes to return to O.

  b Calculate the distance it covers in the first T seconds.

Exercise 204 (Revision)

1 Differentiate the following using the correct notation.

a $y = 3x$

b $y = 10$

c $y = x^3$

d $y = x^4$

e $y = x^5$

f $y = 2x^6$

g $y = 3x^5$

h $y = 20x^8$

2 Differentiate the following using the correct notation.

a $y = 2x^3 + 5x^2$

b $y = 7x^2 - 3x$

c $y = 1 + 5x^3$

d $y = 3x^4 - 5x^2$

e $y = x^2(x + 5)$

f $y = (x - 3)(x + 5)$

g $y = (2x - 1)(x - 4)$

h $y = (x + 2)^2$

3 Find the gradients of the tangents to the following curves at the given points.

a $y = 2x^2$ $(1, 2)$

b $y = 3x - 2x^2$ $(2, -2)$

c $y = 2x^3 + 10x^2$ $(1, 12)$

d $y = (x + 5)(2x + 1)$ $(2, 35)$

4 The flow, $Q\,\text{m}^3/\text{s}$ of a river t hours after midnight is given by the equation
$Q = t^3 - 8t^2 + 14t + 10$.

a Differentiate to find $\dfrac{dQ}{dt}$.

b Find the rate of change of flow of the river at

   i midnight   ii 02:00   iii 05:00.

5 For the curve $y = x^3 - 3x + 2$

a find $\dfrac{dy}{dx}$

b find the co-ordinates of the curve where $\dfrac{dy}{dx} = 0$

c determine whether each point is a maximum or minimum.

6 For the curve $y = x^3 + 3x^2 - 9x + 1$

a find $\dfrac{dy}{dx}$

b find the co-ordinates of the curve where $\dfrac{dy}{dx} = 0$

c determine whether each point is a maximum or minimum.

7 The displacement, s metres, of a ball after t seconds is given by $s = 50 + 12t^2$.

a Find an expression for the ball's velocity, $v = \dfrac{ds}{dt}$ and find its velocity at $t = 2$.

b Find an expression for the ball's acceleration, $a = \dfrac{dv}{dt}$ and find its acceleration at $t = 2$.

8 The displacement, s metres, of a particle after t seconds is given by $s = t^3 + 4t^2 - 3t + 1$.

a Find an expression for the velocity, $v = \dfrac{ds}{dt}$ and find its velocity at $t = 10$.

b Find an expression for the acceleration, $a = \dfrac{dv}{dt}$ and find its acceleration at $t = 10$.

Exercise 204* (Revision)

1 Differentiate the following using correct notation.

a $y = \dfrac{1}{x}$

b $y = \dfrac{1}{x^2}$

c $y = \sqrt{x}$

d $y = x^{-3}$

e $y = 4x^{-4}$

f $y = \dfrac{1}{2x^2}$

g $y = \dfrac{2x^3 + 3x^2 + 4}{x}$

h $y = \dfrac{(x + 3)(2x - 1)}{x}$

2 Find the gradients of the tangents to the following curves at the given points.

   a $y = \dfrac{2}{x}$ $(1, 2)$ **b** $y = 2\sqrt{x} + x$ $(1, 3)$

   c $y = \left(x + \dfrac{1}{x}\right)^2$ $(1, 4)$ **d** $y = \dfrac{(1 - 2x)(1 + 2x)}{2x}$ $(2, -3\tfrac{3}{4})$

3 Find the equation of the tangent to the curve $y = x + \dfrac{1}{x}$ at the point where $x = 2$.

4 Find the equations of the tangents to the curve $y = x^2 - 3x$ at the points where it cuts the x-axis.

5 **a** Find the equation of the tangent to the curve $y = x^3 - 2x^2 + 1$ at the point where $x = 2$.

   b Determine the nature of any stationary points on this curve.

6 If the curve $y = 3x^3 + \dfrac{p}{x} + 3$ has a gradient of 8 where $x = 1$, find the value of p.

7 The temperature, $C°$ Centigrade, of the sea off Tokyo t months after New Year's Day is given by

   $C = 4t + \dfrac{16}{t}$ and is only valid for $1 \leqslant t \leqslant 6$.

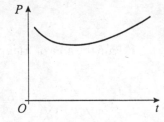

   a Find $\dfrac{dC}{dt}$

   b Find the coolest sea temperature in this period and when it occurs.

   c Find the rate at which the sea temperature is cooling on February 1st.

8 The population, P, of a rare orchid in the Borneo rainforest over t years is given by $P = 5t^2 + \dfrac{10\,000}{t}$ and is valid for $1 \leqslant t \leqslant 5$.

   a Find $\dfrac{dP}{dt}$.

   b Determine when the lowest population occurs and find its value.

9 A catapult projects a small stone vertically upwards such that its height, s m, after t seconds is given by $s = 50t - 5t^2$. Find the greatest height reached by the stone and the time it occurs.

10 The displacement of a comet, s km, after t seconds is given by $s = t(t^2 - 300)$ km.

   a Find expressions for the velocity, $v = \dfrac{ds}{dt}$ and the acceleration, $a = \dfrac{dv}{dt}$.

   b Find the velocity and acceleration of the comet after 5 seconds.

   c Find when the comet is momentarily at rest.

Area of a triangle ($\frac{1}{2}ab$ sin C)

Consider the triangle ABC below, labelled with the usual angle and side convention.

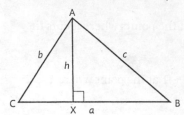

The area of the triangle $= \frac{1}{2} \times$ base $\times$ perpendicular height

$$= \frac{1}{2} \times a \times h$$

$$= \frac{1}{2}a(b \sin C)$$

So the area of any triangle $= \frac{1}{2}ab \sin C$

Simpler to recall is perhaps:

'Area = half the product of two sides $\times$ sine of the included angle'

Example 1

Find the area of the triangle ABC.

Area $= \frac{1}{2}ab \sin C$

$\quad = \frac{1}{2} \times 11.5 \times 7.3 \times \sin 110°$

$\quad = 39.4 \, \text{cm}^2 \, (3 \, \text{s.f.})$

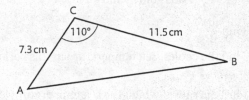

Exercise 205

1 Find the area of triangle ABC.

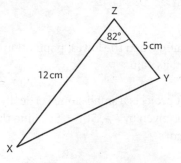

2 Find the area of triangle XYZ.

3 Find the area of triangle PQR.

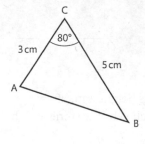

4 Find the area of triangle LMN.

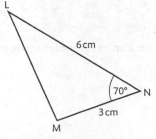

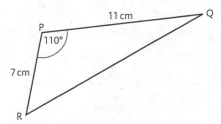

5 Find the area of the triangle ABC if AB = 15 cm, BC = 21 cm and $B = 130°$.

6 Find the area of the triangle XYZ if XY = 10 cm, YZ = 17 cm and $Y = 65°$.

Exercise 205*

1 Find the area of an equilateral triangle of side 20 cm.

2 Find the side length of an equilateral triangle of area 1000 cm².

3 Find side length *a* if the area of triangle ABC is 75 cm².

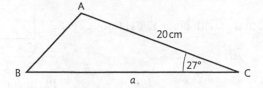

4 Find the angle ABC if the area of the isosceles triangle is 100 cm².

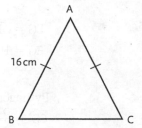

5 Find the area of triangle ABC.

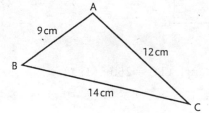

6 Find the perimeter of triangle ABC if its area is 200 cm².

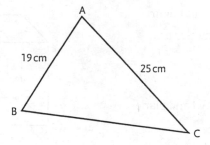

Practical problems in 3D

Place one end of a pencil B on a flat surface and put
your ruler against the other end T as shown.
The angle TBA is the angle between the straight line TB
and the plane WXYZ.

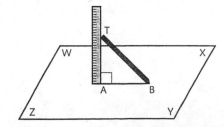

Key Point

● Draw clear, large 3D diagrams including all the facts.

● Redraw the appropriate triangle (usually right-angled) including all the facts. This simplifies a 3D problem into a 2D problem using Pythagoras' Theorem and trigonometry to solve for angles and lengths.

Example 2

ABCDEFGH is a rectangular box.

Find to 3 significant figures

a length EG
b length CE
c the angle CE makes with plane EFGH (angle CEG).

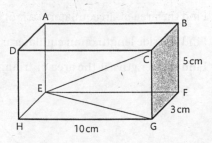

a Draw triangle EGH in 2D.

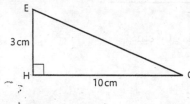

Pythagoras' Theorem

$EG^2 = 3^2 + 10^2$

$\quad = 109$

$EG = \sqrt{109}$

$EG = 10.4\,\text{cm (3s.f.)}$

b Draw triangle CEG in 2D.

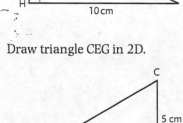

Pythagoras' Theorem

$CE^2 = 5^2 + 109$

$\quad = 134$

$CE = \sqrt{134}$

$CE = 11.6\,\text{cm (3 s.f.)}$

c Let angle CEG = θ

$\tan\theta = \dfrac{5}{\sqrt{109}} \Rightarrow \theta = \text{angle CEG} = 25.6°\ (3\ \text{s.f.})$

Exercise 206

Give all answers to 3 significant figures.

1 ABCDEFGH is a rectangular box.
Find:
  a EG
  b AG
  c the angle between AG and plane EFGH (angle AGE).

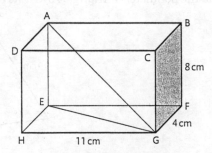

2 STUVWXYZ is a rectangular box.
Find:
  a SU
  b SY
  c the angle between SY and plane STUV (angle YSU).

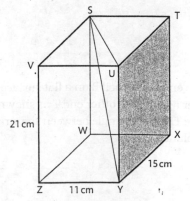

3 LMNOPQRS is a cube of side 10 cm.
Find:
a PR
b LR
c the angle between LR and plane PQRS
(angle LRP).

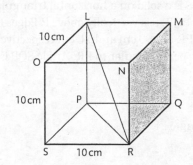

4 ABCDEFGH is a cube of side 20 cm.
Find:
a CF
b DF
c the angle between DF and plane BCGF
(angle DFC)
d the angle MHA, if M is the mid-point of AB.

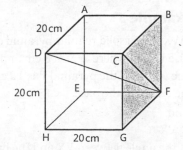

5 ABCDEF is a small ramp where ABCD and
CDEF are both rectangles and perpendicular
to each other.
Find:
a AC
b AF
c angle FAB
d the angle between AF and plane ABCD (angle FAC).

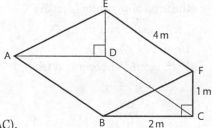

6 PQRSTU is an artificial ski-slope where
PQRS and RSTU are both rectangles and
perpendicular to each other.
Find:
a UP
b PR
c the angle between UP and plane PQRS
(angle UPR)
d the angle between MP and plane PQRS,
if M is the mid-point of TU.

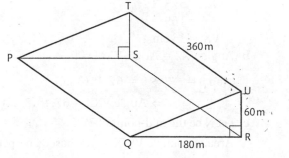

7 ABCD is a solid on a horizontal triangular base
ABC. Edge AD is 25 cm and vertical. AB is
perpendicular to AC. Angles ABD and ACD are
equal to 30° and 20° respectively.
Find:
a AB
b AC
c BC.

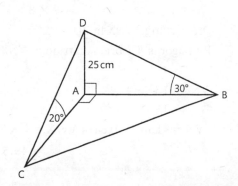

8 PQRS is a solid on a horizontal triangular base PQR. S is vertically above P. Edges PQ and PR are 50 cm and 70 cm respectively. PQ is perpendicular to PR. Angle SQP is 30°. Find:

a SP

b RS

c angle PRS.

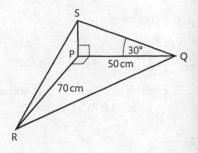

Example 3

VWXYZ is a solid regular pyramid on a rectangular base WXYZ where WX = 8 cm and XY = 6 cm The vertex of the pyramid V is 12 cm directly above the centre of the base.

Find:

a VX

b the angle between VX and the base WXYZ (angle VXZ)

c the area of pyramid face VWX.

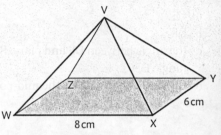

a Let M be the mid-point ZX.

Draw WXYZ in 2D.

Pythagoras' Theorem on triangle ZWX:

$$ZX^2 = 6^2 + 8^2$$
$$= 100$$
$$ZX = 10 \text{ cm} \Rightarrow MX = 5 \text{ cm}$$

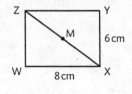

Draw triangle VMX in 2D. Pythagoras' Theorem on triangle VMX:

$$VX^2 = 5^2 + 12^2$$
$$= 169$$
$$VX = 13 \text{ cm}$$

b Angle VXZ = Angle VXM = θ

$$\tan \theta = \frac{12}{5} \Rightarrow \theta = 67.4° \text{ (3 s.f.)}$$
$$\Rightarrow \text{Angle VXZ} = 67.4° \text{ (3 s.f.)}$$

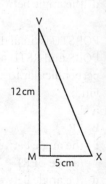

c Let N be the mid-point of WX.

Area of triangle VWX $= \frac{1}{2} \times$ base $\times$ perpendicular height

$$= \frac{1}{2} \times WX \times VN$$
$$= \frac{1}{2} \times 8 \times VN$$

Draw triangle VNX in 2D.

Pythagoras' Theorem on triangle VNX:

$$13^2 = 4^2 + VN^2$$
$$VN^2 = 13^2 - 4^2$$
$$= 153$$
$$VN = \sqrt{153} \Rightarrow \text{Area of VWX} = \frac{1}{2} \times 8 \times \sqrt{153}$$
$$= 49.5 \text{ cm}^2 \text{ (3 s.f.)}$$

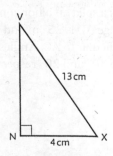

1 PABCD is a solid regular pyramid on a rectangular
base ABCD where AB = 10 cm and BC = 7 cm.
The vertex of the pyramid, P, is 15 cm directly
above the centre of the base.
Find:

a PA

b the angle between PA and base ABCD
(angle PAC)

c the area of pyramid face PBC.

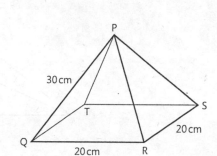

2 PQRST is a solid regular pyramid on a square base
QRST where QR = 20 cm and edge PQ = 30 cm.
Find:

a the height of P above base QRST

b the angle PS makes with the base QRST

c the total external area of the pyramid including
the base.

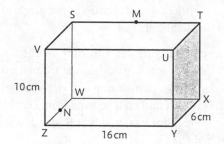

3 STUVWXYZ is a rectangular box.
M and N are the mid-points of
ST and WZ respectively.
Find angle:

a SYW

b TNX

c ZMY.

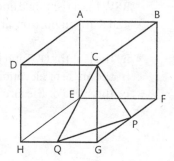

4 ABCDEFGH is a solid cube of volume 1728 cm³.
P and Q are the mid-points of FG and GH respectively.
Find:

a angle QCP

b the total surface area of the solid remaining after
pyramid PGQC is cut off.

5 A church is made from two solid rectangular
blocks with a regular pyramidal roof above the
tower, with V being 40 m above ground level.
Find:

a VA

b the angle of elevation of V from E

c the cost of tiling the tower roof if the church
is charged €250/m².

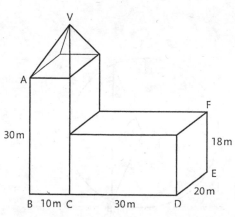

Unit 9 : Trigonometry

6 A hemispherical lampshade of diameter 40 cm is hung from a point by four chains, each of length 50 cm. If the chains are equally spaced on the rim of the hemisphere, find:

a the angle each chain makes with the horizontal

b the angle between two adjacent chains.

7 The angle of elevation to the top of a church tower is measured from A and from B

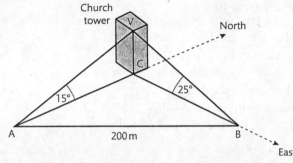

From A, due South of the church tower VC, the angle of elevation ∠VAC = 15°.
From B, due East of the church, the angle of elevation ∠VBC = 25°. AB = 200 m.
Find the height of the tower.

8 An aircraft is flying at a constant height of 2000 m. It is flying due East at a constant speed. At T, the plane's angle of elevation from O is 25°, and on a bearing from O of 310°. One minute later, it is at R and due North of O.

RSWT is a rectangle and the points O, W and S are on horizontal ground.

Find:

a the lengths OW and OS

b the angle of elevation of the aircraft, ∠ROS

c the speed of the aircraft in km/h.

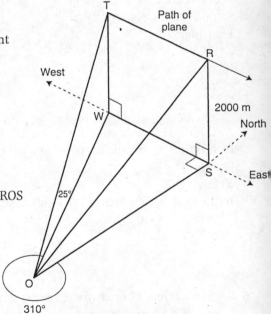

Area of a segment

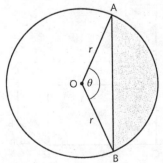

The circle segment (shaded) = Area of sector OAB
− Area of triangle OAB
$$= \frac{\theta}{360} \times \pi r^2 - \frac{1}{2} \times r^2 \sin \theta$$

Example 4

Find the area of the shaded segment.

Area = Area of sector OAB − Area of triangle OAB

$$= \frac{100}{360} \times \pi \times 10^2 - \frac{1}{2} \times 10^2 \times \sin(100°)$$

$$= 38.0 \text{ cm}^2 \text{ (3 s.f.)}$$

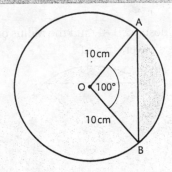

Exercise 207

In Questions 1–6, find the area of the shaded segment.
Give answers to 3 s.f.

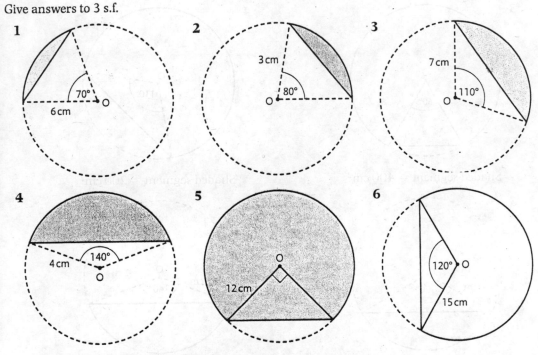

1

70°
6 cm

2

3 cm
80°

3

7 cm
110°

4

4 cm
140°

5

12 cm

6

120°
15 cm

Example 5

Given the shaded segment area is 500 cm², find the
value of r.

Shaded area = Area of sector OAB − Area of triangle OAB

$$500 = \frac{140}{360} \times \pi r^2 - \frac{1}{2} \times r^2 \times \sin(140°)$$

$$= r^2\left(\frac{7}{18} \times \pi - \frac{1}{2} \times \sin(140°)\right)$$

$$= r^2 (0.900337...)$$

$$\frac{500}{(0.900337...)} = r^2$$

$$555.348... = r^2 \Rightarrow r = 23.6 \text{ cm (3 s.f.)}$$

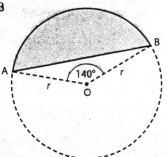

Exercise 207*

In Questions 1–4, find the radius of the circle.
Give answers to 3 s.f.

1

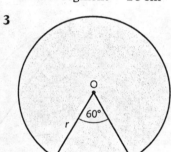

Shaded segment = 10 cm²

2

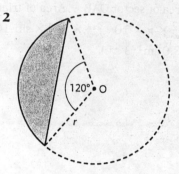

Shaded segment = 100 cm²

3

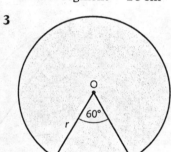

Shaded segment = 400 cm²

4

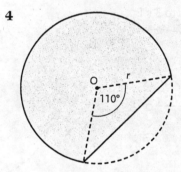

Shaded segment = 200 cm²

5

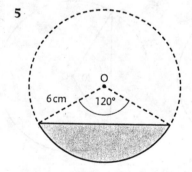

The shaded segment represents a cross-section of a pipe full of water flowing at 10 cm/s. How much water flows through the pipe in 1 hour?

(Answer in m³ in standard form to 3 s.f.)

6

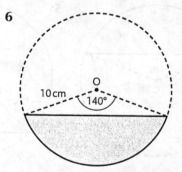

The shaded segment shows a cross-section of a pipe with some rainwater flowing at 4 cm/s into the empty cylindrical tank shown in the diagram.

What is the depth in cm of water in the tank after 1 hour?

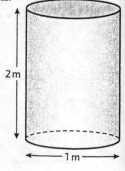

Exercise 208 (Revision)

1 Find the area △XYZ.

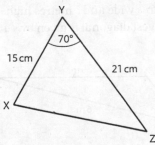

2 Copy this diagram. It shows a pyramid on a square base, where V is vertically above the centre of ABCD. Calculate

  a the length AC

  b the height of the pyramid

  c the angle that VC makes with BC

  d the angle that VC makes with ABCD.

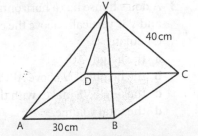

3 Copy this diagram of a ski slope and include these facts:
DE = 300 m, AD = 400 m, CE = 100 m. Find

  a the angle that CD makes with ADEF

  b AE

  c the angle that CA makes with ADEF.

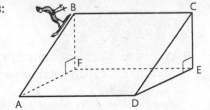

4 Find the area of each triangle ABC.

  a

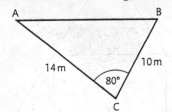

  b

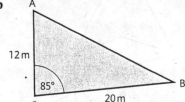

5 ABCDEFGH is a rectangular box. Find

  a FH

  b BH

  c angle BHF

  d the angle between plane ABGH and plane EFGH.

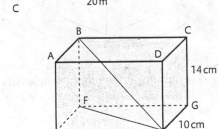

6 Find the area of the shaded segment. Give answers to 3 s.f.

  a

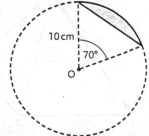

  b

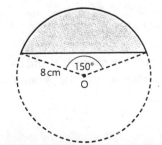

Exercise 208* (Revision)

1 A cuboid is a metres long, b metres wide and c metres high.
Show that the length of the longest diagonal is given by $\sqrt{a^2 + b^2 + c^2}$.

2 Find the area of $\triangle XYZ$.

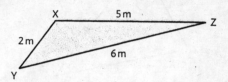

3 A doll's house has a horizontal square base ABCD
and V is vertically above the centre of the base.
Calculate
  a the length AC
  b the height of V above ABCD
  c the angle VE makes with the horizontal
  d the total volume.

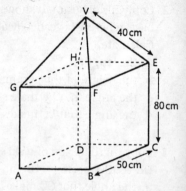

4 A cube of side 8 cm stands on a horizontal table. A hollow cone of height 20 cm is placed
over the cube so that it rests on the table and touches the top four corners of the cube.

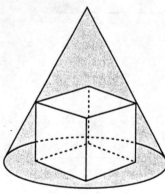

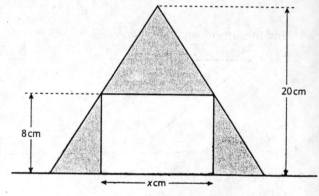

  a Show that $x = 8\sqrt{2}$ cm.   **b** Find the vertical angle of the cone.

5 Find the perimeter of triangle ABC if its area is 250 m²
and angle A is greater than 90°.

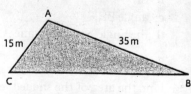

6 VABCD is a right-pyramid on a square base. Find
  a the length AC
  b the pyramid's height
  c angle AVB
  d the angle between plane AVB and base ABCD
  e the total surface area of the pyramid including its base.

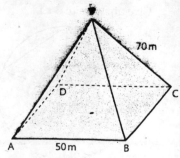

7 The shaded segment area = 100 cm^2.
Find r to 3 s.f.

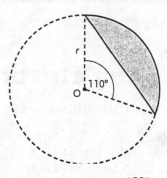

8 The shaded segment shows a cross-section of a pipe carrying some rainwater flowing at x cm/s into an empty cylindrical tank shown.

If the tank is full after 1 hour find x.

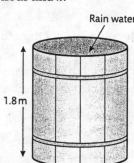

Rain water

1.8 m

1 m

Vector geometry problems

Parallel and equivalent vectors

Two vectors are **parallel** if they have the same direction but not necessarily equal length.

For example, these vectors **a** and **b** are parallel.

$$\mathbf{a} = \begin{pmatrix} 3 \\ 2 \end{pmatrix} \qquad \mathbf{b} = \begin{pmatrix} 6 \\ 4 \end{pmatrix}$$

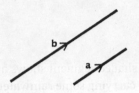

Two vectors are **equivalent** if they have the same direction and length.

Addition of vectors

The result of adding a set of vectors is the vector representing their total effect. This is the **resultant** of the vectors.

To add a number of vectors, they are placed end to end so that the next vector starts where the last one finished. The resultant vector joins the *start* of the first vector to the *end* of the last one.

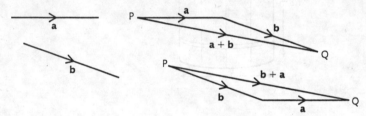

Vector $\overrightarrow{PQ} = \mathbf{a} + \mathbf{b} = \mathbf{b} + \mathbf{a}$

Multiplication of a vector by a scalar

When a vector is multiplied by a scalar, its length is multiplied by this number; but its direction is unchanged, unless the scalar is *negative*, in which case the direction is *reversed*.

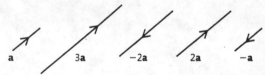

Example 1

Given vectors **a**, **b** and **c** as shown, draw the vector **r** where $\mathbf{r} = \mathbf{a} + \mathbf{b} - \mathbf{c}$.

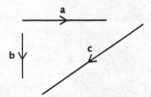

Here is the resultant of $\mathbf{a} + \mathbf{b} - \mathbf{c} = \mathbf{r}$.

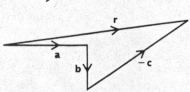

Example 2

ABCDEF is a regular hexagon with centre O.
$\vec{AB} = \mathbf{x}$ and $\vec{BC} = \mathbf{y}$. Express the vectors $\vec{ED}$,
$\vec{DE}$, $\vec{FE}$, $\vec{AC}$, $\vec{FA}$ and $\vec{AE}$ in terms of $\mathbf{x}$ and $\mathbf{y}$.

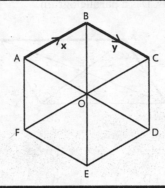

$\vec{ED} = \mathbf{x}$ $\vec{DE} = -\mathbf{x}$

$\vec{FE} = \mathbf{y}$ $\vec{AC} = \mathbf{x} + \mathbf{y}$

$\vec{FA} = \mathbf{x} - \mathbf{y}$ $\vec{AE} = 2\mathbf{y} - \mathbf{x}$

Exercise 209

Use this diagram to answer Questions 1–4.

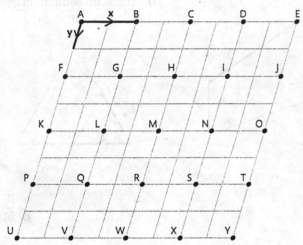

For Questions 1 and 2, express each vector in terms of $\mathbf{x}$ and $\mathbf{y}$.

1   **a** $\vec{XY}$ **b** $\vec{EO}$ **c** $\vec{WC}$ **d** $\vec{TP}$

2   **a** $\vec{KC}$ **b** $\vec{VC}$ **c** $\vec{CU}$ **d** $\vec{AS}$

For Questions 3 and 4, find the vector formed when the vectors given are added to point H.
Write the vector as capital letters (e.g. $\vec{HO}$).

3   **a** $2\mathbf{x}$ **b** $\mathbf{x} + 2\mathbf{y}$ **c** $2\mathbf{x} - \mathbf{y}$ **d** $2\mathbf{x} + 2\mathbf{y}$

4   **a** $4\mathbf{y} + 2\mathbf{x}$ **b** $4\mathbf{y} - 2\mathbf{x}$ **c** $\mathbf{x} - 2\mathbf{y}$ **d** $2\mathbf{x} + 6\mathbf{y}$

In Questions 5–10, express each vector in terms of $\mathbf{x}$ and $\mathbf{y}$.

5 ABCD is a rectangle.   **6** ABCD is a trapezium.

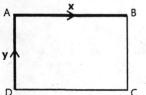   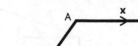

Find   Find

a $\vec{DC}$ **b** $\vec{DB}$ **a** $\vec{AC}$ **b** $\vec{DB}$

c $\vec{BC}$ **d** $\vec{AC}$ **c** $\vec{BC}$ **d** $\vec{CB}$

7 ABCD is a parallelogram.

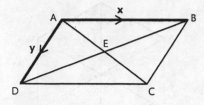

Find

a $\overrightarrow{DC}$ **b** $\overrightarrow{AC}$

c $\overrightarrow{BD}$ **d** $\overrightarrow{AE}$

8 ABCD is a rhombus.

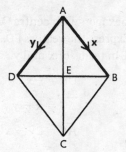

Find

a $\overrightarrow{BD}$ **b** $\overrightarrow{BE}$

c $\overrightarrow{AC}$ **d** $\overrightarrow{AE}$

9

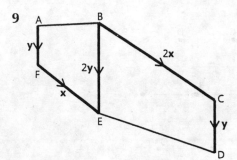

Find

a $\overrightarrow{AB}$ **b** $\overrightarrow{AD}$

c $\overrightarrow{CF}$ **d** $\overrightarrow{CA}$

10 ABC is an equilateral triangle, with $\overrightarrow{AC} = 2\mathbf{x}$, $\overrightarrow{AB} = 2\mathbf{y}$.

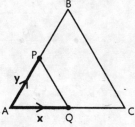

Find

a $\overrightarrow{PQ}$ **b** $\overrightarrow{PC}$

c $\overrightarrow{QB}$ **d** $\overrightarrow{BC}$

11 The gear stick of a car is shown. The lever can only shift along the grooves shown to reach each gear. N is neutral and R is reverse. Express these gear changes in terms of **x** and **y**.

a 1st to 4th   **b** 3rd to 2nd

c N to R   **d** 2nd to 5th

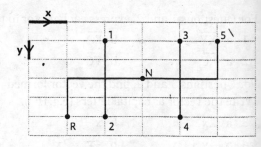

12 A rectangular biscuit tin OABCDEFG is shown. $\overrightarrow{OA} = \mathbf{x}$, $\overrightarrow{OB} = \mathbf{y}$ and $\overrightarrow{OD} = \mathbf{z}$.

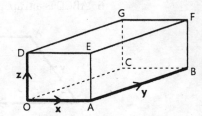

An ant crawls these direct journeys in search of crumbs. Express each journey as a vector in terms of **x**, **y** and **z**.

a $\overrightarrow{OE}$ **b** $\overrightarrow{OB}$ **c** $\overrightarrow{OF}$ **d** $\overrightarrow{EC}$

Example 3

In △OAB, the mid-point on AB is M.
$\overrightarrow{AB} = \mathbf{x}$, $\overrightarrow{OB} = \mathbf{y}$, $\overrightarrow{OD} = 2\mathbf{x}$ and $\overrightarrow{OC} = 2\mathbf{y}$.

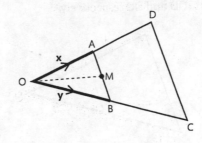

a Express $\overrightarrow{AB}$, $\overrightarrow{OM}$ and $\overrightarrow{DC}$ in terms of $\mathbf{x}$ and $\mathbf{y}$.

$$\overrightarrow{AB} = \overrightarrow{AO} + \overrightarrow{OB} \qquad \overrightarrow{OM} = \overrightarrow{OA} + \overrightarrow{AM}$$

$$= -\mathbf{x} + \mathbf{y} \qquad\qquad = \overrightarrow{OA} + \tfrac{1}{2}\overrightarrow{AB}$$

$$= \mathbf{y} - \mathbf{x} \qquad\qquad = \mathbf{x} + \tfrac{1}{2}(\mathbf{y} - \mathbf{x})$$

$$\overrightarrow{DC} = \overrightarrow{DO} + \overrightarrow{OC} \qquad = \mathbf{x} + \tfrac{1}{2}\mathbf{y} - \tfrac{1}{2}\mathbf{x}$$

$$= -2\mathbf{x} + 2\mathbf{y} \qquad\quad = \tfrac{1}{2}\mathbf{x} + \tfrac{1}{2}\mathbf{y}$$

$$= 2\mathbf{y} - 2\mathbf{x} \qquad\quad = \tfrac{1}{2}(\mathbf{x} + \mathbf{y})$$

$$= 2(\mathbf{y} - \mathbf{x})$$

b How are AB and DC related?

$\overrightarrow{DC} = 2\overrightarrow{AB}$, so AB is parallel to DC and half its magnitude.

Example 4

In the diagram
$$\overrightarrow{AE} = \mathbf{a}, \overrightarrow{EC} = h\mathbf{a}$$
$$\overrightarrow{DE} = \mathbf{b} \text{ and } \overrightarrow{EB} = k\mathbf{b}$$

a **(i)** Find $\overrightarrow{AB}$ in terms of $\mathbf{a}$, $\mathbf{b}$ and k
   (ii) Find $\overrightarrow{DC}$ in terms of $\mathbf{a}$, $\mathbf{b}$ and h

b If $\overrightarrow{AB} = 2\overrightarrow{DC}$ find the values of h and k

a **(i)** $\overrightarrow{AB} = \mathbf{a} + k\mathbf{b}$ **(ii)** $\overrightarrow{DC} = \mathbf{b} + h\mathbf{a}$

b $\mathbf{a} + k\mathbf{b} = 2(\mathbf{b} + h\mathbf{a}) \Rightarrow \mathbf{a} + k\mathbf{b} = 2h\mathbf{a} + 2\mathbf{b}$
Comparing the coefficients of $\mathbf{a}$ gives $1 = 2h \Rightarrow h = \tfrac{1}{2}$
Comparing the coefficients of $\mathbf{b}$ gives $k = 2$

Exercise 209*

In Questions 1–10, express each vector in terms of $\mathbf{x}$ and $\mathbf{y}$, where $\overrightarrow{OA} = \mathbf{x}$ and $\overrightarrow{OB} = \mathbf{y}$.

1 M is the mid-point of AB.

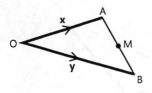

2 The ratio of AM:MB = 1:2.

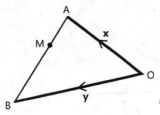

Find
  a $\overrightarrow{AB}$ **b** $\overrightarrow{AM}$ **c** $\overrightarrow{OM}$

Find
  a $\overrightarrow{AB}$ **b** $\overrightarrow{AM}$ **c** $\overrightarrow{OM}$

3 A and B are the mid-points of OD and OC respectively.

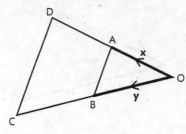

a Find $\overrightarrow{AB}$, $\overrightarrow{OD}$ and $\overrightarrow{DC}$.
b How are AB and DC related?

5 The ratio of $OA:OC = 1:2$ and $OB:OD = 1:2$.

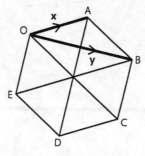

a Find $\overrightarrow{AB}$, $\overrightarrow{OC}$, $\overrightarrow{OD}$ and $\overrightarrow{DC}$.
b How are AB and DC related?

7 OABCDE is a regular hexagon.

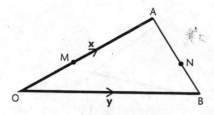

Find $\overrightarrow{AB}$, $\overrightarrow{BC}$, $\overrightarrow{AD}$ and $\overrightarrow{BD}$.

9 $OM:MA = 2:3$ and $AN = \frac{3}{5}AB$.

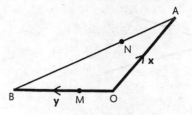

a Find $\overrightarrow{MA}$, $\overrightarrow{AB}$, $\overrightarrow{AN}$ and $\overrightarrow{MN}$.
b How are OB and MN related?

4 The ratio of $OA:AD = 1:2$ and B is the mid-point of OC.

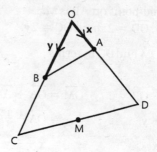

a Find $\overrightarrow{AB}$, $\overrightarrow{OD}$ and $\overrightarrow{DC}$.
b M is the mid-point of CD. Find $\overrightarrow{OM}$.

6 OABC is a parallelogram. The ratio of $OD:DC = 1:2$.

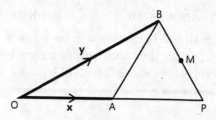

Find $\overrightarrow{AB}$, $\overrightarrow{OD}$, $\overrightarrow{BD}$, $\overrightarrow{OE}$ and $\overrightarrow{DE}$.

8 ABP is an equilateral triangle. OAB is an isosceles triangle, where OA = AB.

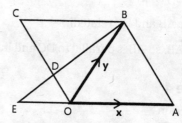

a Find $\overrightarrow{OP}$, $\overrightarrow{AB}$ and $\overrightarrow{BP}$.
b M is the mid-point of BP. Find $\overrightarrow{OM}$.

10 $OM:MB = 1:2$ and $AN = \frac{1}{3}AB$.

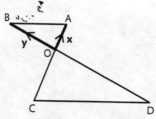

a Find $\overrightarrow{AB}$ and $\overrightarrow{MN}$.
b How are OA and MN related?

Activity 50

A radar tracking station O is positioned at the origin of x and y axes, where the x-axis points due East and the y-axis points due North. All the units are in km.

A helicopter is detected t minutes after midday with position vector $\mathbf{r}$, where

$$\mathbf{r} = \begin{pmatrix} 12 \\ 5 \end{pmatrix} + t\begin{pmatrix} -3 \\ 4 \end{pmatrix}$$

- Copy and complete this table and use it to plot the course of the helicopter.

Time	12:00 $t = 0$	12:01 $t = 1$	12:02 $t = 2$	12:03 $t = 3$	12:04 $t = 4$
$\mathbf{r}$	$\begin{pmatrix} 12 \\ 5 \end{pmatrix}$				

- Calculate the speed of this helicopter in km/hour correct to 1 decimal place, and its bearing.
- An international boundary is described by the line $y = 5x$.
  - Draw the boundary on your graph.
  - Estimate the time the helicopter crosses the boundary by careful inspection of your graph.
  - Considering the helicopter's position vector $\mathbf{r} = \begin{pmatrix} x \\ y \end{pmatrix}$

    express x and y in terms of t and use these equations with $y = 5x$ to find the time when the boundary is crossed, correct to the nearest second.

Transformation matrix

Matrices can be used to describe transformations.

To find the image of the vector $\begin{pmatrix} 1 \\ 1 \end{pmatrix}$ under the transformation

described by $\begin{pmatrix} 1 & 0 \\ 0 & -1 \end{pmatrix}$ do $\begin{pmatrix} 1 & 0 \\ 0 & -1 \end{pmatrix}\begin{pmatrix} 1 \\ 1 \end{pmatrix} = \begin{pmatrix} 1 \\ -1 \end{pmatrix}$ and plot the result.

It is difficult to see if the transformation is a rotation or a reflection, so further vectors need to be transformed. If the shape defined by the vectors has no symmetry it is easier to see what is going on.

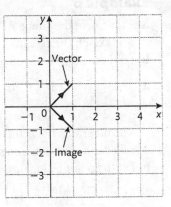

The second graph shows a triangle defined by the vectors $\begin{pmatrix} 1 \\ 1 \end{pmatrix}$, $\begin{pmatrix} 1 \\ 2 \end{pmatrix}$

and $\begin{pmatrix} 3 \\ 1 \end{pmatrix}$, and the images of these vectors under the same transformation.

$$\begin{pmatrix} 1 & 0 \\ 0 & -1 \end{pmatrix}\begin{pmatrix} 1 \\ 1 \end{pmatrix} = \begin{pmatrix} 1 \\ -1 \end{pmatrix}, \begin{pmatrix} 1 & 0 \\ 0 & -1 \end{pmatrix}\begin{pmatrix} 1 \\ 2 \end{pmatrix} = \begin{pmatrix} 1 \\ -2 \end{pmatrix}, \begin{pmatrix} 1 & 0 \\ 0 & -1 \end{pmatrix}\begin{pmatrix} 3 \\ 1 \end{pmatrix} = \begin{pmatrix} 3 \\ -1 \end{pmatrix}$$

The transformation is a reflection in the x-axis.

Drawing the vectors confuses the picture so they are usually left out. Combining the vectors into one matrix shortens the working. If the vertices of the shape are labelled A, B, C,... then the vertices of the image are labelled A', B', C',...

Example 5 shows how this is done.

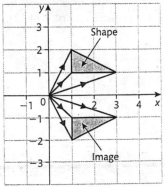

Example 5

The points A(1, 1), B(1, 2) and C(3, 1) are mapped to the points A′, B′ and C′ by the matrix $\begin{pmatrix} 1 & 0 \\ 0 & -1 \end{pmatrix}$. Find the coordinates of A′, B′ and C′ and describe the transformation.

$$\begin{pmatrix} 1 & 0 \\ 0 & -1 \end{pmatrix} \overset{\text{A B C}}{\begin{pmatrix} 1 & 1 & 3 \\ 1 & 2 & 1 \end{pmatrix}} = \overset{\text{A′ B′ C′}}{\begin{pmatrix} 1 & 1 & 3 \\ -1 & -2 & -1 \end{pmatrix}}$$

The diagram shows that $\begin{pmatrix} 1 & 0 \\ 0 & -1 \end{pmatrix}$ represents a reflection in the x-axis.

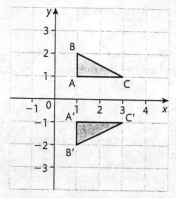

(We have assumed that all the straight lines are still straight after the transformation. This will always be the case.)

Example 6

a Transform the shape A(1, 0), B(1, −1), C(2, −1), D(3, 1) by the matrix $\mathbf{M} = \begin{pmatrix} -1 & 1 \\ 2 & 3 \end{pmatrix}$ and sketch the result.

b Find the point mapped by **M** onto the point (1, 8).

a
$$\begin{pmatrix} -1 & 1 \\ 2 & 3 \end{pmatrix} \overset{\text{A B C D}}{\begin{pmatrix} 1 & 1 & 2 & 3 \\ 0 & -1 & -1 & 1 \end{pmatrix}} = \overset{\text{A′ B′ C′ D′}}{\begin{pmatrix} -1 & -2 & -3 & -2 \\ 2 & -1 & 1 & 9 \end{pmatrix}}$$

The diagram shows this is a combination of reflection, rotation and stretching.

b Let the point be (a, b).

Then $\begin{pmatrix} -1 & 1 \\ 2 & 3 \end{pmatrix}\begin{pmatrix} a \\ b \end{pmatrix} = \begin{pmatrix} 1 \\ 8 \end{pmatrix} \Rightarrow \begin{matrix} -a + b = 1 \\ 2a + 3b = 8 \end{matrix}$

Solving these equations gives $a = 1, b = 2$ so the required point is (1, 2).

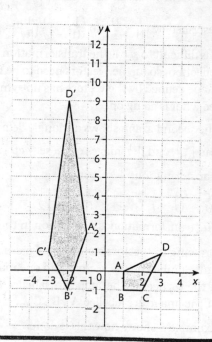

Sometimes a convenient shape to use is the 'unit square'. This is a square with two sides formed by the vectors $\begin{pmatrix} 1 \\ 0 \end{pmatrix}$ and $\begin{pmatrix} 0 \\ 1 \end{pmatrix}$. However, because of all the symmetries of a square it is very important to label all the vertices correctly.

Example 7

Find the image of the 'unit' square under the transformation represented by $\begin{pmatrix} -2 & 0 \\ 0 & 2 \end{pmatrix}$ and describe the transformation.

Point A $\begin{pmatrix} -2 & 0 \\ 0 & 2 \end{pmatrix}\begin{pmatrix} 1 \\ 0 \end{pmatrix} = \begin{pmatrix} -2 \\ 0 \end{pmatrix}$ The answer is **always** the first column of the matrix.

Point B $\begin{pmatrix} -2 & 0 \\ 0 & 2 \end{pmatrix}\begin{pmatrix} 0 \\ 1 \end{pmatrix} = \begin{pmatrix} 0 \\ 2 \end{pmatrix}$ The answer is **always** the second column of the matrix.

Point C $\begin{pmatrix} -2 & 0 \\ 0 & 2 \end{pmatrix}\begin{pmatrix} 1 \\ 1 \end{pmatrix} = \begin{pmatrix} -2 \\ 2 \end{pmatrix}$ The answer is **always** the sum of the rows of the matrix.

Origin $\begin{pmatrix} -2 & 0 \\ 0 & 2 \end{pmatrix}\begin{pmatrix} 0 \\ 0 \end{pmatrix} = \begin{pmatrix} 0 \\ 0 \end{pmatrix}$ The answer is **always** the origin.

The diagram shows the transformation is a reflection in the y-axis and an enlargement of scale factor 2 centre the origin, done in either order.

If the vertices had not been labelled, it would be impossible to say if a reflection or an anticlockwise rotation of 90° had been used.

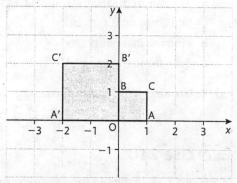

Example 7 shows that the origin never moves under any matrix transformation because $\begin{pmatrix} a & c \\ b & d \end{pmatrix}\begin{pmatrix} 0 \\ 0 \end{pmatrix} = \begin{pmatrix} 0 \\ 0 \end{pmatrix}$. This means that a matrix can only describe a transformation where the origin does not move, so the centre of rotation and the centre of enlargement must be at the origin. **For reflections, the mirror lines must pass through the origin.** As a translation moves every point it cannot be described by a 2 × 2 matrix.

Inverse transformations

The inverse matrix will give the inverse transformation.

Example 8

$\mathbf{R} = \begin{pmatrix} 0 & -1 \\ 1 & 0 \end{pmatrix}$ represents a rotation of 90° anticlockwise about the origin.

a What is the inverse transformation?
b Find the matrix representing the inverse transformation.

a The inverse transformation is a rotation of 90° clockwise about the origin.

b The determinant of $\mathbf{R}$ is $(0 \times 0) - (1 \times -1) = 1$ so $\mathbf{R}^{-1} = \frac{1}{1}\begin{pmatrix} 0 & 1 \\ -1 & 0 \end{pmatrix} = \begin{pmatrix} 0 & 1 \\ -1 & 0 \end{pmatrix}$

The matrix $\begin{pmatrix} 0 & 1 \\ -1 & 0 \end{pmatrix}$ represents a rotation of 90° clockwise about the origin.

Key Points
- A matrix transforms a vector.
- The origin does not move under a matrix transformation.
- The inverse matrix gives the inverse transformation.

Exercise 210

Questions 1–4 all refer to the flag shape F given by the points A(1, 1), B(1, 2), C(1, 3) and D(2, 3) and shown in the diagram. Plot your answers on a copy of the diagram.

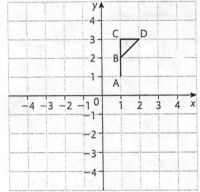

1 Find the image of F under the transformation represented by $\begin{pmatrix} 0 & -1 \\ 1 & 0 \end{pmatrix}$. Describe the transformation.

2 Find the image of F under the transformation represented by $\begin{pmatrix} 1 & 0 \\ 0 & -1 \end{pmatrix}$. Describe the transformation.

3 Find the image of F under the transformation represented by $\begin{pmatrix} 0 & 1 \\ 1 & 0 \end{pmatrix}$. Describe the transformation.

4 Find the image of F under the transformation represented by $\begin{pmatrix} -1 & 0 \\ 0 & -1 \end{pmatrix}$. Describe the transformation.

5 a Transform the triangle A(1, 1), B(1, 2), C(3, 1) by the matrix $\begin{pmatrix} 1 & 2 \\ 1 & 3 \end{pmatrix}$ and show on a sketch.

   b Find the inverse matrix.

   c Transform the image A′B′C′ by the inverse matrix, showing the result on your sketch.

6 The triangle ABC is transformed by the matrix $\begin{pmatrix} 3 & 1 \\ 2 & 1 \end{pmatrix}$ to the triangle A′(2, 1), B′(5, 4), C′(−2, −1). Find the coordinates of A, B and C.

Exercise 210*

Questions 1–4 all refer to the flag shape F given by the points A(1, 0), B(1, 1), C(1, 3) and D(3, 2) and shown in the diagram. Plot your answers on a copy of the diagram.

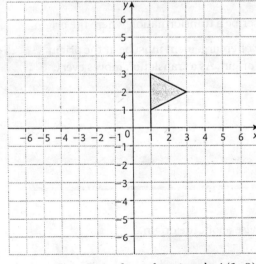

1 Find the image of F under the transformation represented by $\begin{pmatrix} 2 & 0 \\ 0 & 2 \end{pmatrix}$. Describe the transformation.

2 Find the image of F under the transformation represented by $\begin{pmatrix} 2 & 0 \\ 0 & -2 \end{pmatrix}$. Describe the transformation.

3 Find the image of F under the transformation represented by $\begin{pmatrix} 1 & 0 \\ 0 & 2 \end{pmatrix}$. Describe the transformation.

4 Find the image of F under the transformation represented by $\begin{pmatrix} 0.71 & 0.71 \\ -0.71 & 0.71 \end{pmatrix}$. Describe the transformation.

5 a Transform the triangle A(1, 0), B(−1, 1), C(−2, 0) by the matrix $\begin{pmatrix} 2 & 5 \\ 1 & -3 \end{pmatrix}$ and show on a sketch.

   b Find the inverse matrix.

   c Transform the image A′B′C′ by the inverse matrix, showing the result on your sketch.

6 The triangle ABC is transformed by the matrix $\begin{pmatrix} 3 & 2 \\ 4 & 2 \end{pmatrix}$ to the triangle A′(1, 2), B′(2, 2), C′(−1, −2). Find the coordinates of A, B and C.

For each matrix find the determinant. Then transform the triangle A(1, 1), B(2, 3), C(1, 3) by each matrix in turn, sketching the results and finding the areas of the images.

$$\begin{pmatrix} -1 & 0 \\ 0 & -1 \end{pmatrix} \quad \begin{pmatrix} 1 & 0 \\ 0 & -1 \end{pmatrix} \quad \begin{pmatrix} 2 & 0 \\ 0 & 2 \end{pmatrix} \quad \begin{pmatrix} 2 & -2 \\ -1 & 2 \end{pmatrix}$$

Is there a connection between the determinant and the area scale factor of the transformation?

Repeat with some matrices of your choice.

Combined transformation matrix

Multiplying together the matrices representing transformations gives the matrix representing the combined transformation. Because matrix multiplication is seldom commutative, the order of the multiplication is important. If the matrices are **R** and **M**, then **RM** means do **M** first, followed by **R**. (This is the same notation as combined functions.) Similarly **MR** means do **R** first followed by **M**.

Example 9

$\mathbf{M} = \begin{pmatrix} -1 & 0 \\ 0 & 1 \end{pmatrix}$ represents a reflection in the y-axis and $\mathbf{R} = \begin{pmatrix} 0 & -1 \\ 1 & 0 \end{pmatrix}$ represents a rotation of 90° anticlockwise about the origin.

a Calculate **RM** and **MR**.

b Transform the triangle A(1, 1), B(2, 3), C(1, 3) by **RM** and show the result on a sketch. Describe the transformation **RM**.

c Transform the triangle A(1, 1), B(2, 3), C(1, 3) by **MR** and show the result on a sketch. Describe the transformation **MR**.

a $\mathbf{RM} = \begin{pmatrix} 0 & -1 \\ 1 & 0 \end{pmatrix}\begin{pmatrix} -1 & 0 \\ 0 & 1 \end{pmatrix} = \begin{pmatrix} 0 & -1 \\ -1 & 0 \end{pmatrix}$ $\mathbf{MR} = \begin{pmatrix} -1 & 0 \\ 0 & 1 \end{pmatrix}\begin{pmatrix} 0 & -1 \\ 1 & 0 \end{pmatrix} = \begin{pmatrix} 0 & 1 \\ 1 & 0 \end{pmatrix}$

b   A B C   A′ B′ C′

$\begin{pmatrix} 0 & -1 \\ -1 & 0 \end{pmatrix}\begin{pmatrix} 1 & 2 & 1 \\ 1 & 3 & 3 \end{pmatrix} = \begin{pmatrix} -1 & -3 & -3 \\ -1 & -2 & -1 \end{pmatrix}$

c   A B C   A′B′C′

$\begin{pmatrix} 0 & 1 \\ 1 & 0 \end{pmatrix}\begin{pmatrix} 1 & 2 & 1 \\ 1 & 3 & 3 \end{pmatrix} = \begin{pmatrix} 1 & 3 & 3 \\ 1 & 2 & 1 \end{pmatrix}$

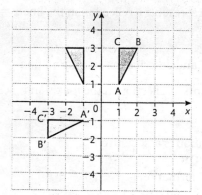

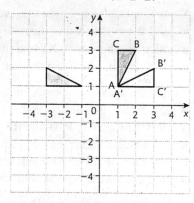

(The pink triangles are to help you see what has happened.)

RM is a reflection in the line $y = -x$ **MR** is a reflection in the line $y = x$

Exercise 211

In this exercise, the centre of all rotations and enlargements is at the origin.

1 **R** represents a 90° clockwise rotation.
  a Find the matrix **R**.
  b Calculate the matrix **RR**.
  c On a sketch, transform the unit square by **RR** and describe the transformation.

2 **X** represents a reflection in the x-axis and **Y** represents a reflection in the y-axis.
  a Find the matrix **X** and the matrix **Y**.
  b Calculate the matrix **XY** and the matrix **YX**.
  c On a sketch, transform the unit square by **XY** and describe the transformation.

3 **R** represents a 180° rotation and **X** represents a reflection in the line $x = y$.
  a Find the matrix **R** and the matrix **X**.
  b Calculate the matrix **RX** and the matrix **XR**.
  c On a sketch, transform the unit square by **XR** and describe the transformation.

4 a Draw a sketch showing the unit square reflected in the x-axis and then rotated 180°.
  b Hence write down the matrix for the combined transformation.
  c Find the matrix for the combined transformation performed in the opposite order.

5 $T = \begin{pmatrix} -2 & 0 \\ 0 & 1 \end{pmatrix}$, $S = \begin{pmatrix} 0 & -1 \\ 1 & 0 \end{pmatrix}$
  a Find **TS** and **ST**.
  b Transform the triangle A(1, 1), B(3, 1), C(1, 2) by **TS** and show the result on a sketch.
  c The triangle FGH is transformed by **ST** to F′(2, 4), G′(2, 2), H′(−1, −2). Find the coordinates of F, G and H.

Exercise 211*

In this exercise, the centre of all rotations and enlargements is the origin.

1 **R** represents a 90° anticlockwise rotation and **Q** represents a 270° clockwise rotation.
  a Find the matrix **R** and the matrix **Q**.
  b Calculate the matrix **RQ** and the matrix **QR**.
  c On a sketch, transform the unit square by **QR** and describe the transformation.

2 **X** represents a reflection in the line $x = y$ and **Y** represents a reflection in the line $x = -y$.
  a Find the matrix **X** and the matrix **Y**.
  b Calculate the matrix **XY** and the matrix **YX**.
  c On a sketch, transform the unit square by **XY** and describe the transformation.

3 **P** represents a 90° clockwise rotation, **Q** represents a reflection in the y-axis and **R** represents a 90° anticlockwise rotation.
  a Find the matrices **P**, **Q** and **R**.
  b Calculate the matrix **PQR**. (Matrix multiplication is associative.)
  c Describe the transformation represented by **PQR**.
  d Find the matrix representing the inverse transformation.

4 a Draw a sketch showing the unit square reflected in the line $x = -y$ and then enlarged (reduced) by a scale factor of $\frac{1}{2}$.
  b Hence write down the matrix for the combined transformation.
  c Find the matrix for the combined transformation performed in the opposite order.

5 $R = \begin{pmatrix} 1 & 1 \\ 0 & 1 \end{pmatrix}, S = \begin{pmatrix} 0 & -2 \\ -2 & 0 \end{pmatrix}$

   a Find **RS** and **SR**

   b Transform the triangle A(2, 3), B(−1, −5), C(−4, 2) by **RS** and show the result on a sketch.

   c The triangle FGH is transformed by **SR** to triangle ABC. Find the coordinates of F, G and H.

Exercise 212 (Revision)

1 In terms of vectors **x** and **y**, find these vectors.

   a $\overrightarrow{AB}$

   b $\overrightarrow{AC}$

   c $\overrightarrow{CB}$

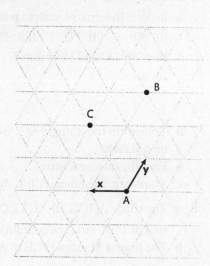

2 $\overrightarrow{OA} = $ **v** and $\overrightarrow{OB} = $ **w** and M is the mid-point of AB. Find these vectors in terms of **v** and **w**.

   a $\overrightarrow{AB}$

   b $\overrightarrow{AM}$

   c $\overrightarrow{OM}$

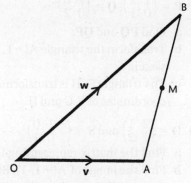

3 ABCDEF is a regular hexagon. Vectors $\overrightarrow{OA} = $ **x** and $\overrightarrow{OB} = $ **y**. Find these vectors in terms of **x** and **y**:

   a $\overrightarrow{AB}$

   b $\overrightarrow{FB}$

   c $\overrightarrow{FD}$

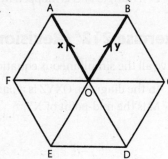

4 In the triangle OPQ, A and B are mid-points of the sides OP and OQ respectively, $\overrightarrow{OA} = $ **a** and $\overrightarrow{OB} = $ **b**.

   a Find in terms of **a** and **b**: $\overrightarrow{OP}$, $\overrightarrow{OQ}$, $\overrightarrow{AB}$ and $\overrightarrow{PQ}$.

   b What can you conclude about AB and PQ?

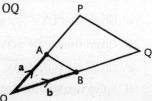

5 OAB is a triangle such that $\overrightarrow{OA} = 2\mathbf{x}$ and $\overrightarrow{OB} = 2\mathbf{y}$ and
AM : MB = 1 : 1. Express the following in terms of $\mathbf{x}$ and $\mathbf{y}$

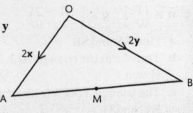

a $\overrightarrow{AB}$

b $\overrightarrow{AM}$

c $\overrightarrow{OM}$

6 OPQ is a triangle such that $\overrightarrow{OP} = \mathbf{p}$ and $\overrightarrow{OQ} = \mathbf{q}$. OM : MP = 3 : 2
and $PN = \frac{2}{5}PQ$.

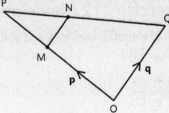

a Express in terms of vectors $\mathbf{p}$ and $\mathbf{q}$

(i) $\overrightarrow{MP}$ (ii) $\overrightarrow{PQ}$ (iii) $\overrightarrow{PN}$ (iv) $\overrightarrow{MN}$

b How are OQ and MN related?

7 a Transform the triangle A(1, 0), B(−3, 3), C(−1, 2) by the matrix $\begin{pmatrix} 2 & -3 \\ -1 & 2 \end{pmatrix}$ and show on a sketch.

b Find the inverse matrix.

c Transform the image A′B′C′ by the inverse matrix, showing the result on your sketch.

8 a The shape A(1, 2), B(−1, 1), C(−1, −2), D(5, 2) is mapped by the matrix $\mathbf{M} = \begin{pmatrix} -1 & 1 \\ 3 & -4 \end{pmatrix}$ to A′B′C′D′. Find the coordinates of A′, B′, C′ and D′, and sketch the results.

b The matrix $\mathbf{M}$ maps EFGH to ABCD. Find the coordinates of E′, F′, G′ and H′.

9 $\mathbf{P} = \begin{pmatrix} 1 & 1 \\ 0 & 1 \end{pmatrix}, \mathbf{Q} = \begin{pmatrix} 2 & 0 \\ 0 & 2 \end{pmatrix}$

a Find $\mathbf{PQ}$ and $\mathbf{QP}$.

b Transform the triangle A(−1, 1), B(2, 3), C(1, −2) by $\mathbf{PQ}$ and show the result on a sketch.

c The triangle FGH is transformed by $\mathbf{QP}$ to F′(1, −1), G′(3, 4), H′(−2, 0). Find the coordinates of F, G and H.

10 $\mathbf{R} = \begin{pmatrix} 2 & 1 \\ 1 & 2 \end{pmatrix}$ and $\mathbf{S} = \begin{pmatrix} -1 & 0 \\ 0 & 1 \end{pmatrix}$

a Find the matrix representing transformation $\mathbf{S}$ followed by $\mathbf{R}$.

b Find the image of A(−1, 1), B(3, 3), C(−2, 4) after transformation by $\mathbf{S}$ followed by $\mathbf{R}$.

c Triangle EFG is mapped by $\mathbf{S}$ followed by $\mathbf{R}$ onto ABC. Find the coordinates of E, F and G.

Exercise 212* (Revision)

Solve all the simultaneous equations using the matrix method.

1 In the diagram, OXYZ is a parallelogram.
M is the mid-point of XY.

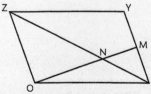

a Given that $\overrightarrow{OX} = \begin{pmatrix} 8 \\ 0 \end{pmatrix}$ and $\overrightarrow{OZ} = \begin{pmatrix} -2 \\ 6 \end{pmatrix}$, write down the vectors $\overrightarrow{XM}$ and $\overrightarrow{XZ}$.

b Given that $\overrightarrow{ON} = v\overrightarrow{OM}$, write down in terms of v the vector $\overrightarrow{ON}$.

c Given that $\overrightarrow{ON} = \overrightarrow{OX} + w\,\overrightarrow{XZ}$, find in terms of w the vector $\overrightarrow{ON}$.

d Solve two simultaneous equations to find v and w.

2 ABCD is a parallelogram in which $\overrightarrow{AB} = \mathbf{x}$ and $\overrightarrow{BC} = \mathbf{y}$.
AE : ED = 1 : 2.

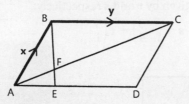

a Express in terms of $\mathbf{x}$ and $\mathbf{y}$, $\overrightarrow{AC}$ and $\overrightarrow{BE}$.

b AC and BE intersect at F, such that $\overrightarrow{BF} = v\overrightarrow{BE}$.

   (i) Express $\overrightarrow{BF}$ in terms of $\mathbf{x}$, $\mathbf{y}$ and v.

   (ii) Show that $\overrightarrow{AF} = (1 - v)\mathbf{x} + \frac{1}{3}v\mathbf{y}$.

   (iii) Use this expression for $\overrightarrow{AF}$ to find the value of v.

3 OPQ is a triangle. OR : RP = 2 : 1 and S is the mid-point of OQ.
M is the mid-point of PQ. If $\overrightarrow{OR} = \mathbf{r}$ and $\overrightarrow{OS} = \mathbf{s}$, express
the following in terms of $\mathbf{r}$ and $\mathbf{s}$

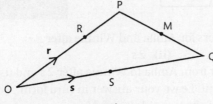

a $\overrightarrow{RS}$ **b** $\overrightarrow{OP}$

c $\overrightarrow{PQ}$ **d** $\overrightarrow{OM}$

4 OPQ is a triangle such that $\overrightarrow{OP} = \mathbf{p}$ and $\overrightarrow{OQ} = \mathbf{q}$. PR : RQ = 1 : 2

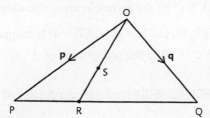

a Express in terms of vectors $\mathbf{p}$ and $\mathbf{q}$

   (i) $\overrightarrow{PQ}$ (ii) $\overrightarrow{PR}$ (iii) $\overrightarrow{OR}$.

b If $\overrightarrow{OS} = k\overrightarrow{OR}$, where k is a constant and OS : SR = 3 : 2, find

   (i) k (ii) $\overrightarrow{OS}$ in terms of $\mathbf{p}$ and $\mathbf{q}$.

5 The position vector $\mathbf{r}$ of a comet with respect to an origin O, t seconds after being detected
is given by $\mathbf{r} = \begin{pmatrix} 1 \\ 5 \end{pmatrix} + t\begin{pmatrix} 2 \\ -1 \end{pmatrix}$, where the units are in km.

a Copy and complete the table for position vector $\mathbf{r}$.

t	0	1	2	3	4	5
$\mathbf{r}$	$\begin{pmatrix} 1 \\ 5 \end{pmatrix}$	$\begin{pmatrix} 3 \\ 4 \end{pmatrix}$				

b Plot the path of the comet for $0 \leqslant t \leqslant 5$.

c Calculate the speed of the comet in km/h and its bearing.

6 Amila and Winnie are playing basketball. During the game, their position vectors on the court are defined relative to the axes on the diagram. At time t seconds after the whistle, their position vectors are given by **r** and **s** respectively:

$$\mathbf{r} = \begin{pmatrix} 2 \\ -1 \end{pmatrix} + t \begin{pmatrix} 1 \\ 2 \end{pmatrix} \qquad \mathbf{s} = \begin{pmatrix} -3 \\ 4 \end{pmatrix} + t \begin{pmatrix} 3 \\ 1 \end{pmatrix}$$

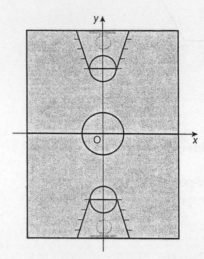

a Find the position vectors for Amila and Winnie after

  (i) 1 s **(ii)** 2 s

b Write down the vector from Amila to Winnie after 2 s and use it to find how far apart they are at this moment. Leave your answer in surd form.

c Calculate the speeds of the two girls in surd form.

7 a Transform the triangle A(1, 0), B(−1, 2), C(−1, 1) by the matrix $\begin{pmatrix} 4 & 3 \\ 3 & 2 \end{pmatrix}$ and show on a sketch.

b Find the inverse matrix.

c Transform the image A′B′C′ by the inverse matrix, showing the result on your sketch.

8 a The shape A(3, −2), B(1, 0), C(21, −16), D(7, −4) is mapped by the matrix
$\mathbf{M} = \begin{pmatrix} -5 & 3 \\ 4 & -2 \end{pmatrix}$ to A′B′C′D′. Find the coordinates of A′, B′, C′ and D′, and sketch the results.

b The matrix **M** maps EFGH to ABCD. Find the coordinates of E′, F′, G′ and H′.

9 $\mathbf{G} = \begin{pmatrix} 2 & 2 \\ 0 & 2 \end{pmatrix}, \mathbf{H} = \begin{pmatrix} 1 & 0 \\ 1 & 1 \end{pmatrix}$

a Find **GH** and **HG**.

b Transform the triangle A(−4, 2), B(−1, 6), C(3, −5) by **GH** and show the result on a sketch.

c The triangle FGH is transformed by **HG** to triangle ABC. Find the coordinates of F, G and H.

10 $\mathbf{Q} = \begin{pmatrix} 5 & -2 \\ -2 & 2 \end{pmatrix}$ and $\mathbf{M} = \begin{pmatrix} 2 & 3 \\ -1 & -1 \end{pmatrix}$

a Find the matrix representing transformation **Q** followed by **M**.

b Find the image of A(−8, 3), B(0, −3), C(8, −3) after transformation by **Q** followed by M

c Triangle EFG is mapped by **Q** followed by **M** onto ABC. Find the coordinates of E, F and C

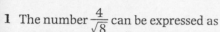

1 The number $\frac{4}{\sqrt{8}}$ can be expressed as

A $\frac{1}{2}$ B 2 C $\frac{1}{\sqrt{2}}$ D $\sqrt{2}$

2 The expression $\frac{x^2 + 2x}{x + 2}$ can be simplified to

A $\frac{x+2}{2}$ B $x + 1$ C x D $\frac{x^2+2}{2}$

3 The value of $\frac{2x}{3} - \frac{x-1}{2}$ expressed as a single fraction is

A $\frac{7x+3}{6}$ B $\frac{x-3}{6}$ C $\frac{x+3}{6}$ D $\frac{x+1}{6}$

4 The displacement x m of an Olympic sprinter t s after the starting pistol is fired is given by $x = 5t^2 + 3t$, valid for $0 \le t \le 1$. His acceleration at $t = \frac{1}{2}$ is:

A 8 B 10 C 5.5 D 13

5 If $x = \begin{pmatrix} 1 \\ 2 \end{pmatrix}$, $y = \begin{pmatrix} -1 \\ 3 \end{pmatrix}$ and $k(x - 2y) = \begin{pmatrix} 15 \\ -20 \end{pmatrix}$, the value of constant k is

A -5 B 5 C -1 D 1

6 The matrix $\begin{pmatrix} 0 & 1 \\ -1 & 0 \end{pmatrix}$ produces which transformation?

A Reflection in the x-axis. B Reflection in the y-axis.
C Rotation of 90° anticlockwise about O. D Rotation of 90° clockwise about O.

7 The area of the shaded segment to 3 sig.figs is

A 14.3 B 9.06
C 27.4 D 34.2

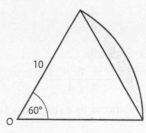

8 The gradient of the curve $y = \frac{1}{x}$ at $x = 2$ is

A $-\frac{1}{4}$ B $\frac{1}{4}$ C 1 D -1

9 The perimeter of the rectangle shown is

A $10\sqrt{5}$ B $6\sqrt{5}$
C 10 D $2\sqrt{5}(\sqrt{3} + 2)$

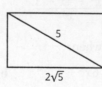

10 A spider is at the top corner of a cube of side π cm. It crawls at π cm/s. What proportion of the cube's surface area can the spider not reach in 1 s?

A $\frac{\pi}{6}$ B $\frac{\pi}{8}$ C $\frac{7\pi}{8}$ D $\frac{5\pi}{6}$

439

1 Which of the following are rational?
   a $0.\dot{8}$ **b** $\sqrt{17}$ **c** $\sqrt{225}$ **d** 2π **e** $\frac{3}{13}$

2 Find a rational number
   a between $\sqrt{5}$ and $\sqrt{7}$ **b** between $\sqrt{7}$ and 3.

3 Find an irrational number
   a between 4 and 5   **b** between 4.5 and 5.

4 Square the following.
   a $\sqrt{6}$ **b** $\sqrt{a}$ **c** $3\sqrt{2}$ **d** $\sqrt{2} \times \sqrt{3}$ **e** $\sqrt{2} \div \sqrt{3}$

5 Write the following as square roots of a single integer.
   a $3\sqrt{3}$ **b** $2\sqrt{7}$ **c** $4\sqrt{5}$ **d** $\sqrt{2} + 2\sqrt{2}$

6 Simplify the following as far as possible.
   a $8\sqrt{5} - 4\sqrt{5}$ **b** $8\sqrt{5} + 4\sqrt{5}$ **c** $8\sqrt{5} \times 4\sqrt{5}$ **d** $8\sqrt{5} \div 4\sqrt{5}$

7 An equilateral triangle has a perimeter of p cm. Its area is $100\sqrt{3}$ cm².
   Find the value of p.

8 Express $\dfrac{2}{\sqrt{8}}$ in the form $a\sqrt{2}$.

9 Find the exact area of the triangle shown, expressing
   your answer in the form $a\sqrt{3}$.

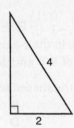

10 Simplify:
   a $\dfrac{x^2 + 4x + 3}{x + 1}$ **b** $\dfrac{x^2 + 4x + 4}{x^2 - 4}$ **c** $\dfrac{x^2 + 5x - 6}{x^2 + 8x + 12}$

11 Express as a single fraction:
   a $\dfrac{2}{x} - \dfrac{3}{2x}$ **b** $\dfrac{1}{x - 2} - \dfrac{1}{x + 3}$ **c** $\dfrac{x}{x^2 - 3x - 4} - \dfrac{1}{x - 4}$

12 Simplify
   a $\dfrac{6x}{3x + 9} \times \dfrac{x^2 + 4x + 3}{x^2}$ **b** $\dfrac{x^2 + x - 6}{x + 1} \div \dfrac{x + 3}{2x^2 + x - 1}$

13 Work out $\dfrac{dy}{dx}$ when:
   a $y = 2x^3 - 6x^2 + 4$ **b** $y = (x^2 + 1)(2x - 1)$ **c** $y = x^2 - x + \dfrac{1}{x^2}$

14 The diagram shows a sketch of the curve
   $y = \dfrac{x^3}{3} + x^2 - 7x - 5$.
   a Work out $\dfrac{dy}{dx}$.
   b Hence find the points on the curve
   where the gradient is parallel
   to the line $y = x$.

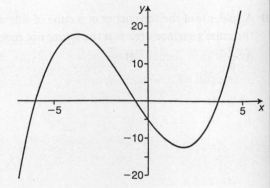

15 The number of hairs H, on the body of a gorilla over a lifetime of 60 years is modelled by the equation $H = 80t^2 - \dfrac{4t^3}{3}$ for $0 \leqslant t \leqslant 60$ where t is time in years.

   a Find $\dfrac{dH}{dt}$

   b Hence calculate

      (i) the rate of hair gain at 20 years

      (ii) the rate of hair loss at 50 years.

   c Calculate the age at which the gorilla has most hair.

16 A serving machine fires a tennis ball vertically into the air. The ball's height, h metres, after t seconds, is given by $h = 40t - 5t^2$ for $0 \leqslant t \leqslant 25$.

   a Work out an expression for the velocity, v, after t seconds.

   b Hence find the time when $v = 0$.

   c Work out the maximum height reached by the ball.

17 A ski-slope is shown.

Calculate

   a the angle that BF makes with the base ADEF

   b the length DF

   c the angle CF makes with the base ADEF

   d the speed of a skier in m/s travelling from M, the mid-point of BC directly to F in 30 seconds.

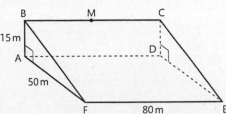

18 The diagram shows the segment of a circle. The segment has an area of 12 cm^2.

   a Calculate the radius of the circle.

   b Calculate the perimeter of the segment.

19 ABCD is a trapezium in which $\overrightarrow{AB} = \mathbf{b}$, $\overrightarrow{BC} = \mathbf{a}$ and $\overrightarrow{AD} = 2\mathbf{a}$. The points P, Q, R and S are the mid-points of the sides AB, BC, CD and AD as shown.

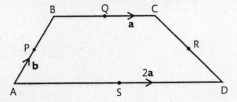

   a Find and simplify the vectors

      (i) $\overrightarrow{AS}$ **(ii)** $\overrightarrow{PQ}$ **(iii)** $\overrightarrow{CD}$ **(iv)** $\overrightarrow{SR}$

   b What can be deduced about PQ and SR?

   c What can be deduced about the quadrilateral PQRS?

20 In the diagram

$\overrightarrow{AX} = \mathbf{a}$, $\overrightarrow{AC} = \mu\mathbf{a}$

$\overrightarrow{MX} = \mathbf{b}$, $\overrightarrow{MB} = \lambda\mathbf{b}$

and M is the mid point of AD.

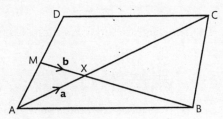

   a Find $\overrightarrow{AM}$

   b **(i)** Find $\overrightarrow{XC}$ in terms of μ and $\mathbf{a}$

      (ii) Find $\overrightarrow{XB}$ in terms of λ and $\mathbf{b}$

   c **(i)** Find $\overrightarrow{AB}$ in terms of λ, $\mathbf{a}$ and $\mathbf{b}$ **(ii)** Find $\overrightarrow{DC}$ in terms of μ, $\mathbf{a}$ and $\mathbf{b}$

$\overrightarrow{AB} = \overrightarrow{DC}$

   d Find μ and λ

21 a Draw the x- and y-axes from $-6 \leqslant x \leqslant 12$ and $-12 \leqslant y \leqslant 12$ and draw the triangle P with vertices (0, 0), (0, 5) and (5, 10).
The matrices **A** and **B** are defined as:

$$\mathbf{A} = \begin{pmatrix} 0.8 & 0.6 \\ 0.6 & -0.8 \end{pmatrix}, \mathbf{B} = \begin{pmatrix} 0 & 1 \\ -1 & 0 \end{pmatrix}$$

b If **AP** = Q. Draw and label triangle Q.

c If matrix **B** represents a rotation, state the centre and angle of the rotation.

d If **BQ** = R. Draw and label triangle R and state what single transformation maps triangle P onto triangle R.

e Simplify the expression $\mathbf{BA}\begin{pmatrix} 4 \\ -2 \end{pmatrix}$ and explain the result.

Consolidation exercise 1

1 Work out these.
   a $2\frac{1}{3} - 1\frac{3}{4}$ b $2\frac{1}{3} \div 1\frac{3}{4}$ c $4 + 2 \times 10$ d $12^0 \div 12^{-1}$

2 Work out these.
   a $4\frac{1}{5} \div \frac{7}{15}$ b $8\frac{3}{7} - 6\frac{4}{5}$

3 Place the following numbers in ascending order.
   14.3, 14.25, 14.532, 14.235

4 Express 945 as a product of prime factors, using indices where necessary.

5 a Express 504 as a product of prime factors, using indices where necessary.
   b Find the LCM and the HCF of 30 and 21.

6 a Express 24 as the product of powers of its prime factors.
   b Express 90 as the product of powers of its prime factors.
   c Find the highest common factor of 24 and 90.
   d Find the lowest common multiple of 24 and 90.

7 a Change $0.3\overset{..}{4}$ to a fraction
   b Change $\frac{3}{8}$ to a decimal.
   c Find 2.8% of $40.
   d What is the reciprocal of 0.5?

8 Change $0.2\overset{..}{1}$ to a fraction in its lowest terms.

9 The surface area of the Earth is 510 million km².
   The surface area of the Pacific Ocean is 180 million km².
   a Express 180 million as a percentage of 510 million
      Give your answer correct to 2 significant figures.

   The surface area of the Arctic Ocean is 14 million km².
   The surface area of the Southern Ocean is 35 million km².
   b Find the ratio of the surface area of the Arctic Ocean to the surface area of the
      Southern Ocean.
      Give your ratio in the form $1 : n$.

10 A mobile phone company makes a special offer.
   Usually one minute of call time costs 5 cents.
   For the special offer, this call time is increased by 20%.
   a Calculate the call time which costs 5 cents during this special offer.
      Give your answer in seconds.
   b Calculate the cost per minute for the special offer.
   c Calculate the percentage decrease in the cost per minute for the special offer.

11 Use standard form to work out an estimate of these, giving your answers in standard form
   correct to 2 significant figures.
   a $(7.5 \times 10^6) \times (7.5 \times 10^{-4})$ b $(7.5 \times 10^6) \div (7.5 \times 10^{-4})$
   c $(7.5 \times 10^{-5}) + (7.5 \times 10^{-6})$ d $\sqrt{(7.5 \times 10^{-5})}$

12 Express the following numbers to the number of decimal places indicated.
   a 37.6245 (3 d.p.) b 37.6235 (2 d.p.)
   c 1.3975 (3 d.p.) d 1.3999 (2 d.p.)

13 Express the following numbers to the number of significant figures indicated.

a 52.35 (3 s.f.)　　　　　　　　　　　**b** 0.5713 (2 s.f.)

c 38 513 (3 s.f.)　　　　　　　　　　　**d** 0.002 627 (2 s.f.)

14 Use standard form to work out the following, and give your answer in standard form correct to 3 significant figures.

a $(7.3 \times 10^7) \times (9.1 \times 10^{-4})$　　　　**b** $(4.5 \times 10^6) \div (8.7 \times 10^4)$

15 Work out these correct to 3 significant figures.

a $\dfrac{12.8}{4.5 \times 8.1}$

b $\sqrt{7.8^2 + 3.7^2 - 2 \times 7.8 \times 3.7 \times \cos 47°}$

c $10^4 \times 9^{-4}$

16 Here is a sequence of blue tiles surrounding white tiles.

a Copy and complete this table.

Number of white tiles (w)	1	2	3	4	5
Number of blue tiles (b)	8				

b Find a formula giving b in terms of w.

c How many blue tiles are needed to surround the pattern with 25 white tiles?

d A pattern has 126 blue tiles. How many white tiles are there?

17 Garden trellis is made from strips of wood joined together with bolts. It comes in different sizes as shown in the diagrams. (Black dots represent the bolts.)

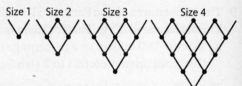

Size 1　Size 2　Size 3　Size 4

a Copy and complete this table.

Size	1	2	3	4	5	6
No. of bolts	1	4				

b Explain how the table could be extended.

c What size of trellis uses 89 bolts?

18 Calculate the following as a power of 2

a $2^3 \times 2^5 \div 2^2$　　　　　　　**b** $(2^3)^7$

19 Calculate the following as a power of 3

a $3^0 \times 3^5 \div 3^{-1}$　　　　　　**b** $3^{-2} \div 3^{-4}$

c $3^m + 3^m + 3^m$　　　　　　　**d** $1 \div 3^{-10}$

20 Simplify these.

a $2\sqrt{3} + 2\sqrt{3}$　　　　**b** $2\sqrt{3} \div 2\sqrt{3}$　　　　**c** $2\sqrt{3} \times 2\sqrt{3}$

21 The world population in 2000 was 6 billion. By 2050, it is expected to have increased by an average of 0.9% per annum.

a Find the expected population in 2050.

b Find the average percentage increase per annum if, by 2050, the population were to have doubled.

22 Blood comprises 8% of the weight of a human. Amy weighs 40 kg.
Find the weight of her blood.

23 The Great Pyramid of Giza took 20 years to build with 2.4 million blocks of stone, each with an average weight of 2.5 tonnes.

   a Find the average number of blocks laid per day.

   b Find the average weight of blocks laid per day, correct to 2 significant figures.

   c How long would it have taken to build another pyramid, with 1.5 million blocks, with the same number of men?

24 In January 1999, by working out $2^{3021377} - 1$, a computer took 46 hours to find the largest prime number, and it contained 909 526 digits. Rebecca tries to write out this number at one digit per second, with each 10 digits taking up 5 cm of space.

   a Find the time it would take her.

   b Find the length of the number, when written out, in kilometres.

25 The largest sewage works in the world is near Washington DC in the US. It treats 740 million gallons per day. Using 1 gallon ≈ 4.55 litres, find the rate of treatment in

   a gallons per second   **b** m^3 per second

26 In the year 2008, 85 people set a new domino record when 4.3 million dominoes fell in 2 hours. On average, how many dominoes fell per second?

27 Lucy earns \$7.50/h for a 35 hour week. Overtime is paid at 25% above the normal rate. How much does she earn for a 50 hour week?

28 Chi-Ho earns \$8.50 per hour for a 40-hour week. Overtime is paid at $1\frac{1}{3}$ of the normal rate. How much does he earn if he works for 46 hours in a week?

29 The exchange rate between dollars and euros is \$1 = €1.196.

   a How many euros can be bought for \$75?

   b How many dollars can be bought for €85?

30 What is the effect of increasing \$80 by 20%, then decreasing this value by 20%?

Consolidation exercise 1*

1 Find the lowest common denominator of the following fractions, and thus list them in ascending order.

$$\frac{3}{4}, \quad \frac{13}{18}, \quad \frac{7}{9}, \quad \frac{2}{3}$$

2 a Nikos drinks $\frac{2}{3}$ of a litre of orange juice each day.

     How many litres does Nikos drink in 5 days?

     Give your answer as a mixed number.

   b (i) Find the lowest common multiple of 4 and 6.

     (ii) Work out $3\frac{3}{4} + 2\frac{5}{6}$.

        Give your answer as a mixed number. You must show all your working.

3 Mortar is made from cement, lime and sand.

   The ratio of their weights is $2:1:9$

   Work out the weight of cement and the weight of sand in 60 kg of mortar.

4 a Express the numbers 180 and 84 as the product of prime factors.

   b Hence work out the LCM and HCF of 180 and 84.

5 Write each of these expressions in the form a^x and hence solve the equation for a.

   a $\dfrac{a^2}{a^{-2}}$, $16 = \dfrac{a^2}{a^{-2}}$ **b** $\dfrac{1}{a^2} + a^{-2}$, $32 = \dfrac{1}{a^2} + a^{-2}$

6 $p = 3^8$

   a Express $p^{\frac{1}{2}}$ in the form 3^k, where k is an integer.

   $q = 2^9 \times 5^{-6}$

   b Express $q^{-\frac{1}{3}}$ in the form $2^m \times 5^n$, where m and n are integers.

7 Simplify $\dfrac{3\sqrt{10}}{\sqrt{40}}$

8 Write down an irrational number between 3 and 4.

9 For $x = 4.385$ and $y = \sqrt{3}$, calculate the value of $\dfrac{x^2 + y^2}{3x - y}$, and express your answer correct to 3 significant figures.

10 Work out the value in standard form to 2 significant figures of $\dfrac{3.23 \times 10^4}{\sqrt{1.81 \times 10^6}}$.

11 Two fruit drinks, *Fruto* and *Tropico*, are sold in cartons.
a *Fruto* contains only orange and mango.
The ratio of orange to mango is 3:2
A carton of *Fruto* contains a total volume of $250\,cm^3$.
Find the volume of orange in a carton of *Fruto*.
b *Tropico* contains only lemon, lime and grapefruit.
The ratios of lemon to lime to grapefruit are 1:2:5.
The volume of grapefruit in a carton of *Tropico* is $200\,cm^3$.
Find the total volume of *Tropico* in a carton.

12 a Write the number 37 000 000 000 in standard form.
b Write 7.5×10^{-5} as an ordinary number.
c Calculate the value of $\dfrac{2.5 \times 10^{-3}}{1.25 \times 10^7}$ in standard form.

13 The cradle used to raise the wreck of the 15th century ship the *Mary Rose*, and the wreck itself, weighed 15 times more out of the water than in. The ratio of the cradle's weight to the wreck's weight was $1:5$. The cradle and wreck weighed 540 tonnes out of the water. What was the apparent weight of the wreck under the water, without the cradle?

14 A car travels from X to Y at an average speed of 60 km/h and returns to X at a speed of 40 kph. What is its average speed for the whole journey?

15 a According to a recent survey, one-third of boys and a quarter of girls exercise regularly by the age of 15. Find, as a percentage, how many more boys exercise than girls.
b Hamial's normal temperature is 37 °C.
(i) He catches an infection and his temperature rises by 4%.
Find his high temperature.
(ii) After taking antibiotics, his temperature drops by 4%.
Find his final temperature, correct to 2 significant figures.

16 The largest
volcanic eruption in the last 10 000 years was at Thera in 1520 BC. It brought about the end of the Minoan civilisation in Greece.

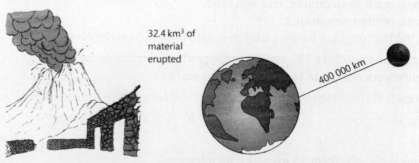

32.4 km³ of material erupted

400 000 km

The volume erupted by Thera equals that of a 400 000 km column that reaches from the Earth to the Moon. If this column has a square cross-section, find its width.

17 At the end of the Ice Age, the water level in the Mediterranean rose by 120 metres in two years. 10 cubic miles of water cut through the Bosphorus into the Black Sea, flooding thousands of square miles.

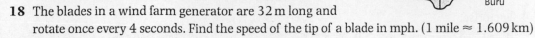

Volume = 10 cubic miles

h

Buru

a In the Mediterranean, what was the average increase in depth, in cm per day?

b Given that 1.609 km = 1 mile, change 10 cubic miles to cubic metres and write your answer in standard form.

c The area of Buru, an island off Borneo, is 9500 km². Find *h*, giving your answer to the nearest cm.

18 The blades in a wind farm generator are 32 m long and rotate once every 4 seconds. Find the speed of the tip of a blade in mph. (1 mile ≈ 1.609 km)

19 The human gut has an area of about two tennis courts and on this live about 100 trillion bacteria. (One trillion is 10^{12}.)

a Taking the area of a tennis court as 260 m², calculate the number of bacteria per cm².

b Assume that each bacterium is in the shape of a cube and that a single layer covers the gut with no gaps. If these cubes were placed, touching each other, in a straight line, how long would the line be?

20 For the sequence 201, 197, 193, 189, ..., find

a the next three terms   **b** the 50th term

c a formula for the *n*th term

21 a Use the difference method to find the next three terms of the sequence 7, 12, 20, 31, 45, ...

b Find the terms *a* and *b* in the sequence ..., *a*, 21, 39, 55, 69, 81, *b*, ...

22 Andy is investigating bees' honeycombs for his biology project. He wants to know the connection between the number of cells and the number of walls for different arrangements of the cells.

He first considers cells in a line.

a Copy and complete this table, and find a formula for *w* in terms of *c*.

He next considers cells arranged in triangles.

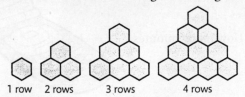

1 row   2 rows   3 rows   4 rows

No. of cells *c*	No. of walls *w*
1	6
2	
3	
4	
5	26

b Copy and complete this table.

No. of rows *r*	1	2	3	4	5
No. of cells *c*	1				15
No. of walls *w*	6			42	
r + *c*	2				

c What formula connects *w* and *r* + *c*?

d Andy's teacher tells him that $c = \frac{1}{2}(r^2 + r)$. Use this to find a formula connecting *w* and *r*.

e How many walls are in a triangular honeycomb with 10 rows?

Consolidation exercise 2

1 Simplify:

  a $x^6 \times x^4$
  b $\dfrac{x^6}{x^4}$
  c $\dfrac{y^4 \times y^5}{y^7}$
  d $(2x^3y^2)^2$

2 Expand and simplify:

  a $(x + 3)(x - 1)$
  b $(2x - 1)(x + 4)$

3 Expand and simplify:

  a $z(z + 3)$
  b $3(x + 4) - 2(2x - 1)$

4 Factorise:

  a $4p - 8$
  b $x^2 + 3x$
  c $3ab^2 + 6a^2b$

5 Factorise:

  a $x^2 - 4$
  b $x^2 + 3x + 2$

6 Simplify:

  a $\dfrac{x}{4} + \dfrac{x}{3}$
  b $\dfrac{x + 1}{2} - \dfrac{x + 1}{3}$
  c $\dfrac{3}{x + 2} + \dfrac{x + 17}{x^2 - x - 6}$

7 a Make u the subject of $m(v - u) = I$.
  b Make r the subject of $\frac{1}{3}\pi r^2 h = 4$.

8 A formula used in mathematics is $v = u + at$.

  a Work out v when $u = 6$, $a = -10$ and $t = 2$.
  b Make a the subject of the formula.

9 A formula used in science is $T = 2\pi\sqrt{\dfrac{l}{g}}$.

  a Work out T when $l = 2.45$ and $g = 9.81$.
  b Make l the subject of the formula.

10 The cost, $\$C$, of using a mobile phone for one month is given by a fixed charge of \$30 plus \$0.15 for every minute spent on calls.

  a One month Ashaf uses her phone for 560 minutes. What is her bill for that month?
  b If Ashaf uses her phone for t minutes, find a formula giving the cost, $\$C$, in terms of t.
  c Make t the subject of your formula.
  d One month Ashaf's bill was \$127.50. How many minutes did she spend on calls that month?

11 Solve:

  a $4x - 7 = 5$
  b $5x - 4 = 2x + 2$
  c $2(x + 3) - 3(x - 3) = 13$

12 Solve:

  a $\dfrac{2 + x}{5} = x - 2$
  b $\dfrac{x + 1}{2} + \dfrac{x - 2}{3} = 4$

13 Tracy thought of a number. She then added 3, multiplied the result by 5 and then subtracted 8.
  Her answer was 42.

  a Let x be the number Tracy first thought of. Form an equation in x.
  b Solve the equation to find the number Tracy first thought of.

14 The angles of a triangle are as shown.
Find the value of a.

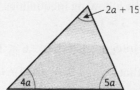

15 A rectangle has sides as shown in the diagram.
Find x and y and the area of the rectangle.

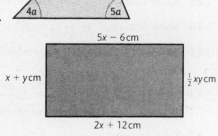

16 y is directly proportional to x. If $y = 12$ when $x = 3$, find
   a the formula for y in terms of x **b** y when $x = 2$ **c** x when $y = 8$

17 y is directly proportional to x^2. If $y = 12$ when $x = 2$, find
   a the formula for y in terms of x **b** y when $x = 4$ **c** x when $y = 27$

18 y is inversely proportional to x. If $y = 8$ when $x = 3$, find
   a the formula for y in terms of x **b** y when $x = 2$ **c** x when $y = 6$

19 y is inversely proportional to $\sqrt{x}$. If $y = 18$ when $x = 4$, find
   a the formula for y in terms of x **b** y when $x = 9$ **c** x when $y = 9$

20 Solve the simultaneous equations $x + y = 8$ and $2x - y = 1$.

21 A farmyard has only horses and ducks in it.
Altogether there are 17 heads and 58 legs in the farmyard.

   a Let x be the number of horses and y be the number of ducks.
      Form two equations involving x and y.

   b Solve the equations to find how many ducks there are.

22 Solve the equations:
   a $x^2 - 16 = 0$ **b** $x^2 - 16x = 0$

23 Solve the equations:
   a $x^2 - 2x - 3 = 0$ **b** $x^2 - 6x + 8 = 0$

24 Solve the equations, giving your anwers to 3 s.f.
   a $x^2 - 4x + 2 = 0$ **b** $2x^2 + 3x - 1 = 0$

25 The length of a rectangle is 4 cm more than the width.
The area of the rectangle is 32 cm².
   a Show that $x^2 + 4x - 32 = 0$.
   b Solve the equation $x^2 + 4x - 32 = 0$.
   c Hence find the perimeter of the rectangle.

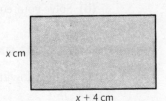

26 The two rectangles shown in the diagram have the same area.
All dimensions are in cm. Find the value of x and the area
of each rectangle.

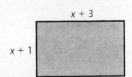

27 Solve the following inequalities, showing the answers on a number line.

   a $4x - 3 > 1$ **b** $2(x + 2) \leqslant 5(x - 1)$

28 n is an integer such that $-6 \leqslant 3n < 6$. Write down the possible values of n.

29 Show that $(x - 1)^2 + (x - 2)^2 = 2x^2 - 6x + 5$

30 Show that $(x - 1)^2 - (x - 2)^2 = 2x - 3$

31 $(x + 3)$ is a factor of $f(x) = x^3 + px^2 - 10x + 24$. Find the value of p and hence factorise $f(x)$ completely.

32 $(x - 1)$ is a factor of $f(x) = 2x^3 + 3x^2 - 3x + q$. Find the value of q, factorise $f(x)$ completely and hence solve $f(x) = 0$.

33 Solve the equations:

   a $1 - \dfrac{x}{2} = \dfrac{x - 17}{4}$ **b** $\dfrac{x + 1}{x + 7} = \dfrac{x - 1}{x + 4}$

34 Rectangular tiles have a width x cm and height $(x + 7)$ cm.

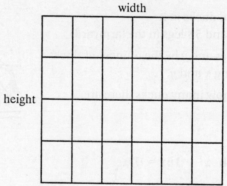

Diagram **NOT** accurately drawn

Some of these tiles are used to form a shape.
The shape is 6 tiles wide and 4 tiles high.

Diagram **NOT** accurately drawn

   a Write down expressions, in terms of x, for the width and height of this shape.

   b The width and height of this shape are equal.

      (i) Write down an equation in x.

      (ii) Solve your equation to find an equation in x.

Consolidation exercise 2*

1 Simplify the following.

   a $\dfrac{a}{4} + \dfrac{2a}{3}$ **b** $\dfrac{a}{4} \div \dfrac{2a}{3}$ **c** $\dfrac{a - b}{a} \times \dfrac{ab}{a - b}$ **d** $\dfrac{x + 2}{5} - \dfrac{x - 3}{3}$

2 Simplify:

   a $\dfrac{x^2 + x}{x}$ **b** $\dfrac{x^2 + x}{x + 1}$

3 a Factorise $x^2 - x - 72$.   **b** Simplify $\dfrac{x^2 - 81}{x^2 - x - 72}$

4 a Factorise $3x^2 + 32x - 11$.   **b** Simplify $\dfrac{3x^2 + 32x - 11}{x^2 - 121}$

5 A formula used in engineering is $l = \frac{1}{3}M(a^2 + b^2)$.

  a Find l to 3 s.f. when $M = 2$, $a = 3.5$ and $b = 5.4$.

  b Make a the subject of the formula.

6 A formula used in mathematics is $S = \frac{n}{2}(2a + (n-1)d)$.

  a Calculate S when $n = 54$, $a = 3$ and $d = 2$.   **b** Make d the subject of the formula.

7 A formula used in science is $\frac{1}{f} = \frac{1}{u} + \frac{1}{v}$.

  a Calculate f when $u = 3$ and $v = 6$.   **b** Make u the subject of the formula.

  c Calculate u when $f = 4$ and $v = 6$.

8 Solve the equation $\frac{x+5}{15} - \frac{x-5}{10} = 1 + \frac{2x}{15}$.

9 When the same number, x, is added to the top and bottom of the fraction $\frac{123}{456}$ the answer is $\frac{1}{2}$. Form an equation in x and solve it to find the number.

10 A triangle has two angles as shown in the diagram.

  a Find the third angle in terms of x.

    There are three different isosceles triangles with these angles.

  b Find the three different values of x which make the triangle isosceles.

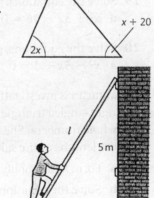

11 A ladder leaning against a vertical wall just reaches a window 5 m above the ground. Let x be the distance of the foot of the ladder from the wall and let l be the length of the ladder.

  a Write down an equation connecting x and l.

  b Chris thinks the ladder is unsafe, so she pulls the base of the ladder out a further 2 metres. The ladder now only reaches 4 m up the wall. Write down another equation involving x and l.

  c Solve your equations to find the length of the ladder.

12 The speed of a skier skiing down a mountain, v m/s, is proportional to the square root of the distance, d m, she has moved from her starting point.

  Given that $v = 20$ when $d = 100$, find

  a the formula for v in terms of d

  b the speed of the skier when she is 49 m from her starting point

  c the distance travelled when the skier's speed is 10 m/s

13 y is inversely proportional to the square of x. Given that $y = 10$ when $x = 2$, find

  a the formula for y in terms of x

  b the value of y when $x = \frac{1}{4}$

  c the value of x when $y = \frac{1}{4}$

14 The temperature of a drink, $T°C$, is inversely proportional to the square root of the time, m minutes, after it has been poured (for $m \geqslant 2$).

  After 4 minutes the temperature is 60 °C.

  a Find the formula for T in terms of m.

  b Find the temperature after 9 minutes.

  c Find how long it takes for the drink to cool to 50 °C.

15 Solve the simultaneous equations $6x - 5y = 7$ and $4x + 3y = 11$.

16 In the triangle shown, AC = CB and AD = AB.

a Find an equation connecting x and y.

Given angle $C = 40°$

b find another equation connecting x and y

c solve these equations to find x and y

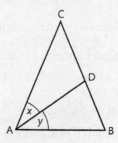

17 Max sold 14 tickets for a concert. He sold x tickets at \$12 each and y tickets at \$18 each. He collected \$204.

a Write down two equations connecting x and y.

b Solve your equations to find how many of each kind of ticket he sold.

18 Solve the equations:

a $x^2 - 100 = 0$ b $x^2 - 100x = 0$

19 Solve the equations:

a $x^2 + 5x - 36 = 0$ b $2x^2 - 14x + 20 = 0$

20 Solve the equations, giving your anwers to 3 s.f.

a $x^2 - 5x + 3 = 0$ b $3x^2 + 3x - 5 = 0$

21 Maurita is investigating whether it is possible to have a right-angled triangle where the sides are three consecutive whole numbers. She lets x, $x + 1$ and $x + 2$ stand for the lengths of the sides.

a Form and simplify an equation for x.

b Solve the equation and show that only one triangle is possible.

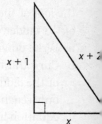

22 The area of the triangle shown is $24\,\text{cm}^2$.

a Show that $x^2 + 6x - 16 = 0$.

b Solve the equation $x^2 + 6x - 16 = 0$.

c Hence find the height of the triangle.

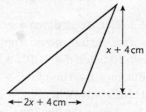

23 The trapezium shown has an area of $21\,\text{cm}^2$.

a Show that $x^2 + 2x - 24 = 0$.

b Solve the equation $x^2 + 2x - 24 = 0$.

c Hence find the length AB.

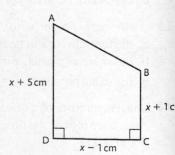

24 n is an integer such that $-4 < 5n + 3 \leqslant 13$. Write down the possible values of n.

25 Solve the inequalities.

a $4 - 2(x + 4) < 10x$ b $\dfrac{22 + x}{4} \geqslant 3x$

26 The sum of any two sides of a triangle must be greater than the third side.

    a Use this to form three inequalities involving x for the triangle shown.

    b Solve these inequalities to find the least and greatest values of x.

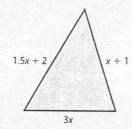

$1.5x + 2$ $x + 1$

$3x$

27 Two chords, AB and CD, of a circle intersect at right angles at X.

AX = 2.8 cm

BX = 1.6 cm

CX = 1.2 cm

Calculate the length AD.

Give your answer correct to 2 significant figures.

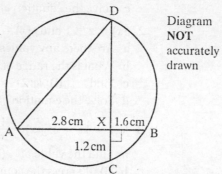

Diagram **NOT** accurately drawn

28 Simplify: $\dfrac{x^2 + 5x + 4}{x^2 - x - 12} \times \dfrac{x^2 + x - 6}{x^2 - x - 2}$.

29 Simplify: $\dfrac{x - 6}{x + 3} \div \dfrac{x^2 - 36}{x^2 + 10x + 21}$.

30 $(x + 2)$ and $(x - 5)$ are factors of $f(x) = x^3 + px^2 + qx + 30$. Find the values of p and q, factorise $f(x)$ completely and hence solve $f(x) = 0$.

31 $(x + 2)$ is a factor of $f(x) = 4x^3 + px^2 - 34x - 8$. Find the value of p, factorise $f(x)$ completely and hence solve $f(x) = 0$.

32 Solve the equations:

    a $\dfrac{2x - 1}{3} - \dfrac{x - 7}{5} = 2$ **b** $\dfrac{1}{x} - \dfrac{1}{x + 4} = \dfrac{1}{x + 1}$

33 The diagram shows one disc with centre A and radius 4 cm and another disc with centre B and radius x cm.

The two discs fit exactly into a rectangular box 10 cm long and 9 cm wide.

The two discs touch at P.

APB is a straight line.

    a Use Pythagoras' Theorem to show that $x^2 - 30x + 45 = 0$

    b Find the value of x.

        Give your value correct to 3 significant figures.

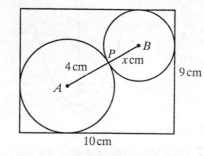

4 cm P B x cm   9 cm

A

10 cm

Diagram **NOT** accurately drawn

Consolidation exercise 3

1 $f(x) = x + 3$ and $g(x) = \dfrac{2}{x}$

   a Find   **(i)** $f(2)$ **(ii)** $fg(2)$ **(iii)** $gf(2)$

   b Which value of x must be excluded from the domain of $gf(x)$?

   c Solve the equation $gf(x) = 1$.

2 $f(x) = x - 1$ and $g(x) = x^2$

   a Are there any values of x that should be excluded from the domain of f?

   b What is the range of   **(i)** f **(ii)** g?

   c Find   **(i)** $fg(2)$ **(ii)** $fg(x)$

   d Solve the equation $gf(x) = f^{-1}(x)$.

3 $f(x) = 3x - 2,\ g(x) = \dfrac{1}{x}$

   a Find the value of   **(i)** $fg(3)$ **(ii)** $gf(3)$

   b **(i)** Express the composite function $fg(x)$ in its simplest form as $fg(x) = \ldots$

      (ii) Which value of x must be excluded from the domain of $fg(x)$?

   c Solve the equation $f(x) = fg(x)$

4 The diagram shows four graphs.

   a $y = 3$

   b $x = 3$

   c $y = \dfrac{1}{2} - 2x$

   d $y = \dfrac{1}{2}x - 2$

   Which graph is which?

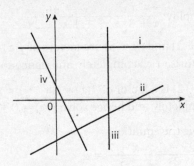

5 Match up the following graphs with their equations.

A   B   C

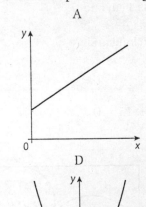
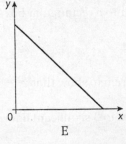
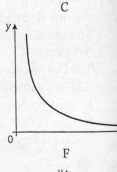

D   E   F

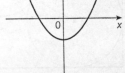
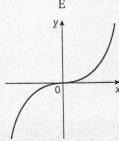
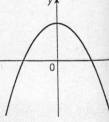

(i) $y = x^2 - 5$ **(ii)** $y = 5 - x$ **(iii)** $y = 5 - x^2$

(iv) $y = x^3$ **(v)** $y = x + 5$ **(vi)** $y = \dfrac{5}{x}$

6 **a** Solve the inequality $4x - 5 > 3$.
b Represent the solution to part **a** on a number line.

7 **a** Draw on one set of axes the graphs of
$x + y = 7$ and $3x - y = 9$ for $0 \leqslant x \leqslant 5$.
b Use these graphs to solve the simultaneous equations $x + y = 7$ and $3x - y = 9$.

8 **a** Draw on one set of axes the graphs of $x + y = 6$, $x = 2$ and $y = 2$.
b By shading the unwanted regions, find the region satisfying $x + y \leqslant 6$, $x > 2$ and $y \geqslant 2$.

9 **a** Find the gradient of the line with equation $3x - 4y = 15$
b Work out the coordinates of the point of intersection of the line with equation
$3x - 4y = 15$ and the line with equation $5x + 6y = 6$

10 **a** Draw the graph of $y = x^2 - 5x + 3$ for $0 \leqslant x \leqslant 5$.
b Use your graph to solve the equation $0 = x^2 - 5x + 3$.

11 **a** Draw an accurate graph of $y = x^2 + x - 1$ for $-3 \leqslant x \leqslant 2$.
b Use your graph to solve the equation $x^2 + x - 1 = 0$.
c Use your graph to solve the equation $x^2 + x = 2$.
d Use your graph to find the smallest value of $f(x) = x^2 + x - 1$.

12 If the graph of $y = x^2 + 2x - 1$ has been drawn, what is the equation of the line that should be drawn to solve these?
a $x^2 + 2x = 2$ **b** $x^2 + x = 3$

13 This diagram shows the graphs of
$y = x^2 - 3x + 1$ and $y = x - 1$.
a What equation in x is solved by the intersection points of these graphs?
b Use the diagram to solve the equation in part **a**.

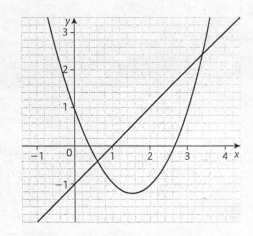

14 A relationship is given by ordered pairs. Find x and y and describe the relationship in words.
a $(0, 0), (1, 2), (x, 4), (3, y), (4, 8)$
b $(0, 0), (x, 1), (2, y), (3, 9), (4, 16)$

15 Draw a mapping diagram for $x \rightarrow 2x + 1$ with domain $\{-1, 0, 1, 2\}$. What is the range?

16 Draw arrows on four copies of the diagram to show the following relationships.

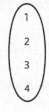

a Many to one   **b** Many to many   **c** One to one   **d** One to many

17 Complete the following mapping diagram.

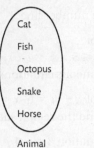

Cat
Fish
Octopus
Snake
Horse

Animal   Number of limbs

Is this mapping a function? Give a reason. Would your conclusions change if the animals changed?

18 Which graph shows a function? Give a reason.

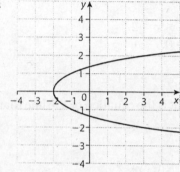

a   b

19 The volume $V(m^3)$ of a hot-air balloon of radius r (m) is given by
$V = \frac{4}{3}\pi r^3$, valid for $2 \leqslant r \leqslant 10$.

a Copy and complete the table to 3 s.f:

r (m)	2	4	6	8	10
V (m³)	33.5				

b Use this table to draw the graph of V (m³) against r (m) for $2 \leqslant r \leqslant 10$.

c Show clearly how you use your graph to estimate.
   (i) V when $r = 7$ m   (ii) r when $V = 500$ m³

20 a Copy and complete the table for the graph $y = \frac{4}{x}$.

x	1	2	3	4	5
y	4				

b Draw the graph of $y = \frac{4}{x}$ for $1 \leqslant x \leqslant 5$.

21 The annual profit P ($\$ \times 10^4$) made by a new sports magazine after t years is given by
$P = \frac{10}{t}$, valid for $2 \leqslant t \leqslant 12$.

a Copy and complete the table.

t (years)	2	4	6	8	10	12
P ($\times 10^4$)	5					0.83

b Use this table to draw the graph of $P($ $\$ \times 10^4$) against t yrs for $2 \leqslant t \leqslant 12$.

c Show clearly how you use your graph to estimate.
   (i) P when $t = 3$ years
   (ii) t when $P = \$14\,000$

2 A cheetah accelerates from rest to 10 m/s for 1 second followed by a further acceleration to 20 m/s in 2 seconds. It then remains at that speed for 30 seconds before retarding at x m/s² to rest.

  a Sketch the speed–time graph for the cheetah's journey.
  b Given that the total distance the cheetah covered was 685 m, find x.
  c Use your speed–time graph to find the cheetah's
      (i) speed after 1 second.
      (ii) acceleration between 1 and 3 seconds.
      (iii) mean speed for the whole journey.

3 Oscar Pistorius held the men's Paralympic 400 m record in 2008 in a time of 46.25 seconds.
  a Calculate his average speed in
      (i) m/s   **(ii)** km/h.
  b Sketch the speed–time graph of Oscar's race assuming he accelerated uniformly for the first 4 seconds and then ran at constant speed.
  c Calculate his maximum speed in m/s.
  d Calculate his initial acceleration in m/s².

4 Part of the graph of $y = x^2 - 2x - 4$ is shown on the grid.

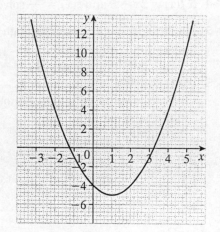

  a Write down the coordinates of the minimum point of the curve.
  b Use the graph to find estimates of the solutions to the equation $x^2 - 2x - 4 = 0$.
    Give your answers correct to 1 decimal place.
  c Draw a suitable straight line on the grid to find estimates of the solutions of the equation
    $x^2 - 3x - 6 = 0$
  d For $y = x^2 - 2x - 4$
      (i) Find $\dfrac{dy}{dx}$
      (ii) find the gradient of the curve at the point where $x = 6$

5 A farmer wants to make a rectangular pen for keeping sheep.

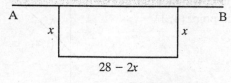

Diagram **NOT** accurately drawn

He uses a wall, AB, for one side. For the other three sides, he uses 28 m of fencing.
He wants to make the area of the pen as large as possible.
The width of the pen is x metres.
The length parallel to the wall is $(28 - 2x)$ metres.
  a The area of the pen is y m². Show that $y = 28x - 2x^2$.
  b For $y = 28x - 2x^2$
      (i) Find $\dfrac{dy}{dx}$
      (ii) Find the value of x for which y is a maximum.
      (iii) Explain how you know that this value gives a maximum.
  c Find the largest possible area of the pen.

26 A particle moves along a line.

For $t \geq 1$, the distance of the particle from O at time t seconds is x metres, where

$$x = 3t - t^2 + \frac{8}{t}$$

a Find an expression for the velocity of the particle.

b Find an expression for the acceleration of the particle.

Consolidation exercise 3*

1 If $f(x) = 2x - 1$ and $g(x) = \frac{1}{2}x$ find

    a $fg(10)$ **b** $gf(10)$ **c** $f^{-1}(x)$ **d** $g^{-1}(x)$

2 If $f(x) = x + 10$ and $g(x) = 10^x$ find

    a $fg(1)$ **b** $gf(1)$ **c** $fg(x)$ **d** $gf(x)$

3 If $f(x) = \dfrac{x+1}{2}$ and $g(x) = \dfrac{6}{x}$ solve for x $f(x) = g(x)$.

4 If $f(x) = 2x - 5$ and $g(x) = \dfrac{25}{x}$ solve for x $f^{-1}(x) = g^{-1}(x)$.

5 The straight line, **L**, passes through the
points $(0, -1)$ and $(2, 3)$.

    a Work out the gradient of **L**.

    b Write down the equation of **L**.

    c Write down the equation of another
line that is parallel to **L**.

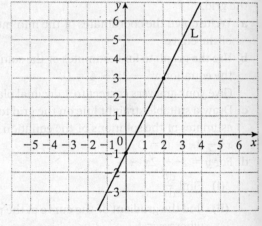

6 a The equation of line L_1 is $y = mx + c$.
State what m and c represent.

    b Find the equation of line L_2 which is parallel
to L_1 and passes through point $(4, 1)$.

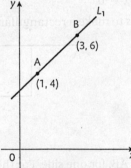

7 Lines L_1 and L_2 intersect at $P(1, 4)$.
The coordinates of P are the solutions
to which two simultaneous equations?

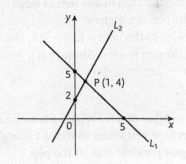

8 The position of a pebble thrown from the top of a cliff, relative to the axes shown, is given by $y = 3x - 0.2x^2$, all units being in metres.

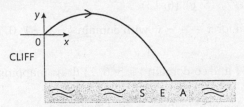

CLIFF

~ S E A ~

a Copy and complete this table and use the values to draw the graph for $0 \leqslant x \leqslant 20$.

x	0	4	8	12	16	20
y	0		11.2		−3.2	

b Use your graph to find the maximum height of the pebble above the sea, given that the cliff is 20 m high.

c What is the horizontal distance from the cliff when the pebble is level with the cliff-top?

9 a Draw an accurate graph of $y = 4x - 1 - x^2$ for $-1 \leqslant x \leqslant 5$.

b Use your graph to solve the equation $4x - x^2 = -1$.

c Use your graph to solve the equation $5x - x^2 = 3$.

10 If the graph of $y = 4 + 2x - x^2$ has been drawn, what is the equation of the line that should be drawn to solve these?

a $1 + 2x - x^2 = 0$ **b** $x^2 - 4x - 2 = 0$

11 a Complete the table of values for $y = x^2 - \dfrac{3}{x}$

x	0.5	1	1.5	2	3	4	5
y	−5.75	−2					24.4

b On a copy of the grid, draw the graph of
$y = x^2 - \dfrac{3}{x}$ for $0.5 \leqslant x \leqslant 5$

c Use your graph to find an estimate for a solution of the equation
$x^2 - \dfrac{3}{x} = 0$

d Draw a suitable straight line on your graph to find an estimate for a solution of the equation
$x^2 - 2x - \dfrac{3}{x} = 0$

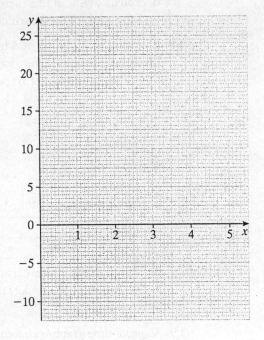

12 The function f is defined as $f(x) = \dfrac{x}{x-1}$.

a Find the value of

    (i) $f(3)$,   **(ii)** $f(-3)$.

b State which value(s) of x must be excluded from the domain of f.

c **(i)** Find $ff(x)$.

    Give your answer in its simplest form.

    (ii) What does your answer to **c (i)** show about the function f?

13 A relationship is given by ordered pairs. Find x and y and describe the relationship in words.

   a $(-2, -8), (x, -1), (0, 0), (2, y), (3, 27)$

   b $(1, 2), (2, 3), (x, 5), (4, 7), (5, y), (6, 13)$

14 Draw a mapping diagram for $x \to 4 - x^2$ with domain $\{-2, -1, 0, 1, 2\}$. What is the range?

15 Using domain $\{0, 1, 2, 3\}$ and co-domain $\{4, 5, 6, 7\}$ draw mapping diagrams to show the following relationships.

   a Many to one   **b** Many to many   **c** One to one   **d** One to many

16 Complete the following mapping diagram. Is this mapping a function? Give a reason. Would your conclusions change if the solids were different?

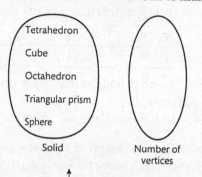

17 Jamal runs a 100 m race in 14.2 s. This is the speed–time graph for Jamal's race.

   a Find the average speed for the race.

   b What is the greatest speed reached by him in the race?

   c Find his acceleration over the first four seconds.

   d Sidd also runs in this race, accelerating constantly from the start for 5 s until he reaches a top speed of 8 m/s, which he maintains until the finish. Does Sidd beat Jamal?

18 Joyce is standing at point A in a field and observes a hare at B, running in a straight line towards a bush at C. The observation angle $x = 60°$ when the hare is first observed, and this changes at a rate of $6°$ per second as the hare runs towards the bush.

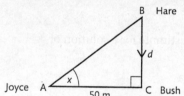

   a If d is the distance of the hare from the bush, copy and complete this table.

t (s)	0	2	4	6	8	10
d (m)	86.6				10.6	

   b Draw a graph of d(m) against time (s) and comment on it.

   c Use the graph to *estimate* the gradient of the graph at $t = 6$ s and state what this represents.

19 A speed–time graph is shown for a journey of a car between two sets of traffic lights. Find:

   a the acceleration over the first 20 s

   b the retardation over the final 10 s

   c the mean speed for the whole journey.

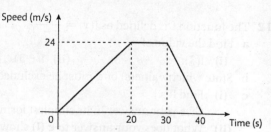

20 The velocity of a firework v m/s, t after launch is given by: $v = 10 + 8t - 2t^2$.
Show by completing the square that the maximum velocity is 18 m/s which occurs
at $t = 2$.

21 The population N of Nightingales in some medieval forests in England t years after the
year 2000 is given by $N = \dfrac{k}{t^2}$, where k is a constant and the formula is valid for $1 \leqslant t \leqslant 10$.
a Use the table below to find the value of the constant k and use the formula to copy and
complete the table:

t (years)	1	2	4	6	8	10
N	1000					10

b Use this table to draw the graph of N against t years for $1 \leqslant t \leqslant 10$.
c Conservationists decide to investigate the decline in numbers when $N = 40$.
Use your graph to estimate in what year this takes place.

22 a For the equation $y = 5000x - 625x^2$, find $\dfrac{dy}{dx}$.

b Find the coordinates of the turning point on the graph of $y = 5000x - 625x^2$.
c **(i)** State whether this turning point is a maximum or a minimum.
   (ii) Give a reason for your answer.
d A publisher has set the price for a new book.
The profit £y, depends on the price of the book, £x, where

$$y = 5000x - 625x^2$$

   (i) What price would you advise the publisher to set for the book?
   (ii) Give a reason for your answer.

Consolidation exercise 4

1 Coventry is 150 km from London on a bearing of 315°. Felixstowe is 120 km from London on a bearing of 060°.

   a Using a scale of 1 cm : 10 km, construct, using ruler and compasses only, a triangle showing the positions of London, Coventry and Felixstowe.

   b From your drawing, estimate the distance between Coventry and Felixstowe.

2 A point P has coordinates (3, 5). State the new position of P after each of these.

   a Reflection in the x-axis   **b** Rotation about O, 90° clockwise

   c Translation along $\begin{pmatrix} -4 \\ 4 \end{pmatrix}$ **d** Reflection in the line $x = 5$

3 Using a scale of 1 cm to 1 unit, draw rectangular axes, labelling both axes from −6 to 6. On your diagram, draw and label the triangle P with vertices (2, 2), (2, 4) and (3, 4).

   a Rotate P −90° about the point (−1, 2). Label the image A.

   b Translate P by the vector $\begin{pmatrix} -6 \\ 1 \end{pmatrix}$. Label the image B.

   c Describe fully the single transformation that maps A to B.

4 A school, S, is 10 km due North of the church, C. A village, V, is 5 km due West of C. Calculate these bearings.

   a S from V   **b** V from S

5 Find lengths of sides p and q and sizes of angles r and s in these triangles.

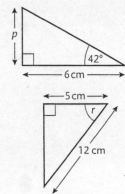
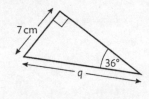
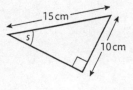

6 Find the sides x and y.

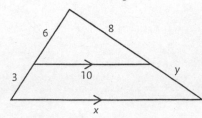

7 A triangle has two angles as shown in this diagram.

   a Find the third angle in terms of x.

   b There are three different isosceles triangles with these angles. Write down three equations giving these angles.

   c Solve the equations to find the three different values of x which make the triangle isosceles.

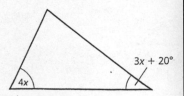

8 a For a regular octagon, work out the size of the exterior angle.

  b Calculate the sum of the interior angles of a regular octagon.

  c Explain why regular octagons will *not* tessellate (meet around a point without any gaps or overlaps).

9 A cylindrical can of beans has a height of 15 cm and base radius of 4 cm. It is put into a cylindrical saucepan of radius 10 cm already fairly full with water, and as a result it is completely covered.

  a By how much does the water level rise? (Assume the saucepan does not overflow)

  b If it was stood on its end, and the water was originally 7 cm deep, by how much would the water level rise now?

10 This figure shows a frustum with a circular base.

  a Use similar triangles to show that $OT = h$.

  b Show that the volume V of the frustum is $V = \dfrac{7\pi r^2 h}{3}$

  c Write the formula in part **b** in the form $V = kr^2 h$, where k is correct to 4 significant figures.

  d The volume of another frustum, where $r = h$, is 35 cm³. Find the value of r.

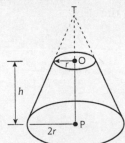

11 Triangle A is similar to triangle B.

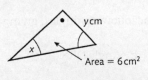

Triangle A

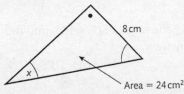

Triangle B

Find the value of y.

12 A floating toy is in the shape of a circular pyramid of height 20 cm. The top section protruding above the water is of height 4 cm and volume 50 cm³, Find the volume of the cone beneath the water level.

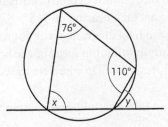

13 Find angles x and y, giving reasons for each step of your calculation.

  a

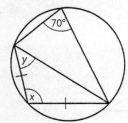

  b

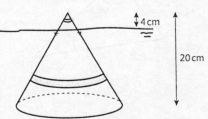

  c

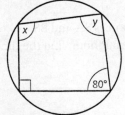

  d

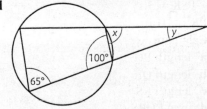

14 **a** Find the values of x, y and z.
   b Find a triangle similar to
   △ABF.

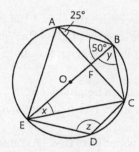

15 In the diagram O is the centre of the circle and the lines PT and QT are tangents to
   the circle.

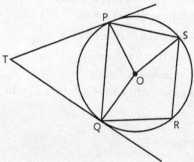

If angle PTQ = 48° and angle PSO = 42°, find the size of the following angles, giving brief
reasons for your answers.

a Angle QPO   **b** Angle QPS   **c** Angle QRS

16 Calculate the value of the side x and angles y and z.

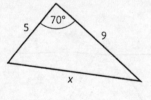 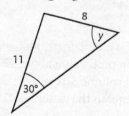 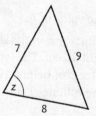

17 A speed boat starts a race at S and travels on a bearing of 060°. After 5 minutes, it rounds
   buoy A and continues on a bearing of 160°. After 6 km, it rounds buoy B and heads back
   to the start on a bearing of 315°.
   a Draw a neat diagram of the course and calculate the distances SA and BS.
   b Assuming that the speed of the boat remains constant throughout, find the time, in
   minutes, it took to complete the course.

18 PQR and PRS are right-angled triangles.
   ∠RPQ = 30°
   ∠PSR = 50°
   PR = 10 m
   Calculate
   a length PQ
   b length QR
   c length RS
   d area PQRS.

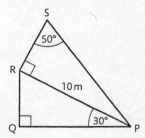

19 A circular big wheel funfair ride of radius 20 m rotates in a clockwise direction. The lowest point is 10 m above the ground and the highest point on the wheel is P.

The wheel rotates at 2° per second.

a Calculate the height of a chair at A above the ground if $x = 30°$.

b Find the height above the ground of the chair 20 s after reaching point A.

c What is the height of the chair 130 s after reaching point A, below point P?

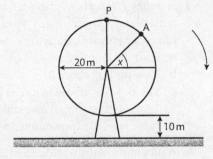

20 ABCDEFGH is a triangular prism.

The base ABCD is a square of length 18 cm.
The face ADEF is perpendicular to the base.
The height DE = 10 cm.

Calculate

a length BD

b length BE

c angle DBE

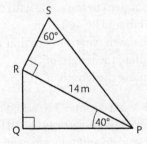

21 Find the area of triangle ABC to 3 s.f.

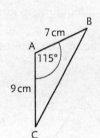

Consolidation exercise 4*

1 PQR and PRS are right-angled triangles.

∠RPQ = 40°

∠PSR = 60°

PR = 14 m

Calculate

a length PQ

b length QR

c length RS

d area PQRS

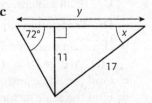

2 Find the angle x and length y in each of these triangles.

a   **b**   **c**

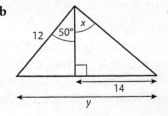

3 A pulley P is vertically above B and PB = 24 m. C is a point level with B and 18 m from it. A 50 m rope has one end pegged at C and the other end at Q, hanging over the pulley as shown.

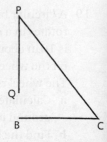

  a Find the length CP.
  b How far is Q above B?
  c The peg C is moved to a new position 14 m further from B and still on the same level. By how much does Q rise?

4 A mechanical drawbridge 5 m above a river consists of two equal spans of length 15 m both rising up at the same rate of 5° per second.

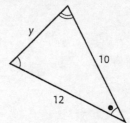
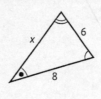
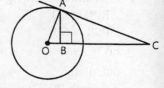

  a When $b = 20°$, find distance BC and the height of B above the river.
  b At the moment when $b = 20°$, a stuntman runs up the slope AB starting at A at a constant speed of 5 m/s attempting to jump across the gap. He can clear a horizontal distance of 5 m.
    Does he make the jump across the gap successfully?

5 Find the sides x and y.

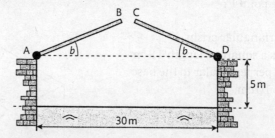

6 AC is a tangent to the circle with centre at O.
  AB = 5 cm and BC = 12 cm.
  a Write down two triangles that are similar to △ABC.
  b Calculate the length of OB.
  c Calculate the length of AC.

7 a Draw x-axis from −3 to 5 and y-axis from −5 to 6 and on these axes draw and label the triangle P with vertices (1, 1), (1, 3) and (2, 1).
    Four transformations are defined.
    A: Reflection in $y = −1$
    B: Reflection in $y + x = 0$
    C: Enlargement, centre 0, scale factor 2
    D: Translation along $\begin{pmatrix} -3 \\ -1 \end{pmatrix}$

  b Draw, on the same axes, triangles Q, R, S and T where
    (i) A(P) = Q   **(ii)** B(Q) = R   **(iii)** C(P) = S   **(iv)** D(S) = T
  c Find the angle and centre of rotation that takes P onto R.
  d Find the scale factor and centre of enlargement that takes P onto T.

8 This diagram shows a hollow metal pipe.

a Show that the volume V of metal is $V = \pi h(R - r)(R + r)$.

b Find the volume of metal in a pipe where $R = 12.5\,\text{cm}$, $r = 8.5\,\text{cm}$ and $h = 20\,\text{cm}$. Take $\pi = 3\frac{1}{7}$.

c Another pipe has a cross-sectional area of $28\,\text{cm}^2$ and volume of $126\,\text{cm}^3$. Find h.

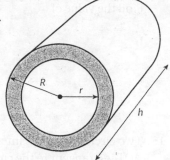

9 This figure shows three identical spherical billiard balls of diameter $6\,\text{cm}$ inside an equilateral triangular frame. Work out the perimeter of the frame, leaving your answer in surd form.

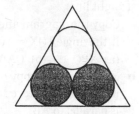

10 A sector with an angle of $144°$ is cut out of a circle of radius $10\,\text{cm}$. The remainder is turned into a cone by bringing two radii together.

a Work out the circumference of the base of the cone in terms of π.

b Find the base radius of the cone.

c Calculate the height of the cone.

d If the volume of the cone is the same as a sphere, find the radius of the sphere, leaving your answer as a cube root. (Vol. sphere $= \frac{4}{3}\pi r^3$)

11 A right-circular cone of base radius $10\,\text{cm}$ has a volume of $500\,\text{cm}^3$. Given that the volume $= \frac{\pi r^2}{3} \times h$ and the curved surface area $= \pi rl$, find the cone's total surface area.

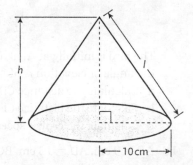

12 If PA and PB are tangents, PA is parallel to BC and angle APB is $78°$, calculate the angles x, y and z, being careful to state reasons for each step of your working.

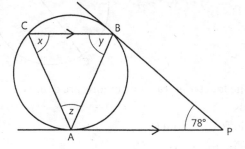

13 In the figure, the perimeter of the sector is $4r$.

a Find the arc length AB, in terms of r.

b Show that the angle s is $114.6°$, correct to 4 significant figures.

c Find, in terms of r, the shaded segment area.

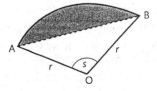

14 The cone P is similar to cone Q.
Find the value of
a x
b y

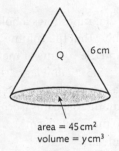

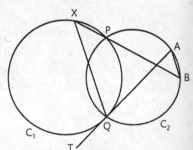

area = x cm²
volume = 80 cm³

area = 45 cm²
volume = y cm³

15 Two circles C_1 and C_2 intersect at P and Q. The line AQT is a tangent to the circle C_1 at Q. XPB is a straight line.

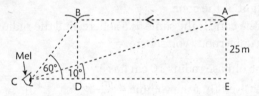

a Copy the diagram and join PQ.
If the angle TQX = 100°, calculate the angles
(a) XPQ   (b) QAB

b Make a second copy of the given diagram and prove that, whatever the size of angle TQX, XQ is parallel to AB.

16 Find the total perimeter of an equilateral triangle of area $100\sqrt{3}$ cm².

17 Mel is a birdwatcher who spots a crow flying directly towards her at a constant height of 25 m above her eye level.

The bird is initially at A, at an angle of elevation of 10°. Ten seconds later, it is at B, where the angle of elevation is 60°.
a Calculate the distance CD.
b Calculate the distance CE.
c Calculate the crow's speed.

18 AP = 9 cm, AD = 13 cm, BC = 15 cm,
PB = x and CP = y.
  Find x and y.

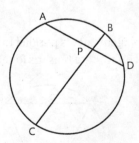

19 Point P is at (2, 5). The following transformations are defined as
A Reflection in x-axis
B Rotation of 90° clockwise about O
C Translation along vector $\begin{pmatrix} -3 \\ 7 \end{pmatrix}$
D Enlargement scale factor +2 about O
Find the image of P after it has undergone the transformations
a A(P)   **b** BA(P)   **c** C(P)   **d** DC(P)

20 A Swing-Ball tennis game consists of a tennis ball attached to a 1 m string, one end of which is fixed to a vertical pole of height h metres.

$\sin 30° = \cos 60° = \frac{1}{2}$

$\sin 60° = \cos 30° = \frac{\sqrt{3}}{2}$

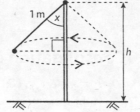

a Given that, when angle $x = 60°$, the height of the ball above the ground is 1.5 m, find h.

b When $x = 60°$, the ball takes $\sqrt{3}$ seconds to perform one revolution of its circular path. Find the speed of the ball in terms of π.

c When $x = 30°$, the ball takes 2 seconds to perform one revolution of its circular path. Find the percentage change in the ball's speed from that at $x = 60°$.

21 A regular tetrahedron of side 10 cm is shown. P is directly below O on horizontal base ABC and Q is the mid-point of AB.

a Find the length of AP, PQ and OQ.

b What angle does OA make with ABC?

c What angle does OAB make with ABC?

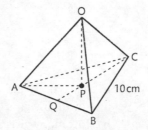

Consolidation exercise 5

1 A = {*a, c, e, g, i, k*} and B = {*d, e, f, g*}.
   a List the members of the set.
      (i) A ∩ B   **(ii)** A ∪ B
   b Show sets A and B in a Venn diagram.

2 Copy the Venn diagram and shade the region representing (P ∪ R) ∩ Q'.

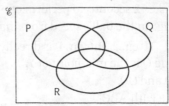

3 In a class of 30 girls, 18 play netball, 12 do gymnastics and 4 do not do games at all.
   a How many girls do both gymnastics and netball?
   b How many girls play netball, but not gymnastics?

4 From a pack of 52 playing cards
   H = {Hearts}
   R = {Red cards}
   P = {Picture cards}
   Copy the Venn diagram and label the three sets, H, R and P.

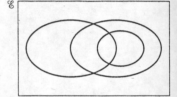

5 A person is selected at random from the group shown.
   Work out these probabilities.
   a The chosen person wears glasses.
   b He/she does not wear glasses.
   c The person is bald.
   d The person is bald or wears glasses.

6 A letter is chosen at random from the word 'CARIBBEAN'.
   Calculate the probability that it is
   a an R   **b** not an A   **c** a vowel

7 The ages of the players at a football club are given in this table.

17	23	25	30	24	18
18	36	20	20	27	18
24	34	32	32	22	20
26	25	27	23	24	21
21	19	33	29	25	25

   a Work out the median age of the players.
   b Calculate the mean age of the players.

8 The total number of points scored in a series of
basketball matches is given in the frequency table.
  a How many matches are included in the survey?
  b Calculate an estimate for the mean number
    of points scored per match.
  c Estimate the probability that the number of
    points scored in a match is between 125
    and 155.

Points x	Frequency f
121–130	7
131–140	11
141–150	15
151–160	13
161–170	9
171–180	5

9 The frequency table shows the distribution of
heights of a batch of laurel shrubs grown
at a garden centre.

Height (cm)	f
20–40	15
40–50	18
50–55	22
55–60	19
60–70	18
70–90	8

  a Calculate the frequency density for each group.
  b Draw a histogram of the data.
  c Calculate, to one decimal place, an estimate of the mean height of a shrub.

10 In an experiment the lengths of 80 daisy stalks were measured.
The results are shown in the table.

Length l (mm)	Frequency f
$0 < l \leqslant 5$	5
$5 < l \leqslant 10$	10
$10 < l \leqslant 20$	22
$20 < l \leqslant 30$	25
$30 < l \leqslant 50$	18

  a Construct a histogram for these results.
  b Calculate an estimate of the number of daisies in this group that have a stalk length
    between 4 mm and 8 mm.
  c Calculate an estimate of the mean length of the daisy stalks.

11 40 people were asked how long it takes them to travel to work.
The results are shown in the table.

Time t (min)	Frequency f
$15 < t \leqslant 20$	12
$20 < t \leqslant 30$	12
$30 < t \leqslant 40$	10
$40 < t \leqslant 70$	6

  a Construct a histogram for these results
  b Calculate an estimate of the number of people who took more than 60 minutes to travel
    to work.
  c Calculate an estimate of the number of people who took less than 24 minutes to travel
    to work.

12 A farmer checks the masses of a sample of apples for quality control. The unfinished table and histogram show the results.

Mass m (g)	Frequency f
$60 < m \leqslant 70$	
$70 < m \leqslant 80$	22
$80 < m \leqslant 100$	40
$100 < m \leqslant 120$	20
$120 < m \leqslant 160$	

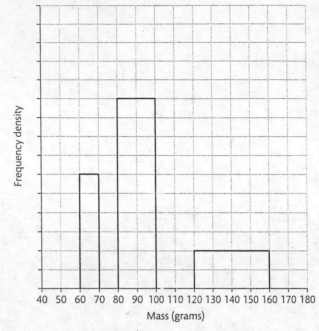

a Use the histogram to complete the table.
b Use the table to complete the histogram.
c Calculate an estimate of the number of apples with mass between 75 and 95 grams.

13 The unfinished table and histogram give the waiting times of patients at a doctor's surgery.
The number of patients waiting 10–12 minutes is 8 more than the number waiting 0–6 minutes.

Time t (min)	Frequency
$0 < t \leqslant 6$	
$6 < t \leqslant 10$	26
$10 < t \leqslant 12$	
$12 < t \leqslant 14$	18
$14 < t \leqslant 22$	24

a Use the histogram to complete the table.
b Use the table to complete the histogram.
c Calculate an estimate of the percentage of patients who wait between 8 and 16 minutes.
d Calculate an estimate of the median waiting time.

14 The probability that a person chosen at random has brown eyes is 0.45.
The probability that a person chosen at random has green eyes is 0.12.
a Work out the probability that a person chosen at random has either brown eyes **or** green eyes.
250 people are to be chosen at random.
b Work out an estimate for the number of people who will have green eyes.

15 The diagram shows six counters.

Each counter has a letter on it.
Bishen puts the six counters in a bag. He takes a counter at random from the bag.
He records the letter which is on the counter and replaces the counter in the bag.
He than takes a second counter at random and records the letter which is on the counter.

a Calculate the probability that the first letter will be A and the second letter will be N.

b Calculate the probability that both letters will be the same.

16 In order to start a course, Bae has to pass a test.
He is allowed only two attempts to pass the test.
The probability that Bae will pass the test at his first attempt is $\frac{2}{5}$
If he fails at his first attempt, the probability that he will pass at his second attempt is $\frac{3}{4}$

a Complete the probability tree diagram.

b Calculate the probability that Bae will be allowed to start the course.

First attempt   Second attempt

.... Pass

.... Fail

17 $\mathbf{p} = \begin{pmatrix} 3 \\ 4 \end{pmatrix}$

a Find the magnitude of $\mathbf{p}$.

b Find the angle $\mathbf{p}$ makes with the x-axis.

18 The vector $\mathbf{q}$ has length 6 and makes an angle of 30° with the x-axis as shown.
Express $\mathbf{q}$ as a column vector.

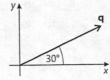

19 $\mathbf{u} = \begin{pmatrix} 2 \\ 3 \end{pmatrix}$ and $\mathbf{v} = \begin{pmatrix} -1 \\ 5 \end{pmatrix}$. Find $2\mathbf{v}$, $\mathbf{u} + \mathbf{v}$, $\mathbf{u} - \mathbf{v}$ and the length of $2\mathbf{u} - 3\mathbf{v}$.

20 If $\mathbf{A} = \begin{pmatrix} 3 & 1 \\ 3 & 2 \end{pmatrix}$ and $\mathbf{B} = \begin{pmatrix} 0 & -4 \\ -5 & 2 \end{pmatrix}$ find the value of

    a $\mathbf{A} + \mathbf{B}$ **b** $\mathbf{A} - \mathbf{B}$ **c** $2(\mathbf{A} + 2\mathbf{B})$ **d** $3(\mathbf{A} - 4\mathbf{B})$

21 If $\mathbf{M} = \begin{pmatrix} -1 & 1 \\ 0 & 1 \end{pmatrix}$ and $\mathbf{N} = \begin{pmatrix} 1 & -4 \\ 3 & 1 \end{pmatrix}$ find the value of

    a $\mathbf{MN}$ **b** $\mathbf{NM}$ **c** $\mathbf{M}^2$ **d** $\mathbf{N}^{-1}$

22 In the diagram, M, N and O are the mid-points of AB, BC and CA respectively.
$\overrightarrow{OA} = \mathbf{x}$, $\overrightarrow{OB} = \mathbf{y}$.
Find in terms of vectors $\mathbf{x}$ and $\mathbf{y}$, $\overrightarrow{AB}$, $\overrightarrow{OM}$, $\overrightarrow{ON}$ and $\overrightarrow{MN}$.
What is the relationship between MN and OA?

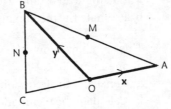

23 In the triangle OPQ, OA is one third of OP and OB is one third of OQ.

    a Find the vector AB.

    b Find the vector PQ.

    c What is the relationship bewteen the lines AB and PQ?

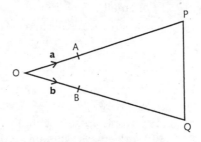

24 $R = \begin{pmatrix} 2 & 3 \\ 0 & -1 \end{pmatrix}$, $S = \begin{pmatrix} 1 & 4 \\ -2 & -1 \end{pmatrix}$

   a Find the matrix **T** representing the transformation **R** followed by transformation **S**.

   b Find the inverse of **T**.

   c The triangle ABC is mapped by **T** onto the triangle A′(−1, −1), B′(5, 2), C′(−1, 1). Find the coordinates of A, B and C.

Consolidation exercise 5*

1 Copy the Venn diagram and shade the region representing $(A' \cup B') \cap C$.

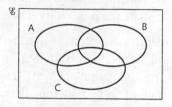

2 The universal set, $\mathscr{E} = \{2, 3, 4, \ldots, 12\}$.

   A = {factors of 24}

   B = {multiples of 3}

   C = {even numbers}

   On the diagram, draw a ring to represent the set C, and write the members of $\mathscr{E}$ in the appropriate regions.

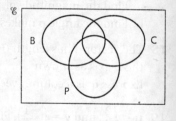

3 All boys in a class of 30 study at least one of the three sciences: Physics, Chemistry and Biology.

   14 study Biology.

   15 study Chemistry.

   6 study Physics and Chemistry.

   7 study Biology and Chemistry.

   8 study Biology and Physics.

   5 study all three.

   Use the Venn diagram to work out how many boys study Physics.

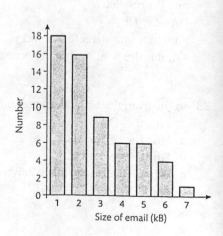

4 This frequency table gives the ages of the employees at a small company.

Group	20–30	30–40	40–50	50–60
Frequency	12	9	6	3

   a Calculate an estimate for the mean age.

   b Estimate the median age.

5 This bar chart shows the size of the emails in a phone-in mailbox.

   a Construct the frequency table.

   b Calculate the mean, median and mode of the size of emails.

   c Display the results in a pie chart.

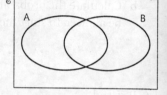

6 A = {Factors of 24} and B = {multiples of 3}

A whole number from 1 to 24 inclusive is randomly chosen.

Find the probability that the number is in the set.

a A **b** B′ **c** A∩B **d** A∪B

7 Two normal six-sided dice are thrown.
X is the difference between the scores.

a Copy and fill in this table to show all the possible values of X.

b Calculate p($X = 1$).

c Calculate p($X > 2$).

X	1	2	3	4	5	6
1						
2						
3						
4						
5						
6						

8 A football team of eleven players has a mean height of 1.83 m. One player is injured and is replaced by a player of height 1.85 m. The new mean height of the team is now 1.84 m. What is the height of the injured player?

9 A box contains twelve roses. Four are white, two are red and six are pink.
Sacha picks out one rose at random. What is the probability that it is

a pink **b** not red

c white, red or pink **d** yellow?

10 Ellis receives 25 text messages. Six are from his mother, 16 are from his friend Leo and the rest are from his sister Laura. He chooses one at random to read first. What is the probability that it is

a from his father **b** from his mother or his sister

c from Leo **d** not from Leo?

11 Here is a four sided spinner.

Its sides are labelled 1, 2, 3 and 4.

The spinner is biased.

The probability that the spinner lands on each of the numbers 1, 2 and 3 is given in the table.

The spinner is spun once.

a Work out the probability that the spinner lands on 4.

b Work out the probability that the spinner lands on either 2 or 3.

Number	Probability
1	0.25
2	0.25
3	0.1
4	

12 $\frac{1}{3}$ of the people in a club are men.

The number of men in the club is n.

a Write down an expression, in terms of n, for the number of people in the club.

Two of the people in the club are chosen at random.

The probability that both these people are men is $\frac{1}{10}$.

b Calculate the number of people in the club.

13 There are 48 beads in a bag.
Some of the beads are red and the rest of the beads are blue.
Shan is going to take a bead at random from the bag.
The probability that she will take a red bead is $\frac{3}{8}$.
a Work out the number of red beads in the bag.
Shan adds some **red** beads to the 48 beads in the bag.
The probability that she will take a red bead is now $\frac{1}{2}$.
b Work out the number red beads she adds.

14 In an experiment the lengths of 100 cats' whiskers
were measured. The results are shown in the table.
a Construct a histogram for these results.
b Calculate an estimate of the number of whiskers
in this group that have a length between 5.5 and
8.5 mm.
c Calculate an estimate of the mean length of the
whiskers.

Length l (mm)	Frequency f
$5 < l \leqslant 6$	18
$6 < l \leqslant 6.5$	15
$6.5 < l \leqslant 7$	21
$7 < l \leqslant 8$	26
$8 < l \leqslant 10$	20

15 The table gives the times of 50 cyclists in a race.
a Construct a histogram of these results.
b Calculate an estimate of the percentage of cyclists
in this group that had a time of less than
48 minutes.
c Calculate an estimate for the median time.

Time t (min)	Frequency f
$20 < t \leqslant 30$	5
$30 < t \leqslant 35$	4
$35 < t \leqslant 40$	9
$40 < t \leqslant 50$	22
$50 < t \leqslant 70$	7
$70 < t \leqslant 100$	3

16 A manufacturer checks how long the batteries last
in a new portable CD player using a random sample
of 200 players. The unfinished table and histogram
show the results.

a Use the histogram to complete the table.
b Calculate the value of x.
c Use the table to complete the histogram.

Life t (h)	Frequency
$4 < t \leqslant 6$	40
$6 < t \leqslant 7$	
$7 < t \leqslant 7.5$	25
$7.5 < t \leqslant 8$	
$8 < t \leqslant 10$	50
$10 < t \leqslant x$	

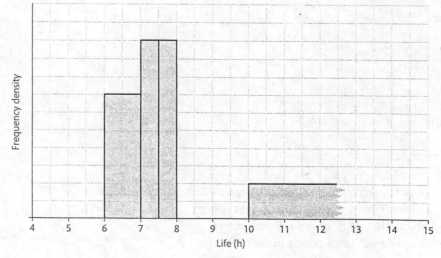

17 A cat breeder records the birth masses of 200 kittens. The unfinished table and histogram show the results. The difference between the number of kittens in the 100 – 105 g class and the 110 – 120 g class is 30.

Mass *m* (grams)	Frequency
$85 < m \leqslant 90$	
$90 < m \leqslant 100$	
$100 < m \leqslant 105$	
$105 < m \leqslant 110$	30
$110 < m \leqslant 120$	

  a Use the histogram to complete the table.

  b Use the table to complete the histogram.

  c Calculate the probability that the birth mass is between 95 grams and 110 grams.

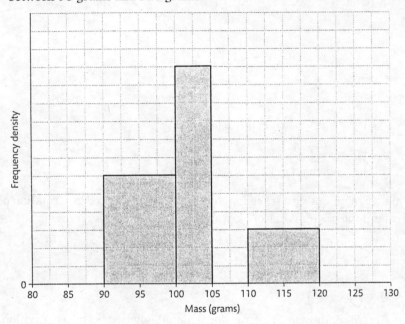

18 You are given these vectors:

$$\mathbf{x} = \begin{pmatrix} 1 \\ 1 \end{pmatrix} \qquad \mathbf{y} = \begin{pmatrix} 2 \\ -3 \end{pmatrix} \qquad \mathbf{z} = \begin{pmatrix} -5 \\ 5 \end{pmatrix}$$

  a If $\mathbf{x} + \mathbf{y} + \mathbf{z} = \mathbf{w}$, find the length of $\mathbf{w}$.

  b If $p\mathbf{x} + q\mathbf{y} + \mathbf{z} = \mathbf{0}$, find the values of p and q.

19 **a** Express OA as a column vector.

  b Express AB as a column vector.

  c Express OB as a column vector.

  d Hence find the length OB and the angle OB makes with the *x*-axis.

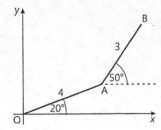

20 Vectors are defined as in the diagram. S is the mid-point of RT.

  a Find the vector $\overrightarrow{PU}$.

  b Find the vector $\overrightarrow{TR}$.

  c Find the vector $\overrightarrow{US}$.

  d What is the relationship between the lines PU and US?

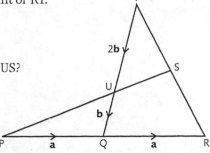

21 In the diagram

$\overrightarrow{OA} = \mathbf{a}$, $\overrightarrow{OC} = \mu\mathbf{a}$

$\overrightarrow{OB} = \mathbf{b}$ and $\overrightarrow{OD} = \lambda\mathbf{b}$

(i) Find AB

(ii) Find CD

$\overrightarrow{CD} = 3\overrightarrow{AB}$

(iii) Find the values of μ and λ

(iv) What is the relation between AB and CD?

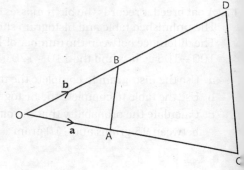

22 If $X = \begin{pmatrix} -1 & 1 \\ 3 & 2 \end{pmatrix}$, $Y = \begin{pmatrix} 1 & 0 \\ -2 & -2 \end{pmatrix}$ find the value of:

a $(XY)^{-1}$ **b** X^2 **c** $X^2 + Y^2$ **d** X^4

23 If $A = \begin{pmatrix} 1 & -1 \\ -1 & 2 \\ 3 & 5 \end{pmatrix}$, $B = \begin{pmatrix} 0 & 1 & -2 \\ 1 & 3 & 1 \end{pmatrix}$ find the value of

a AB   **b** BA

c What does this tell you about matrix multiplication?

24 a Find the matrix representing a reflection in the line $y = x$ followed by a rotation of 90° clockwise about the origin. What single transformation does this represent?

b The matrix $\begin{pmatrix} a & b \\ c & d \end{pmatrix}$ maps the point (1, 5) to (−8, 8) and the point (4, −1) to (−11, −10). Find the values of a, b, c and d.

25 $P = \begin{pmatrix} 2 & 3 \\ 0 & -1 \end{pmatrix}$, $Q = \begin{pmatrix} 1 & 4 \\ -2 & -1 \end{pmatrix}$

a Find the matrix **M** representing the transformation **P** followed by transformation **Q**.

b The triangle ABC is mapped by **M** onto the triangle A′(1, −9), B′(4, −8), C′(2, −18). Find the coordinates of A, B and C.

Answers

Exercise 1

1 $\frac{2}{3}$ **3** $\frac{1}{2}$ **5** $\frac{1}{4}$, 0.25
7 $\frac{7}{20}$, 35% **9** $3\frac{2}{5}$ **11** $\frac{11}{6}$

Exercise 1*

1 $\frac{2}{7}$ **3** $\frac{1}{2}$ **5** $\frac{5}{6}$ **7** $\frac{4}{9}$
9 680 **11** 39 **13** 26.25

Exercise 2

1 5 **3** 13 **5** -12 **7** -2 **9** 10

Exercise 2*

1 -2 **3** 18 **5** 12 **7** 36 **9** 144

Exercise 3

1 1.8 **3** 0.684 **5** £45.50 **7** 10%
9 Percentage increase in value $= \frac{468}{7800} \times 100 = 6\%$

Exercise 3*

1 0.75 **3** 52.5 **5** $80.04
7 Profit $= €4.25$; percentage profit $= \frac{4.25}{34} \times 100 = 12.5\%$
9 11.1%; 12.5%

Exercise 4

1 10^5 **3** 10^3 **5** 10^2
7 10^3 **9** 4.56×10^2 **11** 5.68×10^2
13 4000 **15** 560 **17** 10000

Exercise 4*

1 4.5089×10^4 **3** 10^3 **5** 10^{21}
7 6.16×10^6 **9** 9.1125×10^{16} **111** 9.653×10^8
13 Saturn 10 cm, Andromeda Galaxy 1 million km,
Quasar 1000 million km

Exercise 5

1 800 **3** 0.44 **5** 34.78 **7** 1×10^5

Exercise 5*

1 10 **3** 0.067 **5** 9.00 **7** 1.06×10^5

Exercise 6 (Revision)

1 $\frac{1}{3}$ **2** $\frac{2}{7}$ **3** $\frac{1}{5}$ **4** $\frac{1}{8}$
5 $\frac{1}{2}$ **6** $1\frac{1}{3}$ **7** $1\frac{1}{5}$ **8** $\frac{3}{4}$
9 $\frac{1}{10}$ **10** $\frac{3}{10}$ **11** $\frac{2}{5}$ **12** $\frac{3}{4}$
13 50 **14** 60 **15** 300 **16** 8
17 -16 **18** -48 **19** $-\frac{1}{3}$ **20** 48
21 $\frac{1}{4}$ **22** $\frac{1}{10}$ **23** $\frac{3}{4}$ **24** $\frac{3}{5}$
25 $\frac{7}{20}$ **26** 150 m **27** $360 **28** 1650 m
29 £2040 **30** $90 **31** $897 **32** £56.25
33 6×10^{10} **34** 4×10^2 **35** 5.6×10^{16} **36** 1230
37 1240 **38** 1240 **39** 54300 **40** 54400
41 1.234 **42** 1.235 **43** 1.231 **44** 1.204
45 1.201

Exercise 6* (Revision)

1 $\frac{1}{11}$ **2** $\frac{3}{20}$ **3** $\frac{9}{10}$ **4** $\frac{1}{8}$
5 42 g **6** 37.5% **7** 12%
8 a 56.7 s **b** 19%
9 a 154 m **b** 23.2%
10 a €118 800 **b** $€120\,000\left(1 - \frac{x^2}{10^4}\right)$
11 7450 **12** 0.0745 **13** 74 500 **14** 74 500 000
15 5.3×10^6 **16** 8.8×10^{14} **17** 5.0×10 **18** 3.7×10^8
19 3.5×10^8 **20** 6.3×10^{10} **21** 0.201 **22** 0.00201
23 3080 **24** 47 600 **25** 0.079 **26** 0.072
27 0.063 **28** 0.111

Exercise 7

1 $5a$ **3** $2a + 3b$ **5** $4ab$ **7** $-3pq$
9 $-6x + 2$ **11** $-4xy$ **13** 0

Exercise 7*

1 $5ab$ **3** $3x + 3$ **5** $-xy$ **7** $6ab$
9 $3ab + 3bc$ **11** $x + 1$ **13** $h^3 + h^2 + 3h + 4$

Exercise 8

1 $6a$ **3** $3x^3$ **5** $6st$ **7** $2a^2b^2$ **9** $12x^3$

Exercise 8*

1 $10xy$ **3** $8a^3$ **5** $15x^4y^2$ **7** $18y^3$ **9** $30a^3b^3c^5$

Exercise 9

1 $4x + 4y$ **3** $10 + 15a$ **5** $-6a - 24$
7 $-a + 2b$ **9** $3t - 18$

Exercise 9*

1 $5x + 10y$ **3** $12m - 8$ **5** $15a + 5b - 20c$
7 $3y - x$ **9** $-1.4x - 3.8$

Exercise 10

1 4 **3** 25 **5** 12
7 2.4 **9** 26.6 **11** 0.985

Exercise 10*

1 99.9 **3** 40.7 **5** 8.49

Exercise 11

1 $x = 3$ **3** $x = -2$ **5** $x = 8$
7 $x = -6$ **9** $x = -2$ **11** $x = \frac{5}{9}$
13 $x = -1$ **15** $x = 10$; 40, 80, 60
17 -15

Exercise 11*

1 $x = 4$ **3** $x = -2$ **5** $x = -4$
7 $x = 0$ **9** 11, 44, 67 kg **11** $x = 2.5$
13 4.20 cm

Exercise 12

1 $x = 1$ **3** $x = 4$ **5** $x = 1$
7 $x = 0$ **9** $x = 2, 38$

Exercise 12*

1 $x = 4$	**3** $x = 1\frac{1}{2}$	**5** $x = \frac{7}{9}$	**7** $x = 5$
9 $x = 0.576$ (3 sig figs)	**11** 42 years		

Exercise 13

1 $x = 13$	**3** $x = 2$	**5** $x = \frac{5}{2}$

Exercise 13*

1 $x = 8$	**3** $x = 4$	**5** $x = \frac{3}{4}$	**7** 15

Exercise 14 (Revision)

1 $3x - 2$	**2** ab	**3** $6a$	**4** $2a^2$
5 a^3	**6** $2a^4$	**7** $4a^4$	**8** $-4ab - 5a$
9 $x + 7y$	**10** $x = 7$	**11** $x = 4.8$	**12** $x = 2$

13 145, 146, 147

14 a $4x + 12 = 54$ **b** $x = 10.5, 10.5, 16.5$

Exercise 14* (Revision)

1 $4xy^2 - 3x^2y$	**2** $2x^3y^3$	**3** 1
4 $2x^3y + xy^3 + x^4$	**5** $x = 20$	**6** $x = \frac{5}{4}$
7 $x = -6$	**8** $x = 2$	**9** $x = 4$
10 $72\,m^2$	**11** 11 years old	**12** 6 m/s

13 $12\,800

Exercise 15

1 1	**3** 3	**5** $-\frac{1}{4}$
7 1.5 m	**9** 2 m	
11 a 2	**b** 159 m	

Exercise 15*

1 $\frac{3}{8}$	**3** 52	**5** Yes

Exercise 16

1 1, $(0, 1)$	**3** 3, $(0, 5)$	**5** $\frac{1}{3}$, $(0, 2)$
7 $-\frac{1}{3}$, $(0, -2)$	**9** $y = 2x + 1$	**11** $y = 2x + 4$

13 For example:

a $y = x - 2$ **b** $y = 5 - x$ **c** $y = 2$

Exercise 16*

1 3 $(0, 2)$	**3** 5 $(0, \frac{1}{2})$	**5** -3 $(0, \frac{5}{2})$

7 Infinity (vertical line)

9 $y = 2.5x - 2.3$

11 $2y = 5x - 7$

13 For general equation $y = mx + c$

a $m = 1$ $c < 0$ **b** $m < 0$ $c > 0$ **c** $m = 0$ $c > 0$

Exercise 17

1 $(5, 0)$ $(0, 5)$	**3** $(3, 0)$ $(0, 6)$
5 $(5, 0)$ $(0, 4)$	**7** $(-8, 0)$ $(0, 6)$

Exercise 17*

1 $(3, 0)$ $(0, 12)$	**3** $(6, 0)$ $(0, 12)$
5 $(7.5, 0)$ $(0, -6)$	**7** $(-3.5, 0)$ $(0, 3)$

Exercise 18 (Revision)

1 a 2 **b** -1

2 4.5 m

3 a 3, -2 **b** $-2, 5$

4 a $y = 2x - 1$ **b** $y = -3x + 2$

5 a Graph through $(0, -3)$ $(1, -1)$

b Graph through $(0, 4)$ $(4, 0)$

c Graph through $(5, 0)$ $(0, 2)$

6 $y = 2x + 4$; $4x = 2y + 7$

$x - 3y = 1$; $9y = 3x + 4$

$4x - 3y = 12$; $3y = 4x - 1$

$3x - 4y = 12$; $4x - 4y = 12$; $4y = 3x + 7$

7 a $\frac{2}{3}$ **b** $-\frac{1}{2}$

Exercise 18* (Revision)

1 a $\frac{1}{2}$ **b** -2.5

2 5 m

3 a Graph through $(0, -2)$ $(1, 1)$ **b** Graph through $(0, 3)$ $(1, 1)$

c Graph through $(0, 2.5)$ $(1, 2)$ **d** Graph through $(2, 0)$ $(0, 3\frac{1}{3})$

4 $b = \frac{3}{2}$

5 $3y = x + 6$

6 c $-40°$

7 $y = x + 2, y = x - 2, y = -x + 2, y = -x - 2$

8 $y = 3, y = -3, x = 3, x = -3, y = x, y = -x$

Exercise 19

1 $a = 102°, b = 78°$	**3** $a = 73°, b = 34°$
5 $a = 31°, b = 31°$	**7** $a = 58°, b = 32°$
9 9 sides	

Exercise 19*

1 $a = 137°, b = 43°$	**3** $34°, 85°$
5 $x = 50°$	**7** 13 $a = 40°, b = 113°$
9 a $a = 56°, b = 38°$	

Exercise 20

1 5.9 cm	**3** 150 m	**5** 152 m

Exercise 20*

1 5.4 cm **3,5** Student's own diagrams

Exercise 21 (Revision)

1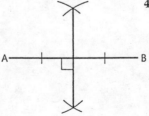

2

3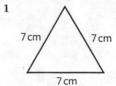

4 Measure and check arcs are included

5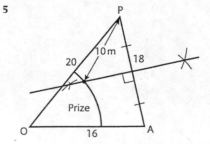
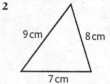

6 $x = 143°, y = 37°$ **7** $x = 30°, y = 60°$
8 $x = 69°, y = 42°$ **9** $x = 65°, y = 115°$
10 $x = 40°, y = 140°$ **11** $x = 77°, y = 103°$
12 $x = 60°, y = 120°$ **13** $x = 25°, y = 115°$
14 $x = 100°, y = 75°, z = 135°$

Exercise 21* (Revision)

1   **a** 36   **b** $144°$ **c** $1440°$
2   **a** 22 sides   **b** $16\frac{4}{11}°$
3   **b** 30 m approx   **4** —
5 $x = 46\frac{2}{3}°, y = 133\frac{1}{3}°$ **6** $x = 66°$
7 $x = 160°$ **8** $x = 36°, y = 106°, z = 38°$
9 $x = 70°, y = 60°$ **10** $x = 60°, y = 30°$
11 $x = 30°, y = 120°$ **12** $x = 36°$
13 $x = 20°$ **14** $x = y = 33°, z = 83°$
15 $x = 150°$ **16** $x = 36°$

Exercise 22

1   **a** Any two vegetables   **b** Any two colours
    c Any two letters   **d** Any two odd numbers
3   **a** {the first four letters of the alphabet}
    b {days of the week beginning with T}
    c {first four square numbers}   **d** {even numbers}
5   b and c

Exercise 22*

1   **a** Any two planets   **b** Any two polygons
    c Any two elements   **d** Any two square numbers
3   **a** {seasons of the year}   **b** {conic sections}
    c {first five powers of 2}   **d** {Pythagorean triples}
5   a, c and d

Exercise 23

1   **a** 16   **b** $n(T) = 14$; 14 pupils like toffee
    c $n(C \cap T) = 12$; 12 pupils like both chocolate and toffee
    d 21
3   **a** 35   **b** 3   **c** 11   **d** 2   **e** 64

Exercise 23*

1   **a**

(Venn diagram: universal set $\mathscr{E}$, set A and set B. A only: 2, 10; intersection: 4, 6, 8; B only: 5, 7; outside: 1, 3, 9, 11)

    b {4, 6, 8}, 3   **c** Yes
    d {1, 2, 3, 5, 7, 9, 10, 11}   **e** Yes

3   **a**

(Venn diagram: universal set $\mathscr{E}$, three sets V, A, B. i, o in V; a in V∩A; b, c in A; u in V∩B; e in centre; d in A∩B; f, g, h, i, k, l, m, n, p, q, r, s, t, v, w, x, y, z outside)

    b {a, e}, {a, i, o}, {u}   **c** {e}
5   {a, b, c} {a, b} {a, c} {b, c} {a} {b} {c}, ∅, 2^n

Exercise 24

1   **a**

(Venn diagram: universal set $\mathscr{E}$, sets A and B. A only: 1, 7, 9; intersection: 3, 5; B only: 4, 6; outside: 2, 8)

    b {1, 3, 4, 5, 6, 7, 9}, 7   **c** Yes
    d {2, 8}   **e** No

3   **a**

(Venn diagram: universal set $\mathscr{E}$, sets R and I overlapping; E inside I)

    b An isosceles right-angled triangle.
    c Isosceles triangles, triangles that are isosceles or right-angled or both.
    d Equilateral triangles, ∅

Exercise 24*

1   **a**

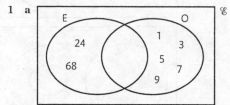

(Venn diagram: universal set $\mathscr{E}$, sets E and O. E: 24, 68; intersection: 5, 9; O: 1, 3, 7)

    b {1, 2, 3, 4, 5, 6, 7, 8, 9}   **c** $E \cap O = ∅$
    d $E \cap O = \mathscr{E}$
3 $B \subset A$ or $B = ∅$

Exercise 25 (Revision)

1   **a** any 2 spices   **b** any 2 pets
    c any 2 fruits   **d** any 2 colours
2   **a** {4, 9, 16, 25}   **b** {1, 2, 3, 4, 6, 8, 12, 24}
    c {a, e, i}   **d** {april, june, sept, nov}
3   **a** {prime numbers less than 10}
    b {even numbers between 31 and 39}
    c {days in weekend}   **d** {vowels}
4   **a** T   **b** F   **c** T   **d** T
5   **a**

(Venn diagram: universal set $\mathscr{E}$, set A containing set B. B: 4, 8; A outside B: 2, 6, 10; outside A: 1, 3, 5, 7, 9)

    b $A' = \{1, 3, 5, 7, 9\}$ add numbers
    c $n(B') = 8$ **d** Yes

6

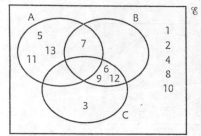

a A ∪ B = {5, 6, 7, 9, 11, 12, 13}

b B ∩ C = {6, 9, 12}　　　　**c** A ∩ C = all odd numbers 3–13

7 **a** 　　**b**

c　　**d**

8 **a** {females born in Africa}

b ∅

c She was born in Africa *or* China

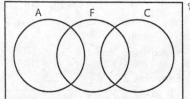

9 **a** There are no red dresses

b All dresses are green

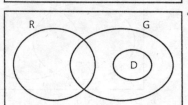

10 **a** 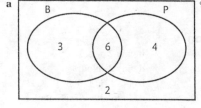　　**b** 15

Exercise 25* (Revision)

1 **a** {4, 8, 12, 16}　　**b** {R, O, Y, G, B, I, V}

　　c {CAT, CTA, ATC, ACT, TAC, TCA}

　　d {1, 2, 3, 4, 6, 9}

2 **a** {factors of 12}　　**b** {1st five fibonacci}

　　c {suits of playing cards} **d** {3D shapes}

3 **a** 9　　**b**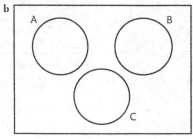

4 **a** {ace of diamonds}

b ∅

c {all diamonds plus 3 other aces}

d 2

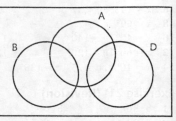

5 **a** {right-angled isosceles triangles}　　**b** E (or ∅)

c ∅

d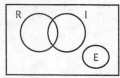

6 **a** {20}　　　　**b** {12, 24}　　　　**c** Yes

7 8

8 **a** 8　　　　　　　　　　　　　　**b** 2

9 **a** 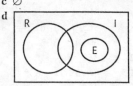　　**b** 19

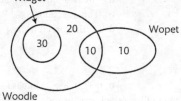

10 **a** 　　　　**b** 20

Multiple choice 1

1 C	**2** C	**3** B	**4** A	**5** A
6 D	**7** D	**8** A	**9** B	**10** D

Self-assessment 1

1 **a** 3.85×10^5　　　　**b** 3250

2 **a** 36.58　　　　　　　　**b** 37

3 **a** 4.00×10^2　　　　**b** 4.81×10^5

4 **a** 1.20×10^6　　　　**b** 2.60×10^3

5 **a** £4.62　　　　　　　　**b** £42.77

6 **a** $8ab$　　　**b** $3xy + 3x$　　　**c** $3a^3b$

7 **a** $x = 3$　　**b** $a = 3.5$　　　**c** $x = 3$

8 **a** $x + 1$　　**b** $x + x + 1 + x + 2 = 525$　　**c** $x = 174$

9 $x - 3y = 12$ and $6y - 2x = 7$

10 **a** 2　　　　　　**b** −4

11 **a** 400 m　　　　**b** $q = 7$

12 **b** AC = 8.6 cm

13 **c** RS = 7.9 cm, area = 31.6 cm²

14 $p = 80$

15 **d** $x = 0.6, y = 3.71$

16 A gradient = $\frac{1}{2}$　intercept = 4　equation is $y = \frac{1}{2}x + 4$

　　B gradient = −2　intercept = 8　equation is $y = -2x + 8$

17 **a** A ∩ C = ∅　　**b** C ∪ D = C　　**c** A ∩ B ≠ ∅

18 **a** 13　　**b** 7

19 **a** and **c**

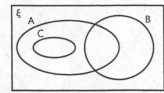

b 90°, 45°, 45°

20 **a** {6, 12, 18}   **b** {multiples of 6}   **c** {odd numbers}
    d {3, 9, 15}   **e** {odd multiples of 3}

Exercise 26

1 10^{-1} **3** 10^{-3} **5** 0.001
7 0.000 001   **9** 5.43×10^{-1} **11** 6.7×10^{-1}
13 100   **15** 128   **17** 0.018

Exercise 26*

1 10   **3** 0.01   **5** 1000
7 10^4 **9** 5000 viruses
11 **a** 10^{27}, 27 zeros
    c 10^7
    d 2×10^{23}, 2×10^{16} cm, $(2 \times 10^{16}) \div (4 \times 10^9) \approx 5 \times 10^6$ times!

Exercise 27

1 $\frac{6}{7}$ **3** $1\frac{2}{9}$ **5** $\frac{23}{24}$ **7** $\frac{5}{18}$
9 $\frac{3}{5}$ **11** 3   **13** $4\frac{7}{12}$ **15** $4\frac{4}{5}$

Exercise 27*

1 $\frac{3}{4}$ **3** $\frac{19}{20}$ **5** $\frac{1}{5}$ **7** $\frac{2}{9}$
9 $7\frac{2}{3}$ **11** $6\frac{7}{9}$ **13** $2\frac{2}{3}$ **15** $4\frac{1}{2}$

Exercise 28

1 $168 $224   **3** 574, 410

Exercise 28*

1 $45 : $75   **3** $32   **5** 1 mg

Exercise 29

1 $2^4 = 16$ **3** $2^2 = 4$ **5** $3^6 = 729$
7 $(0.1)^3 = 0.001$ **9** $4^4 = 256$

Exercise 29*

1 $8^3 = 512$ **3** $5^3 = 125$ **5** $5^3 = 125$
7 $2^0 = 1$ **9** 2 097 152, 524 288

Exercise 30

1 Yen 180   **3** NZ$2.75
5 **a** 105 mm/h   **b** 1.75 mm/min

Exercise 30*

1 $15 ¥2667   **3** ≈ 50
5 **a** 94.4 m/s   **b** 0.00106 s

Exercise 31

1 5000, 500 000, 5 000 000
3 0.05, 50, 50 000

Exercise 31*

1 $2.5 \times 10^7, 2.5 \times 10^9, 2.5 \times 10^{10}$
3 $5 \times 10^{-4}, 0.5, 500$
5 **a** 2×10^{14} **b** 10^6 km

Exercise 32

1 $2 \times 10^6, 2 \times 10^{10}, 2 \times 10^{12}$ **2** $6 \times 10^{-4}, 6 \times 10^2, 6 \times 10^8$

Exercise 32*

1 $6 \times 10^{-3}, 6 \times 10^7, 6 \times 10^9$ **3** $2 \times 10^9, 2 \times 10^{15}, 2 \times 10^{19}$

Exercise 33

1 $1 \times 10^6, 1 \times 10^{15}, 1 \times 10^{18}$ **3** $4 \times 10^{-12}, 4 \times 10^{-3}, 4 \times 10^6$
5 1000

Exercise 33*

1 $6 \times 10^{-7}, 6 \times 10^8, 6 \times 10^{11}$ **3** $5 \times 10^7, 5 \times 10^{16}, 5 \times 10^{22}$
5 5.12×10^5 **7** 10^{45}

Exercise 34 (Revision)

1 $\frac{29}{35}$ **2** $\frac{1}{35}$ **3** $\frac{6}{35}$
4 $1\frac{1}{14}$ **5** $3\frac{26}{35}$ **6** $1\frac{16}{35}$
7 $2\frac{34}{35}$ **8** $2\frac{11}{40}$ **9** 12 m : 24 m
10 45 kg : 60 kg   **11** $160 : $240
12 160 mins : 200 mins   **13** £19 : £38 : £76
14 64 km : 96 km : 192 km   **15** 1.23×10^{-2}
16 1.24×10^{-2} **17** 1.60×10^{-4}
18 8.89×10^{-3} **19** 4.31×10^3
20 1.02×10^8 **21** 2.50×10^{-12}
22 2.93×10^{-8} **23** 3.62×10^{-8}
24 5.61×10^{-17} **25** 2.10×10^7
26 2.46×10^{-5}
27 **a** $150   **b** $1650   **c** $75
28 **a** €1.20   **b** €6   **c** €14.40
29 **a** 162   **b** 180   **c** 12 960
30 **a** **(i)** 46.7 km   **(ii)** 17 056 km
    b **(i)** 0.214 days   **(ii)** 0.0749 days
    c 5.41×10 mm/sec

Exercise 34* (Revision)

1 $\frac{3}{44}$ **2** $\frac{27}{35}$ **3** $\frac{3}{5}$
4 $\frac{25}{63}$ **5** 4   **6** $2\frac{89}{100}$
7 10   **8** $\frac{2m}{5}$
9 $X : Y : Z = £3000 : £6000 : £2000$
10 **a** 500 mm   **b** 20 m²
11 1.85×10^{-6} **12** 2.66×10^{-22}
13 2.41×10^{-6} **14** 1.62×10^{-3}
15 **a** 4.08×10^{-7} **b** 1.76×10^{-9}
    c 3.87×10^{-11} **d** 4.83×10^{-4}
16 2.90×10^{-5} km
17 **a** 3.02×10^{30} mm²   **b** 1.69×10^{-8} %
18 6.32×10^{-13} km/s

Exercise 35

1 4   **3** 2   **5** $\frac{b}{2}$ **7** $4c$ **9** $\frac{4}{x}$ **11** $\frac{1}{5b^2}$

Exercise 35*

1 $\frac{1}{2}$ **3** $\frac{x}{4}$ **5** $\frac{2}{b}$ **7** $3y$ **9** $\frac{3}{abc}$ **11** $\frac{1}{a^2b^2}$

Exercise 36

1 $\frac{5x^2}{4}$ **3** 1   **5** 6   **7** $\frac{b}{6}$ **9** $\frac{2}{y}$

Exercise 36*

1. $\dfrac{x^2}{6}$ 3. $\dfrac{3x}{z}$ 5. $\dfrac{1}{x}$ 7. y 9. $\dfrac{3x^4}{8y}$

Exercise 37

1. $\dfrac{3x}{4}$ 3. $\dfrac{a}{12}$ 5. $\dfrac{5x}{12}$

7. $\dfrac{3a + 4b}{12}$ 9. $\dfrac{a}{b}$

Exercise 37*

1. $\dfrac{7x}{18}$ 3. $\dfrac{14x + 20y}{35}$ 5. $\dfrac{17}{6b}$

7. $\dfrac{7 - 3x}{10}$ 9. $\dfrac{3x + 5}{12}$

Exercise 38

1. $x = \pm 3$ 3. $x = \pm 4$ 5. $x = \pm 3$
7. $x = 13$ 9. $x = 16$

Exercise 38*

1. $x = \pm 5$ 3. $x = \pm 7$ 5. $x = 81$
7. $x = \pm 4$ 9. $x = 10 \text{ or } -16$

Exercise 39

1. $4.5\,\text{cm}$ 3. $h = 8\,\text{cm}$
5. a. $9.42 \times 10^8\,\text{km}$ b. $110\,000\,\text{km/h}$
7. $1.5 \times 10^9\,\text{km}$

Exercise 39*

1. $14\,\text{cm}$ 3. $h = 5.5\,\text{cm}$ 5. $5.30\,\text{cm}^2$
7. $15.9\,\text{km}$ 9. $8.37\,\text{cm}$

Exercise 40

1. $2^{10} = 1024$ 3. $2^6 = 64$ 5. $2^{12} = 4096$ 7. a^5
9. e^6 11. c^5 13. $6a^5$

Exercise 40*

1. $6^{12} = 2.18 \times 10^9$
3. $8^{12} = 6.87 \times 10^{10}$
5. a^{12} 7. $2e^8$ 9. $48j^{12}$
11. $27a^6$ 13. $8b^4$

Exercise 41

1. $<$ 3. $>$ 5. $x \le 0, x > 2$
7. $x > 5$ 9. $x < 3$ 11. $x \ge 9$
13. $x < 0$ 15. $x < -2$ 17. $x < -1$
19. $x \ge -\dfrac{2}{3}$ 21. $\{5, 6\}$ 23. $\{0, 1\}$
25. $\{2, 3\}$

Exercise 41*

1. $x \le 0 \text{ or } x > 3; 0 \ge x > 3 \to 0 \ge 3!$
3. $x < 5\frac{1}{3}$ 5. $x < -3\frac{1}{5}$ 7. $-1 < x \le 3$
9. 23

Exercise 42 (Revision)

1. 3 2. x 3. $3x$
4. 4 5. a 6. $6x$
7. $\dfrac{9y}{20}$ 8. $\dfrac{2x}{15}$ 9. $\dfrac{4a + b}{10}$
10. $x = \pm 4$ 11. $x = \pm 6$ 12. $x = 20$
13. a^{10} 14. b^2 15. c^{12}
16. $>$ 17. $<$ 18. $<$
19. $=$ 20. $-3 < x \le 2$ 21. $x > 5$

22. $x \le 4.5$ 23. $x \ge 2$ 24. $x \ge 1$
25. $\{3, 4\}$ 26. $3.24\,\text{cm}$ 27. $11.34\,\text{km}$

Exercise 42* (Revision)

1. $\dfrac{4a}{b}$ 2. $\dfrac{5x}{y}$ 3. $\dfrac{b}{4a}$
4. $\dfrac{b}{2}$ 5. $\dfrac{5}{xy}$ 6. $\dfrac{18b}{a}$
7. $\dfrac{8a}{5}$ 8. $\dfrac{7}{12b}$ 9. $\dfrac{2x + 4}{21}$
10. $x = \pm 3$ 11. $x = 2$ 12. $x = \pm 4$
13. a^4 14. $4b^6$ 15. $81c^7$
16. $-3 < x \le 0, -2$ 17. $x < -4.4$ 18. $x > -4$
19. $x \le 4.5$ 20. 37 21. $\{-3, -2, -1, 0\}$
22. $50.1\,\text{cm}$ 23. $0.39\,\text{s}$
24. $1130\,\text{km/s} = 4.07 \times 10^6\,\text{km/h}$

Exercise 43

1. $1, 3, 5$ and $-2, 2, 6$ $(3, 4)$ 3. $(3, 8)$
5. $(4, 4)$

Exercise 43*

1. $(6, 13)$ 3. $(2.57, 0.29)$
5. a. $C = 0.25t + 15, C = 0.14t + 100$
   c. $773\,\text{min}$

Exercise 44

1. $x \le 2$ 3. $y \le -2$
5. 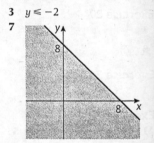   7.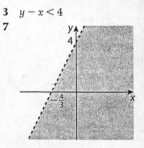

Exercise 44*

1. $y > -2$ 3. $y - x < 4$
5. 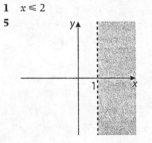   7.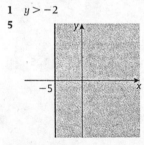

Exercise 45

1. $2 < x < 5$ 3. $x \le -3, x \ge 4$
5. $x + y > 3$ and $x - y \le 2$
7. 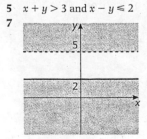   9.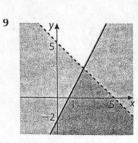

Exercise 45*

1 $-3 \leqslant x < 4$ **3** $y \geqslant 0, y < 2x + 4, 4x + 3y \leqslant 12$

5

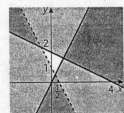

7   **b** $y < x + 2, y < 2 - 4x, 2y > -x - 2$ **c**   1

Exercise 46 (Revision)

1 $(2, 3)$ **2** $(2, 1)$ **3** $(2, 1)$ **4** $(4, 3)$

5 $(-1, 3)$ **6** $(-2, -2)$ **7** $(2, 2)$ **8** $\left(\frac{6}{5}, \frac{2}{5}\right)$

9 $\left(\frac{14}{3}, \frac{4}{3}\right)$ **10** $\left(\frac{7}{3}, -\frac{5}{3}\right)$ **11** $x \geqslant 1$ **12** $x < 3$

13 $y \geqslant 2$ **14** $y < 1$ **15** $x + y \leqslant 4$ **16** $y < x + 1$

17

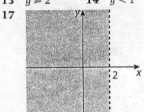

18

19

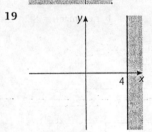

20

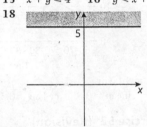

21 $-1 < x \leqslant 3$ **22** $-1 \leqslant y \leqslant 2$

23

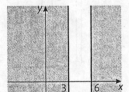

24

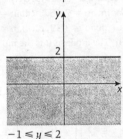

Exercise 46* (Revision)

1 $\left(\frac{3}{2}, \frac{9}{2}\right)$ **2** $\left(\frac{7}{3}, \frac{2}{3}\right)$ **3** $(2, 1)$ **4** $(4, 2)$

5 $\left(\frac{16}{3}\right), \left(\frac{10}{3}\right)$ **6** $\left(\frac{16}{5}, -\frac{2}{5}\right)$ **7** $\left(\frac{16}{7}, \frac{15}{7}\right)$ **8** $\left(\frac{30}{13}, \frac{42}{13}\right)$

9 $\left(-\frac{8}{5}, \frac{6}{5}\right)$ **10** $\left(-\frac{7}{5}, -\frac{4}{5}\right)$ **11** $x \geqslant 1$ **12** $y < 3$

13 $x > -2$ **14** $y \leqslant -1$ **15** $x + y > 6$ **16** $y \leqslant 2x - 1$

17

18

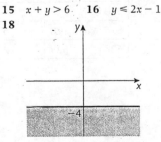

19

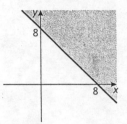

20

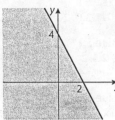

21 $x + y \leqslant 5, y \geqslant 0$ **22** $y < x + 4, x \geqslant 0$

23

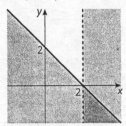

24

Exercise 47

1 x: hyp, y: opp, z: adj   **3** $\frac{4}{3}$

5   87   **7**   6.66 cm   **9**   11.3 cm   **11**   100 m²

Exercise 47*

1   14.4 cm   **3**   8.45 m   **5**   22.4 m

7 $x = 10.9$ cm, $y = 6.4$ cm

Exercise 48

1   45°   **3**   70.0°   **5**   75°   **7**   28.2°   **9**   15°

Exercise 48*

1   **a**   69°   **b**   139°

3   160°   **5**   36.4 m   **7**   13.9°

Exercise 49 (Revision)

1   7.00   **2**   6.71   **3**   6.99

4   11.0   **5**   8.57   **6**   6.93

7   59.0°   **8**   32.5   **9**   58.0°

10   5.19 cm²   **11**   30°   **12** $\theta = 46.3°$

13 $\theta = 5.20°$ **14** $\theta = 59.6°$ **15** $\theta = 16.1°$

Exercise 49* (Revision)

1 $x = 6.53$ cm, $y = 1.55$ cm   **2** $x = 34.6$ cm, $y = 29.1$ cm

3 $x = 8.39$ m, $y = 3.53$ m   **4** $x = 12.1$ cm, $y = 5.12$ cm

5   549 m

6   **a**   0.63.4°   **b**   243°

7   **a**   1.01 m   **b**   Undesirable to have too large a blind distance.

8   18.4°   **9**   71.6°

10   56.3°   **11**   144 cm

12   23.3 m   **13**   50.4 m

14   correct proof   **15**   correct proof

Exercise 50

1

Score	1	2	3	4	5	6
f	9	10	7	5	4	5

3   Mean = 10, Median = 12, Mode = 14

Exercise 50*

1   Mean = 48.9 s, median = 45 s.

3   10.5 years

Exercise 51

1 204°, 32°, 20°, 56°, 48°; 29%

Exercise 51*

1 **a** 2.6; 2
3 14 min 41 s

Exercise 52 (Revision)

1 mean = 21, median = 21, mode = 20
2 **a** mean = 1.69, median = 1.72   **b** new mean 1.76
3 **a**

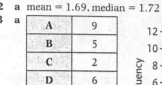

A	9
B	5
C	2
D	6
E	8

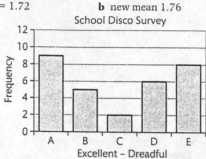

School Disco Survey

b Most felt strongly one way or the other, not so many thought it was average.

4 Strawberries 125°, bananas 75°, yoghurt 100°, iced water 60°.
5 **a**

Score	Frequency
1	7
2	10
3	5
4	10
5	6
6	12

b

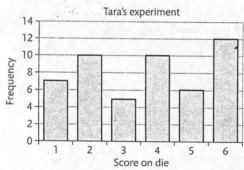

Tara's experiment

c No clear bias. (Sample is too small to draw conclusions from.)
6 **a**

Times	Frequency
40–45	12
45–50	5
50–55	8
55–60	4
60–65	2
65–70	10
70–75	4
75–80	5

b

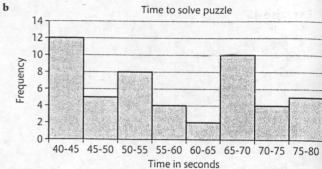

Time to solve puzzle

7 **a** 85   **c** 23.5%
   b

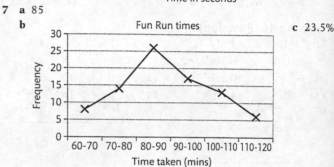

Fun Run times

8 **a** 14.2 m   **b** 1.40 m
9 **a** mean = 0.8,   median = 0.5, mode = 0   **b** 0.909

Exercise 52* (Revision)

1 **a** mean = 7.45, median = 7.45   **b** new mean = 8.2
2 **a** Mean = 2.43, median = 2, mode = 0
   b The mode as it is the most flattering figure.
   c

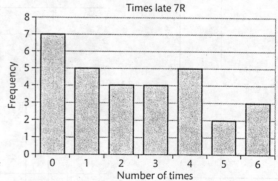

Times late 7R

3 540
4 **a**

Weight g	Frequency
$490 \leqslant w < 495$	1
$495 \leqslant w < 500$	3
$500 \leqslant w < 505$	10
$505 \leqslant w < 510$	18
$510 \leqslant w < 515$	1
$515 \leqslant w < 520$	5
$520 \leqslant w < 525$	2

b

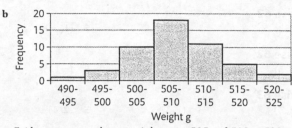

c Evidence suggests the mean is between 505 and 510 so 500 is probably minimum weight.

5 a 126

b

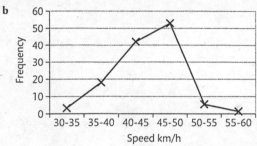

c Speed limit is probably 50 km/h as there is a sharp cut off at that speed.

6 a 15

b

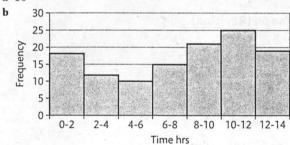

c 46.7%

7 1.74 m

8 a 83 cm   **b** 9.37 cm

Multiple choice 2

1 A   **2** D   **3** A   **4** B   **5** D
6 C   **7** B   **8** B   **9** C   **10** A

Self-assessment 2

1 a $^1 0.0504$ **b** 0.00002

2 a $\frac{\cancel{2}^1}{\cancel{5}} \times \frac{\cancel{15}^3}{\cancel{2}\cancel{1}}_7 = \frac{3}{7}$ **b** $\frac{39}{12} + \frac{28}{12} = \frac{67}{12} = 5\frac{7}{12}$

   c $\frac{^2\cancel{14}}{_3\cancel{15}} \times \frac{\cancel{25}^5}{7} = \frac{10}{3} = 3\frac{1}{3}$

3 a 295 kg : 472 kg   **b** \$0.64 : \$2.24 : \$1.60

4 108

5 78.4 km/h

6 27.8 seconds

7 a $3a^2$ **b** $\frac{1}{2x}$ **c** $3b^2$

8 $\frac{7a+2b}{14}$ **b** $\frac{y}{10}$ **c** $\frac{35x-12y-81}{20}$

9 a ± 5 **b** 16   **c** ± 6

10 a 68 °F   **b** 37.8 °C   **c** −40°

11 a y^{13} **b** b^9 **c** $16a^2b^8$

12 a b^9 **b** $49a^8$ **c** p^8

13 a $\frac{7d}{16b}$ **b** $\frac{7a^2c^3}{b}$ **c** 3

14 a $x > 12$ **b** $x \leqslant 7$

15 $-5, -4, -3, -2, -1, 0, 1, 2$

16 a, b, d   **c** (4, 7)

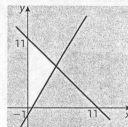

17 a $x = 4.37$ **b** $y = 2.33$ **c** $z = 67.4°$

18 125°

19 a mean = 23.4, median = 23, mode = 27   **b** 23

20 $p = 80, q = 48, r = 56, s = 64, t = 50$

Exercise 53

1 a 1.1   **b** 1.2   **c** 1.30
   d 1.01   **e** 1.15   **f** 1.25

3 660 pupils

5 €515.21

Exercise 53*

1 a 1.125   **b** 1.04   **c** 1.05

3 \$436.80   **5** €692.84

7 £8870, £3940, £1750   **9** $P = \$614$

11 $n = 14.2$ years

Exercise 54

1 7, 14, 21, 28, 35   **3** 1, 2, 3, 4, 6, 12
5 $2 \times 2 \times 7$ **7** No
9 $2 \times 2 \times 2 \times 11$

Exercise 54*

1 No   **3** 3, 7, 11
5 1, 3, 5, 15, 25, 75   **7** $3 \times 5 \times 11$
9 59, 61   **11** $2^4 \times 3^2 \times 7$

Exercise 55

1 2   **3** 22   **5** 30   **7** $2x$
9 $4y^2$ **11** $6ab$ **13** $\frac{3}{4}$ **15** $\frac{1}{2}$
17 $\frac{1}{5}$ **19** $\frac{17}{140}$ **21** $\frac{7}{12}$

Exercise 55*

1 HCF = 6, LCM = 36   **3** HCF = y, LCM = $6xyz$
5 HCF = xy, LCM = x^2yz **7** HCF = $3xyz$, LCM = $18x^2y^2z^2$
9 $\frac{13}{36}$ **11** $\frac{29}{72}$

Exercise 56

1 6.96   **3** 6.96   **5** 134   **7** 12.9
9 2.58   **11** 2.69   **13** 11.3   **15** 625
17 191   **19** 1.75×10^{10}

Exercise 56*

1 3.43   **3** −1.01   **5** −0.956   **7** 0.103   **9** 3.60

Exercise 57 (Revision)

1 1.12, 1.25, 75%, 99%

2 0.88, 0.75, 75%, 99%

3	**a** $535	**b** $572.45	**c** $612.52
4	**a** £104	**b** £108.16	**c** £121.67
5	**a** €26 250	**b** €27 562.50	**c** €31 907.04
6	**a** £41 400	**b** £38 088	**c** £29 658.67
7	**a** 2	**b** 5	**c** 8 **d** 2
8	**a** 12	**b** 42	**c** 120 **d** 630

9 2.64   **10** 4.37   **11** 0.245   **12** 6.75

13 5.94   **14** 0.314   **15** 27.3   **16** 2 400 000

17 26 200   **18** 15.6   **19** 755 000   **20** 25.5

Exercise 57* (Revision)

1 **a** €3345.56   **b** 12 years

2 **a** $134 391.64   **b** 14 years

3 £7366.96

4 **a** $425 570.14   **b** $597 895.41

5 €652.70

6 **a** £16 769.97   **b** £13 887.21

7 **a** HCF = 10, LCM = 420

b HCF = 28, LCM = 420

c HCF = 14 , LCM = 210

8 **a** HCF = $2xyz$, LCM = $12x^2y^2z^2$

b HCF = $5pq$, LCM = $140pq$

c HCF = $6a^2b^2c^2$, LCM = $36a^4b^3c^4$

9 862 000   **10** 2.79   **11** 2.31

12 5.68   **13** 14.7   **14** 0.104

Exercise 58

1 $x(x + 3)$ **3** $5(a - 2b)$ **5** $2x(x + 2)$

7 $ax(x - a)$ **9** $3pq(3p + 2)$

Exercise 58*

1 $5x^3(1 + 3x)$ **3** $3x^2y^2(3x - 4y^2)$ **5** $ab(c^2 - b + ac)$

7 $3x(10x^2 + 4y - 7z)$ **9** $\frac{xy}{16}(2x^2 - 4y + xy)$

11 $4pqr(4pr^2 - 7 - 5p^2q)$ **13** $(x - y)^2(x + y - 1)$

Exercise 59

1 $x + 1$ **3** 2 **5** $\frac{t}{r}$

Exercise 59*

1 $x + y$ **3** $2 + 3x^2$ **5** y

7 1 **9** 5

Exercise 60

1 $x = 8$ **3** $x = 2$ **5** $x = -6$

7 $x = -4$ **9** $x = 14$ **11** $x = 0$

13 90 cm

Exercise 60*

1 $x = 9$ **3** $x = 9$ **5** $x = 0$

7 $x = 3$ **9** $x = 7$ **11** 60 km, 3 h

Exercise 61

1 $x = 2$ **3** $x = \frac{3}{5}$ **5** $x = 10$

7 $x = 50$ **9** $x = \frac{5}{3}$

Exercise 61*

1 $x = 4$ **3** $x = \frac{1}{6}$ **5** $x = 4$

7 $x = \pm 2$ **9** $x = \frac{5}{6}$

Exercise 62

1 (5, 3)   **3** (1, 2)   **5** (1, −1)

Exercise 62*

1 (8, 3)   **3** (1, 5)   **5** (−1, 5)

Exercise 63

1 (2, 5)   **3** (1, 3)   **5** (2, 1)

Exercise 63*

1 (3, −1)   **3** (−0.4, 2.6)   **5** (0.5, 0.75)

7 (−0.6, −0.8)

Exercise 64

1 $x = 3, y = 1$ **3** $x = 1, y = 6$ **5** $x = 3, y = -1$

Exercise 64*

1 $x = 1, y = 2$ **3** $x = 2, y = 1$ **5** $x = -3, y = \frac{1}{2}$

Exercise 65

1 29, 83   **3** 9, 4   **5** Burger 99p, cola 49p

7 11

Exercise 65*

1 (2, 3)   **3** 1.5 m s⁻¹   **5** 7.5 km   **7** 50 m

Exercise 66 (Revision)

1 $x(x - 8)$ **2** $3x(x + 4)$ **3** $6xy(y - 5x)$

4 $3x(4x^2 + 3x - 5)$ **5** $x - 1$ **6** $\frac{(x + y)}{(x - y)}$

7 $x = 4$ **8** 6   **9** −4

10 2   **11** 24   **12** (−1, 3)

13 (0, 3)   **14** (2, 2)   **15** (1, 3)

16 CD £7.50, Tape £3.50   **17** 19 @ 10c, 11 @ 20c

Exercise 66* (Revision)

1 $3x^3(x - 4)$ **2** $\frac{2}{3}\pi r^2(2r + 1)$

3 $6x^2y(4xy - 3)$ **4** $3a^2b^2c^2(5b - 3a + 7c)$

5 $\frac{x}{y}$ **6** x **7** $x = \frac{1}{3}$

8 $x = -4$ **9** $x = 6$ **10** $x = \frac{1}{2}$

11 70 years   **12** (2, 3)   **13** (4, 1)

14 (4, 1.5)   **15** $\left(3\frac{1}{3}, 2\right)$ **16** $a = \frac{3}{11}, b = \frac{2}{11}$

17 Mike is 38, Ben is 14

Exercise 67

1 **a** 65 km/h   **b** 50 km/h   **c** 12:00

   d 72.5 km   **e** 11:08 approx

3 **a**   **b** 14:00

Exercise 67*

1 a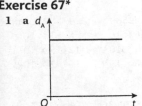

b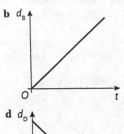

c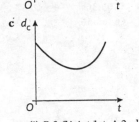

d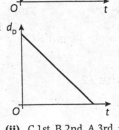

3 a (i) B & C joint 1st, A 2nd
(iii) A 1st, B 2nd, C 3rd
b 28.5 s   c B

(ii) C 1st, B 2nd, A 3rd

d (i) A   (ii) C

Exercise 68

1 a $2\,m/s^2$ b $4\,m/s^2$ c 150 m   d 10 m/s
3 a $2\,m/s^2$ b $1\,m/s^2$ c 8000 m   d 50 m/s
5 11:40

Exercise 68*

1 a $0.6\,m/s^2$ b $-0.5\,m/s^2$ c 4.43 m/s (3 s.f.)
3 a $t = 10\,s$, so distance = 1900 m
b $-3\,m/s^2$ c 47.5 m/s
5

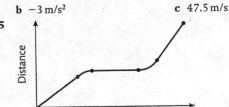

Exercise 69 (Revision)

1 a 20 min   b 10:00   c 10 km/h   d $3\frac{1}{3}$ km
2 a 0.4 m/s   b 10 min   c 0.2 m/s
3

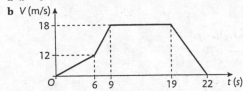

a $\frac{2}{15}\,m/s^2$ b $0\,m/s^2$ c $\frac{4}{15}\,m/s^2$ d $3\frac{1}{7}\,m/s$
4 a 400 min   b 1050 m   c 10.5 m/s   d 0.33 m/s²
5 a $3\,m/s^2$ b $0\,m/s^2$ c 43.3 m/s

Exercise 69* (Revision)

1 a 50 m/s   b 0.5 s at 30 m/s approx
2 a 32 m   b $\frac{2}{3}\,m/s^2$ c 3.2 m/s

3 a $x = 6$
b V (m/s)

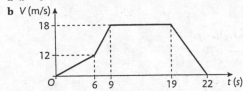

c (i) $2\,m/s^2$ (ii) $6\,m/s^2$ (iii) 13.1 m/s
4 a (i) 6.4 m/s   (ii) 23.0 km/h
b 7.08 m/s   c 0.590 m/s²
5 a False; it is constant at $\frac{2}{3}\,m/s^2$ b True
c True   d False; it is 72 km/h

Exercise 70

1 2.46   3 8.09   5 8.76   7 67.6 m

Exercise 70*

1 6.57 m
3 a 107 m   b 79.7 m
5 a 2.25 km   b 3.90 km   c 4.5 km

Exercise 71

1 48.6°   3 78.9°   5 70.5°

Exercise 71*

1 37.8°   3 57.3°   5 72.7°

Exercise 72

1 5   3 5.18   5 60°   7 32.2°
9 1.38 m   11 62.3 m   13 10.0°

Exercise 72*

1 18   3 7.96   5 16.8°
7 a 4.66 km north   b 17.4 km west
9 22.2 m   11 7.99 km

Exercise 73 (Revision)

1 $x = 14.1\,cm$, $d = 70.5°$ 2 $x = 7.87\,m$, $d = 10.2°$
3 $x = 16.7\,km$, $d = 39.9°$ 4 $x = 11.7\,cm$, $d = 31.2°$
5 $x = 2.38\,m$, $d = 4.62°$ 6 $x = 14.3\,km$, $d = 79°$
7 43.4 cm²   8 33.7°
9 a 0.5   b $f = 30°$
10 a 20.5 m   b 19.1 m   c 20.7 m

Exercise 73* (Revision)

1 Ascends in 3 min 52 s, so reaches surface with 8 seconds to spare.
2 a 17.2 km, 284°   b 18:11:10
3 3.56 m
4 a 16.2 m   b 16.2 s   c 432 m
5 $p = 25$ 6 $q = 5$
7 2.5 km   8 1.79 m
9 $p = 25$ 10 $q = 5$

Exercise 74

1 a

Score	1	2	3	4	5	6
f	7	8	5	4	3	3

b 2.9   c 2.5
3 a 80   b 2.4   c 40%

Exercise 74*

1 197 cm   **3** 91.5 s

Exercise 75 (Revision)

1 a 20   **b** 1   **c** 0   **d** 1.4
2 a 40   **b** 2   **c** 1   **d** 2.05
3 a 50   **b** 16–20   **c** 15.3
4 a 24   **b** $4 < w \leqslant 6$ **c** 4.75
5 a 25   **b** $17.5 < t \leqslant 20.5$ **c** 17.92 mins
6 a 32   **b** $14 \leqslant a < 15$ **c** 14.3
     d Decrease as below the mean

Exercise 75* (Revision)

1 a 8   **b** 73   **c** 74   **d** 73.2
2 a 7   **b** 2   **c** 26.7%
3 a 28 days ⇒ February   **b** 16–20   **c** 15.3
4 a 1000   **b** 601–800   **c** 770.5
5 a 540   **b** 3000–3300   **c** 3360
6 a $22 - x$ **b** 9

Multiple choice 3

1 D   **2** B   **3** C   **4** B   **5** C
6 B   **6** B   **6** D   **6** C   **10** C

Self-assessment 3

1 a £466.19   **b** £473.89
2 £10 675
3 9.45×10^{12} km = a light year.
4 a 0.09215784314   **b** 9.22×10^{-2}
5 HCF = 6, LCM = 60
6 a $\dfrac{5}{xy}$ **b** $\dfrac{2x + 6}{21}$
7 $x = \dfrac{1}{9}$
8 $x = -0.4, y = -9.2$
9 b (i) $0.5\,\text{m/s}^2$ (ii) $-1\,\text{m/s}^2$ (iii) 7.5 m/s
10 b Eliza home 12:00, Albert home 12:00
     c Eliza 1.48 m/s, Albert 2.22 m/s
11 a 6.43 cm   **b** 7.66 cm   **c** 3.66 cm
     d 60.3°   **e** 120°   **f** 12.9 cm²
12 a 5 m   **b** 30.0°   **c** 14.1 m
13 $x = 25.8°, y = 10.3$ cm
14 2.6 goals/game
15 a $\Sigma fx = 72$ goals   **b** 1.44 goals/game

Exercise 76

1 $442   **3** $40   **5** 13.3%   **7** €74.10

Exercise 76*

1 $44   **3** $60   **5** €73 000   **7** $450

Exercise 77

1 150   **3** 300   **5** 200   **7** 4×10^3
9 8.7×10^4 **11** 2×10^6 **13** 7×10^6 **15** 8×10^2

Exercise 77*

1 4   **3** 100 cm²   **5** 1.2×10^9 **7** 7.06×10^8
9 0.2   **11** 10 000   **13** 3×10^{-1} **15** 1×10^{-3}

Exercise 78 (Revision)

1 $120   **2** $20   **3** £2500   **4** 80 seconds
5 1×10^8 **6** 4×10^{13} **7** 6×10^{15} **8** 1×10^{11}
9 2×10^2 **10** 2×10^3 **11** 5×10^7 **12** 5×10^8
13 4×10^{14} **14** 3×10^{15} **15** 1×10^3 **16** 2×10^{10}

Exercise 78* (Revision)

1 $4329   **2** $1811.59   **3** $409.36   **4** $2573.53
5 $1283.76   **6** $888 889   **7** €6863.56   **8** $2326.24
9 2×10^6 **10** 4×10^{-3} **11** 2×10^{-12}
12 7×10^4 **13** 2×10 **14** 8×10^{-3}

Exercise 79

1 $x = a - 2$ **3** $x = c - a$ **5** $x = \dfrac{(b - a)}{3}$ **7** $x = \dfrac{(4 - b)}{a}$
9 $x = \dfrac{c}{(a + b)}$ **11** $x = \dfrac{(a - 3b)}{3}$ **13** $x = ab$ **15** $x = r(p + q)$

Exercise 79*

1 $x = \dfrac{(c - b)}{a}$ **3** $x = \dfrac{cd}{b}$ **5** $x = \dfrac{P - b^2}{\pi}$
7 $x = \pi - b$ **9** $x = \dfrac{a}{b}$ **11** $x = \dfrac{s}{(p - q)}$
13 $h = \dfrac{3V}{\pi r^2}$ **15** $s = \dfrac{(v^2 - u^2)}{2a}$ **17** $a = \dfrac{S(1 - r)}{(1 - r^n)}$

Exercise 80

1 $x = \sqrt{\left(\dfrac{b}{a}\right)}$ **3** $x = \sqrt{(2D - C)}$ **5** $x = \sqrt{\left(\dfrac{(c - 2b)}{a}\right)}$
7 $x = \dfrac{ab}{a - 1}$ **9** $r = \sqrt{\left(\dfrac{A}{4\pi}\right)}$ **11** $r = \sqrt[3]{\dfrac{3V}{4\pi}}$

Exercise 80*

1 $x = \sqrt{\left(\dfrac{S}{R}\right)}$ **3** $x = \dfrac{c}{b - a}$ **5** $x = \dfrac{\tan b + ac}{(1 - a)}$
7 $x = \sqrt{(Ab - Da)}$ **9** $v = \sqrt{(2gh)}$ **11** $a = b - \sqrt{12s}$
13 $d = \left(\dfrac{k}{F}\right)^3$ **15** $x = \dfrac{p(y + 1)}{y - 1}$

Exercise 81

1 a 155 min   **b** 2 kg
3 a 0.15   **b** 200
5 $A = \dfrac{\pi r^2}{4}, P = r\left(2 + \dfrac{\pi}{2}\right)$
7 a $A = 19.6$ cm², $P = 17.9$ cm
     b 11.28 cm   **c** 14.0 cm   **d** 4.55

Exercise 81*

1 a 22   **b** 400
3 a 209   **b** 3   **c** 8
5 a $A = r^2\left(\dfrac{\pi}{4} + 1\right)$ $P = r\left(\dfrac{\pi}{2} + 4\right)$ $A = 28.6$ cm² $P = 22.3$ cm
     b 4.10 cm   **c** 12.6 cm   **d** 3.12 cm

Exercise 82 (Revision)

1 $x = \dfrac{b}{a}$ **2** $x = ac$ **3** $y = \dfrac{a - c}{b}$ **4** $y = \sqrt{\dfrac{d}{b}}$
5 $y = \dfrac{b^2}{a}$ **6** $y = \dfrac{d}{a - c}$ **7** $y = \dfrac{bc}{c - 1}$
8 a 45 m/s   **b** $t = \dfrac{v - 20}{2}$ **c** 2 s
9 a 26   **b** 100   **c** $1800
10 a $A = 4n + 2$ **b** 402 cm²
     c $n = \dfrac{A - 2}{4}$ **d** 53

Exercise 82* (Revision)

1 $x = \dfrac{c - b}{a}$ **2** $x = \dfrac{b}{a - d}$ **3** $x = \dfrac{ab - \tan c}{a}$

4 $y = \sqrt{\dfrac{a}{b - c}}$ **5** $y = \dfrac{ac - d}{a - b}$ **6** $y = b - d(c - a)^2$

7 a 20 **b** $A = \dfrac{N - 2}{0.4}$ **c** 70 m²

8 a 2.81 s **b** $\ell = 10\left(\dfrac{t}{2\pi}\right)^2$ **c** 1.58 m

9 a \$14 **b** $n = \dfrac{200}{C - 10}$ **c** 160

10 a 291 cm² **b** $h = \dfrac{A}{2\pi r} - r$ **c** 15.9 cm

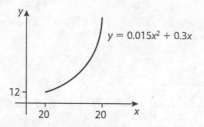

b 61.9 m **c** 48.6 mph **d** 0.7 s

Exercise 83

1 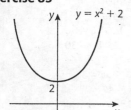 **3**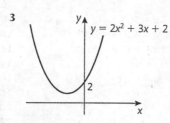

5 a

t	0	1	2	3	4	5	6
y	5	2.5	1	0.5	1	2.5	5

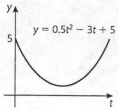

b 5 m **c** 0.5 m, 3 s
d 5 m **e** $0 \leqslant t \leqslant 0.76$, $5.24 \leqslant t \leqslant 6$

Exercise 83*

1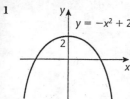

3 a

t	0	1	1.5	2	3	4
y	4	6	6.25	6	4	0

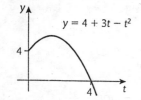

b 4 m **c** 4 p.m.
d 6.25 m, 1.30 p.m. **e** Between 12.23 p.m. and 2.37 p.m.

5 a

x	20	30	40	50	60	70	80
y	12	22.5	36	52.5	72	94.5	120

Exercise 84 (Revision)

1 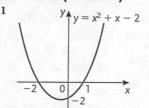 **2**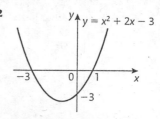

3

x	−2	−1	0	1	2	3
y	5	3	3	5	9	15

4

x	−2	−1	0	1	2	3
y	−3	−5	−5	−3	1	7

5

x	−3	−2	−1	0	1	2	3	4
y	10	7	6	7	10	15	22	31

6

x	−3	−2	−1	0	1	2	3	4
y	−4	−6	−6	−4	0	6	14	24

7 a 4.9 cm² **b** 4.5 cm **c** 12.9 cm²

8 a

t	0	1	2	3	4	5
y	0	5	20	45	80	125

b i 61 m **ii** 2.2 s

Exercise 84* (Revision)

1

x	−3	−2	−1	0	1	2	3
y	23	10	1	−4	−5	−2	5

2

x	−3	−2	−1	0	1	2	3
y	−23	−10	−1	4	5	2	−5

3

x	−4	−3	−2	−1	0	1	2	3	4
y	13	0	−9	−14	−15	−12	−5	6	21

4

x	0	1	2	3	4
y	9	1	1	9	25

5 b 28.3 m

6 a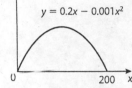

$y = 0.2x - 0.001x^2$

b 10 m   **c** $29 \leqslant x \leqslant 170$

7 a $k = 3$

t	0	1	2	3	4
p	0	7	8	3	-8

c (i) £8333 @ $t = 1.7$ months   (ii) $t > 3.3$ months

8 a 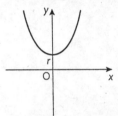   **b**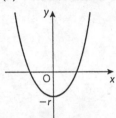

c 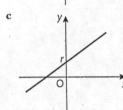   **d**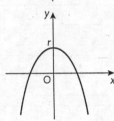

Exercise 85
1 $a = 230°, b = 25°$ **3** $a = 100°, b = 260°$
5 $a = 110°, b = 35°$ **7** $a = 108°, b = 72°, c = 54°$
9 $a = 55°$ **11** $a = 90°, b = 32°, c = 58°$

Exercise 85*
1 a $37°$ **b** $53°$
3 a $22.5°$ **b** $45°$
5 $OBA = 90° - x$; $AOB = 2x$; $COB = 180° - 2x$
7 a 12 cm   **b** 54 cm²   **c** 7.2 cm

Exercise 86
1 $a = 124°$ **3** $a = 102°$
5 $a = 56°$ **7** $a = 50°, b = 130°$
9 $a = 100°, b = 160°$ **11** $a = 67°, b = 85°$

Exercise 86*
1 Correct proof   **3** $126°$
5 $x = 116°, y = 64°$ **7** Correct proof

Exercise 87
1 T_1, T_2 **3** $x = 5\, y = 8$
5 a 7.5 cm   **b** 10.5 cm

Exercise 87*
1 E(4, 2), F(4, 4)
3 a 11.25 cm   **b** 5 cm
5 18 m   **7** 239 litres

Exercise 88
1 10.3 cm   **3** 11.8 m   **5** 3.16 m

Exercise 88*
1 12.4 cm   **3** 17:28   **5** 27.5 m

Exercise 89
1 Yes SAS   **3** No   **5** Yes RHS

Exercise 89*
1 Correct proof   **3** Correct proof
5 Correct proof

Exercise 90 (Revision)
1 $a = 50°, b = 280°$ **2** $a = 90°, b = 30°$
3 $a = 70°, b = 20°$ **4** $a = 55°, b = 70°$
5 $a = 60°$ **6** $a = 140°$
7 $a = 50°$ **8** $a = 140°$
9 $a = 40°, b = 20°$ **10** $a = 120°, b = 30°$
11 $a = 65°, b = 115°$ **12** $a = 50°, b = 130°$
13 $a = 18, b = 14$ **14** $a = 10.5$
15 $a = 12, b = 12$ **16** $a = 16, b = 3$
17 $a = 6.40$ **18** $b = 4.47$
19 $c = 15.0$ **20** AC = 36.6 cm
21 a No   **b** Yes SSS   **22** SAS

Exercise 90* (Revision)
1 $a = 60°, b = 300°$ **2** $a = 90°, b = 45°$
3 $2a = 36°, 3a = 54°$ **4** $a = 55°, b = 35°$
5 $a = 100°$ **6** $a = 80°$
7 $a = 290°$ **8** $a = 102°$
9 $a = 40°, b = 60°$ **10** $a = 35°, b = 25°$
11 $a = 110°, b = 70°$ **12** $a = 60°, b = 60°$
13 $a = 4.5$ **14** $a = 6, b = 4.5$
15 $a = 4.5, b = 2.5$
16 a The angles of both triangles are the same: 55°, 60°, 65°
   b $a = 3.61, b = 2.87$
17 $a = 5.39$ **18** $a = 5.20$
19 a $r = 11.7$ **b** $a = 18.7$ **20 a** XC = 2   **b** AC = 4.47
21 see online pdf   **22** see online pdf

Exercise 91
1 a $p(\text{odd}) = \frac{6}{10}$
   b Inconclusive. More trials would improve the experiment.
3 $p(\text{vowel}) = \frac{9}{20}$

Exercise 91*
1 a $p(L\ \text{success}) = \frac{8}{12}$ $p(R\ \text{success}) = \frac{4}{12}$
   b Learning curve, so warm up before playing. Practise more from RHS.
3 a $p(W) = \frac{12}{20} = \frac{3}{5}$; $p(P) = \frac{8}{20} = \frac{2}{5}$
   b No. of $W = \frac{3}{5} \times 100 = 60 \rightarrow$ No. of $P = 40$

Exercise 92
1 a $p(g) = \frac{4}{10} = \frac{2}{5}$ **b** $p(a) = \frac{3}{10}$
   c $p(t) = 0$ **d** $p(S) = \frac{9}{10}$
3 a $p(R) = \frac{1}{2}$ **b** $p(K) = \frac{1}{13}$
   c $p(\text{mult of } 3) = \frac{3}{13}$ **d** $p(AJQK) = \frac{4}{13}$
5 a $\frac{1}{10}$ **b** $\frac{1}{2}$
   c $\frac{3}{10}$ **d** $\frac{2}{5}$
7 $p(S) = \frac{3}{5}$
9 a $p(P) = \frac{1}{2}$ **b** $p(R') = \frac{2}{3}$
   c $p(WRP) = 1$ **d** $p(Y) = 0$

Exercise 92*

1 a (i) $\frac{5}{36}$ **(ii)** $\frac{1}{12}$ **(iii)** $\frac{1}{12}$ **(iv)** $\frac{15}{36}$
b 7

3 a $\frac{9}{25}$ **b** $\frac{14}{25}$ **c** $\frac{6}{25}$ **d** $\frac{9}{25}$

5 a $\frac{1}{2}$ **b** $\frac{3}{18} = \frac{1}{6}$ **c** $\frac{13}{18}$ **d** $\frac{5}{18}$

7 a $\frac{1}{36}$ **b** 0 **c** $\frac{11}{36}$ **d** $\frac{8}{36}$

9 $f = 5$

Exercise 93 (Revision)

1 $\frac{2}{3}$; More trials for a better estimate

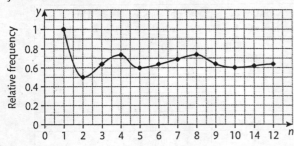

2 $\frac{13}{15}$

3 a $\frac{7}{51}$ **b** $\frac{1}{17}$ **c** $\frac{1}{3}$ **d** 0

4 HH, HT, TH, TT **a** $\frac{1}{4}$ **b** $\frac{1}{2}$

5 a $\frac{1}{8}$ **b** $\frac{1}{8}$ **c** 0 **d** $\frac{15}{16}$

6 $\frac{10}{494} = \frac{5}{247}$

7 a 0 **b** $\frac{9}{25}$ **c** $\frac{16}{25}$ **d** $\frac{9}{25}$

8 a $\frac{1}{13}$ **b** $\frac{2}{13}$ **c** $\frac{3}{4}$ **d** $\frac{3}{26}$

9 a $\frac{1}{8}$ **b** $\frac{1}{2}$ **c** $\frac{3}{4}$ **d** $\frac{5}{8}$

10 a RG GR GG **b** $\frac{2}{3}$

Exercise 93* (Revision)

1 a 2002, $\frac{7}{10}$; 2003, $\frac{13}{20}$; 2004, $\frac{17}{30}$
b Decrease in numbers from 2002 is suggested by the data

2 a $\frac{1}{12}$ **b** $\frac{3}{4}$ **c** $\frac{11}{36}$

3 a (i) $\frac{1}{11}$ **(ii)** $\frac{2}{11}$ **(iii)** 0
b Z or U, $\frac{2}{11}$

4 a $\frac{6}{25}$ **b** $\frac{19}{25}$ **c** $\frac{3}{25}$ **d** $\frac{9}{25}$

5 £45

6 HHH, HHT, HTH, HTT, THH, TTH, TTT
a $\frac{1}{8}$ **b** $\frac{3}{8}$ **c** $\frac{1}{2}$

7 a $\frac{25}{28}$ **b** $\frac{1}{6}$ **c** 102

8 a

	Pink Box		
	△	○	☆
□	□△	□○	□☆
Blue Box □	□△	□○	□☆
☆	☆△	☆○	☆☆

b (i) $\frac{5}{9}$ **(ii)** $\frac{4}{9}$

9 10

10 $\frac{8}{245}$

Multiple choice 4

1 D **2** A **3** B **4** C **5** A
6 D **7** A **8** A **9** C **10** D

Self-assessment 4

1 $14\,286 **2** €22\,727
3 a $\approx 625\,000$ **b** ≈ 500
4 $x = ab - Vg$ **5** $x = \frac{cd + ab}{a + c}$
6 $x = (PR)^2 - Q$
7 a, b $x < 20.3$ or 1.8 for both equations.

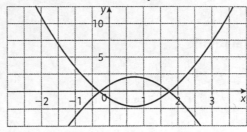

8 a $x = 13$ **b** $y = 13.2$
9 12.0 units **10** $x = 6\frac{2}{7}, y = 12$
11 $x = 15, y = 9$ **12** $x = 130°, y = 25°, z = 65°$
13 $x = 70°, y = 55°, z = 35°$ **14** $x = 130°, y = 65°, z = 115°$
15 $x = 124°, y = 34°, z = 62°$

16 a

	1	2	3	4	5	6
1	2	2	4	5	6	7
2	3	4	5	6	7	8
3	4	5	6	7	8	9
4	5	6	7	8	9	10
5	6	7	8	9	10	11
6	7	8	9	10	11	12

b (i) $\frac{1}{12}$ **(ii)** $\frac{7}{36}$ **(iii)** $\frac{5}{6}$ **(iv)** 0

17 a $\frac{3}{26}$ **b** $\frac{3}{26}$ **c** $\frac{11}{26}$ **d** $\frac{8}{13}$

18 a HHH HHT HTH THH TTH THT HTT TTT
b (i) $\frac{1}{8}$ **(ii)** $\frac{3}{8}$ **(ii)** $\frac{7}{8}$

19 a $e = \dfrac{\left(t\sqrt{\frac{5}{h}} - 1\right)}{\left(t\sqrt{\frac{5}{h}} + 1\right)}$ **b** $\frac{2}{3}$

20 BC common to both △s so SSS

Exercise 94

1 a 4 days **b** 2 days **c** $2\frac{2}{3}$ days
3 a 60 years **b** 15 years **c** 1200 years
5 a 32 km/litre **b** 20 litres

Exercise 94*

1 a

Number of light bulbs (N)	Power of each bulb (P)
6	500
5	600
2	1500
30	100

b $NP = 3000$

3

Number of men	Number of tunnels	Time in years
100 000	4	4
100 000	2	2
20 000	8	40
400 000	2	0.5

5 a 16 000 kg
b (i) 4 min **(ii)** 0.6 min **(iii)** 1 min **(iv)** 8 sec

Exercise 95

1	0.375	3	0.28125	5	$0.1\dot{8}$	7	$0.3\dot{8}$
9	$\frac{3}{20}, \frac{5}{64}$	11	$\frac{5}{9}$	13	$\frac{7}{90}$	15	$\frac{5}{90} = \frac{1}{18}$

Exercise 95*

1	$0.4\dot{6}$	3	$2.3\dot{0}$	5	$\frac{11}{16}, \frac{7}{40}$	7	$\frac{24}{99} = \frac{8}{33}$
9	$9\frac{19}{990}$	11	$\frac{412}{999}$	13	$\frac{11}{90}$	15	$0.0\dot{3}\dot{7}$

Exercise 96 (Revision)

1 **a** 2 days **b** 1 day **c** $\frac{1}{2}$ day

2 **a** $2\frac{1}{2}$ days **b** $1\frac{1}{4}$ days **c** $\frac{1}{2}$ day

3

Number of years, n	Number of men, m
1	12 000
2	6000
3	4000
4	3000
6	2000
$12\,000 \div m$	m
n	$12\,000 \div n$

4 **a** (i) $2\frac{1}{2}$ days (ii) $1\frac{1}{4}$ days (iii) 1 day (iv) $\frac{5}{p}$ days

b (i) 10 people (ii) 5 people (iii) 4 people (iv) $\frac{20}{d}$ people

5 **a** 9 hours **b** 90 km/hr

6 **a** $0.\dot{3}$ **b** $0.\dot{1}$ **c** $0.08\dot{3}$ **d** $0.0\dot{6}$

7 **a** $\frac{4}{9}$ **b** $\frac{7}{9}$ **c** $\frac{53}{99}$ **d** $\frac{82}{99}$

8 **a** $\frac{301}{999}$ **b** $\frac{707}{999}$ **c** $\frac{7}{198}$ **d** $\frac{409}{99\,900}$

Exercise 96* (Revision)

1

Number of women, w	Length of dry-stone wall, x m	Time of construction, t days
1	8	8
3	18	6
4	24	6
6	12	2
$2\frac{2}{3}$	32	12
w	x	$x \div w$
w	wt	t
$x \div t$	x	t

2

Number of bees, b	Length of bee's journey, x km	Mass of honey, m g
10	150	10
20	375	50
100	750	500
500	300	1000
10^6	1000	6.67×10^7
b	x	$bx \div 150$
b	$150m \div b$	m
$150m \div x$	x	m

3

Number of grass-cutters, n	Number of rugby pitches, r	Time, t hours
2	1	0·5
$1\frac{1}{3}$	2	1.5
4	4	1
$r \div t$	r	t
n	tn	t
n	r	$r \div n$

4

Number of houses, h	Mass of waste, w kg	Time, t weeks
1	20	1
100	20 000	10
9615	10^6	52
h	w	$w \div 20h$
$w \div 20t$	w	t
h	$20ht$	t

5 **a** $\frac{107}{333}$ **b** $\frac{34}{45}$ **c** $\frac{25}{66}$ **d** $\frac{203}{198}$

6 $\frac{254}{713}$

Exercise 97

1 $x^2 + 5x + 4$ 3 $x^2 - 4x - 12$ 5 $x^2 + 6x + 9$

7 $x^2 - 25$ 9 $15x^2 - 7x - 2$

Exercise 97*

1 $x^2 + 4x - 21$ 3 $x^2 + 24x + 144$

5 $x^2 + x(b - a) - ab$ 7 $15x^3 + 21x^2 + 5x + 7$

9 $\frac{a^2}{4} - \frac{ab}{5} + \frac{b^2}{25}$ 11 4

13 $a = 3, b = 1$

Exercise 98

1 **a** $x^2 + 3x + 2$ **b** $3x + 2$ **c** $x = 3$

3 **a** $5x^2 + 25x + 30$ **b** $2x^2 + 30x + 62$

5 $x = 6$

Exercise 98*

1 **a** $\pi(x^2 + 12x + 36)$ **b** $x = 0.75$

3 $x = 6$

5 **a** $2n$ is divisible by 2 and so is even; $2n + 1$ is then odd.

b $(2n + 1)(2m + 1) = 4mn + 2n + 2m + 1$; $4mn, 2n$ and $2m$ are even so this is odd.

c $(2n - 1)(2n + 1) + 1 = 4n^2 = (2n)^2$

Exercise 99

1 $x(x - 3)$ 3 $x(x - 31)$ 5 $(x - 4)(x + 4)$

Exercise 99*

1 $x(x - 312)$ 3 $(x - 8)(x + 8)$ 5 $(x - 15)(x + 15$

Exercise 100

1 $a = 1$ 3 $a = -1$ 5 $a = 2$

Exercise 100*

1 $a = 3$ 3 $a = -7$ 5 $a = -8$

Exercise 101

1 $(x - 2)(x - 1)$ **3** $(x - 4)(x - 3)$ **5** $(x - 1)(x - 8)$

Exercise 101*

1 $(x + 7)(x + 3)$ **3** $(x - 8)(x - 8)$ **5** $(x + 9)(x + 5)$

Exercise 102

1 $(x + 3)(x - 2)$ **3** $(x + 2)(x - 6)$ **5** $(x + 7)(x - 2)$

Exercise 102*

1 $(x + 6)(x - 5)$ **3** $(x + 12)(x - 5)$ **5** $(x + 8)(x - 15)$

Exercise 103

1 $(x - 1)(x - 2)$ **3** $(x + 1)(x + 12)$ **5** $(x - 4)(x - 4)$

Exercise 103*

1 $(x + 10)(x - 2)$ **3** $(x + 9)(x + 4)$ **5** $(x + 12)(x - 4)$

Exercise 104

1 $x = -1$ or $x = -2$ **3** $x = 7$ or $x = 2$ **5** $x = 0$ or $x = 10$

Exercise 104*

1 $x = -8$ or $x = 4$ **3** $x = 0$ or $x = 8$

5 $x = -1$ or $x = 1$ or $x = -\frac{5}{2}$

Exercise 105

1 $x = 1$ or $x = -2$ **3** $x = -2$ or $x = -4$ **5** $x = 5$ or $x = 3$

Exercise 105*

1 $x = 4$ or $x = 5$ **3** $x = -9$ or $x = -12$

5 $x = -16$ or $x = -6$ **7** $x = -15$ or $x = 8$

Exercise 106

1 $x = 0$ or $x = 2$ **3** $x = 0$ or $x = 25$ **5** $x = -2$ or $x = 2$

Exercise 106*

1 $x = 0$ or $x = 125$ **3** $x = -8$ or $x = 8$ **5** $x = \pm\sqrt{7}$

Exercise 107

1 $3, -4$ **3** $-4, 5$ **5** $5, -7$

7 a $x^2 + 3x$ **b** $x = 3$

9 10 cm by 4 cm   **11** $x = 3$

Exercise 107*

1 11, 13 or $-13, -11$ **3** 30 cm by 40 cm

5 1 s and 2 s   **7** 8, 9 or $-9, -8$

9 20   **11** 4

Exercise 108 (Revision)

1 $x^2 - 10x - 21$ **2** $x^2 + 4x + 4$

3 $2x^2 - 7x - 15$

4 a $x^2 + 5x + 6$ **b** $5x + 6$ **c** $x = 4$

5 $(x - 6)(x + 6)$ **6** $(x + 3)(x + 1)$

7 $(x + 4)(x - 2)$ **8** $x = 6$ or $x = -1$

9 $x = 0$ or $x = 5$ **10** $x = -6$ or $x = 6$

11 $x = -4$ or $x = 5$ **12** 20 cm by 30 cm

Exercise 108* (Revision)

1 $x^2 - 3x - 108$ **2** $4x^2 - 12x + 9$

3 $6x^2 + 7x - 3$ **4** 4.25 m

5 27 cm   **7** $x = -11$ or $x = 11$

7 $x = 0$ or $x = 7$ **8** $x = -7$ or $x = 8$

9 $x = 9$ or $x = 6$ **10** 4, 10

11 b $x = 30$ **12** 400 cm²

Exercise 109

1 $x = 3$ or $x = 2$ **3** $x = -1$ or $x = 3$ **5** $x = 0$ or $x = 3$

Exercise 109*

1 $x = \frac{1}{2}$ or $x = 2$ **3** $x = \frac{2}{3}$ or $x = 2$

5 $y = x^2 - 6x + 5$ **7** $y = x^2 - 6x + 9$

9 Two solutions   One solution   No solutions

Exercise 110

1 $x = -2.2$ or $x = 2.2$ **3** $x = -1$ or $x = 2$

5 $x = -3.8$ or $x = 1.8$ **7** $x = 0.6$ or $x = 3.4$

9 $x = -2.9$ or $x = 3.4$

Exercise 110*

1 $x = -1.3$ or $x = 2.3$ **3** $x = -2.6$ or $x = -0.4$

5 $x = 2$ **7** $x = -2.7$ or $x = 2.2$

9 $x = -2.8$ or $x = 3.2$

Exercise 111

1 a $x = 0$ or $x = 3$ **b** $x = -0.56$ or $x = 3.56$

   c $x = 0.38$ or $x = 2.62$ **d** $x = -0.24$ or $x = 4.24$

   e $x = -0.79$ or $x = 3.79$ **f** $x = 0.21$ or $x = 4.79$

3 a $x = 1$ or $x = 3$ **b** $x = -0.65$ or $x = 4.65$

   c $x = 0.70$ or $x = 4.30$ **d** $x = -0.56$ or $x = 3.56$

5 $(2.71, 3.5)$, no

Exercise 111*

1 a $x = 5$ or $x = 0$ **b** $x = 4.30$ or $x = 0.70$

   c $x = 3.73$ or $x = 0.27$ **d** $x = 0.76$ or $x = 5.24$

3 a $x = -1.78$ or $x = 0.28$ **b** $x = -2.35$ or $x = 0.85$

   c $x = -2.28$ or $x = -0.22$

7 a $y = x + 2$ **b** $y = 3x - 1$

Exercise 112 (Revision)

1 a $-2.6, 2.6, y = 7$ **b** $0, 1, y = x$

   c $-1.3, 2.3, y = x + 3$ **d** $-2.6, 1.6, y = 4 - x$

   e $1, y = 2x - 1$ **f** $-2.4, 0.4, y = 1 - 2x$

2 a $-1.6, 0.6, y = 0$ **b** $-1, 0, y = -1$

   c $-2, 1, y = 1$ **d** $-2.6, 1.6, y = 3$

   e $-3.3, 0.3, y = -2x$ **f** $-1, 1, y = x$

3 a $y = 2$ **b** $y = x$

   c $y = 1 - x$ **d** $y = 3 + 2x$

4 a $y = 0$ **b** $y = -3$

   c $y = 3$ **d** $y = x$

5 a $x^2 + 1 = 0$ **b** $2x^2 + x - 3 = 0$

   c $x^2 - 4x + 3 = 0$ **d** $2x^2 + 2x - 4 = 0$

6 a $-1.4, 0, 1.4, y = 2x$ **b** $0, y = -3x$

   c $1.4, y = 4 - x$ **d** $-1.9, 0.3, 1.5, y = 3x - 1$

Exercise 112* (Revision)

1 **a** $0.4, 2.6, y = 0$ **b** $-0.8, 3.8, y = 4$
   c $0.3, 3.7, y = x$ **d** $-0.2, 4.2, y = x + 2$
   e $-1.4, 3.4, y = 6 - x$ **f** $-1.3, 2.3, y = 4 - 2x$
   g $k = -2.25$

2 **a** $y = -1$ **b** $y = 6$
   c $y = x$ **d** $y = 2x + 2$
   e $y = -4x + 3$ **f** $y = -\frac{1}{}$

3 $(2, 3)$

4 $(72, 14)$

5 $x = 1.13$ or -2.64

6 $x = 1.13$ or -2.64

Exercise 113

1

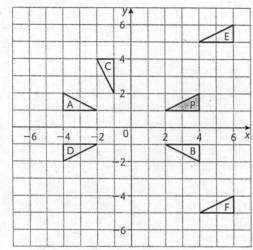

3

Object	Reflection in line	Image
A	$x = 5$	B
F	$x = 5$	G
G	$x = 5$	F
A	$x = 7$	C
D	$x = 12$	B
G	$x = 9$	H
K	$x = 24$	J
I	$x = 12$	G
H	$x = 18$	K
E	$x = 10$	A
J	$x = 15$	G
D	$x = 17$	E

Exercise 113*

1 **a** $(1, -2)$ **b** $(1, 2)$ **c** $(1, 10)$
   d $(1, 18)$ **e** $(4, 1)$ **f** $(5, 2)$
   g $(6, 3)$ **h** $(12, 9)$

3 **a**

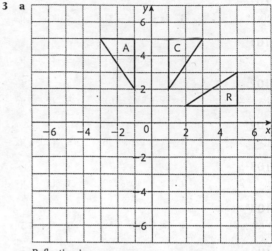

   c Reflection in $y = x$

Exercise 114

1 $(4, 0)$ $(4, -4)$ $(0, -4)$

3 B to A: $(10, 8)$ SF $= 2$; A to E: $(8, 0)$ SF $= \frac{1}{2}$; D to B: $(12, 4)$ SF $= \frac{3}{2}$;
   E to D: $(14, 12)$ SF $= \frac{2}{3}$; D to C: $(4, 4)$ SF $= 2$

Exercise 114*

1 $(0, 0)$ $(-2, 0)$ $(-4, 4)$

3 A to C: $(2, 11)$ SF $= 2$; A to D: $(3, 11)$ SF $= 4$; C to D: $(5, 11)$ SF $= 2$;
   C to E: $(4, 10)$ SF $= 3$; B to C: $(4, 5)$ SF $= \frac{1}{2}$; E to D: $(1, 7)$ SF $= \frac{2}{3}$

Exercise 115

1 Reflection in $y = 2$; reflection in x-axis; translation $(0, -4)$

3 Reflection in y-axis; reflection in $y = 2$; rotation $180°$ about $(0, 2)$

5 Reflection in y-axis; reflection in $x + y = 0$; rotation $90°$ about $(0, 0$

7 Reflection in $x = -4$; reflection in $y = 2$; rotation $180°$ about $(-4, 2$

Exercise 115*

1 Reflection in $x + y = 6$; rotation $-90°$ about $(2, 4)$

3 Rotation $143°$ about $(-2, 2)$; reflection in $y = 2x + 6$

Exercise 116 (Revision)

1 **a** $(3, -4)$ **b** $(-3, 4)$ **c** $(4, -3)$ **d** $(10, -2)$

2 **a** $(-3, -5)$ **b** $(3, 5)$ **c** $(-5, -3)$ **d** $(-8, 8)$

3 **a** $A'(1, -2), B'(1, -6), C'(8, -2)$
   b $A'(-2, 1), B'(-6, 1), C'(-2, 8)$
   c $A'(-4, 6), B'(-4, 10), C'(3, 6)$
   d $A'(2, 0), B'(2, 8), C'(16, 0)$

4

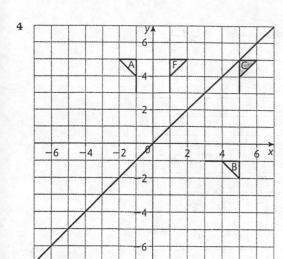

c Rotation of 90° clockwise around O.

e Translation along $\begin{pmatrix} 4 \\ 0 \end{pmatrix}$

5 $x = -11, y = -1$

Exercise 116* (Revision)

1 A(1, −5), B(−1, −5), C(1, −9)

2 d Translation along vector $\begin{pmatrix} -2 \\ 2 \end{pmatrix}$

3 e Rotation of −90° about (0, −1)

    f Enlargement of scale factor +2 centre (−1, −4)

4 a Reflection in x-axis   **b** Reflection in y-axis

    c Reflection in $y = x + 2$ **d** Reflection in $y = x - 2$

    e Rotation 90° about (−2, 0)   **f** Rotation 90° about (0, −2)

5

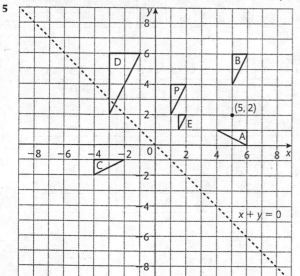

Exercise 117

1 f.d.: 0.20, 0.40, 0.90, 0.60, 0.45, 0.05

3 a f.d.: 3.5, 7, 10, 24, 38, 16, 9

    b 6.5–7 kg

    c 55%

5 a f.d.: 3.5, 9.5, 12. 13.6, 10.4, 2.5

    c $\bar{x} = 28.8$ years

Exercise 117*

1 a f.d.: 0.04, 0.07, 0.087, 0.113, 0.024, 0.012

    b 51.4%

    c $\bar{x} = 368.5$

    d 350

3 a f.d.: 10, 13, 15, 15, 13, 7, 7

    b $\bar{x} = 9.77$ years

    c 6.5 cm, 7.5 cm, 7.5 cm, 6.5 cm, 3.5 cm, 3.5 cm

5 a 522 customers

    b £1566

    c $\bar{x} = 52.5$ customers per hour. Not a useful statistic

    d 10.00–12.00, 1 staff; 12.00–14.00, 4 staff; 14.00–18.00, 2 staff; 18.00–20.00, 3 staff

7 a 6, 8

    b f.d.: 36, 17, 6, 1

    c $\bar{x} = 97.7$ min

Exercise 118 (Revision)

1 a

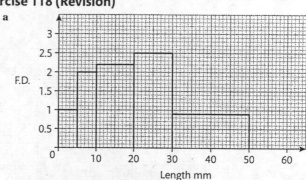

Length, l mm	F.D.
$0 < l \le 5$	1
$5 < l \le 10$	2
$10 < l \le 20$	2.2
$20 < l \le 30$	2.5
$30 < l \le 50$	0.9

    b 7   **c** 22.0 mm

2 a

Time, t mins	F.D.
$15 < t \le 20$	2.4
$20 < t \le 30$	1.2
$30 < t \le 40$	1
$40 < t \le 70$	0.2

    b 2   **c** 17 (16.8)   **d** $20 < t \le 30$

3 a b

Mass, m g	Frequency	F.D.
$60 < m \leq 70$	12	1.2
$70 < m \leq 80$	22	2.2
$80 < m \leq 100$	40	2
$100 < m \leq 120$	20	1
$120 < m \leq 160$	16	0.4

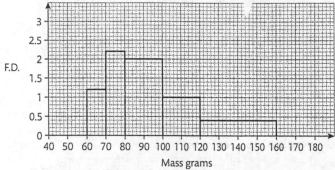

c 41

4

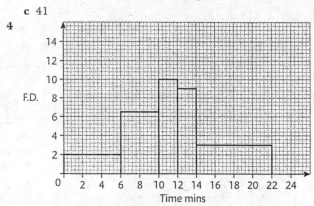

Time, t mins	Frequency	F.D.
$0 < t \leq 6$	12	2
$6 < t \leq 10$	26	6.5
$10 < t \leq 12$	20	10
$12 < t \leq 14$	18	9
$14 < t \leq 22$	24	3

c 60%   **d** 11.2 mins

Exercise 118* (Revision)

1 a

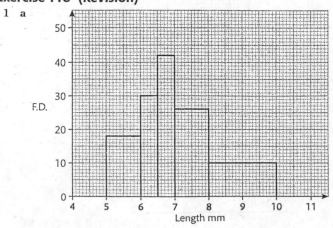

Length, l mm	F.D.
$5 < l \leq 6$	18
$6 < l \leq 6.5$	30
$6.5 < l \leq 7$	42
$7 < l \leq 8$	26
$8 < l \leq 10$	10

b 76   **c** 7.10 mm

2 a

Time, t mins	F.D.
$20 < t \leq 30$	0.5
$30 < t \leq 35$	0.8
$35 < t \leq 40$	1.8
$40 < t \leq 50$	2.2
$50 < t \leq 70$	0.35
$70 < t \leq 100$	0.1

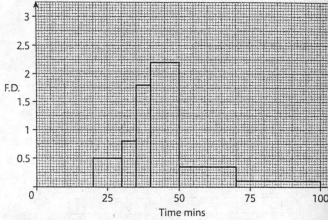

b 71.2%   **c** 43.2 mins   **d** $40 < t \leq 50$

3 a, c

Life, t hrs	Frequency	F.D.
$60 < t \leq 80$	24	1.2
$80 < t \leq 90$	26	2.6
$90 < t \leq 95$	20	4
$95 < t \leq 100$	25	5
$100 < t \leq 115$	54	3.6
$115 < t \leq x$	51	1

b $x = 166$

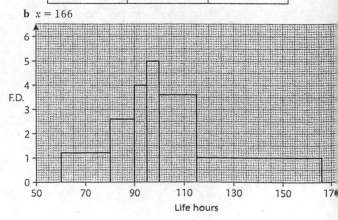

a, b

Mass, m kg	Frequency	F.D.
$2 < m \leqslant 3$	52	52
$3 < m \leqslant 3.25$	34	136
$3.25 < m \leqslant 3.5$	30	120
$3.5 < m \leqslant 4$	60	120
$4 < m \leqslant 4.75$	24	32

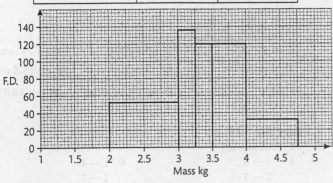

c 0.83　　　　**d** 3.37 kg

Multiple choice 5

1 A	2 D	3 C	4 C	5 D
6 C	7 B	8 D	9 A	10 D

Self-assessment 5

1 **a** 4 girls　　　　　　　　　**b** 40 cars

2

Number of men, m	Number of humous tubs, h	Time, t mins
2	120	5
4	480	10
6	1080	15
m	h	$h \div (12m)$
$h \div (12t)$	h	t
m	$12mt$	t

3 **a** $\frac{5}{9}$ 　**b** $\frac{37}{990}$ 　**c** $\frac{2908}{4995}$ 　**d** $\frac{35117}{33300}$

4 $\frac{263}{329}$

5 **a** $x^2 + 7x + 10$ 　**b** $y^2 + 2y - 99$ 　**c** $9 - 6p + p^2$

6 **a** $15x^2 + x - 6$ 　**b** $a^2 - 6ab + 9b^2$

7 **a** $x(x - 3)$ 　　　　　**b** $(x - 7)(x + 7)$
　c $(x + 1)(x + 2)$ 　　　**d** $(x - 8)(x + 1)$

8 **a** $(x - 7)(x - 3)$ 　　　**b** $3p^2(p + 1)(p - 1)$

9 **a** $x = -3$ or -4 　　　**b** $x = -2$ or 3
　c $x = 0$ or 3 　　　　　**d** $x = 2$ or 10

10 **b** $x = 6$

11 **a** $y = 6$ 　　**b** $y = 2x + 1$ 　　**c** $y = 2 - x$

12 **d** Translation $\binom{10}{10}$ 　　**e** $x + y = 14$
　f 180° 　　　　　　　　　**g** (11, 3)
　h (15, 15), $\frac{1}{3}$

13

Consumption, m, ml	Frequency
$0 < m \leqslant 50$	9
$50 < m \leqslant 100$	42
$100 < m \leqslant 125$	30
$125 < m \leqslant 150$	25
$150 < m \leqslant 200$	25
$200 < m \leqslant 300$	22
$300 < m \leqslant 500$	27

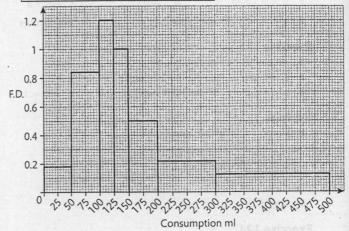

c 134 ml　　　**d** 50.1%　　　**e** 171.5 ml

Exercise 119

1 $270　　3 $234　　5 $319.20　　6 $18.38

Exercise 119*

1 $549　　3 $9
5 **a** $6 240 000　　　　　**b** $11.90
7 B better than A in year 3

Exercise 120

1 **a** Aus $277.50 　　　　**b** R$457.50
　c CNY 1707.50 　　　　**d** €185
3 **a** $300 　　　　　　　　**b** $900
5 $37 736
7 Jamaica (500 000) by $10 000

Exercise 120*

1 **a** £24.15 　　　　　　　**b** €27.92
　c Aus $41.89 　　　　　**d** R$69.06
3 41.4 million
5 MXN 1941, R$414
7 JMD 4 240 000 = U$ 160 000

Exercise 121 (Revision)

1 $555 Australian　　　　　2 £283.78
3 €27.92　　　　　　　　　4 Jamaican $206.07
5 Jam$53 000, Pesos 25 720, Reais 10 980
6 £13 663　　　　　　　　　7 £3055
8 Gerard: €865.38/week, Marcelle: €860/week. Gerard by €5.38/week.

Exercise 121* (Revision)

1 1596 Ringitts and 4098 Yuan　　2 4160 Rupees and 1438 Yuan
3 $x = 47 619$ 　　　　　　　　4 $y = 10 841 270$

5 $y = 20$ **6** $y = 14$

7 $2.40   **8** €3.69

Exercise 122

1 **a** $y = 5x$ **b** 30   **c** 5

3 **a** $e = \dfrac{M}{20}$ **b** 5 m   **c** 120 kg

5 Yes, as 210 people would turn up to swim

Exercise 122*

1 **a** $v = 9.8t$ **b** 49 m/s   **c** 2.5 s

3 **a** $d = 150m$ **b** 1500 km   **c** 266.7 g

5 **a** $h = \dfrac{3y}{2}$ **b** 0.75 m   **c** 4 months

Exercise 123

1 **a** $y = 4x^2$ **b** 144   **c** 4

3 **a** $v = 2w^3$ **b** 54   **c** 4

5 **a** $y = 5t^2$ **b** 45 m   **c** $\sqrt{20} \approx 4.5$ s

Exercise 123*

1

g	2	4	6
f	12	48	108

3 **a** $R = \left(\dfrac{5}{256}\right)s^2$ **b** 113 km/h

5 $x = 10\sqrt{2}$

Exercise 124

1 **a** $y = \dfrac{12}{x}$ **b** $y = 6$ **c** $x = 4$

3 **a** $m = \dfrac{36}{n^2}$ **b** $m = 9$ **c** $n = 6$

5 **a** $I = \dfrac{4 \times 105}{d^2}$ **b** 0.1 candle power

Exercise 124*

1

b	2	5	10
a	50	8	2

3 **a** $R = \dfrac{2}{r^2}$ **b** $\dfrac{2}{9}$ ohm

5 **a**

Day	N	T
Mon	400	25
Tues	447	20
Wed	500	16

b 407 approx.

Exercise 125 (Revision)

1 **a** $y = 6x$ **b** $y = 42$ **c** $x = 11$

2 **a** $p = 5q^2$ **b** $p = 500$ **c** $q = 11$

3 **a** $c = \dfrac{3}{4}a^2$ **b** $675   **c** 28.3 m²

4

a	1	2	4	8
t	80	40	20	10

5 $a = 20b$

b	10	5	30
a	200	300	600

6 **a** $d = 5t^2$ **b** $d = 20$ m   **c** $t \approx 4.24$ s

7 **a** $C = 1.5d$ **b** $97.50   **c** 53.3 cm

8 **a** $p = 1.5n^2$ **b** p = £216   **c** $n = 20$

Exercise 125* (Revision)

1 **a** $y^2 = 50z^2$ **b** 56.6   **c** 5.85

2 **a** $m = \dfrac{8839}{\sqrt{n}}$ **b** 3.23×10^5 **c** 7.81×10^{-5}

3 1500 m

4

x	0.25	1	4	25
y	20	10	5	2

5

b	125	8	1
a	2	5	10

6 **a** $e = 0.5v^2$ **b** $e = 1250$ kJ   **c** $v = 1414$ m/s

7 **a** $n = \dfrac{250}{t^2}$ **b** $n = 62.5$ **c** $t = 15.8$ yrs

8 **a** $v = 3.1\sqrt{d}$ **b** (i) $v = 21.9$ m/s   (ii) $v = 98.0$ m/
     c 0.104 m   **d** $d = 5011$ m

Exercise 126

1 One to many

3
$1 \rightarrow 4$
$2 \rightarrow 5$
$3 \rightarrow 6$
$13 \rightarrow 16$

5
$1 \rightarrow 4$
$2 \rightarrow 7$
$3 \rightarrow 10$
$5 \rightarrow 16$

7 $a = 5$ $b = 2$ $x \rightarrow x + 3$

9 $a = 1$ $b = 9$ $x \rightarrow x - 4$

11 $a = -2$ $b = -1$ $c = 0$

13 $a = 5$ $b = 2$ $c = 1$

Exercise 126*

1 $1, 3, -1, \frac{1}{2}$ one to one

3
$-1 \rightarrow -1$
$0 \rightarrow -\frac{1}{2}$
$5 \rightarrow 2$
$15 \rightarrow 7$

5
$-1 \rightarrow -1$
$0 \rightarrow -2$
$5 \rightarrow 23$
$3 \rightarrow 7$

7 $a = 4$ $b = 5$ $x \rightarrow 2x$

9 $a = 1$ $b = 18$ $x \rightarrow \frac{1}{3}x$

11 $a = 0$ $b = 3$ $c = 1$

13 $a = 0$ $b = 1, -1$ $c = 2, -2$

Exercise 127

1 Yes   **3** Yes   **5** Yes   **7** Yes   **9** No

Exercise 127*

1 Yes   **3** Yes   **5** No   **7** No

Exercise 128

1 **a** 5   **b** -5 **c** 2   **d** 1

3 **a** 8   **b** 3   **c** $1\frac{1}{4}$ **d** 0

5 **a** 2   **b** 0   **c** 0

7 **a** $1\frac{1}{2}$ **b** $2\frac{1}{2}$ **c** $2 - \dfrac{1}{y}$

9 3

11 **a** $1 - 2x$ **b** $2x + 5$ **c** $2x + 3$

Exercise 128*

1 **a** 12   **b** 7   **c** 8   **d** $8 - 2p$

3 **a** 0   **b** $1\frac{1}{4}$ **c** 0   **d** $p(p + 2)$

5 **a** $-\frac{1}{3}$ **b** $\frac{1}{5}$ **c** $-\dfrac{1}{195}$

7 **a** -4 **b** $\frac{2}{3}$ **c** $\dfrac{(9y + 2)}{(3y - 4)}$

9 $-2, 3$
11 **a** $2 + x$ **b** $2 + 4x$ **c** $2x - 4$

Exercise 129

1 **a** $0 \leqslant x \leqslant 2$ **b** $-2 \leqslant x \leqslant 2$
3 **a** $-1 \leqslant x \leqslant 0$ **b** $-2 \leqslant x \leqslant 0$
5 **a** $-0.5 \leqslant x \leqslant 1.5$ **b** $-4.5 \leqslant x \leqslant 5.5$

Exercise 129*

1 **a** $\{8, 5, 2, -1\}$ **b** $\{\mathbb{R}\}$
3 **a** $\{0, 8, 24\}$ **b** $\{x : x \geqslant 0 \; x \in \mathbb{R}\}$
5 **a** $\left\{1, \frac{1}{2}, \frac{1}{3}, \frac{1}{4}\right\}$ **b** $\{x : 0 < x \leqslant 1\}$

Exercise 130

1 -1 **3** $\{x : x < 2, x$ a real number$\}$
5 0 **7** None   **9** ± 2

Exercise 130*

1 $\frac{1}{2}$ **3** $\{x : x > 9, x$ a real number$\}$
5 -1 **7** ± 1
9 $\{x : x > -2, x$ a real number$\}$

Exercise 131

1 $fg(3) = 6, gf(3) = 6$ **3** $fg(4) = 5, gf(4) = \frac{4}{5}$
5 **a** $2x + 4$ **b** $2x + 2$ **c** $4x$ **d** $x + 4$
7 **a** x **b** x **c** $x - 12$ **d** $x + 12$

Exercise 131*

1 $fg(-3) = 19, gf(-3) = 8$ **3** $fg(-3) = -4\frac{1}{2}, gf(-3) = -\frac{3}{7}$
5 **a** $2(x - 2)^2$ **b** $2x^2 - 2$ **c** $8x^4$ **d** $x - 4$
7 **a** $4\sqrt{\left(\frac{x}{4} + 4\right)}$ **b** $\sqrt{(x + 4)}$
    c $16x$ **d** $\sqrt{\left[\frac{1}{4}\sqrt{\left(\frac{x}{4 + 4}\right) + 4}\right]}$
9 **a** $\{x : x \neq 5, x$ a real number$\}$ **b** $\{x : x \neq \pm 2, x$ a real number$\}$

Exercise 132

1 7 **3** $\frac{(x - 4)}{6}$ **5** $\frac{x}{3 + 6}$ **7** $\frac{3}{(4 - x)}$
9 **a** 4 **b** $\frac{5}{2}$ **c** 1

Exercise 132*

1 17 **3** $\frac{4}{3} - \frac{x}{24}$ **5** $\frac{7}{(4 - x)}$ **7** $\sqrt{\frac{(x - 16)}{2}}$
9 **a** 4 **b** 7 **c** 0
11 $x = 1$ or $x = 2$

Exercise 133 (Revision)

1 **a** $x = 6 \, y = 7, \quad y = x + 1$ **b** $x = 4 \, y = 0, \quad y = 2x$
2 $-1 \to 3 \quad \{$range $-3, -1, 1, 3\}$
     $0 \to -1$
     $1 \to 1$
     $2 \to 3$
3 Many possible answers
4 mango $\to 5$, coconut $\to 7$, yam $\to 3$, pumpkin $\to 7$, guava $\to 5$
5 **a** function   **b** not function
6 **a** 13 **b** -2 **c** 7
7 **a** $x = \frac{-1}{2}$ **b** $x = \frac{5}{4}$
8 **a** $5x - 1$ **b** $5x + 3$

9 **a** $x = 1$ **b** $x = \frac{1}{2}$ **c** $x < -1$ **d** $x < 2$
10 **a** $f(x) \in \mathbb{R}$ **b** $g(x) \geqslant 1$ **c** $h(x) \geqslant 0$ **d** $f(x) \in \mathbb{R}$

Exercise 133* (Revision)

1 **a** $x = 4 \, y = 4, \quad y = x^2$ **b** $x = 2 \, y = 9, \quad y = 2x + 1$
2 $-2 \to 4 \quad \{$range $-3, -1, 1, 3\}$
     $-1 \to -2$
     $0 \to -4$
     $1 \to -2$
     $2 \to 4$
3 Many possible answers
4 terahedron $\to 4$, cube $\to 6$, octahedron $\to 8$, triangular prism $\to 5$, dodecahedron $\to 12$
5 **a** not function   **b** function
6 **a** ± 4 **b** ± 3 **c** 0
7 **a** $3, -2$ **b** $8, -7$
8 **a** $4 - 2x$ **b** $7 - 2x$
9 **a** $x = \frac{4}{3}$ **b** $x = -2$ **c** $x < \frac{2}{5}$ **d** $-3 < x < 3$
10 **a** $f(x) \geqslant 3$ **b** $g(x) \geqslant 0$ **c** $h(x) \in \mathbb{R}$ **d** $f(x) \in \mathbb{R}$

Exercise 134

1 $100°$ **3** $45°$ **5** $280°$ **7** $60°$
9 ADB and BCA are angles in the same segment

Exercise 134*

1 $140°$ **3** $110°$ **5** $76°$ **7** $3x$
9 BEC = CDB (angles in the same segment), so CEA = BDA

Exercise 135

1 $70°$ **3** $80°$ **5** $100°$
7 **a** NTM = NPT (Alternate segment)
    b PLT = NTM (Corresponding angles)

Exercise 135*

1 $65°$
3 ATE = $55°$ (alternate segment)
    TBC = $125°$ (angles on straight line)
    BTC = $35°$ (angle sum of triangle)
    ATB = $90°$ (angles on straight line)
    AB is a diameter
5 **a** $55°$ **b** $35°$
7 see online pdf

Exercise 136

1 22.5 cm   **3** 18 cm   **5** 12 cm

Exercise 136*

1 3 cm   **3** 8 cm   **5** 4.5 cm

Exercise 137

1 18.8 cm, 28.3 cm^2 **3** 22.3 cm, 30.3 cm^2
5 50.8 cm, 117 cm^2 **7** 37.7 cm, 37.7 cm^2

	Radius in cm	Circumference in cm	Area in cm^2
9	2.11	13.3	14
11	5.17	32.5	84

13 7.54 km

Exercise 137*

1 20.5 cm, 25.1 cm²   3 43.7 cm, 99.0 cm²
5 37.7 cm, 56.5 cm²   7 $r = 3.19$ cm, $P = 11.4$ cm
9 569 m²
11 a 40 100 km   b 464 m/s
13 $r = 1.79$ cm, $A = 7.53$ cm²

Exercise 138

1 8.62 cm   3 38.4 cm   5 34.4°   7 14.3 cm

Exercise 138*

1 3.55 cm   3 25.1°   5 13.4 cm
7 33.0 cm   9 4.94 cm

Exercise 139

1 12.6 cm²   3 170 cm²   5 76.4°   7 5.86 cm

Exercise 139*

1 15.8 cm²   3 53.3°   5 4.88 cm
7 11.5 cm   9 7.31 cm²   11 2.58 cm²

Exercise 140 (Revision)

1 62°   2 55°   3 124°   4 132°
5 54°   6 66°   7 30°   8 222°
9 12   10 3   11 8   12 8
13 4   14 3   15 5   16 4
17 a 80°   b 100°   c 50°   d 50°
18 a $A = 45.9$, $P = 35.4$ b $A = 1.93$, $P = 10.7$
    c $A = 3.43$, $P = 12.6$ d $A = 17.7$, $P = 22.3$
19 a $A = 19.5$, $P = 20.9$ b $A = 43.3$, $P = 29.6$
    c $A = 6.98$, $P = 11.0$ d $A = 13.1$, $P = 28.2$

Exercise 140* (Revision)

1 20°   2 65°   3 65°   4 45°
5 a 55°   b 35°
    c ∠TDC = 35° (Angles in the same segment)
    ∴ ∠EDC = 90° and EC is the diameter
6 a 40°   b 50°
    c ∠ZXT = ∠WVT (Angle in alternate segment)
    ∴ XZ is parallel to WV (Alternate angles)
7 6   8 3   9 8   10 5
11 4   12 4   13 $\sqrt{2}$ 14 4.85
15 a $A = 96$, $P = 49.1$ b $A = 113$, $P = 49.7$
    c $A = 5.37$, $P = 25.7$ d $A = 49.1$, $P = 28.6$
16 a $x = 45.8°$, $A = 10$ b $x = 251°$, $P = 39$
17 See online pdf

Exercise 141

1 $\mathbf{p} + \mathbf{q} = \binom{6}{8}$; $\mathbf{p} - \mathbf{q} = \binom{-2}{-2}$; $2\mathbf{p} + 3\mathbf{q}\binom{16}{21}$

3 $\mathbf{p} + \mathbf{q} = \binom{4}{6}$; $\mathbf{p} - \mathbf{q} = \binom{-2}{-2}$; $2\mathbf{p} + 5\mathbf{q} = \binom{17}{24}$

5 $\mathbf{v} + \mathbf{w} = \binom{4}{5}$, $\sqrt{41}$; $2\mathbf{v} - \mathbf{w} = \binom{5}{-2}$, $\sqrt{29}$; $\mathbf{v} - 2\mathbf{w} = \binom{-1}{-7}$, $\sqrt{50}$

Exercise 141*

1 $\mathbf{p} + \mathbf{q} = \binom{5}{0}$, 5, 090°

    $\mathbf{p} - \mathbf{q} = \binom{-1}{2}$, $\sqrt{5}$, 333°

    $2\mathbf{p} - 3\mathbf{q} = \binom{-5}{5}$, $\sqrt{50}$, 315°

3 $m = -1$, $n = -2$

5 a $\binom{-5}{8.7}$ km   b $\binom{-8.5}{-3.1}$ km

Exercise 142

1 a 2×1 b 1×2

3 $\begin{pmatrix} 3 & 4 \\ -4 & 4 \\ 6 & 18 \end{pmatrix}$ 5 $\binom{2}{2}$ 7 Not possible

Exercise 142*

1 $p = 2$ 3 $t = -2$ 5 $a = 12$ $b = 10$

Exercise 143

1 $\binom{1}{2}$ 3 $\begin{pmatrix} 0 & 1 & 3 \\ 2 & 2 & 0 \end{pmatrix}$ 5 $\begin{pmatrix} 8 & 2 \\ 15 & -8 \end{pmatrix}$ 7 $\begin{pmatrix} 13 & 24 \\ 18 & 61 \end{pmatrix}$

Exercise 143*

1 $x = 10$ $y = 30$ 3 $x = 2$ $y = 1$
5 $x = 10$ $y = 90$ 7 $m = -2$

Exercise 144

1 $\frac{1}{7}\begin{pmatrix} 1 & 1 \\ -4 & 3 \end{pmatrix}$ 3 $-\frac{1}{8}\begin{pmatrix} -2 & -1 \\ 2 & 3 \end{pmatrix}$ 5 $\frac{1}{12}\begin{pmatrix} 1 & 1 \\ -7 & 5 \end{pmatrix}$

7 $-1\begin{pmatrix} 2 & 1 \\ 5 & 2 \end{pmatrix}$ 9 $\frac{1}{222}\begin{pmatrix} 9 & 12 \\ -11 & 10 \end{pmatrix}$

Exercise 144*

1 $x = 1.5$ 3 Reflection in x axis
5 $\begin{pmatrix} 3 & 1 \\ 5 & 2 \end{pmatrix}$ 7 $\begin{pmatrix} -7 & -2 \\ -4 & -1 \end{pmatrix}$

Exercise 145 (Revision)

1 $\mathbf{p} + \mathbf{q} = \binom{1}{5}$, $\sqrt{26}$; $\mathbf{p} - \mathbf{q} = \binom{5}{3}$, $\sqrt{34}$; $2\mathbf{p} - 2\mathbf{q} = \binom{13}{10}$, $\sqrt{269}$

2 $\mathbf{r} + \mathbf{s} = \binom{5}{-1}$, $\sqrt{26}$; $\mathbf{r} - \mathbf{s} = \binom{-1}{-9}$, $\sqrt{82}$; $3\mathbf{s} - 2\mathbf{r} = \binom{5}{22}$, $\sqrt{509}$

3 a $\binom{5}{1}$ b $\binom{6}{-4}$ c $\sqrt{29}$ d $v = 1$ $w = 1$

4 a $n = 1$ $m = -3$ 5 $m = -2$ $s = 5$

6 a $\begin{pmatrix} 2 & -3 \\ 0 & 2 \end{pmatrix}$ b $\begin{pmatrix} 2 & 5 \\ 6 & 0 \end{pmatrix}$ c $\begin{pmatrix} 8 & -28 \\ -12 & 12 \end{pmatrix}$ d $\begin{pmatrix} 4 & 32 \\ 30 & -6 \end{pmatrix}$

7 a $\begin{pmatrix} -1 & 5 \\ 2 & -7 \end{pmatrix}$ b $\begin{pmatrix} -9 & -3 \\ 2 & 1 \end{pmatrix}$ c $\begin{pmatrix} 3 & 0 \\ 0 & 3 \end{pmatrix}$ d $\begin{pmatrix} 1 & 4 \\ 0 & 1 \end{pmatrix}$

Exercise 145* (Revision)

1 $m = 3$, $n = 1$ 2 $m = -2$, $n = 5$

3 a $\binom{-7}{19}$ b $\sqrt{140}$ 340°   c 5.1 km/h

4 a (i) $\binom{6}{-3}$ (ii) $\binom{3}{9}$ b $m = 1$ $n = 3$
    c $r = 10$ $s = 33$ d $v = 2$ $u = -1$
5 $n = -1$ $m = 2$

6 a $\begin{pmatrix} 9 & 0 \\ -31 & -28 \end{pmatrix}$ b $-\frac{1}{42}\begin{pmatrix} 16 & 6 \\ 1 & -3 \end{pmatrix}$

    c $-\frac{1}{42}\begin{pmatrix} -6 & -9 \\ 8 & 19 \end{pmatrix}$ d $\begin{pmatrix} 41 & 15 \\ 75 & 86 \end{pmatrix}$

7 a $x = 2$ $y = 3$ b $\begin{pmatrix} 8 & -8 \\ 2 & 9 \end{pmatrix}$

Multiple choice 6

1 C   2 D   3 B   4 B   5 A
6 A   7 D   8 A   9 B   10 D

Self-assessment 6

1 $546
2 ¥4098 : R1598 : R9222
3 a $y = 8x$ b $y = 80$ c $x = 5$
4 a $x = 4\sqrt[3]{y}$ b $y = 512$
5 a $y = \frac{48}{x}$ b $y = 6$ c $x = 4$

6 a $N = \dfrac{9000}{d^2}$ b $N = 2250$ c $d = 3$

7 a $A = 15h^2$ b $A = 135\,\text{m}^2$ c $h = 6\,\text{m}$

8 a $x = 60°, y = 60°, z = 55°$ b $x = 40°, y = 70°, z = 40°$

9 a $BP = 5.4\,\text{cm}$ b $CQ = 4\,\text{cm}$

10 20.1 cm

11 a $28.6°$ b 3.02 cm

12 $\{22, 17, 12, 7, 2\}$

13 a 99 b 4 c $x = 10$

14 a 263 b 598

c $2x^2 + 6x + 3$ d $(2x + 3)(2x + 6)$

15 a $f^{-1}(x) = \dfrac{x + 1}{3}$ b $g^{-1}(x) = \dfrac{1}{x}$

c $x = 1.3$ or -2.3 d $\left(\dfrac{3}{x} - 1\right)^2$

16 a $\begin{pmatrix} -1 \\ 5 \end{pmatrix}$ b $\begin{pmatrix} -3 \\ -1 \end{pmatrix}$ c $\sqrt{41}$

17 $a = 1\ b = -3$

18 a $\begin{pmatrix} 2 & -8 \\ 4 & 2 \end{pmatrix}$ b $\begin{pmatrix} -7 & -8 \\ 4 & -7 \end{pmatrix}$ c $\dfrac{1}{9}\begin{pmatrix} 1 & 4 \\ -2 & 1 \end{pmatrix}$

Exercise 146

1 $\frac{1}{9}$ 3 $\frac{1}{8}$ 5 9 7 3 9 $\frac{1}{16}$

11 $\frac{1}{16}$ 13 16 15 2.84 17 0.0123 19 0.0370

21 64 23 c^{-3} 25 a^{-2}

Exercise 146*

1 $\frac{1}{64}$ 3 1 5 12.5 7 10

9 $\frac{1}{2}$ 11 0.364 13 0.00137 15 $2c^{-4}$

17 $12a^{-1}$ 19 c^{-1} 21 2 23 $k = 2\frac{1}{3}$

25 b 27 $5a^{-2}$ 29 $-9a^6$

Exercise 147

1 $\frac{1}{3}$ 3 $\frac{1}{9}$ 5 27 7 0.024

9 36.5 11 b^{-2} 13 f^{-1}

Exercise 147*

1 $\frac{1}{6}$ 3 16 5 1.29 7 2.85

9 a^{-2} 11 e 13 $\frac{1}{36}$

Exercise 148 (Revision)

1 $\frac{1}{16}$ 2 10 3 4 4 8 5 5

6 5 7 $\frac{1}{64}$ 8 $\frac{1}{4}$ 9 $\frac{1}{11}$ 10 9

11 $\frac{1}{81}$ 12 27 13 $36^{-\frac{1}{2}} = \dfrac{1}{36^{\frac{1}{2}}} = \dfrac{1}{\sqrt{36}} = \dfrac{1}{6}$

14 $9^{\frac{3}{2}} = \left(9^{\frac{1}{2}}\right)^3 = (\sqrt{9})^3 = 3^3 = 27$ 15 a^2 16 a^4

17 d^{-2} 18 $b^{-\frac{1}{2}}$ 19 $b^{\frac{9}{2}}$ 20 c

Exercise 148* (Revision)

1 $\frac{1}{32}$ 2 6 3 1 4 3

5 4 6 $\frac{1}{7}$ 7 64 8 8

9 $\frac{1}{9}$ 10 $\frac{1}{625}$

11 $(0.125)^{-\frac{2}{3}} = \left(\dfrac{1}{8}\right)^{-\frac{2}{3}} = \left(\dfrac{8}{1}\right)^{\frac{2}{3}} = \left(8^{\frac{1}{3}}\right)^2 = (\sqrt[3]{8})^2 = 2^2 = 4$

12 $\left(4^{\frac{1}{3}}\right)^{-1\frac{1}{2}} = \left(4^{\frac{1}{3}}\right)^{-\frac{3}{2}} = 4^{-\frac{1}{2}} = \dfrac{1}{4^{\frac{1}{2}}} = \dfrac{1}{\sqrt{4}} = \dfrac{1}{2}$

13 $3c$ 14 $5b^{-2}$ 15 a^2 16 $3c^{-2}$

17 $2a^{-1}$ 18 $d^{\frac{7}{6}}$ 19 $-9a^6$ 20 2

Exercise 149

1 $x = -2$ or $x = -1$ 3 $x = -5$ or $x = -2$

5 $x = 3$ 7 $x = -1$ or $x = 0$

9 $x = \pm 2$

Exercise 149*

1 $x = -5$ or $x = -1$ 3 $x = -8$ or $x = -7$

5 $x = 7$ 7 $x = 0$ or $x = 13$

9 $x = \pm 9$

Exercise 150

1 $x = \pm\frac{7}{2}$ 3 $x = -2$ or $x = 0$

5 $x = 2$ or $x = 3$ 7 $x = -1.5$ or $x = -1$

9 $x = -3$ or $x = 3$ 11 $x = -2$ or $x = -\frac{1}{3}$

13 $x = -2$ or $x = 3$ 15 $x = -4$ or $x = \frac{2}{3}$

Exercise 150*

1 $x = \pm\frac{5}{7}$ 3 $x = -2$ or $x = 0$

5 $x = 1$ or $x = 2$ 7 $x = -9$ or $x = -\frac{4}{3}$

9 $x = -\frac{1}{2}$ or $x = -\frac{1}{4}$ 11 $x = 0.8$ or $x = 1.5$

13 $x = -4$ or $x = 4$ 15 $x = -5$ (repeated root)

17 $x = \frac{2}{3}$ or $x = 1.5$

Exercise 151

1 $x = -3.45$ or $x = 1.45$ 3 $x = -5.46$ or $x = 1.46$

5 $x = -14.2$ or $x = 0.21$ 7 $x = -2.90$ or $x = 6.90$

9 $x = -3.56$ or $x = 0.561$

Exercise 151*

1 $x = 0.171$ or $x = 5.83$ 3 $x = -7.58$ or $x = 1.58$

5 $x = -0.41$ or $x = 2.41$ 7 $x = -1.00$ or $x = 3.50$

9 $x = 0.27$ or $x = -2.77$

Exercise 152

1 $x = -3.45$ or $x = 1.45$ 3 $x = -5.46$ or $x = 1.46$

5 $x = -14.2$ or $x = 0.211$ 7 $x = -2.90$ or $x = 6.90$

9 $x = -3.56$ or $x = 0.561$ 11 $x = -3.37$ or $x = 2.37$

Exercise 152*

1 $x = 0.171$ or $x = 5.83$ 3 $x = -7.58$ or $x = 1.58$

5 $x = 1.59$ or $x = 4.41$ 7 $x = -1.69$ or $x = 7.69$

9 $x = 0.258$ or $x = 7.74$ 11 $x = 0.105$ or $x = 5.37$

Exercise 153

1 1 3 0 5 0 7 2 9 0

Exercise 153*

1 2 3 2 5 1 7 2 9 2

Exercise 154 (Revision)

1 a $x = \pm 4$ b $x = \pm 5$

c $x = 0$ or -4 d $x = 0$ or 3

2 a $x = 1$ or 3 b $x = -3$ or -4

c $x = -4$ or 3 d $x = -2$ or 4

3 a $x = -2$ or 3 b $x = -1$ or -2

c $x = -1$ or $\frac{2}{3}$ d $x = 0$ or -1

4 a $x = -3.73$ or -0.27 b $x = 0.7$ or 4.30

c $x = -1.24$ or 3.24 d $x = -2.77$ or 0.27

5 a $x = -2$ or 5 b $x = -3$ or 2

c $x = 0$ or 4 d $x = 0.23$ or 1.43

Exercise 154* (Revision)

1 a $x = \pm 13$ b $x = \pm 2$

c $x = 0$ or 9 d $x = \pm 6$

2 **a** $x = -9$ or 8 **b** $x = -1$ or 4
  c $x = -6$ or 8 **d** $x = -6$ or 3
3 **a** $x = -3$ or -4 **b** $x = 2$
  c $x = -4$ or 6 **d** $x = -\frac{5}{2}$ or $\frac{3}{4}$
4 **a** $x = -0.44$ or 3.44 **b** $x = -1.37$ or 0.37
  c $x = -0.57$ or 2.91 **d** $x = -4.46$ or 0.46
5 **a** $x = -7$ or 2 **b** $x = -2$ or 4
  c $x = \frac{1}{3}$ or 2 **d** $x = -0.38$ or 3.30

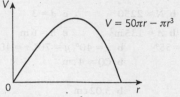

c $V_{\max} = 427.3 \text{ cm}^3$ **d** $d = 8.16 \text{ cm}, h = 8.2 \text{ cm}$

Exercise 155

1 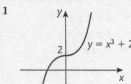   3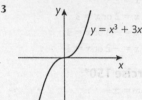

$y = x^3 + 2$ $y = x^3 + 3x$

5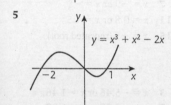

$y = x^3 + x^2 - 2x$

7 **a** $V = x^2(x - 1) = x^3 - x^2$
  c 48 m^2
  d $4.6 \text{ m} \times 4.6 \text{ m} \times 3.6 \text{ m}$

Exercise 155*

1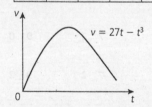

$y = 2x^3 - x^2 + x - 3$

3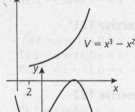

$y = -2x^3 + 3x^2 + 4x$

5 **a**

t	0	1	2	3	4	5
v	0	26	46	54	44	10

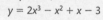

$v = 27t - t^3$

  b $v_{\max} = 54 \text{ m/s}$ and occurs at $t = 3 \text{ s}$
  c $v \geq 30 \text{ m/s}$ when $1.2 \geq t \geq 4.5$ so for about 3.3 s
7 **a** $A = 100\pi = 2\pi r^2 + 2\pi r h$
    $\rightarrow 100\pi - 2\pi r^2 = 2\pi r h$
    $\rightarrow \frac{50}{r} - r = h$
    $\rightarrow V = \pi r^2 h = \pi r^2 \left(\frac{50}{r} - r\right) = 50\pi r - \pi r^2$

  b

r	0	1	2	3	4	5		
V	0	153.9	289.0	386.4	427.3	392.7	263.9	22.0

$V = 50\pi r - \pi r^3$

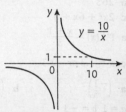

Exercise 156

1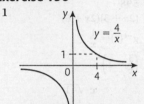

$y = \frac{4}{x}$

3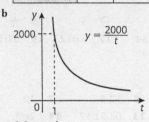

$y = \frac{10}{x}$

5 **a**

t (months)	1	2	3	4	5	6
y	2000	1000	667	500	400	333

  b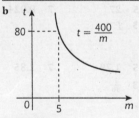

$y = \frac{2000}{t}$

  c 3.3 months   **d** 2.7 months approx.
7 **a** $k = 400$

m (min)	5	6	7	8	9	10
t (°C)	80	67	57	50	44	40

  b

$t = \frac{400}{m}$

  c 53 (°C)   **d** $6 \text{ min } 40 \text{ s}$ **e** $5.3 \leq m \leq 8$

Exercise 156*

1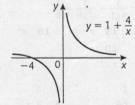

$y = 1 + \frac{4}{x}$

3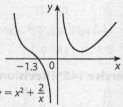

$y = x^2 + \frac{2}{x}$

5 **a**

x (°)	30	35	40	45	50	55	60
d (m)	3.3	7.9	10	10.6	10	8.6	6.7

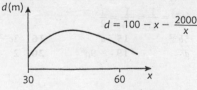

$d = 100 - x - \frac{2000}{x}$

b 10.6 m at $x = 45°$
$37° \leqslant x \leqslant 54$

7 b

r (m)	1	2	3	4	5
A (m²)	106	75	90	126	177

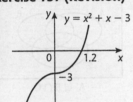

$A = 2\pi r^2 + \dfrac{100}{r}$

c $A = 75\,m^2$ at $r = 2.0\,m$

Exercise 157 (Revision)

1

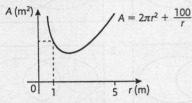

$y = x^2 + x - 3$

2

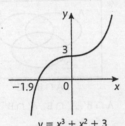

$y = x^3 + x^2 + 3$

3

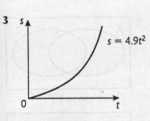

$s = 4.9t^2$

4

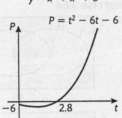

$P = t^2 - 6t - 6$

5 a $k = 2800$

b

t (weeks)	30	32	34	36	38	40
x (kg)	93	88	82	78	74	70

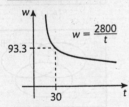

$w = \dfrac{2800}{t}$

c 35 weeks

d Clearly after 500 weeks, Nick cannot weigh 5.6 kg. So there is a domain over which the equation fits the situation being modelled.

6 a

m	5	6	7	8	9	10
t	85	70.8	60.7	53.1	47.2	42.5

b $6 \leqslant m \leqslant 8.4$

Exercise 157* (Revision)

1

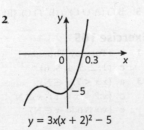

$y = 2x^3 - x^2 + 3x$

2

$y = 3x(x + 2)^2 - 5$

3

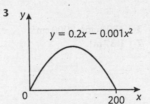

$y = 0.2x - 0.001x^2$

b 10 m   **c** $29 \leqslant x \leqslant 170$

4 a

t	0	1	2	3	4	5
Q	10	17	14	7	2	5

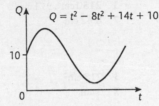

$Q = t^2 - 8t^2 + 14t + 10$

b $Q_{max} = 17.1\,m^3/s$ at 01.66   **c** Between midnight and 02.35

5 a $\dfrac{600}{x}$

c

x	5	10	15	20	25	30	35	40
L	130	80	70	70	74	80	87	95

d 69.3 m at $x = 17.3$ m

e $11.6 < x < 25.9$

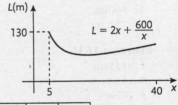

$L = 2x + \dfrac{600}{x}$

6 b

x	−1	0	1	2	3	4
y	−6	5	6	3	2	9

b R(0.6, 6.4) S(2.7, 1.7)

c

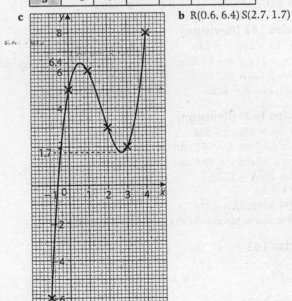

Exercise 158

1 120 cm³

3 48 cm², 108 cm²

5 1.57 m³, 7.85 m²

7 800 m³

9 18.0 m

Exercise 158*

1 4800 cm³
5 7151 cm³, 2260 cm²
9 approx 30 m

3 3800 cm³, 8000 cm²
7 1.37 m³

Exercise 159

1 2.57×10^6 m³
3 Volume = 7069 cm³; Surface area = 1414 cm²
5 396 m³, 311 m²
7 61 cm

Exercise 159*

1 $83\frac{1}{3}$ mm³
5 1089 cm³
9 12 cm

3 2150 cm³, 971 cm²
7 0.42 cm

Exercise 160

1 16 cm²
5 6 cm

3 213 cm²
7 3 cm

Exercise 160*

1 675 cm²
5 10 cm
9 44%

3 7.5 cm
7 1000 cm²
11 19%

Exercise 161

1 800 cm³
5 15.1 cm

3 14.1 cm³
7 5.06 cm

Exercise 161*

1 135 cm³
3 33.4 cm
5 72.8%
7 a 270 g
9 a 2000
   c 810 cm²

b 16 cm
b 22.5 m
d 240 g

Exercise 162 (Revision)

1 a SA = 207, V = 226
2 $r = 1.91$ cm, A = 11.5 cm²
4 P = 20, A = 18.75
6 V = 246, A = 62.5

b SA = 152, V = 96
3 $r = 4.57$ cm, SA = 263 cm²
5 $x = 12$
7 $h = 10$, SA = 120

Exercise 162* (Revision)

1 a SA = 452, V = 509
2 $r = 4.67$ cm, A = 34.2 cm²
3 $r = 2.84$ cm, V = 48.0 cm³
4 P = 20, A = 31.36
5 $x = 4.5$
6 V = 2048, SA = 1875
7 Diameter of Moon = 3479 km, SA Earth = 5.09×10^8 km²

b SA = 223, V = 192

Exercise 163

1 6
5 a 15

3 22
b 8
c 5

Exercise 163*

1 23
5 $8 \leq x \leq 14, 0 \leq y \leq 6$

3 100

Exercise 164

1

A ∩ B'

A ∪ B'

A' ∩ B'

A' ∪ B'

3

A ∩ B ∩ C

A' ∪ B ∩ C

5 A ∩ B', A' ∩ B', A' ∩ B'

Exercise 164*

1

(A' ∩ B').

(A ∪ B')'

(A ∩ B')'

(A ∪ B)'

3

A ∩ B ∩ C

(A ∩ B ∩ C)'

(A ∩ B') ∪ C

(A ∪ B)' ∩ C

5 B ∩ (A ∪ C)', (B' ∩ C), (B' ∩ C) ∪ (A ∩ B ∩ C')

Exercise 165

1 a {Tuesday, Thursday}
   c {1, 2, 3, 4, 5, 6}
3 a {x:x < 7, x ∈ N}
   c {x:2 ≤ x ≤ 11, x ∈ N}
   e {x:x is odd, x ∈ N}

b {Red, Amber, Green}
d {−1, 0, 1, 2, 3, 4, 5, 6}
b {x:x > 4, x ∈ N}
d {x:−3 < x < 3, x ∈ N}
f {x:x is prime}

Exercise 165*

1 **a** {2, 4, 6, 8, 10, 12} **b** {3, 7, 11, 15, 19, 23}
  c {2, 4, 6}
  d {Integers between 1 and 12 inclusive}
3 **a** ∅ **b** $(1, \frac{1}{2}, \frac{1}{4}, \frac{1}{8}, \frac{1}{16})$
  c {2} **d** {−3, 2}
5

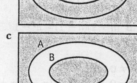

Exercise 166 (Revision)

1 **a**

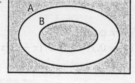

  b 17 **c** 30
2 **a** 6 **b** 2 **c** 10
3 **a** 17% **b** 52% **c** 31%
4 **a** **b**

5 A′ ∪ B′
6 **a** {−2, −1, 0, 1, 2, 3} **b** {1, 2, 3, 4}
7 **a** {$x:x$ is even, $x \in$ N}
  b {$x:x$ is a factor of 24, $x \in$ N}
  c {$x:-1 \leq x \leq 4, x \in$ N}

Exercise 166* (Revision)

1 34 **2** 10 **3** 2
4 **a** **b**

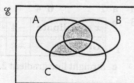

5 **a** (A ∪ B′) ∩ C **b** A ∪ B ∪ C′
6 **a** {−1, 1} **b** {0, −4} **c** ∅
7 **a** {$x:x > -5, x \in$ N}
  b {$x:4 < x < 12, x \in$ N}
  c {$x: x$ is a multiple of 3, $x \in$ N} or {$x:x = 3y, y \in$ N}

Multiple choice 7

1 C **2** C **3** B **4** A **5** D
6 D **7** A **8** D **9** D **10** A

Self-assessment 7

1 **a** x^7 **b** x^3 **c** x^{10} **d** x^{-3}

2 **a** $\frac{1}{9}$ **b** 2 **c** $\frac{1}{2}$ **d** $\frac{1}{25}$ **e** $\frac{1}{16}$
3 **a** b **b** c^{-3} **c** b^{-2} **d** c **e** 1
4 **a** $x = 7$ **b** $x = -1$ **c** $x = \frac{1}{2}$ **d** $x = \frac{1}{3}$
5 **a** $x = 3$ or 4 **b** $x = -5$ or -2
  c $x = 0$ or 7 **d** $x = 0$ or 5
6 **a** $x = \frac{1}{2}$ or 2 **b** $x = -4$ or $\frac{-2}{3}$
  c $x = \frac{2}{3}$ or -1 **d** $x = \frac{-5}{2}$ or 2
7 **a** $x = -2$ or 8 **b** $x = -5.46$ or 1.46
  c $x = 0.84$ or 7.16 **d** $x = -3$ or 1
8 **a** $x = -0.79$ or 2.12 **b** $x = 0.31$ or 1.29
  c $x = -3.91$ or 1.41 **d** $x = 0.21$ or 4.79
9 $a = 15$
10 $a = 60$ $b = 3$
11 1560 cm² (3 sf)
12 158 cm (3 sf)
13 **a** SA = 452 V = 509 **b** SA = 223 V = 192
14

m	5	6	7	8	9	10
t	80	66.7	57.1	50	44.4	40

  c 53°C **d** 6.7 min **e** 5.3–8 min
15 **a** 135 cm³ **b** 10.8 cm **c** 12 cm³ 55.7 cm³
16 **a** 9 **b** 5 **c** 7
17 **a** **b**

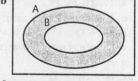

  c **d**

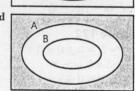

Exercise 167

1 2, 4, 6, 8 **3** 15, 10, 5, 0
5 12, 6, 3, 1.5 **7** Subtract 5; −7, −12, −17
9 Halve: 4, 2, 1

Exercise 167*

1 −1, 0.5, 2, 3.5 **3** 1, 2.5, 6.25, 15.625
5 1, 1, 2, 3, 5, 8 **7** Divide by 3; 3, 1, $\frac{1}{3}$
9 Multiply by $-\frac{1}{2}$; $\frac{1}{16}$, $-\frac{1}{32}$, $\frac{1}{64}$

Exercise 168

1 3, 5, 7, 9 **3** 30, 27, 24, 21 **5** 3, 9, 27, 81
7 8 **9** 7

Exercise 168*

1 −1, 4, 9, 14 **3** 1, $\frac{3}{2}$, 2, $\frac{5}{2}$ **5** 2, 5, 10, 17
7 8 **9** 10

Exercise 169

1 17, 20, 23 **3** −7, −10, −13 **5** 56, 76, 99
7 0, 8, 19 **9** −4, −11, −20

Exercise 169*

1 58, 78, 101 **3** 1, −6, −15 **5** 1, 8, 17
7 71, 101, 139 **9** −6, −19, −38

Exercise 170

1 $3n + 1$ **3** $34 - 4n$

5 a $1, 3, 5, 7, 9, 11, \ldots$ **b** $c = 2l - 1$

c c is always odd, 50 layers.

Exercise 170*

1 $4n - 1$ **3** $9 - 3n$

5 a $6, 10, 14, 18, 22, 26, \ldots$ **b** $s = 4n + 2$

c 202

Exercise 171 (Revision)

1 $-2, 43, 493$ **2** $5, 8, 11$

3 a $15 - 3n$ **b** $6n + 4$

4 a $1800\,\text{m}$ **b** $800 + 200n$ **c** 36 days

5 a Odd numbers **b** $4, 9, 16, 25$ **c** 121

d n^2 **e** 29

6 a $n + n + 1$ **b** $2n + 1$

c $n + n + 1 = 2n + 1$ **d** $2n$ is always even, so $2n + 1$ is odd

Exercise 171* (Revision)

1 $5, -58, -688$ **2** $-2, -5, -8$

3 $4n - 10$ **4** $18 - 5n$

5 a £2.90 **b** $50 + 20n$ **c** 12 years

6 a $17, 21, 25, 29$ **b** $7, 13$

c $1, 3, 7, 13, \ldots$; 21st term

7 a $n(n + 1)$ **b** $n^2 + n$ **c** $n(n + 1) = n^2 + n$

d Either n or $n + 1$ is even, so their product is even.

8 a $n + 1$ **b** $n - 1$

c $(n + 1)(n - 1)$ **d** $(n + 1)(n - 1) = n^2 - 1$.

Exercise 172

1 $f(-2) = 0, f(4) = 0$ **3** $f(1) \neq 0$ so no

5 $f(6) = 0, (x - 3)$ **7** $f(1) = 0, f(-9) = 0, (x - 1)(x + 9)$

9 $f(-3) = 0 \Rightarrow p = -1$

Exercise 172*

1 $f(-9) \neq 0$ so no **3** $f(-7) \neq 0$ so no

5 $f(2) = 0, f(19) = 0, (x - 2)(x - 19)(x - 3)$

7 $f(-12) = 0 \Rightarrow p = 36$ **9** $f(-5) = 0 \Rightarrow r = -35$

Exercise 173

1 $p = 3$ **3** $r = 24$

5 $x = -1, -5,$ or 4 **7** Other factor is $(x + 1), x = 2, -4, -1$

9 Other factors are $(x + 1), (x - 2), x = 4, -1$ or 2

Exercise 173*

1 $p = -24$ **3** $p = 7, q = -15$

5 Other factor is $(x - 2), x = -3, 4$ or 2

7 Other factors are $(x - 2)$ and $(x - 3), x = -6, 2$ or 3

9 $(x + 1)(x + 2)(x - 2), x = -1, -2$ or 2

Exercise 174

1 $x(x + 1)(x - 2)$ **3** $(x + 2)(x - 1)^2$

5 $2(x + 1)(x + 2)(x + 3)$

7 $p = 0, (x - 1)(x + 3)(2x + 3), x = 1, -3$ or -1.5

Exercise 174*

1 $x(2x + 1)(x - 3)$ **3** $(x + 1)(x - 2)^2$

5 $3(x - 3)(x + 1)^2$

7 $p = 4, (x + 4)(x - 2)(3x - 2), x = -4, 2$ or $2/3$

Exercise 175

1 $12, 8$ or $-8, -12$ **3** 2.61

5 3.22 **7** 7 and 8

Exercise 175*

1 4 **3** $8, 9$ or $-9, -8$

5 $4.40\,\text{m}$ by $11.6\,\text{m}$

7 a 38 sides **b** 30 sides too few, 31 sides too many

9 $14.8\,\text{cm}$

Exercise 176

1 $p = 0$

2 $(x - 3), (x + 1); x = -3, 3$ or -1

3 $(x + 2)(x - 2)(x - 1), x = -2, 2$ or 1

4 $x(x + 4)(x - 7); x = 0, -4, 7$

5 $p = -5, (x + 4)(x - 1)(2x + 1), x = -4, 1$ or $\frac{1}{2}$

6 $2.5\,\text{cm}$

7 $\frac{1}{2}x(x + 3) = 25, 5.73\,\text{cm}$

8 $x^2 + (x + 1)^2 = (x + 2)^2, x = 3$

9 $x^2 + (x + 1)^2 = 145, 8$ and 9

10 $x(30 - x) = 210, 11.1\,\text{cm}$ by $18.9\,\text{cm}$

Exercise 176*

1 $p = 3, q = -6$

2 $(x + 5)(x - 1)(x + 3); x = -5, 1$ or -3

3 $x = 7, -1$ or 2

4 $x = 1$ or 3

5 $p = -1, (x + 5)(x + 2)(2x - 3), x = -5, -2$ or $-\frac{3}{2}$

6 $4.63\,\text{cm}$

7 $6\,\text{m} \times 5\,\text{m}$

8 $5, 12, 13$

9 9 and 11

10 $x^2 + (5 - x)^2 = 16, 1.18\,\text{cm}$ and $3.82\,\text{cm}$

11 $\pi(x + 2)^2 - \pi 2^2 = \pi 2^2, x = 0.828\,\text{m}$

12 $\left(\frac{1200}{x} - 20\right)(x + 3) = 1200, x = 12$ days

Exercise 177

1 a 2 **b** 3 **c** -1 **d** -2 **e** $x = 1$

3 a 4 **b** 2 **c** -4 **d** $x = 3$

Exercise 177*

1 a 2 **b** 4 **c** -1 **d** -3 **e** $x = 1.5$

3 b

x coordinate	-4	-3	-2	-1	0	1	2	3	4
Gradient	-8	-6	-4	-2	0	2	4	6	8

c Straight line gradient 2 passing through the origin

Exercise 178

1 a (i) $1\,\text{m/s}$ **(ii)** $0\,\text{m/s}$ **(iii)** $2\,\text{m/s}$

b 0–20 s gradually increased speed then slowed down to a stop

20–30 s stationary

30–40 s speed increasing

40–50 s travelling at a constant speed of $2\,\text{m/s}$

50–60 s slowing down to a stop

3 a

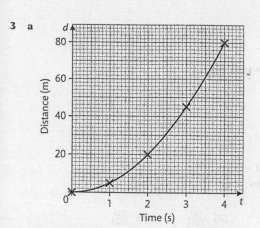

b (i) 10 m/s² **(ii)** 20 m/s² **(iii)** 30 m/s²

Exercise 178*

1 a

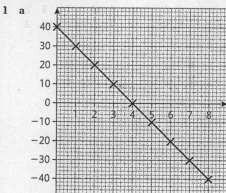

b

t (s)	0	1	2	3	4	5	6	7	8
V (m/s)	40	30	20	10	0	−10	−20	−30	−40

c Straight line graph passing through $(0, -40)$ and $(8, -40)$

d Acceleration is constant (-10 m/s), i.e. constant deceleration (10 m/s)

3 a

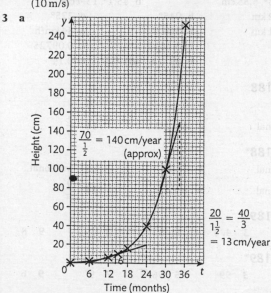

$\dfrac{70}{\frac{1}{2}} = 140\text{ cm/year}$ (approx)

$\dfrac{20}{1\frac{1}{2}} = \dfrac{40}{3} = 13\text{ cm/year}$

b $t = 15$, 13 cm/year (approx.) $t = 30$, 140 cm/year (approx.)

Exercise 179 (Revision)

1 a (i) 1 **(ii)** −2 **(iii)** 3
  b $x = -2$ **c** $y = 2x - 2$
2 b (i) 8 ms⁻² **(ii)** 4 ms⁻² **c** 2.5 s
3 a 2 ms⁻¹ **(ii)** 0 ms⁻¹ **(iii)** −1.6 ms⁻¹
  b 23.5 s at −2.8 ms⁻¹
  c Increases velocity for 5 secs, slows down for 5 secs, on flower for 10 secs, returns faster than outward journey.

4 a

Time, t, in days	0	1	2	3	4	5	6	7	8
Depth, d, in cm	20	17	14.45	12.28	10.44	8.87	7.54	6.41	5.45

b 4.27 days **c** −2.35 cm/day
d −0.051 cm/hr **e** −3.25 cm/day at $t = 0$

Exercise 179* (Revision)

1 a (i) 1 **(ii)** −2 **(iii)** 3
  b −10 **c** Negative of each other.
  d $y = 2x + 6$ **e** ±1.414
2 b (i) 5 **(ii)** 10 **(iii)** 15 **iv** 20 **v** 0
  c Should be $v = 5t$ **d** Constant acceleration of 5 ms⁻²
3 b Runs off for about 1.7 secs, returns more slowly, at 5 secs runs off again
  c (i) 1.6 ms⁻¹ **(ii)** −1.6 ms⁻¹ **d** 6.9 s

4 a

Time, t (years)	0	10	20	30	40	50	60	70
Number, N, of eagles	100	150	225	338	506	759	1139	1709

b (i) 13.7 **(ii)** 30.8 **c** After 39 years

Exercise 180

1 $x = 17.5°, 163°$ **3** $x = 72.5°$ **5** $x = 71.6°$
7 $x = 107°$ **9** $x = 198°$ **11** $x = 108°$

Exercise 180*

1 $x = 27.2°, 153°$ **3** $x = 207°, 333°$
5 $x = 75.7°, 284°$ **7** $x = 112°, 248°$
9 $x = 67.9°, 248°$ **11** $x = 125°, 305°$

Exercise 181

1 $x = 5.94$ **3** MN = 39.0 cm **5** AC = 37.8 cm
7 $x = 37.3°$ **9** ∠ABC = 38.8° **11** ∠ACB = 62.2°

Exercise 181*

1 $x = 29.7$ **3** EF = 10.4 cm, ∠DEF = 47.5°, ∠FDE = 79.0°
5 13 m **7** BC = 261 km
9 PR = 115 m, 112 m

Exercise 182

1 $x = 7.26$ **3** AB = 39.1 cm **5** RT = 24.2 cm
7 Y = 70.5° **9** ∠XYZ = 109.6°

Exercise 182*

1 $x = 9.34$ **3** ∠BAC = 81.8°
7 QR = 4.18 cm, ∠PQR = 39.2°, ∠QRP = 62.8°
9 11.6 km

Exercise 183

1 a 9.64 **b** 38.9°
3 a 54.9° **b** 88.0° **c** 37.1°

5 a 4.1 b 4.9
7 a 16.8 km b 168°

Exercise 183*
1 247 km, 280°
3 a 50.4° b 7.01 m c 48.4°
5 BC = 23.4 km, 186.3°
7 a 38.1° b 29.4 cm

Exercise 184 (Revision)
1 9.22 cm 2 18.0 cm 3 15.6 cm 4 13.5 cm
5 11.9 cm 6 19.4 cm 7 46.5° 8 38.2°
9 55.1° 10 36.2° 11 20.7° 12 59.0°

Exercise 184* (Revision)
1 BH = 506 m
2 6.32 cm and 9.74 cm
3 a 60°, 50°, 70° b 4.9 km
4 22.3°
5 a C = 42.2°, a = 6.96 m b C = 44.7°, a = 5.84 m
6 a 25.5 km b 022.7° c 202.7°
7 a 42.2 km b 022.7° c 14.1 km/h
8 82.8°, 41.4°, 55.8°

Exercise 185
1 a $\frac{1}{36}$ b $\frac{25}{36}$ c $\frac{5}{36}$ d $\frac{5}{18}$
3 a $\frac{4}{25}$ b $\frac{9}{25}$ c $\frac{6}{25}$ d $\frac{12}{25}$
5 a $\frac{4}{9}$ b $\frac{4}{9}$ c $\frac{1}{9}$
7 a $\frac{1}{169}$ b $\frac{1}{4}$ c $\frac{30}{169}$ d $\frac{1}{8}$

Exercise 185*
1 a $\frac{4}{15}$ b $\frac{8}{15}$ c $\frac{3}{5}$
3 a $\frac{1}{9}$ b $\frac{4}{9}$ c $\frac{5}{9}$
5 a $\frac{1}{4}$ b $\frac{1}{2}$ c $\frac{15}{32}$ d $\frac{1}{16}$
7 a $\frac{1}{5}$ b $\frac{13}{35}$ c $\frac{4}{5}$

Exercise 186
1 a $\frac{1}{9}$ b $\frac{4}{9}$
3 a (i) $\frac{43}{63}$ (ii) $\frac{20}{63}$ b $\frac{2}{7}$
c Let X be the number of beads added to the box:
$$p(W_2) = \frac{2}{7} \times \frac{(2+X)}{(7+X)} + \frac{5}{7} \times \frac{2}{(7+X)}$$
$$= \frac{2}{[7(7+X)]} \times (2+X) + 5] = \frac{2}{7} \quad \text{Therefore true!}$$

Exercise 186*
1 a 0.0034 b 0.0006 c 0.0532
3 a $\frac{1}{8}$ b $\frac{8}{15}$ c $\frac{13}{60}$
5 a $p(H_1) = \frac{1}{4}$
b $p(H_2) = p(HH) + p(\overline{H}H) = \frac{1}{4} \times \frac{12}{51} + \frac{3}{4} \times \frac{13}{51} = \frac{1}{4}$
c $p(H_3) = p(HHH) + p(\overline{H}\overline{H}HH) + p(\overline{H}HH) + p(\overline{H}\overline{H}H)$
$= \frac{1}{4} \times \frac{12}{51} \times \frac{11}{50} + \frac{1}{4} \times \frac{39}{51} \times \frac{12}{50} + \frac{3}{4} \times \frac{13}{51} \times \frac{12}{50} + \frac{3}{4} \times \frac{38}{51} \times \frac{13}{50}$
$= \frac{1}{4}$

Exercise 187 (Revision)
1 a $\frac{9}{25}$ b $\frac{12}{25}$
2 a $\frac{1}{16}$ b $\frac{3}{8}$
3 a $\frac{9}{25}$ b $\frac{12}{25}$
4 a $\frac{4}{5}$ b $\frac{6}{25}$

5 a $\frac{1}{36}$ b $\frac{5}{18}$
6 a 0.1 b 0.7 c 0.15

Exercise 187* (Revision)
1 a $\frac{1}{6}$ b $\frac{5}{18}$ c $\frac{13}{18}$
2 a $\frac{2}{9}$ b $\frac{8}{45}$ c $\frac{2}{45}$ d $\frac{43}{45}$
3 a $\frac{1}{32}$ b $\frac{5}{16}$ c $\frac{3}{16}$
4 $\frac{n}{25} \times \frac{(n-1)}{24} = 0.07$
$n^2 - n - 42 = 0 \rightarrow n = 7$
p(diff colours) $= \frac{7}{25} \times \frac{18}{24} + \frac{18}{25} \times \frac{7}{24} = \frac{21}{50}$
5 a 1:3:5 b $\frac{4}{45}$
c (i) $\frac{16}{2025}$ (ii) $\frac{164}{2025}$ (iii) $\frac{344}{2025}$
6 a $\frac{6}{25}$ b $\frac{19}{25}$ c $\frac{12}{43}$ d $\frac{31}{43}$ e 0.320

Multiple choice 8
1 B 2 B 3 D 4 A 5 B
6 D 7 C 8 D 9 A 10 A

Self-assessment 8
1 a $4n+1$ b 401 c 50
2 a $18-6n$ b -582 c 25
3 $f(2) = 0$
4 $p = 6$
5 $(x+1)(x-1)(x-5)$; $x = -1, 1$ or 5
6 $a = -2$ or 2
7 $p = -14$ $(x+3)(x-2)(3x+4)$; $x = -3, 2$ or $-\frac{4}{3}$
8 a $x(x+2) = 6$ b $x = 1.65$
9 2.19×3.19 m
10 9 cm
11 4.83 (3 sf)
12 11 and 12
13 a 1.2 b -3.5 c 0.67
14 a $-1.8, 1.1$ b $-2.8, 2.4$ c $-2.4, 1.7$
15 (1.1, 1.8) (−1.8, −3.2)
16 a 31.3°, 5.55 cm b 45.1°, 13.4 cm
17 a 82.2 km b 045° c 225°
18 a 36.1 km b 064.6° c 34 km/h
19 a 0.2 b 0.1625 c 0.09725
20 a 0.1 b (i) 0.8 (ii) 0.5 c 0.14

Exercise 188
1 $\frac{57}{100}$ 3 7 5 $\frac{3}{1}$
7 Irrational 9 e.g. 2.5 11 $\frac{2}{\pi}$

Exercise 188*
1 Irrational 3 $\frac{3}{5}$ 5 Irrational
7 Irrational 9 e.g. $\sqrt{7}$ 11 $\frac{9}{\pi}$

Exercise 189
1 $6\sqrt{5}$ 3 $4\sqrt{3}$ 5 8 7 $2\sqrt{2}$ 9 8

Exercise 189*
1 $5\sqrt{11}$ 3 99 5 $56\sqrt{7}$ 7 9 9 6

Exercise 190
1 $2\sqrt{3}$ 3 $4\sqrt{3}$ 5 $3\sqrt{3}$ 7 $\sqrt{50}$
9 $\sqrt{54}$ 11 $\frac{2}{5}$ 13 $12, 10\sqrt{2}, \sqrt{26}$

Answers

Exercise 190*

1. $2\sqrt{7}$ 3. $4\sqrt{5}$ 5. $5\sqrt{3}$ 7. $\sqrt{75}$
9. $\sqrt{63}$ 11. $\frac{9}{10}$ 13. $8\text{ cm}^2, 4\sqrt{8}\text{ cm}$

Exercise 191

1. $\frac{\sqrt{3}}{3}$ 3. $\sqrt{3}$ 5. $\frac{2\sqrt{3}}{3}$ 7. $\frac{3\sqrt{2}}{4}$ 9. $\frac{(2+\sqrt{2})}{2}$

Exercise 191*

1. $\frac{\sqrt{13}}{13}$ 3. $\sqrt{a}$ 5. $2\sqrt{3}-1$ 7. $\frac{\sqrt{3}}{3}$ 9. $2+\sqrt{5}$

Exercise 192 (Revision)

1. $0.\dot{3}$ and $\sqrt{25}$ 2. 2, for example (answers may vary)
3. e.g. $\sqrt{11}$ 4. $\sqrt{45}$ 5. $5\sqrt{3}$ 6. $\sqrt{3}$
7. 18   8. 1.5   9. $2\sqrt{2}$ 10. $3\sqrt{7}$
11. $3\sqrt{3}$ 12. $\frac{2}{3}$ 13. $\sqrt{5}$ 14. $2\sqrt{3}$
15. $\frac{3}{2}$ 16. $1+\sqrt{3}$ 17. $16\sqrt{2}, 30, 2\sqrt{17}$

Exercise 192* (Revision)

1. $\sqrt{3}^2$ and $0.\dot{2}\dot{3}$ 2. 3, for example (answers may vary)
3. e.g. 3   4. $\sqrt{176}$ 5. $\sqrt{176}$ 6. $2\sqrt{5}$
7. $2\sqrt{5}$ 8. $\frac{5}{3}$ 9. $\frac{5}{3}$ 10. $11\sqrt{12}$
11. $11\sqrt{12}$ 12. $\frac{3}{5}$ 13. $\frac{\sqrt{5}}{10}$ 14. $2\sqrt{6}$
15. $2+\sqrt{6}$ 16. 2   17. $\frac{(2+3\sqrt{2})}{4}$ 18. $5\sqrt{3}$
19. $\sqrt{3}, \frac{1}{2}, \frac{\sqrt{3}}{2}$

Exercise 193

1. $\frac{3}{2}$ 3. $\frac{x}{y}$ 5. $x+2$ 7. $\frac{1}{(x+2)}$ 9. $\frac{(x+y)}{(x-y)}$

Exercise 193*

1. $\frac{3}{5}$ 3. $\frac{x}{y}$ 5. $\frac{(x-4)}{(x+3)}$ 7. $\frac{(x+3)}{(x+4)}$ 9. $\frac{(r-3)}{(r+1)}$

Exercise 194

1. $\frac{5x+3}{6}$ 3. $\frac{x-3}{12}$ 5. $\frac{5x+3}{4}$
7. $\frac{3-x}{2}$ 9. $\frac{5x+1}{6}$

Exercise 194*

1. $\frac{9x+13}{10}$ 3. $\frac{1-12x}{15}$ 5. $\frac{2x+9}{8}$
7. $\frac{84-7x}{24}$ 9. $\frac{7x+1}{100}$

Exercise 195

1. $\frac{5}{6x}$ 3. $\frac{(x-4)}{2(x-2)}$ 5. $\frac{(x+8)}{(x-1)(x+2)}$
7. $\frac{(x^2+x+2)}{x(x+2)}$ 9. $\frac{(2x+3)}{(x+2)(x+1)}$

Exercise 195*

1. $\frac{17}{15x}$ 3. $\frac{x^2+x+1}{1+x}$ 5. $\frac{1}{(x+1)}$
7. $\frac{2x+3}{(x+1)(x+2)}$ 9. $\frac{x+2}{(x+1)^2}$

Exercise 196

1. $2x$ 3. $\frac{y}{x}$ 5. $\frac{(p-1)}{(p-2)}$
7. $\frac{(x-2)}{(x-4)}$ 9. $\frac{(x+3)}{(x+4)}$

Exercise 196*

1. $\frac{2(x-3)}{(x+2)}$ 3. $\frac{1}{(x+1)}$ 5. $\frac{(x+2)}{(x-2)}$
7. $\frac{(p+4)}{(p-5)}$ 9. $\frac{y}{(x+3y)}$ 11. $-(x+4)(x-3)$

Exercise 197

1. 21   3. $\frac{1}{2}$ 5. -8 7. $\frac{1}{2}$ 9. -6

Exercise 197*

1. $\frac{2}{5}$ 3. $\frac{1}{3}$ 5. 15   7. 1   9. $-\frac{8}{3}$

Exercise 198

1. $-7, 2$ 3. 6   5. $-4, 5$ 7. $-1, 4$

Exercise 198*

1. $-3.5, 1$ 3. 4   5. $-\frac{7}{3}, 2$
7. $-2, 6$ 9. 10.4   11. $-\frac{2}{5}, 2$

Exercise 199 (Revision)

1. 3   2. $x+2$ 3. $\frac{(x+3)}{(x-3)}$
4. $\frac{(x-1)}{(x+3)}$ 5. $\frac{(x-3)}{6}$ 6. $\frac{3x-21}{20}$
7. $\frac{-3}{(x+1)(x-2)}$ 8. $\frac{x+5}{(x-1)(x+1)}$ 9. $\frac{x^2+2x-4}{x(x-2)}$
10. $\frac{4x+10}{(x+2)(x+4)}$ 11. $\frac{x(x-1)}{(x+1)}$ 12. $\frac{x-2}{x-3}$
13. $\frac{8}{3}$ 14. -3 15. -7
16. -2 17. $-2, 1$ 18. $\frac{1}{2}$
19. $-8, 2$ 20. 1

Exercise 199* (Revision)

1. $\frac{2}{3}$ 2. $\frac{(x-11)}{(x+5)}$ 3. $\frac{(x+4)}{(x-7)}$
4. $\frac{(x-1)}{(x+1)}$ 5. $\frac{x-11}{12}$ 6. $\frac{5x-13}{18}$
7. $\frac{5x+7}{12}$ 8. $\frac{7x+6}{10}$ 9. $\frac{2x+3}{(x+1)(x+2)}$
10. $\frac{-1}{(x-4)(x+1)}$ 11. $x+1$ 12. 1
13. -1 14. 5   15. $\frac{15}{16}$
16. 5   17. $-\frac{1}{3}$ or 2   18. $-\frac{2}{3}$ or 1
19. -0.464 or 6.46 20. -9.16 or 3.16

Exercise 200

1. $\frac{dy}{dx}=3x^2$ 3. $\frac{dy}{dx}=5x^4$ 5. $\frac{dy}{dx}=8x^7$
7. $\frac{dy}{dx}=10x^9$ 9. $\frac{dy}{dx}=-x^{-2}$ 11. $\frac{dy}{dx}=-x^{-2}$
13. 4   15. 32   17. 448

Exercise 200*

1. $\frac{dy}{dx}=12x^{11}$ 3. $\frac{dy}{dx}=1$ 5. $\frac{dy}{dx}=-3x^{-4}$
7. $\frac{dy}{dx}=-6x^{-7}$ 9. $\frac{dy}{dx}=\frac{1}{2}x^{-\frac{1}{2}}$ 11. $\frac{dy}{dx}=\frac{1}{2}x^{-\frac{1}{2}}$
13. $-\frac{3}{16}$ 15. $-\frac{1}{4}$ 17. $\frac{1}{4}$
19. 0

Exercise 201

1 $\dfrac{dy}{dx} = 5x^4 + 2x$ **3** $\dfrac{dy}{dx} = 6x^2 + 4$ **5** $\dfrac{dy}{dx} = 9x^2 + 8x^3$

7 $\dfrac{dy}{dx} = 4x$ **9** $\dfrac{dy}{dx} = -x^{-2}$ **11** $\dfrac{dy}{dx} = -6x^{-4}$

13 $\dfrac{dy}{dx} = 2x - 2x^{-3}$ **15** $\dfrac{dy}{dx} = 3x^2 + 3x^{-4}$

Exercise 201*

1 $\dfrac{dy}{dx} = 3x^2 + 4x$ **3** $\dfrac{dy}{dx} = 2x + 4$ **5** $\dfrac{dy}{dx} = 2x + 6$

7 $\dfrac{dy}{dx} = 8x - 4$ **9** $\dfrac{dy}{dx} = 8x - 2x^{-3}$ **11** $\dfrac{dy}{dx} = 5$

13 $(3, -9)$ **15**

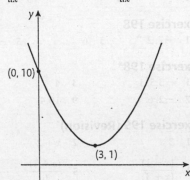

Exercise 202

1 **a** $2x + 4$ **b** $(-2, 6)$
3 **a** $x = 3$ **b** $y = 24$
5 **a** 15 **b** 6 **c** $y = 6x + 9$
7 **a** $8t + 8$ **b** $16°$ per minute   **c** $48°$ per minute
9 **a** $3x^2 - 24x$ **b** $x = 0$ or 8 **c** $(0, 5), (8, -251)$
     d Maximum: $(0, 5)$, minimum: $(8, -251)$;
11 **a** $6 - 2x$ **b** 20
13 **a** $-2 - 2x$ **b** 9

Exercise 202*

1 $6x - 7$ **b** $y - 5x - 7$
3 Minimum at $(-3, -54)$, maximum at $(3, 54)$
5 **a** $11\,150$ **b** 325 per year
7 **a** $80 - 40t$ **b** $t = 2$ **c** $350°$ **d** $120°$ per hour
9 **a** $\dfrac{dy}{dx} = 6x^2 - 2x - 4$ **b** $x = \dfrac{-2}{3}$
     c $(1, 7) \left(\dfrac{-2}{3}, 11.63\right)$
11 **a** $2x - 16x^{-2}$ **b** $x = 2$ **c** $(2, 12)$
13 $3x - y = -4$ and $3x - y = 4$

Exercise 203

1 $10t$
3 **a** $40 + 10t$ **b** $70\,\text{m/s}$
5 32
7 **a** $v = 3t^2 + 8t - 5$ **b** $a = 6t + 8$ **c** $v = 6\,\text{m/s}; a = 14\,\text{m/s}^2$
9 **a** $2t + 10$ **b** $14\,\text{m/s}$

Exercise 203*

1 **a** $v = 8t + \dfrac{2}{t^2}$ **b** $a = 8 - \dfrac{4}{t^3}$
3 **a** $20 + 10t$ **b** $30\,\text{m/s}, 40\,\text{m/s}, 50\,\text{m/s}$ **c** $10\,\text{m/s}^2$
5 **a** $40 - 10t$ **b** $30\,\text{m/s}, 20\,\text{m/s}, 10\,\text{m/s}, 0\,\text{m/s}$ **c** $80\,\text{m}$
7 **a** $\dfrac{ds}{dt} = -2 + \dfrac{18}{t^2}; \dfrac{dv}{dt} = -\dfrac{36}{t^3}$ **b** $3\,\text{s}$ **c** $28\,\text{m}$
9 **a** $8\,\text{s}$ **b** $3\,\text{s}$ **c** $170\,\text{m}$

Exercise 204 (Revision)

1 **a** $\dfrac{dy}{dx} = 3$ **b** $\dfrac{dy}{dx} = 0$ **c** $\dfrac{dy}{dx} = 3x^2$
    d $\dfrac{dy}{dx} = 4x^3$ **e** $\dfrac{dy}{dx} = 5x^4$ **f** $\dfrac{dy}{dx} = 12x^5$
    g $\dfrac{dy}{dx} = 15x^4$ **h** $\dfrac{dy}{dx} = 160x^7$
2 **a** $\dfrac{dy}{dx} = 6x^2 + 10x$ **b** $\dfrac{dy}{dx} = 14x - 3$ **c** $\dfrac{dy}{dx} = 15x^2$
    d $\dfrac{dy}{dx} = 12x^3 - 10x$ **e** $\dfrac{dy}{dx} = 3x^2 + 10x$ **f** $\dfrac{dy}{dx} = 2x + 2$
    g $\dfrac{dy}{dx} = 4x - 9$ **h** $\dfrac{dy}{dx} = 2x + 4$
3 **a** $\dfrac{dy}{dx} = 4$ **b** $\dfrac{dy}{dx} = -5$ **c** $\dfrac{dy}{dx} = 26$
    d $\dfrac{dy}{dx} = 19$
4 **a** $\dfrac{dQ}{dt} = 3t^2 - 16t + 14$
    b **(i)** $14\,\text{m}^3/\text{s}^2$ **(ii)** $-6\,\text{m}^3/\text{s}^2$ **iii** $9\,\text{m}^3/\text{s}^2$
5 **a** $\dfrac{dy}{dx} = 3x^2 - 3$ **b** $(1, 0), (-1, 4)$
    c $(-1, 4)$ is max, $(1, 0)$ is min.
6 **a** $\dfrac{dy}{dx} = 3x^2 + 6x - 9$ **b** $(-3, 28), (1, -4)$
    c $(-3, 28)$ is max, $(1, -4)$ is min.
7 **a** $v = 24t\,\text{m/s}, 48\,\text{m/s}$ **b** $a = 24\,\text{m/s}^2, 24\,\text{m/s}^2$
8 **a** $v = 3t^2 + 8t - 3\,\text{m/s}, 377\,\text{m/s}$ **b** $a = 6t + 8\,\text{m/s}^2, 68\,\text{m/s}^2$

Exercise 204* (Revision)

1 **a** $\dfrac{dy}{dx} = -x^{-2} = -\dfrac{1}{x^2}$ **b** $\dfrac{dy}{dx} = -2x^{-3} = -\dfrac{2}{x^3}$
    c $\dfrac{dy}{dx} = \dfrac{1}{2}x^{-\frac{1}{2}} = \dfrac{1}{2\sqrt{x}}$ **d** $\dfrac{dy}{dx} = -3x^{-4} = -\dfrac{3}{x^4}$
    e $\dfrac{dy}{dx} = -16x^{-5} = -\dfrac{16}{x^5}$ **f** $\dfrac{dy}{dx} = -x^{-3} = -\dfrac{1}{x^3}$
    g $\dfrac{dy}{dx} = 4x + 3 - 4x^{-2}$ **h** $\dfrac{dy}{dx} = 2 + 3x^{-2}$
2 **a** -2 **b** 2 **c** 0 **d** $-2\frac{1}{8}$
3 $4y = 3x + 4$ **4** $y = -3x, y = 3x - 9$
5 **a** $y = 4x - 7$ **b** $x = 0$ is a max, $x = \frac{4}{3}$ is a min.
6 $p = 1$
7 **a** $\dfrac{dC}{dt} = 4 - 16t^{-2} = 4 - \dfrac{16}{t^2}$ **b** $t = 2, C = 16$
    c $-12\,°\text{C/month}$
8 **a** $\dfrac{dP}{dt} = 10t - 10\,000t^{-2} = 10t - \dfrac{10\,000}{t^2}$
    b $t = 10, P = 1500$
9 $t = 5, s = 125\,\text{m}$
10 **a** $v = \dfrac{ds}{dt} = 3t^2 - 300\,\text{km/s}, a = \dfrac{dv}{dt} = 6t\,\text{km/s}^2$
    b at $t = 5$, $v = -225\,\text{km/s}, a = 30\,\text{km/s}^2$
    c $t = 10\,\text{s}$

Exercise 205

1 $7.39\,\text{cm}^2$ **3** $36.2\,\text{cm}^2$ **5** $121\,\text{cm}^2$

Exercise 205*

1 $173\,\text{cm}^2$ **3** $16.5\,\text{cm}$ **5** $53.5\,\text{cm}^2$

Exercise 206

1 **a** $11.7\,\text{cm}$ **b** $14.2\,\text{cm}$ **c** $34.4°$

Answers

3   **a**   14.1 cm   **b**   17.3 cm   **c**   35.4°
5   **a**   4.47 m   **b**   4.58 m   **c**   29.2°   **d**   12.6°
7   **a**   43.3 cm   **b**   68.7 cm   **c**   81.2 cm

Exercise 206*

1   **a**   16.2 cm   **b**   67.9°   **c**   55.3 cm²
3   **a**   30.3°   **b**   31.6°   **c**   68.9°
5   **a**   15 m   **b**   47.7°   **c**   €91 300
7   46.5 m

Exercise 207

1   5.08 cm²   **3**   24.0 cm²   **5**   411 cm²

Exercise 207*

1   6.97 cm   **3**   11.5 cm   **5** 7.96×10^{-1} m³

Exercise 208 (Revision)

1   148 cm²
2   **a**   AC = 42.4 cm   **b**   33.9 cm   **c**   68.0°   **d**   58.0°
3   **a**   18.4°   **b**   500 m   **c**   11.3°
4   **a**   68.9 m²   **b**   120 m²
5   **a**   22.4 cm   **b**   26.4 cm   **c**   32.1°   **d**   35.0°
6   **a**   14.1 cm²   **b**   67.8 cm²

Exercise 208* (Revision)

1   Pythagoras proof
2   4.68 m²
3   **a**   AC = 70.7 cm   **b**   98.7 cm
    c   27.9°   **d**   216 000 cm²
4   **a** x is the length of the diagonal of the
     square that is the bottom face of the cube.
     Using Pythagoras
     $x^2 = 8^2 + 8^2$
        $= 128$
        $x = \sqrt{128} = \sqrt{64 \times 2} = \sqrt{64} \times \sqrt{2} = 8\sqrt{2}$

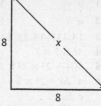

    b   36.9°
5   92.1 m
6   **a**   70.7 m   **b**   60.4 m   **c**   41.8°   **d**   67.5°   **e**   9038 m²
7   14.3 cm
8   5.2 cm/s

Exercise 209

1   **a** $\overrightarrow{XY} = x$ **b** $\overrightarrow{EO} = 4y$ **c** $\overrightarrow{WC} = -8t$ **d** $\overrightarrow{TP} = -4x$
3   **a** $\overrightarrow{HJ}$ **b** $\overrightarrow{HN}$ **c** $\overrightarrow{HL}$ **d** $\overrightarrow{HO}$
5   **a** $\overrightarrow{DC} = x$ **b** $\overrightarrow{DB} = x + y$ **c** $\overrightarrow{BC} = -y$ **d** $\overrightarrow{AC} = x - y$
7   **a** $\overrightarrow{DC} = x$ **b** $\overrightarrow{AC} = x + y$ **c** $\overrightarrow{BD} = y - x$ **d** $\overrightarrow{AE} = \frac{1}{2}(x + y)$
9   **a** $\overrightarrow{AB} = x - y$ **b** $\overrightarrow{AD} = 3x$ **c** $\overrightarrow{CF} = 2y - 3x$ **d** $\overrightarrow{CA} = y - 3x$
11   **a** $2x + 4y$ **b** $4y - 2x$ **c** $2y - 2x$ **d** $3x - 4y$

Exercise 209*

1   **a** $\overrightarrow{AB} = y - x$ **b** $\overrightarrow{AM} = \frac{1}{2}(y - x)$ **c** $\overrightarrow{OM} = \frac{1}{2}(x + y)$
3 $\overrightarrow{AB} = y - x; \overrightarrow{OD} = 2x; \overrightarrow{DC} = 2y - 2x$
    d   DC = 2AB and they are parallel lines
5   **a** $\overrightarrow{AB} = y - x; \overrightarrow{OC} = -2x; \overrightarrow{OD} = -2y; \overrightarrow{DC} = 2y - 2x$
    b   DC = 2AB and they are parallel lines
7 $\overrightarrow{AB} = y - x; \overrightarrow{BC} = y - 2x; \overrightarrow{AD} = 2y - 4x; \overrightarrow{BD} = y - 3x$
9   **a** $\overrightarrow{MA} = \frac{3}{5}x; \overrightarrow{AB} = y - x; \overrightarrow{AN} = \frac{3}{5}(y - x); \overrightarrow{MN} = \frac{3}{5}y$
    b   OB and MN are parallel; $\overrightarrow{MN} = \frac{3}{5}\overrightarrow{OB}$

Exercise 210

1   Rotation 90° anticlockwise about (0, 0)
3   Reflection in $y = x$
5   **a**   A′(3, 4)   B′(5, 7)   C′(5, 6)   **b** $\begin{pmatrix} 3 & -2 \\ -1 & 1 \end{pmatrix}$

Exercise 210*

1   Enlargement SF2 centre (0, 0)
3   Stretch × 2 in y direction
5   **a**   New image A′(2, 1)   B′(3, −4)   C′(−4, −2)   **b** $\frac{1}{11}\begin{pmatrix} 3 & 5 \\ 1 & -2 \end{pmatrix}$

Exercise 211

1   **a** $\begin{pmatrix} 0 & 1 \\ -1 & 0 \end{pmatrix}$ **b** $\begin{pmatrix} -1 & 0 \\ 0 & -1 \end{pmatrix}$
    c   Rotation 180° centre (0, 0)
3   **a** $R = \begin{pmatrix} -1 & 0 \\ 0 & -1 \end{pmatrix}$ $X = \begin{pmatrix} 0 & 1 \\ 1 & 0 \end{pmatrix}$
    b $RX = \begin{pmatrix} 0 & -1 \\ -1 & 0 \end{pmatrix}$ $XR = \begin{pmatrix} 0 & -1 \\ -1 & 0 \end{pmatrix}$
    c   Reflection in $y = -x$
5   **a** $TS = \begin{pmatrix} 0 & 2 \\ 1 & 0 \end{pmatrix}$ $ST = \begin{pmatrix} 0 & -1 \\ -2 & 0 \end{pmatrix}$ **b**   A′(2, 1)   B′(2, 3)   C′(4, 1)
    c   F(−2, −2)   G(−1, −2)   H(−1, 1)

Exercise 211*

1   **a** $R = \begin{pmatrix} 0 & -1 \\ 1 & 0 \end{pmatrix}$ $Q = \begin{pmatrix} 0 & -1 \\ 1 & 0 \end{pmatrix}$
    b $RQ = QR \begin{pmatrix} -1 & 0 \\ 0 & -1 \end{pmatrix}$ **c**   Rotation 180°
3   **a** $P = \begin{pmatrix} 0 & 1 \\ -1 & 0 \end{pmatrix}$ $Q = \begin{pmatrix} -1 & 0 \\ 0 & 1 \end{pmatrix}$ $R = \begin{pmatrix} 0 & -1 \\ 1 & 0 \end{pmatrix}$
    b $PQR = \begin{pmatrix} 1 & 0 \\ 0 & -1 \end{pmatrix}$ **c**   Reflection in x axis
    d $(PQR)^{-1} = \begin{pmatrix} 1 & 0 \\ 0 & -1 \end{pmatrix}$
5   **a** $RS = \begin{pmatrix} -2 & -2 \\ -2 & 0 \end{pmatrix}$ $SR = \begin{pmatrix} 0 & -2 \\ -2 & -2 \end{pmatrix}$
    b   A′(−10, −4)   B′(12, 2)   C′(4, 8)
    c   F$\left(-\frac{1}{2}, -1\right)$ G$\left(2, \frac{1}{2}\right)$ H(−3, 2)

Exercise 212 (Revision)

1   **a** $\overrightarrow{AB} = 3y + x$ **b** $\overrightarrow{AC} = 2y + 2x$ **c** $\overrightarrow{CB} = -x + y$
2   **a** $\overrightarrow{AB} = w - v$ **b** $\overrightarrow{AM} = \frac{1}{2}(w - v)$ **c** $\overrightarrow{OM} = \frac{1}{2}(v + w)$
3   **a** $\overrightarrow{AB} = y - x$ **b** $\overrightarrow{FB} = 2y - x$ **c** $\overrightarrow{FD} = y - 2x$
4   **a** $\overrightarrow{OP} = 2a$ $\overrightarrow{OQ} = 2b$ $\overrightarrow{AB} = b - a$ $\overrightarrow{PQ} = 2b - 2a$
    b   AB parallel to PQ, PQ = 2AB
5   **a** $\overrightarrow{AB} = 2y - 2x$ **b** $\overrightarrow{AM} = y - x$ **c** $\overrightarrow{OM} = x + y$
6   **a**   (i) $\overrightarrow{MP} = \frac{2}{5}p$ (ii) $\overrightarrow{PQ} = q - p$
      (iii) $\overrightarrow{PN} = \frac{2}{5}(q - p)$ (iv) $\overrightarrow{MN} = \frac{2}{5}q$
    b   OQ parallel to MN, MN = $\frac{2}{5}$ OQ
7   **a**   A′(2, −1)   B′(3, 3)   C′(−8, 1)
    b $\begin{pmatrix} 2 & 3 \\ 1 & 2 \end{pmatrix}$ **c**   A′B′C′ → ABC
8   **a**   A′(1, −5)   B′(2, −7)   C′(−1, 5)   D′(−3, 7)
    b
9   **a** $PQ = \begin{pmatrix} 2 & 2 \\ 0 & 2 \end{pmatrix}$ $QP = \begin{pmatrix} 2 & 2 \\ 0 & 2 \end{pmatrix}$
    b   A′(0, 2)   B′(10, 6)   C′(−2, −4)
    c   F$\left(1, -\frac{1}{2}\right)$ G$\left(-\frac{1}{2}, 2\right)$ H(−1, 0)
10   **a** $RS = \begin{pmatrix} -2 & 1 \\ -1 & 2 \end{pmatrix}$ **b**   A′(−1, 1)   B′(−3, 3)   C′(8, 10)
    c

Exercise 212* (Revision)

1 a $\overrightarrow{XM}\begin{pmatrix}-1\\3\end{pmatrix}$

$\overrightarrow{XZ}=\begin{pmatrix}-10\\6\end{pmatrix}$

b $v\begin{pmatrix}7\\3\end{pmatrix}$ **c** $\begin{pmatrix}8\\0\end{pmatrix}+w\begin{pmatrix}-10\\6\end{pmatrix}$ **d** $v=\frac{2}{3},w=\frac{1}{3}$

2 a $\overrightarrow{AC}=x+y;\overrightarrow{BE}=\frac{1}{3}y-x$

b (i) $\overrightarrow{BF}=v\left(\frac{1}{3}y-x\right)$

(ii) $\overrightarrow{AF}=x+\overrightarrow{BF}=x+v\left(\frac{1}{3}y-x\right)$

(iii) $v=\frac{3}{4}$

3 a $\overrightarrow{RS}=-r+s$ **b** $\overrightarrow{OP}=\frac{3}{2}r$

c $\overrightarrow{PQ}=-\frac{3}{2}r+2s$ **d** $\overrightarrow{OM}=s-\frac{3}{4}r$

4 a (i) $\overrightarrow{PQ}=-p+q$ (ii) $\overrightarrow{PR}=\frac{1}{3}(q-p)$ (iii) $OR=\frac{1}{3}(2p+q)$

b (i) $k=\frac{3}{5}$ (ii) $OS=\frac{1}{5}(2p+q)$

5 a

t	0	1	2	3	4	5
r	$\begin{pmatrix}1\\5\end{pmatrix}$	$\begin{pmatrix}3\\4\end{pmatrix}$	$\begin{pmatrix}5\\3\end{pmatrix}$	$\begin{pmatrix}7\\2\end{pmatrix}$	$\begin{pmatrix}9\\1\end{pmatrix}$	$\begin{pmatrix}11\\0\end{pmatrix}$

b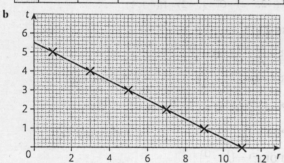

c 8050 km/hr, 117°

6 a (i) $r=\begin{pmatrix}3\\1\end{pmatrix},s=\begin{pmatrix}0\\5\end{pmatrix}$ (ii) $r=\begin{pmatrix}4\\3\end{pmatrix},s=\begin{pmatrix}3\\6\end{pmatrix}$

b $\begin{pmatrix}3\\1\end{pmatrix};\sqrt{10}$ **c** $\sqrt{5};\sqrt{10}$

7 a A'(4, 3) B'(2, 1) C'(−1, −1)

b $\begin{pmatrix}-2&3\\3&-4\end{pmatrix}$ **c** A'B'C' → ABC

8 a A'(−21, 16) B'(−5, 4) C'(−137, 116) D'(−47, 36)

b

9 a $GH=\begin{pmatrix}4&2\\2&2\end{pmatrix}$ $HG=\begin{pmatrix}2&2\\2&4\end{pmatrix}$

b A'(−12, 4) B'(8, 10) C'(2, −4)

c

10 a $MQ=\begin{pmatrix}4&2\\-3&0\end{pmatrix}$ **b** A'(−26, 24) B'(−6, 0) C'(26, −24)

c E(−1, −2) F(1, −2) G(1, 2)

Multiple choice 9

1 D **2** C **3** B **4** A **5** B
6 D **7** B **8** A **9** B **10** A

Self-assessment 9

1 Rational **a**, **c** and **e**

2 a e.g. 2.3 **b** e.g. 2.8

3 a e.g. $\sqrt{17}$ **b** e.g. $\sqrt{23}$

4 a 6 **b** a **c** 18 **d** 6 **e** $\frac{2}{3}$

5 a $\sqrt{27}$ **b** $\sqrt{28}$ **c** $\sqrt{80}$ **d** $\sqrt{18}$

6 a $4\sqrt{5}$ **b** $12\sqrt{5}$ **c** 160 **d** 2

7 $p=20$

8 $\frac{1}{2}\sqrt{2}$

9 $2\sqrt{3}$

10 a $x+3$ **b** $\dfrac{(x+2)}{(x-2)}$ **c** $\dfrac{(x-1)}{(x+2)}$

11 a $\frac{1}{2}x$ **b** $\dfrac{5}{(x-2)(x+3)}$ **c** $\dfrac{(x-1)}{(x+2)}$

12 a $\dfrac{2(x+1)}{x}$ **b** $(x-2)(2x-1)$

13 a $\frac{dy}{dx}=6x^2-12x$ **b** $\frac{dy}{dx}=6x^2-2x+2$ **c** $\frac{dy}{dx}=2x-1-2/x^3$

14 a $\frac{dy}{dx}=x^2+2x-7$ **b** $\left(-4,17\frac{2}{3}\right),\left(2,-12\frac{1}{3}\right)$

15 a 1600 hairs/yr **b** 2000 hairs/yr **c** 40 yrs

16 a $v=40-10t$ **b** $t=4$ s **c** 80 m

17 a 16.7° **b** DF = 94.3 m

c 9.03° **d** 2.19 m/s

18 31.9 cm

19 a (i) a (ii) $\frac{1}{2}$(a + b) (iii) a − b (iv) $\frac{1}{2}$(a + b)

b Parallel and equal lengths

c PQRS is a parallelogram

20 a a − b

b (i) $(\mu-1)$a (ii) $(\lambda-1)$b

c (i) a + $(\lambda-1)$b (ii) $(\mu-2)$a + 2b

d $\mu=2$ $\lambda=3$

21 b Q points (0, 0) (3, −4) (10, −5)

c 90 degrees clockwise about the origin

d R points (0, 0) (−4, −3) (5, −10)

Consolidation exercise 1

1 a $\frac{7}{12}$ **b** $1\frac{1}{3}$ **c** 24 **d** 12

2 a 9 **b** $1\frac{22}{35}$

3 14.235, 14.25, 14.3, 14.532

4 $3^3\times5\times7$

5 $504=2^3\times3^2\times7$

6 a $24=2^3\times3$ **b** $90=2\times3^2\times5$

c 6 **d** 360

7 a $\frac{34}{99}$ **b** 0.375 **c** \$1.12 **d** 2

8 $\frac{181}{333}$

9 a 35% **b** 1 : 2.5

10 a 72 s **b** 4.17c **c** 16.7%

11 a 10^3 **b** 2×10^{10} **c** 7×10^{-5} **d** 6×10^{-3}

12 a 37.625 **b** 37.62 **c** 1.398 **d** 1.40

13 a 52.4 **b** 0.0026 **c** 38 500 **d** 0.57

14 a 5×10^{-3} **b** 5×10^1

15 a 0.351 **b** 5.93 **c** 1.52

16 a 10, 12, 14, 16 **b** $b=2w+6$

c 56 **d** 60

17 a 8, 13, 19, 26 **c** 12

18 a 2^6 **b** 2^{21}

19 a 3^6 **b** 3^2 **c** 3^{3m} **d** 3^{10}

20 a $4\sqrt{3}$ **b** 1

21 a 9.39 billion people **b** 1.40%

22 3.2 kg

23 a 329 blocks **b** 820 tonnes **c** 12.5 ears

24 a 10.5 days **b** 4.55 km

25 a 8565 gallons/s **b** 39 m²/s

26 597 dominos

27 \$403.13

28 \$408

29 a $89.7 b $71.07
30 £76.80

Consolidation exercise 1*

1 $\frac{2}{3} = \frac{24}{36}, \frac{13}{18} = \frac{26}{36}, \frac{3}{4} = \frac{27}{36}, \frac{14}{18} = \frac{28}{36}$

2 a $3\frac{1}{3}$ b (i) 12 (ii) $6\frac{7}{12}$

3 cement 10 kg sand 45 kg

4 a $180 = 2^2 \times 3^2 \times 5, 84 = 2^2 \times 3 \times 7$

5 a $a^4, a = 2$ b $a^{-\frac{1}{3}}, a = 8$ c $2a^{-2}, a = \frac{1}{4}$

6 a 3^4 b $2^{-3} \times 5^2$

7 $\frac{3}{2}$

8 $\sqrt{5}$ or $\sqrt{6}$ or $\sqrt{7}$ or $\sqrt{8}$

9 1.95

10 2.4×10^1

11 a $150\,\text{cm}^3$ b $320\,\text{cm}^3$

12 a 3.7×10^{10} b 0.000 075 c 2×10^{-10}

13 30 tonnes

14 48 km/h

15 a 33.3% b (i) 38.5 °C c 37°C

16 9 metres

17 a 16.4 cm/day b $4.17 \times 10^{10}\,\text{m}^3$

18 160

19 a 19 230 769 b 520 m

20 a 185, 181 b 5 c $205 - 4n$

21 a 11, 16, 21 b $r + c = 2, 5, 9, 14, 20$
  c $w = 3(r + c)$ d $2w = 9r + 3r^2$ e 195

22 a 11, 16, 21 b $r + c$: 2, 5, 9, 14, 20
  c $w = 3(r + c)$ d $w = \frac{3}{2}(3r + r^2)$ e 195

Consolidation exercise 2

1 a x^{10} b x^2
  c y^2 d $4x^6 y^4$

2 a $x^2 + 2x - 3$ b $2x^2 + 7x - 4$

3 a $z^2 + 3z$ b $14 - x$

4 a $4(p - 2)$ b $x(x + 3)$ c $3ab(b + 2a)$

5 a $(x + 2)(x - 2)$ b $(x + 1)(x + 2)$

6 a $\frac{7x}{12}$ b $\frac{(x + 1)}{6}$ c $\frac{4}{(x - 3)}$

7 a $v - 1\,\text{m}$ b $\sqrt{(12\pi h)}$

8 a -14 b $\frac{(v - u)}{t}$

9 a 3.14 b $g\left(\frac{T}{2\pi}\right)^2$

10 a 114 b $C = 30 + 0.15t$

11 a 3 b 2 c 2

12 a 3 b 5

13 a $5(x + 3) - 8 = 42$ b 7

14 15

15 $x = 6, y = 3$; 216 cm²

16 a $y = 4x$ b 8 c 2

17 a $y = 3x^2$ b 48 c 3

18 a $y = \frac{24}{x}$ b 12 c 4

19 a $y = \frac{36}{\sqrt{x}}$ b 12 c 16

20 (3, 5)

21 a $x + y = 17, 4x + 2y = 58$ b 5

22 a $x = -4$ or 4 b $x = 0$ or 16

23 a $x = -1$ or 3 b $x = 2$ or 4

24 a $x = 0.59$ or 3.41 b $x = -1.78$ or 0.281

25 b $-8, 4$ c 24
26 $x = 5$, area 48
27 a $x > 1$ b $x \geqslant 3$
28 $-2, -1, 0, 1$
31 $p = -3$ $(x + 3)(x - 2)(x - 4)$
32 $q = -2$ $(2x + 1)(x + 2)(x - 1)$ $x = -\frac{1}{2}, -2$ or 1
33 a $x = 7$ b $x = 11$

Consolidation exercise 2*

1 a $\frac{11a}{12}$ b $\frac{3}{a}$ c b d $\frac{(21 - 2x)}{15}$

2 a $x + 1$ b x

3 a $(x - 9)(x + 8)$ b $\frac{(x + 9)}{(x + 8)}$

4 a $(3x - 1)(x + 11)$ b $\frac{(3x - 1)}{(x - 11)}$

5 a 27.6 b $\frac{\sqrt{3I}}{M - b^2}$

6 a 3024 b $\frac{2(S - an)}{n(n - 1)}$

7 a $\frac{15}{8}$ b $\frac{fv}{(v - f)}$ c 12

8 -1

9 $\frac{x + 123}{x + 456} = \frac{1}{2}$, 210

10 a $(160° - 3x)$ b $x = 20°, 32°, 35°$

11 a $x^2 + 25 = I^2$ b $(x + 2)^2 + 16 = I^2$
  c 5.15 m to 3 s.f.

12 a $v = 2\sqrt{d}$ b 14 m/s c 25 m

13 a $y - \frac{40}{x^2}$ b 640 c 12.6

14 a $T = \frac{120}{\sqrt{m}}$ b 40 c 13.8 min

15 $x = 2$ $y = 1$

16 a $2x + 3y = 180$ b $2x + 2y = 140$
  c $x = 30, y = 40$

17 a $x + y = 14, 12x + 18y = 204$
  b $x = 8, y = 6$

18 a $x = -10$ or 10 b $x = 0$ or 100

19 a $x = -9$ or 4 b $x = 2$ or 5

20 a $x = 0.697$ or 4.30 b $x = -1.88$ or 0.884

21 a $x^2 - 2x - 3 = 0$ b $x = -1$ or $x = 3, \therefore 3, 4, 5$ triangle

22 a $-8, 2$ b 6

23 b $-6, 4$ c 5

24 $-1, 0, 1, 2$

25 a $x > -\frac{1}{3}$ b $x \leqslant 2$

26 a $2.5x + 3 < 3x, 4x + 1 > 1.5x + 2, 4.5x + 2 > x + 1$
  b $\frac{2}{5} < x < 6$

27 4.7 cm

28 $\frac{x + 4}{x - 4}$

29 $\frac{x + 7}{x + 6}$

30 $p = -6, q = -1, x = -2, 5$ or 3

31 $p = -7, (4x + 1)(x - 4)(x + 2)$ $x = -\frac{1}{4}, 4$ or -2

32 a $x = 2$ b $x = 2$ or -2

33 1.58 cm

Consolidation exercise 3

1 a (i) 5 (ii) 4 (iii) $\frac{2}{5}$
  b $x = 0$ c $x = -1$

2 a No
  b (i) $\mathbb{R}$ **(ii)** $x \geq 0$
  c (i) 3 **(ii)** $x^2 - 1$
3 a (i) -1 **(ii)** $\frac{1}{7}$
  b (i) $\frac{3}{x} - 2$ **(ii)** 0
  c ± 1
4 a i **b** iii **c** iv **d** ii
5 A **(v)** B **(ii)** C **(vi)** D **(i)** E **(iv)** F **(iii)**
6 $x > 2$
7 a **b** $x = 4, y = 3$

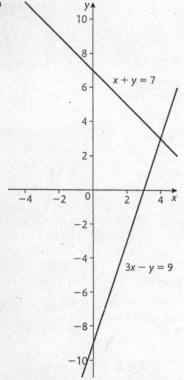

8

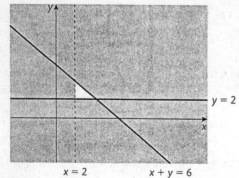

9 a $\frac{3}{4}$ **b** $(3, -1.5)$
10 a **b** $x \approx 0.7, x \approx 4.3$

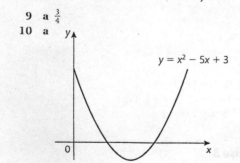

11 a **b** $x = -1.6$ or 0.6

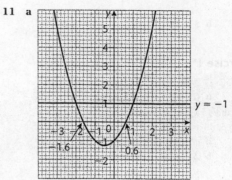

  c $x = -2, x = 1$ **d** -1.25
12 a $y = 1$ **b** $y = x + 2$
13 a $x^2 - 4x + 2 = 0$ **b** $x = 0.6$ or $x = 3.4$
14 a $x = 2$ $y = 6$ **b** $y = 2x$
15 $-1 \rightarrow -1$ range $\{-1, 1, 3, 5\}$
  $0 \rightarrow 1$
  $1 \rightarrow 3$
  $2 \rightarrow 5$
16 Various possible answers.
17 Yes it is a function

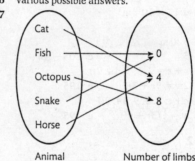

18 a not a function **b** a function
19 c $1440 \,\text{m}^3$, $4.9 \,\text{m}$
20 a

x	1	2	3	4	5
y	4	2	$1\frac{1}{3}$	1	0.8

  b

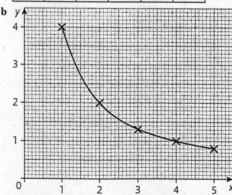

21 c (i) $\$3.3 \times 10^4$ **(ii)** 7.1 years
22 b 3 seconds
  c (i) 10 mls **(ii)** 5 m/s² **(iii)** 19.0 m/s
23 a (i) 8.65 m/s = 31.1 km/h
  c 9.04 m/s **d** 2.26 m/s²
24 a $(1, -5)$ **b** $-1.2, 3.2$
  c Line is $y = x + 2$, solutions 4.4 or -1.4
  d (i) $\frac{dy}{dx} = 2x - 2$ **(ii)** Gradient is 10

25 b (i) $28 - 4x$ **(ii)** $x = 7$ **(iii)** Graph is a negative parabola
   c $98\,\text{m}^2$

26 a $v = 3 - 2t - \dfrac{8}{t^2}$ **b** $a = -2 + \dfrac{16}{t^3}$ **c** $t = 2$

Consolidation exercise 3*

1 a 9   **b** 9.5   **c** $\dfrac{x+1}{z}$ **d** $2x$

2 a 20   **b** 10^{11} **c** $10^x + 10$ **d** 10^{x+10}

3 $x = -4$ or 3

4 $x = -10$ or 5

5 a 2   **b** $y = 2x - 1$
   c $y = 2x \pm c$ (c positive integer)

6 a $a = -1, b = 1, c = 3$ **b** $y = x + 3$

7 $x + y = 5$ and $y - 2x = 2$

8 a

x	0	4	8	12	16	20
y	0	8.8	11.2	7.2	-3.2	-20

   b $31.3\,\text{m}$ **c** $15\,\text{m}$

9 a $x^3 - 3x^2 - x + 2 = 0$
   b $x = -0.86$ or $x - 0.75$ or $x = 3.12$

10 a $y = 3$ **b** $y = 2 - 2x$

11 a

x	0.5	1	1.5	2	3	4	5
y	-5.75	-2	0.25	2.5	8	15.25	24.4

   b

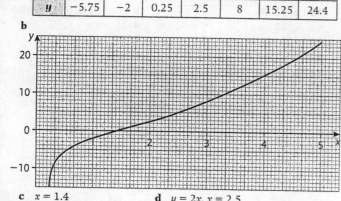

   c $x = 1.4$ **d** $y = 2x,\ x = 2.5$

12 a (i) 1.5   **(ii)** 0.75   **b** $x = 1$
   c (i) $ff(x) = x$ **(ii)** it is its own inverse

13 a $x = -1$ $y = 8$ $y = x^3$
   b $x = 3$ $y = 9$ $y = 2x - 1$

14 $-2 \to 0$ {range 0, 3, 4}
      $-1 \to 3$
       $0 \to 4$
       $1 \to 3$
       $2 \to 0$

15 Various possible answers

16 Yes it is a function

17 a $7.04\,\text{m/s}$ **b** $8.20\,\text{m/s}$ **c** $2.05\,\text{m/s}$
   d No. Sidd takes 15 s to run 100 m so gets beaten by 0.8 s.

18 a

t (s)	0	2	4	6	8	10
d (m)	86.6	55.5	36.3	22.3	10.6	0

   b Graph shows that hare's speed is very gradually decreasing as it reaches the bush.
   c Gradient at $t = 6\,\text{s}$ is approx. $-6.25\,\text{m/s}$: Velocity.

19 a $1.2\,\text{m/s}^2$ **b** $2.4\,\text{m/s}^2$ **c** $15\,\text{m/s}$

20 a $v = -2(t-2)^2 - 18 \therefore v_{max} = 18$

21 $t = 5$

---

22 a $\dfrac{dy}{dx} = 5000 - 1250x$ **b** $(4, 10\,000)$
   c max   **d** £4

Consolidation exercise 4

1 b $215\,\text{km}$

2 $(3, -5)$ **b** $(5, -3)$ **c** $(-1, 9)$ **d** $(7, 5)$

3 a $(-1, -1), (1, -1), (1, -2)$
   b $(-4, 3), (-4, 5), (-3, 5)$
   c Rotation $+90°$ about the point $(-3, -2)$

4 a $0.26.6°$ **b** $206.6°$

5 $p = 5.40\,\text{cm}, q = 11.9\,\text{cm}, r = 65.4°, s = 41.8°$

6 $x = 15; y = 4$

7 a $160° - 3x$
   b $2x = x + 20°, 2x = 160° - 3x$
   c $x = 20°, 32°, 35°$

8 a $45°$ **b** $1080°$
   c each interior angle has size $135°$, $\dfrac{360}{135}$ does not give a whole number

9 a $2.4\,\text{cm}$ **b** $1\frac{1}{3}\,\text{cm}$

10 a $\dfrac{h + OT}{OT} = \dfrac{2r}{r}$ $\therefore 2 \times OT = h + OT$ and $OT = h$
   b $V = \left(\frac{1}{3} \times \pi 4r^2 \times 2h\right) - \left(\frac{1}{3} \times \pi r^2 \times h\right)$
       $= \dfrac{8\pi r^2 h}{3} - \dfrac{\pi r^2 h}{3} = \dfrac{7\pi r^2 h}{3}$
   c $V = 7.3304r^2 h$ **d** $r = 1.68\,\text{cm}$

11 $y = 4\,\text{cm}$

12 $6200\,\text{cm}^3$

13 a $x = 100°, y = 90°$ **b** $x = 70°, y = 76°$
   c $x = 110°, y = 35°$ **d** $x = 65°, y = 35°$

14 a $x = 25°, y = 65°, z = 115°$ **b** triangle ECF

15 a $\angle TPQ = 66°$ (triangle TPQ is isosceles)
     $\angle OPT = 90°$ (TP is tangent)
     $\therefore \angle QPO = 24°$
   b $OPS = 42°$ (isosceles triangle)
     $\therefore \angle QPS = 66°$
   c $\angle QRS = 114°$ (opposite angles of a cyclic quadrilateral)

16 $x = 8.7; y = 43.4°, z = 73.4°$

17 a $SA = 2.63\,\text{km}, BS = 6.12\,\text{km}$

18 a $PQ = 8.7\,\text{m}$ **b** $QR = 5\,\text{m}$ **c** $RS = 8.4\,\text{m}$

19 a $40\,\text{m}$ **b** $26.5\,\text{m}$ **c** $4.68\,\text{m}$

20 a $25.5\,\text{cm}$ **b** $27.3\,\text{cm}$ **c** $21.4°$

21 $28.5\,\text{cm}^2$

Consolidation exercise 4*

1 a $PQ = 8.7\,\text{m}$ **b** $QR = 5\,\text{m}$ **c** $RS = 8.4\,\text{m}$
   d Area PQRS $= 63.8\,\text{m}^2 = 64\,\text{m}^2$ (to 2 s.f.)

2 a $x = 29.7°, y = 1.2$ **b** $x = 61.1°, y = 23.2$
   c $x = 40.3°, y = 16.5$

3 a $CP = 30\,\text{m}$ **b** $QB = 4\,\text{m}$ **c** $10\,\text{m}$

4 a $BC = 1.81\,\text{m}; 10.1\,\text{m}$
   b No, because the gap when $b = 35°$ is $5.43\,\text{m}$

5 $x = 6\frac{2}{3}, y = 9$

6 a OBA, OAC   **b** $2\frac{1}{2}$ (2.08) **c** 13

7 a –
   b (i) $(1, -3), (1, -5), (2, -3)$
     (ii) $(3, -1), (5, -1), (3, -2)$
     (iii) $(2, 2), (2, 6), (4, 2)$
     (iv) $(-1, 1), (-1, 5), (1, 1)$
   c 90 degrees clockwise about $(1, -1)$
   d Scale factor 2 centre $(3, 1)$

8 **a** $V = \pi R^2 h - \pi r^2 h = \pi h(R^2 - r^2) = \pi h(R - r)(R + r)$
  b $5280\,\text{cm}^3$ **c** $h = 4.5\,\text{cm}$

9 Perimeter $= 18(1 + \sqrt{3}\,\text{cm})$

10 **a** 8π **b** $4\,\text{cm}$
  c $9.2\,\text{cm}$ **d** $r = \sqrt[3]{36.8}\,\text{cm}$

11 $3795\,\text{cm}^2$

12 $x = y = 51°, z = 78°$

13 **a** $AB = 2r$ **b** $\frac{s}{360} \times 2\pi r$, so $s = 114.6°$
  c $0.545r^2$

14 **a** 20 **b** 270

15 **a** **(i)** $100°$ **b** $80°$
    (ii) $\angle TQX = \angle QPX$ (angles in alternate segments)
    $\therefore \angle XQA = \angle QPB$
    $\angle QAB = \angle QPB$ (angles in same segment)
    $\therefore XQ$ is parallel to AB

16 $60\,\text{cm}$

17 **a** $CD = 14.4\,\text{m}$ **b** $CE = 142\,\text{m}$ **c** $12.7\,\text{m/s}$

18 $x = 3\,\text{cm}\quad y = 12\,\text{cm}$

19 **a** $(2, -5)$ **b** $(-2, 5)$
  c $(-1, 12)$ **d** $(-2, 24)$

20 **a** $h = 2\,\text{m}$ **b** $\pi\,\text{m/s}$
  c $\frac{1}{2}\pi\,\text{m/s}$, so 50% decrease

21 **a** $AP = \frac{10\sqrt{3}}{3}\quad PQ = \frac{5\sqrt{3}}{3}\quad OQ = 5\sqrt{3}$
  b $54.7°$ **c** $70.5°$

Consolidation exercise 5

1 **a** **(i)** $A \cap B = \{e, g\}$
    (ii) $A \cup B = \{a, c, d, e, f, g, i, k\}$
  b

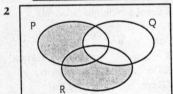

2

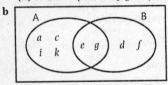

3 **a** 4 **b** 14

4

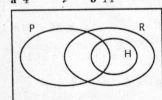

5 **a** $\frac{1}{3}$ **b** $\frac{2}{3}$ **c** $\frac{1}{12}$ **d** $\frac{1}{3}$

6 **a** $\frac{1}{9}$ **b** $\frac{7}{9}$ **c** $\frac{4}{9}$

7 **a** 24 years   **b** 24.6 years

8 **a** rotation 90° clockwise centre $(2, 0)$
  b **(i)** false   **(ii)** true   **(iii)** true

9 **a** triangle vertices are: $(0, 6), (2, 6), (2, 10)$
  b triangle vertices are: $(6, 3), (6, 4), (8, 3)$

10 **a** 60 matches   **b** 149 points   **c** 0.6

11 **a** 0.75, 1.8, 4.4, 3.8, 1.8, 0.4
  b

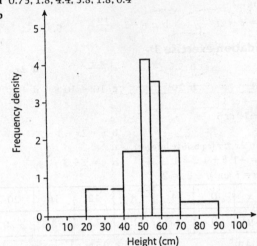

  c $53.2\,\text{cm}$

12 **b** 4 **c** $22.0\,\text{cm}$

13 **b** 2 **c** 17

14 **a** bars at 11 units and 5 units
  b $11, 16$ **c** 41

15 **a** $12, 20$ **b** heights 6.5, 9 and 3 units
  c 57 **d** $11.2\,\text{min}$

16 **a** 0.57 **b** 30

17 **a** $\frac{1}{6}$ **b** $\frac{7}{18}$

18 **a** First attempt   Second attempt   **b** $\frac{17}{20}$

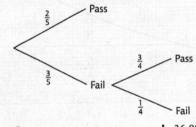

19 **a** 5 **b** $36.9°$

20 $\begin{pmatrix} 6\cos 30 \\ 6\sin 30 \end{pmatrix}$

21 $2\mathbf{v} = \begin{pmatrix} -2 \\ 10 \end{pmatrix}\quad \mathbf{u} + \mathbf{v} = \begin{pmatrix} 1 \\ 8 \end{pmatrix}\quad \mathbf{u} - \mathbf{v} = \begin{pmatrix} 3 \\ -2 \end{pmatrix}\quad \sqrt{130}$

22 **a** $A + B = \begin{pmatrix} 3 & -3 \\ -2 & 4 \end{pmatrix}$ **b** $A - B = \begin{pmatrix} 3 & 5 \\ 8 & 0 \end{pmatrix}$
  c $2(A + 2B) = \begin{pmatrix} 6 & -14 \\ -14 & 12 \end{pmatrix}$ **d** $3(A + 4B) = \begin{pmatrix} 9 & 51 \\ 69 & -18 \end{pmatrix}$

23 **a** $\begin{pmatrix} 2 & 5 \\ 3 & 1 \end{pmatrix}$ **b** $\begin{pmatrix} -1 & -3 \\ -3 & 4 \end{pmatrix}$ **c** $\begin{pmatrix} 1 & 0 \\ 0 & 1 \end{pmatrix}$ **d** $\frac{1}{13}\begin{pmatrix} 1 & 4 \\ -3 & 1 \end{pmatrix}$

24 **a** $\overrightarrow{AB} = \mathbf{y} - \mathbf{x}\quad \overrightarrow{OM} = \frac{1}{2}(\mathbf{x} + \mathbf{y})\quad \overrightarrow{ON} = \frac{1}{2}(\mathbf{y} - \mathbf{x})\quad \overrightarrow{MN} = -\mathbf{x}$
  MN is parallel to OA and the same length.

25 **a** $\overrightarrow{AB} = \mathbf{b} - \mathbf{a}$ **b** $\overrightarrow{PQ} = 3(\mathbf{b} - \mathbf{a})$
  c AB is parallel to PQ; $PQ = 3AB$

26 **a** $T = \begin{pmatrix} 2 & -1 \\ -4 & -5 \end{pmatrix}$ **b** $T^{-1} = \frac{1}{14}\begin{pmatrix} 5 & -1 \\ -4 & -2 \end{pmatrix}$
  c $A\left(\frac{-2}{7}, \frac{3}{7}\right)\quad B\left(\frac{23}{14}, \frac{-12}{7}\right)\quad C\left(\frac{-3}{7}, \frac{1}{7}\right)$

Consolidation exercise 5*

1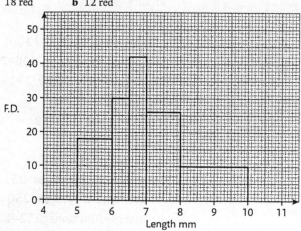

2

3 17

4 a 34.6   **b** 33.5

5 a

size (kb)	1	2	3	4	5	6	7
f	18	16	9	6	6	4	1

   b mean = 2.7 KB, median = 2 KB, mode = 1 KB

   c Angle sizes shown in table

size (KB)	1	2	3	4	5	6	7
Angle size	108°	96°	54°	36°	36°	24°	6°

6 a $\frac{8}{24} = \frac{1}{3}$ **b** $\frac{2}{3}$ **c** $\frac{4}{24} = \frac{1}{6}$ **d** $\frac{12}{24} = \frac{1}{2}$

7 a

	1	2	3	4	5	6
1	0	1	2	3	4	5
2	1	0	1	2	3	4
3	2	1	0	1	2	3
4	3	2	1	0	1	2
5	4	3	2	1	0	1
6	5	4	3	2	1	0

   b $\frac{10}{36} = \frac{5}{18}$ **c** $\frac{12}{36} = \frac{1}{3}$

8 1.74 m

9 a $\frac{1}{2}$ **b** $\frac{5}{6}$ **c** 1   **d** 0

10 a 0   **b** $\frac{9}{25}$ **c** $\frac{16}{25}$ **d** $\frac{9}{25}$

11 a 0.4   **b** 0.35

12 a $3n$ **b** 21 people

13 a 18 red   **b** 12 red

14 a

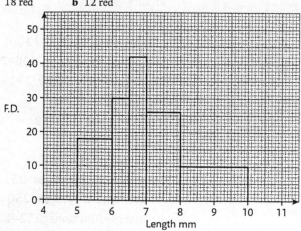

b 76   **c** 7.10 mm

Length, l mm	F.D.
$5 < l \leqslant 6$	18
$6 < l \leqslant 6.5$	30
$6.5 < l \leqslant 7$	42
$7 < l \leqslant 8$	26
$8 < l \leqslant 10$	10

15 a

Time, t mins	F.D.
$20 < t \leqslant 30$	0.5
$30 < t \leqslant 35$	0.8
$35 < t \leqslant 40$	1.8
$40 < t \leqslant 50$	2.2
$50 < t \leqslant 70$	0.35
$70 < t \leqslant 100$	0.1

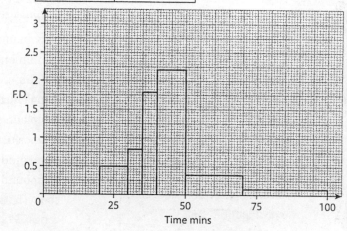

b 71.2%   **c** 43.2 mins

16 a 40, 35, 25, 25, 50, 25   **b** $x = 12.5$
   c —

17 a 20, 60, 60, 30, 30   **c** 0.6

18 a $\sqrt{13}$ **b** $p = 1$ $q = 2$

19 a $\begin{pmatrix} 4\cos 20 \\ 4\sin 20 \end{pmatrix}$ **b** $\begin{pmatrix} 3\cos 50 \\ 3\sin 50 \end{pmatrix}$

   c $\begin{pmatrix} 4\cos 20 + 3\cos 50 \\ 4\sin 20 + 3\sin 50 \end{pmatrix}$ **d** 6.77 units 32.8°

20 a **(i)** $\overrightarrow{OD} = \frac{1}{2}\mathbf{b}$ $\overrightarrow{OQ} = \frac{1}{2}\mathbf{a}$ **(ii)** $\overrightarrow{PQ} = \frac{1}{2}(\mathbf{a} - \mathbf{b})$

   (iii) $\overrightarrow{OR} = \mathbf{a} + \frac{1}{2}\mathbf{c}$ $\overrightarrow{OS} = \frac{1}{2}(\mathbf{a} + \mathbf{b} + \mathbf{c})$

   (iv) $\overrightarrow{SR} = \frac{1}{2}(\mathbf{a} - \mathbf{b})$

   b $\overrightarrow{PQ} = \overrightarrow{SR}$

21 a $\overrightarrow{PU} = \mathbf{a} - \mathbf{b}$ **b** $\overrightarrow{TR} = 3\mathbf{b} + \mathbf{a}$ **c** $\overrightarrow{US} = \frac{1}{2}(\mathbf{a} - \mathbf{b})$
   d PUS a sraight line PU = 2US

22 **(i)** $\overrightarrow{AB} = \mathbf{b} - \mathbf{a}$ **(ii)** $\overrightarrow{CD} = \lambda\mathbf{b} - \mu\mathbf{a}$
   (iii) $\lambda = 3$ $\mu = 3$ **(iv)** CD parallel to AB; CD = 3AB

23 a $\frac{1}{10}\begin{pmatrix} -4 & 2 \\ 1 & -3 \end{pmatrix}$ **b** $\begin{pmatrix} 4 & 1 \\ 3 & 7 \end{pmatrix}$

   c $\begin{pmatrix} 5 & 1 \\ -3 & -3 \end{pmatrix}$ **d** $\begin{pmatrix} 19 & 11 \\ 33 & 52 \end{pmatrix}$

24 a $\begin{pmatrix} -1 & -2 & -3 \\ 2 & 5 & 4 \\ 5 & 18 & -1 \end{pmatrix}$ **b** $\begin{pmatrix} -7 & 8 \\ 1 & 10 \end{pmatrix}$ **c** not commutative

25 a $\begin{pmatrix} 1 & 0 \\ 0 & -1 \end{pmatrix}$ **b** reflection in x axis
   b $a = 3$ $b = -1$ $c = \frac{-6}{21}$ $d = \frac{-34}{21}$

Index

Index